Integral Transforms and Engineering

With the aim to better understand nature, mathematical tools are being used nowadays in many different fields. The concept of integral transforms, in particular, has been found to be a useful mathematical tool for solving a variety of problems not only in mathematics but also in various other branches of science, engineering, and technology.

Integral Transforms and Engineering: Theory, Methods, and Applications presents a mathematical analysis of integral transforms and their applications. The book illustrates the possibility of obtaining transfer functions using different integral transforms, especially when mapping any function into the frequency domain. Various differential operators, models, and applications are included such as classical derivative, Caputo derivative, Caputo-Fabrizio derivative, and Atangana-Baleanu derivative.

This book is a useful reference for practitioners, engineers, researchers, and graduate students in mathematics, applied sciences, engineering, and technology fields.

Integral Transforms and Engineering

Theory, Methods, and Applications

Abdon Atangana

Ali Akgül

CRC Press
Taylor & Francis Group
Boca Raton London New York

CRC Press is an imprint of the
Taylor & Francis Group, an **informa** business

First edition published 2023
by CRC Press
6000 Broken Sound Parkway NW, Suite 300, Boca Raton, FL 33487-2742

and by CRC Press
4 Park Square, Milton Park, Abingdon, Oxon, OX14 4RN

CRC Press is an imprint of Taylor & Francis Group, LLC

ISBN: 978-1-032-41683-0 (hbk)
ISBN: 978-1-032-41820-9 (pbk)
ISBN: 978-1-003-35986-9 (ebk)

DOI: 10.1201/9781003359869

Typeset in Nimbus Roman
by KnowledgeWorks Global Ltd.

Dedication

_____

First and foremost, we would like to praise and thank God, the Almighty, who has granted countless blessings, knowledge, and opportunity to the writers, so that we have finally been able to reach the accomplishment of completing our book.
We wish to show our greatest appreciation to our families. We can't thank them enough for their tremendous support and help. Without their encouragement and guidance, this book would not have materialized.

Contents

Preface

To better understand nature, researchers have provided some mathematical tools that are used currently in many branches of science, technology, and engineering. In particular, the concept of integral transformation was suggested and has been found as a useful mathematical tool for solving a range of problems in mathematics and applied mathematics. It is worth noting that a mathematical operator is called an integral transform if it maps a function from its original function into another function space through an integral. However, it is possible that some properties of the initial functions could be easier to characterize and manipulate than the original function space. Usually, the transformed function can be mapped back to the initial function space using the inverse transform. In mathematics and applied mathematics, several problems have been found difficult to be solved in their original presentations. Thus, an integral transform maps an equation from its original domain into another. However, controlling and solving a differential or integral equation in the target domain can be easier than manipulating and finding the solution in the original space. The obtained solution is mapped back into the original space using the inverse integral transform. The available literature shows that there are several applications of probability that are connected to integral transforms, for instance, the pricing kernel, also known as the stochastic discount factor. Another important field where these mathematical operators are applied is the control theory. This theory comprises two principal approaches for continuous time and discrete linear time-invariant systems. We will be interested only in the continuous time, which generates frequency-domain techniques, relying on the notions of the transfer function and the frequency response. The electric circuits are in this case the principal step toward understanding complex electrical engineering notions. More importantly, circuit analysis creates the progressive methods that are moving the industry forward. Laplace transform is, therefore, an important mathematical operator used to obtain the transfer function that it in turn is used to obtain the Bode diagram. Besides the Laplace transform, several different integrals have been suggested in the last decades and have been found to have some interesting properties comparable to those of the Laplace transform. We can list the Sumudu transform, Mohand transforms, Sawi transform, Elzaki transform, Aboodh transform, Pourreza transform, α integral Laplace transform, Kamal transform, G-Transform, and Natural transform. These integral transforms have played a significant role in solving differential equations with integer and non-integer orders in the last decades. In particular, fractional linear differential equations have been acknowledged as powerful mathematical tools to replicate complex phenomena. The tools of fractional calculus have played a significant role in enhancing the modeling methods for several real-world problems. In this book, information on the mathematical analysis of integral transform and their applications in control theory are presented. We discuss the possibility of obtaining transfer functions using different integral transforms especially when they map any function into the frequency domain.

We applied these integral transforms for many electric circuit models. Different differential operators are considered including classical derivative, Caputo derivative, Caputo-Fabrizio derivative, and Atangana-Baleanu derivative in the models. This book could be convenient for graduate students and investigators in pure and applied mathematics and engineering.

Authors

Abdon Atangana works at the Institute for Groundwater Studies, University of the Free State, Bloemfontein, South Africa, as a full-time professor. His research interests are, but are not limited to: fractional calculus and applications, numerical and analytical methods, and modeling. He is the author of more than 270 research papers and five books in top-tier journals of applied mathematics and groundwater modeling. He is the founder of numerous mathematical operators, for example, fractional differential and integral operators with singular and nonlocal kernels (Atangana-Baleanu derivatives and integral), fractal-fractional calculus, and piecewise calculus, and some numerical methods. He serves as an editor in esteemed journals in various fields of study. He has been invited as a plenary, and keynote speaker at more than 30 international conferences. He was elected Highly Cited Mathematician in 2019, Highly Cited Mathematician with Crossfield Impact in 2020, and Highly Cited Mathematician in 2021. He is a recipient of the Obada prize 2020, the TWAS-Hamdan Award 2020, and many others. He is also a fellow of the World Academia of Science.

Ali Akgül[1,2,3] works at the Department of Mathematics, Siirt University, Turkey, as a full-time professor. He is head of the department of Mathematics. His research interests are, but are not limited to, fractional calculus and applications, numerical and analytical methods, and modeling. He is the author of more than 250 research papers in respected journals. He has been invited as a keynote speaker at more than 15 international conferences. His name has been enlisted in the 'World Ranking Top Two Percent Scientists' list, 2020, 2021, and 2022. He is also a recipient of the Obada prize 2022 (Young Distinguished Researchers).

[1]Department of Computer Science and Mathematics, Lebanese American University

[2]Siirt University, Art and Science Faculty, Department of Mathematics

[3]Near East University, Mathematics Research Center, Department of Mathematics

1 Sumudu and Laplace Transforms

The Laplace transform is perhaps one of the most used integral transforms in the field of mathematics, technology, and engineering. In general, the Laplace transform was named after Pierre-Simon Laplace as he introduced it as an integral that covers a function of a real variable to a function of a complex variable also known as complex frequency. The operator has found applications in all fields of science and engineering. In particular, this transform is used to solve linear ordinary and partial differential and integral equations. One of its great achievements is to transform a convolution into a product; therefore, the theorem of convolution is used to obtain the solution of linear equations with integer and non-integer orders. It is worth noting that the current widespread application of the Laplace transform especially in engineering can be traced to World War II as it was replacing the earlier Heaviside operational calculus. Gustav [6] provided the advantages of the Laplace transform. Indeed, there are some questions raised around some properties of the Laplace transform; for example, it fails to preserve parity of the function and the units. However, the operator has been used to solve problems in mechanical engineering and electrical engineering since it has the great property of reducing a linear differential equation to an algebraic equation, which can be later solved by a recognized routine of algebra. The original solution can be obtained by the mean of the inverse Laplace transform. On the other hand, by the year 1990 Gamage K Watugala suggested an integral transform similar to the Laplace transform to solve differential equations and control engineering problems the operator was called Sumudu transform and has been applied in many real-world problems with great success. In particular, it was observed that the transform has many interesting properties that over-performed those of the Laplace transform. For example, unlike the Laplace transform, the Sumudu transform preserve units and parity of the function, the differentiation, and integration in the t-domain is equivalent to division and multiplication of the transformed function $F(u)$ by u in the u-domain. Where u- domain is a complex number.

1.1 DEFINITIONS

DEFINITION 1.1 Let $f(t)$ be defined for $t \geq 0$. The Laplace transform of $f(t)$ defined by $F(s)$ or $L\{f(t)\}$ is an integral transform given by the Laplace integral [9]:

$$L\{f(t)\} = F(s) = \int_0^\infty \exp(-st)f(t)dt. \qquad (1.1)$$

DOI: 10.1201/9781003359869-1

1

Provided that this (improper) integral exists, i.e., that the integral is convergent. The Laplace transform is an operation that transforms a function of t (i.e., a function of time domain), defined on $[0, \infty)$, to a function of s (i.e., of frequency domain). $F(s)$ is the Laplace transform, or simply transform, of $f(t)$. Together, the two functions $f(t)$ and $F(s)$ are called a Laplace transform pair.

DEFINITION 1.2 Over the set of functions,

$$A = \{f(t) \mid \exists M, \tau_1, \tau_2 > 0, \mid f(t) \mid < M \exp(\mid t \mid / \tau_j, \text{ if } t \in (-1)^j \times [0, \infty)\}, \quad (1.2)$$

the Sumudu transform is presented as [10–12]

$$G(u) = S[f(t)] = \int_0^\infty f(ut) \exp(-t) dt, \quad u \in (-\tau_1, \tau_2). \quad (1.3)$$

DEFINITION 1.3 Let $u \in C_{-1}^m[0, T]$. The Caputo fractional derivative of order α of the function u is defined below [13, 14]:

$$
{}_0^C D_t^\alpha u(t) =
\begin{cases}
\frac{1}{\Gamma(m-\alpha)} \int_0^t \frac{1}{(t-\eta)^{1+\alpha-m}} \frac{d^m u(\eta)}{d\eta^m} d\eta, & m-1 < \alpha < m, \quad m \in \mathbb{N}, \\
\frac{d^m u(t)}{dt^m}, & \alpha = m, \quad m \in \mathbb{N}.
\end{cases}
\quad (1.4)
$$

DEFINITION 1.4 Let $0 < \alpha < 1$ and $u \in H^1(0, T), T \in \mathbb{R}_*^+$. We define the Caputo-Fabrizio fractional derivative of order α of a function u by [15]

$$
{}_0^{ABC} D_t^\alpha u(t) = \frac{M(\alpha)}{1-\alpha} \int_0^t u'(\eta) \exp\left(-\frac{\alpha}{1-\alpha}(t-\eta)\right) d\eta, \quad (1.5)
$$

where $H^1(0, T)$ denotes the Sobolev space, and $M(\alpha)$ is a normalization function in which $M(0) = M(1) = 1$.

DEFINITION 1.5 Let $\alpha \in [0, 1]$ and $u \in H^1(0, T), 0 < T$. The Atangana-Baleanu fractional derivative in Caputo sense of order α of a function u is given by [15]

$$
{}_0^{ABC} D_t^\alpha u(t) = \frac{AB(\alpha)}{1-\alpha} \int_0^t E_\alpha\left(-\frac{\alpha}{1-\alpha}(t-\eta)^\alpha\right) u'(\eta) d\eta, \quad (1.6)
$$

where $H^1(0, T)$ denotes the Sobolev space and $AB(\alpha) = 1 - \alpha + \frac{\alpha}{\Gamma(\alpha)}$ is named the normalization function in which $AB(0) = AB(1) = 1$.

1.2 PROPERTIES OF LAPLACE AND SUMUDU TRANSFORMS

We present some important properties of these two integral transforms.

1.2.1 PROPERTIES OF LAPLACE

The Laplace transform has several properties that have been found useful in many theoretical problems as well as applications. For readers that are not aware of these properties, in this section we present some useful properties of the Laplace transform. Let g and f be two functions for which their Laplace transform exists [15]

$$\mathscr{L}\left(af\left(t\right)+bg\left(t\right)\right)=a\mathscr{L}\left(f\left(t\right)\right)+b\mathscr{L}\left(g\left(t\right)\right)$$

$$=af\left(s\right)+bG(s)$$

The above property is known as linearity

$$\forall n \geq 1$$

$\mathscr{L}\left\{t^{n} f\left(t\right)\right\} = (-1)^{n} F^{n}\left(s\right)$, where

$$F^{n}\left(s\right)\ is\ the\ n.\ derivative\ of\ F\left(s\right)$$

$$\mathscr{L}\left(f^{(n)}\left(t\right)\right) = s^{n} f\left(s\right) - \sum_{k=1}^{n} s^{n-k} f^{(k-1)}(0)$$

$$\mathscr{L}\left(e^{at} f\left(t\right)\right) = f\left(s-a\right)$$

$$\mathscr{L}\left(f\left(t-a\right) U\left(t-a\right)\right) = e^{-as} f\left(s\right), \quad a > 0$$

$$\mathscr{L}\left(f\left(at\right)\right) = \frac{1}{a} F\left(\frac{s}{a}\right), \quad a > 0$$

$$\mathscr{L}\left(\left(f \ast g\right)\left(t\right)\right) = F\left(s\right) G(s)$$

The above is known as convolution theorem

$$f\left(0^{+}\right) = \lim_{s \to \infty} sf(s)$$

which is known as initial value theorem

$$\lim_{s \to \infty} sf(s)$$

if all poles of $sf(s)$ are in the left half plane.

1.2.2 PROPERTIES OF SUMUDU

We present some important properties of the Sumudu transform.
Let f and g be two functions such that their Sumudu transform exists. Then, [11]

$$\forall n \geq 1,$$

$$S\left(t^{n} f^{(n)}\left(t\right)\right) = u^{n} f^{(n)}\left(u\right)$$

$$S\left(t^{n+1} f\left(t\right)\right) = u^{n+1} \sum_{k=0}^{n+1} a_{k}^{n+1} u^{k} f^{(n)}\left(u\right)$$

$$S\left(f^{(n)}(t)\right) = \frac{F(u)}{u^n} - \frac{F(0)}{u^n} - \cdots \frac{f^{n-1}(0)}{u}$$

$$S\left(\int_0^t f(\tau)\,d\tau\right) = uF(u)$$

$$S(f(at)) = F(au)$$

$$S\left(\frac{1}{t}\int_0^t f(\tau)\,d\tau\right) = \frac{1}{u}\int_0^u f(v)\,dv$$

$$S((f * g)(t)) = uF(u)G(u)$$

The above is the equivalent convolution

$$S((f * g)^n(t)) = u^n F(u) G(u)$$

The Sumudu transform may be used to solve problems without resorting to a new frequency domain. In fact, the Sumudu transform which is itself linear preserves linear functions, and hence in particular does not change units [11].

1.2.3   SOME EXAMPLES OF SUMUDU AND LAPLACE TRANSFORMS

We give some examples of the Laplace transforms and Sumudu transforms as [6, 11]

$$L\{t^{-\alpha}\} = s^{\alpha-1}\Gamma(1-\alpha). \tag{1.7}$$

$$L\{sin(t)\} = \frac{1}{s^2+1}. \tag{1.8}$$

$$L\{cos(t)\} = \frac{s}{s^2+1}. \tag{1.9}$$

$$S[t^{-\alpha}] = s^{-\alpha}\Gamma(1-\alpha). \tag{1.10}$$

$$S[cos(t)] = \frac{1}{s^2+1}. \tag{1.11}$$

$$S[sin(t)] = \frac{s}{s^2+1}. \tag{1.12}$$

We have the following relations for the Caputo, Caputo-Fabrizio, and Atangana-Baleanu derivatives:

$$_0^C D_t^\alpha u(t) = \frac{du(t)}{dt} * \frac{t^{-\alpha}}{\Gamma(1-\alpha)}. \tag{1.13}$$

$$_0^{CF} D_t^\alpha u(t) = \frac{du(t)}{dt} * \frac{M(\alpha)}{1-\alpha}\exp\left(-\frac{\alpha}{1-\alpha}t\right). \tag{1.14}$$

$$_0^{ABC} D_t^\alpha u(t) = \frac{du(t)}{dt} * \frac{AB(\alpha)}{1-\alpha}E_\alpha\left[\frac{-\alpha}{1-\alpha}t^\alpha\right]. \tag{1.15}$$

The Laplace transform of Caputo derivative is given as

$$L\{{}_0^C D_t^\alpha u(t)\} = (sL\{u(t)\} - u(0)) s^{\alpha-1}. \tag{1.16}$$

The Laplace transform of Caputo-Fabrizio derivative is given as

$$L\{{}_0^{CF} D_t^\alpha u(t)\} = -(sL\{u(t)\} - u(0)) \frac{M(\alpha)}{s\alpha - s - \alpha}. \tag{1.17}$$

The Laplace transform of Atangana-Baleanu derivative is given as

$$L\{{}_0^{ABC} D_t^\alpha u(t)\} = (sL\{u(t)\} - u(0)) \frac{AB(\alpha)s^{\alpha-1}}{s^\alpha(1-\alpha)+\alpha}. \tag{1.18}$$

The Sumudu transform of Caputo derivative is given as

$$S[{}_0^C D_t^\alpha u(t)] = \frac{S[u] - u(0)}{s^\alpha}. \tag{1.19}$$

The Sumudu transform of Caputo-Fabrizio derivative is given as

$$S[{}_0^{CF} D_t^\alpha u(t)] = (S[u] - u(0)) \frac{M(\alpha)}{\alpha s + 1 - \alpha}. \tag{1.20}$$

The Sumudu transform of Atangana-Baleanu derivative is given as

$$S[{}_0^{ABC} D_t^\alpha u(t)] = (S[u] - u(0)) \frac{AB(\alpha)}{1 - \alpha + \alpha s^\alpha}. \tag{1.21}$$

For $1 < \alpha \le 2$, we defined the Caputo Derivative as

$$\begin{aligned} {}_0^C D_t^\alpha f(t) &= \frac{1}{\Gamma(2-\alpha)} \int_0^t \frac{d^2 f(\tau)}{d\tau^2}(t-\tau)^{1-\alpha} d\tau \\ &= \frac{1}{\Gamma(2-\alpha)} \frac{d^2 f(t)}{dt^2} * t^{1-\alpha} \end{aligned}$$

Then, we have the Laplace transform of the above equation as

$$L\left({}_0^C D_t^u f(t)\right) = \frac{1}{\Gamma(2-\alpha)} L\left(\frac{d^2 f(t)}{dt^2}\right) L\left(t^{1-\alpha}\right)$$

$$L\left({}_0^C D_t^\alpha f(t)\right) = \left(s^2 L(f(t)) - sf(0) - f'(0)\right) s^{\alpha-2}$$

For $1 < \alpha < 2$, we defined the Caputo-Fabrizio Derivative as

$$\begin{aligned} {}_0^{CF} D_t^\alpha f(t) &= \frac{M(\alpha)}{2-\alpha} \int_0^t \frac{d^2 f(\tau)}{d\tau^2} \exp\left(\frac{-\alpha}{2-\alpha}(t-\tau)\right) d\tau \\ &= \frac{M(\alpha)}{2-\alpha} \frac{d^2 f(t)}{dt^2} * \exp\left(\frac{-\alpha}{2-\alpha}t\right) \end{aligned}$$

Then, we have the Laplace transform of the above equation as

$$L\left({}^{CF}_0 D^\alpha_t f(t)\right) = \frac{M(\alpha)}{2-\alpha} L\left(\frac{d^2 f(t)}{dt^2}\right) L\left(\exp\left(\frac{-\alpha}{2-\alpha}t\right)\right)$$

$$L\left({}^{CF}_0 D^\alpha_t f(t)\right) = \frac{M(\alpha)}{\alpha+s(2-\alpha)} \left(s^2 L(f(t)) - sf(0) - f'(0)\right)$$

For $1 < \alpha \le 2$, we defined the ABC derivative as

$$
\begin{aligned}
{}^{ABC}_0 D^\alpha_t f(t) &= \frac{AB(\alpha)}{2-\alpha} \int_0^t \frac{d^2 f(\tau)}{d\tau^2} E_\alpha\left(\frac{-\alpha}{2-\alpha}(t-\tau)^\alpha\right) d\tau \\
&= \frac{AB(\alpha)}{2-\alpha} \frac{d^2 f(t)}{dt^2} * E_\alpha\left(\frac{-\alpha}{2-\alpha}t^\alpha\right)
\end{aligned}
$$

Then, we have the Laplace transform of the above equation as

$$L\left({}^{ABC}_0 D^\alpha_t f(t)\right) = \frac{AB(\alpha)}{2-\alpha} L\left(\frac{d^2 f(t)}{dt^2}\right) L\left(E_\alpha\left(\frac{-\alpha}{2-\alpha}t^\alpha\right)\right)$$

$$L\left({}^{ABC}_0 D^\alpha_t f(t)\right) = \frac{AB(\alpha)s^{\alpha-1}}{\alpha+(2-\alpha)s^\alpha} \left(s^2 L(f(t)) - sf(0) - f'(0)\right)$$

For $1 < \alpha \le 2$, we defined the Caputo Derivative as

$$
\begin{aligned}
{}^C_0 D^\alpha_t f(t) &= \frac{1}{\Gamma(2-\alpha)} \int_0^t \frac{d^2 f(\tau)}{d\tau^2}(t-\tau)^{1-\alpha} d\tau \\
&= \frac{1}{\Gamma(2-\alpha)} \frac{d^2 f(t)}{dt^2} * t^{1-\alpha}
\end{aligned}
$$

Then, we have the Sumudu transform of the above equation as

$$
\begin{aligned}
S\left({}^C_0 D^\alpha_t f(t)\right) &= S\left(\frac{1}{\Gamma(2-\alpha)} \frac{d^2 f(t)}{dt^2} * t^{1-\alpha}\right) \\
&= sS\left(\frac{d^2 f(t)}{dt^2}\right) S\left(\frac{1}{\Gamma(2-\alpha)} * t^{1-\alpha}\right) \\
&= s\left(\frac{S[f] - f(0)}{s^2} - \frac{f'(0)}{s}\right)(s^{1-\alpha}) \\
&= s^{2-\alpha}\left(\frac{S[f] - f(0)}{s^2} - \frac{f'(0)}{s}\right)
\end{aligned}
$$

For $1 < \alpha \leq 2$, we defined the Caputo-Fabrizio Derivative as

$$
\begin{aligned}
{}^{CF}_{0}D^{\alpha}_t f(t) &= \frac{M(\alpha)}{2-\alpha} \int_0^t \frac{d^2 f(\tau)}{d\tau^2} \exp\left(\frac{-\alpha}{2-\alpha}(t-\tau)\right) d\tau \\
&= \frac{M(\alpha)}{2-\alpha} \frac{d^2 f(t)}{dt^2} * \exp\left(\frac{-\alpha}{2-\alpha}t\right)
\end{aligned}
$$

Then, we have the Sumudu transform of the above equation as

$$
\begin{aligned}
S\left({}^{CF}_{0}D^{\alpha}_t f(t)\right) &= S\left(\frac{M(\alpha)}{2-\alpha} \frac{d^2 f(t)}{dt^2} * \exp\left(\frac{-\alpha}{2-\alpha}t\right)\right) \\
&= sS\left(\frac{d^2 f(t)}{dt^2}\right) S\left(\frac{M(\alpha)}{2-\alpha} \exp\left(\frac{-\alpha}{2-\alpha}t\right)\right) \\
&= s\left(\frac{S[f]-f(0)}{s^2} - \frac{f'(0)}{s}\right)\left(\frac{M(\alpha)}{2-\alpha+\alpha s}\right) \\
&= \frac{sM(\alpha)}{2-\alpha+\alpha s}\left(\frac{S[f]-f(0)}{s^2} - \frac{f'(0)}{s}\right)
\end{aligned}
$$

For $1 < \alpha \leq 2$, we defined the ABC derivative as

$$
\begin{aligned}
{}^{ABC}_{0}D^{\alpha}_t f(t) &= \frac{AB(\alpha)}{2-\alpha} \int_0^t \frac{d^2 f(\tau)}{d\tau^2} E_\alpha\left(\frac{-\alpha}{2-\alpha}(t-\tau)^\alpha\right) d\tau \\
&= \frac{AB(\alpha)}{2-\alpha} \frac{d^2 f(t)}{dt^2} * E_\alpha\left(\frac{-\alpha}{2-\alpha}t^\alpha\right)
\end{aligned}
$$

Then, we have the Sumudu transform of the above equation as

$$
\begin{aligned}
S\left({}^{ABC}_{0}D^{\alpha}_t f(t)\right) &= \frac{AB(\alpha)}{2-\alpha} S\left(\frac{d^2 f(t)}{dt^2}\right) S\left(E_\alpha\left(\frac{-\alpha}{2-\alpha}t^\alpha\right)\right) \\
&= s\left(\frac{S[f]-f(0)}{s^2} - \frac{f'(0)}{s}\right)\left(\frac{AB(\alpha)}{2-\alpha+\alpha s^\alpha}\right) \\
&= \frac{sAB(\alpha)}{2-\alpha+\alpha s^\alpha}\left(\frac{S[f]-f(0)}{s^2} - \frac{f'(0)}{s}\right)
\end{aligned}
$$

Theorem 1.1

We obtain the Sumudu transform of the Mittag-Leffler function as [4]

$$
S\left[E_\alpha\left(-\frac{\alpha}{1-\alpha}t^\alpha\right)\right] = \frac{1}{1+\frac{\alpha s^\alpha}{1-\alpha}}, \tag{1.22}
$$

where

$$|s|^\alpha < \frac{1-\alpha}{\alpha}$$

$$|s| < \left(\frac{1-\alpha}{\alpha}\right)^{\frac{1}{\alpha}}$$

$$a^2+b^2 < \left(\frac{1-\alpha}{\alpha}\right)^{\frac{2}{\alpha}}.$$

∎

PROOF We have the following relation between Sumudu transform and Laplace transform.

$$S[t^\alpha] = s^{2\alpha+1}L[t^\alpha]. \tag{1.23}$$

We obtain

$$
\begin{aligned}
S\left[E_\alpha\left(-\frac{\alpha}{1-\alpha}t^\alpha\right)\right] &= S\left[\sum_{k=0}^{\infty}\frac{(\frac{-\alpha t^\alpha}{1-\alpha})^k}{\Gamma(\alpha k+1)}\right] \\
&= \sum_{k=0}^{\infty}\frac{(\frac{-\alpha}{1-\alpha})^k}{\Gamma(\alpha k+1)}S[t^{\alpha k}] \\
&= \sum_{k=0}^{\infty}\frac{(\frac{-\alpha}{1-\alpha})^k}{\Gamma(\alpha k+1)}s^{2\alpha k+1}L[t^{\alpha k}] \\
&= \sum_{k=0}^{\infty}\frac{(\frac{-\alpha}{1-\alpha})^k}{\Gamma(\alpha k+1)}s^{2\alpha k+1}s^{-1-\alpha k}\Gamma(\alpha k+1) \\
&= \sum_{k=0}^{\infty}\left(\frac{-\alpha}{1-\alpha}s^\alpha\right)^k \\
&= \frac{1}{1+\frac{\alpha s^\alpha}{1-\alpha}},
\end{aligned}
$$

by using the properties of Sumudu transform and the relation between Sumudu and Laplace transform.   ∎

REMARK 1.1 The Sumudu transform of even functions is even, and the Sumudu transform of odd functions is odd.   ∎

PROOF Let $f(t)$ be an even function. Thus, we have $f(t) = f(-t)$. Therefore, we obtain

$$S[-u] = \int_0^\infty \exp(-t)f(-tu)dt = \int_0^\infty \exp(-t)f(tu)dt = S[u]. \tag{1.24}$$

Now let $f(t)$ be an odd function. Thus, we have $f(t) = -f(-t)$. Finally, we get

$$S[-u] = \int_0^\infty \exp(-t)f(-tu)dt = -\int_0^\infty \exp(-t)f(tu)dt = -S[u]. \qquad (1.25)$$

The above cannot be proven using Laplace transform. For instance, the Laplace transform of $sin(t)$ is $\frac{1}{1+s^2}$. This function is even function. This is very important as the Sumudu transform conserves the parity. ∎

We shall use some properties of the Laplace and the Sumudu transforms to derive the solution of the below decay equations. We have stressed the fact that these equations may not have exact solutions, or if they are obtained, they may not satisfy the initial conditions due to the memory effect. We consider the below decay equation with different differential operators.

We consider the decay equation with the classical derivative as

$$Df = \lambda f. \qquad (1.26)$$

We consider the decay equation with the Caputo-Fabrizio derivative as

$$_0^{CF}D_t^\alpha f = \lambda f. \qquad (1.27)$$

We consider the decay equation with the Caputo derivative as

$$_0^C D_t^\alpha f = \lambda f. \qquad (1.28)$$

We consider the decay equation with the Atangana-Baleanu derivative as

$$_0^{ABC}D_t^\alpha f = \lambda f. \qquad (1.29)$$

We find the Laplace transform of these equations. We apply the Laplace transform to both sides of Eq. 1.26, and we obtain

$$L(Df) = L(\lambda f) \qquad (1.30)$$
$$sL(f) - f(0) = \lambda L(f) \qquad (1.31)$$

After rearranging the above equation, we get

$$L(f) = \frac{f(0)}{s - \lambda}. \qquad (1.32)$$

We apply the Laplace transform to both sides of Eq. 1.27, and we obtain

$$L(_0^{CF}D_t^\alpha f) = L(\lambda f) \qquad (1.33)$$
$$-(sL(f) - f(0))\frac{M(\alpha)}{s\alpha - s - \alpha} = \lambda L(f). \qquad (1.34)$$

After rearranging the above equation, we get

$$L(f) = \frac{f(0)M(\alpha)}{\lambda(s\alpha - s - \alpha) + sM(\alpha)}. \tag{1.35}$$

We apply the Laplace transform to both sides of Eq. 1.28, and we obtain

$$L({}_0^C D_t^\alpha f) = L(\lambda f) \tag{1.36}$$

$$(sL(f) - f(0))s^{\alpha-1} = \lambda L(f). \tag{1.37}$$

After rearranging the above equation, we get

$$L(f) = \frac{f(0)s^{\alpha-1}}{s^\alpha + \lambda}. \tag{1.38}$$

We apply the Laplace transform to both sides of Eq. 1.29, and we obtain

$$L({}_0^{ABC} D_t^\alpha f) = L(\lambda f) \tag{1.39}$$

$$(sL(f) - f(0))\frac{AB(\alpha)s^{\alpha-1}}{s^\alpha(1-\alpha) + \alpha} = \lambda L(f). \tag{1.40}$$

After rearranging the above equation, we get

$$L(f) = \frac{f(0)AB(\alpha)s^{\alpha-1}}{-\lambda(s^\alpha(1-\alpha) + \alpha) + AB(\alpha)s^\alpha}. \tag{1.41}$$

We find the Sumudu transform of these equations. We apply the Sumudu transform to both sides of Eq. 1.26, and we obtain

$$S(Df) = S(\lambda f) \tag{1.42}$$

$$\frac{S(f) - f(0)}{s} = \lambda S(f). \tag{1.43}$$

After rearranging the above equation, we get

$$S(f) = \frac{f(0)}{1 - \lambda s}. \tag{1.44}$$

We apply the Sumudu transform to both sides of Eq. 1.27, and we obtain

$$S({}_0^{CF} D_t^\alpha f) = S(\lambda f) \tag{1.45}$$

$$(S(f) - f(0))\frac{M(\alpha)}{\alpha s + 1 - \alpha} = \lambda S(f). \tag{1.46}$$

After rearranging the above equation, we get

$$S(f) = \frac{f(0)M(\alpha)}{M(\alpha) - \lambda(s\alpha + 1 - \alpha))}. \tag{1.47}$$

We apply the Sumudu transform to both sides of Eq. 1.28, and we obtain

$$S({}^{C}_{0}D^{\alpha}_{t}f) = S(\lambda f) \tag{1.48}$$

$$\frac{S(f) - f(0)}{s^{\alpha}} = \lambda S(f). \tag{1.49}$$

After rearranging the above equation, we get

$$S(f) = \frac{f(0)}{1 - \lambda s^{\alpha}}. \tag{1.50}$$

We apply the Sumudu transform to both sides of Eq. 1.29, and we obtain

$$S({}^{ABC}_{0}D^{\alpha}_{t}f) = S(\lambda f) \tag{1.51}$$

$$(S(f) - f(0))\frac{AB(\alpha)}{1 - \alpha + \alpha s^{\alpha}} = \lambda S(f). \tag{1.52}$$

After rearranging the above equation, we get

$$S(f) = \frac{f(0)AB(\alpha)}{AB(\alpha) - \lambda(1 - \alpha + \alpha s^{\alpha})}. \tag{1.53}$$

2 Transfer Functions and Diagrams

Laplace transform appears to be one of the most used integral transforms, which could be applied in many fields of science. For instance, linear ordinary and partial differential equations can be solved using the Laplace transform. In signal analysis, a very important field, the Laplace transform is an essential mathematical tool that is being used very intensively to calculate the well-known transfer function. This function can be used to estimate the Bode diagram. We shall recall that the transfer function is a suitable depiction of a linear time-invariant dynamical system. We shall note that, mathematically, the function called transfer is a function of complex variables. In the case of finite dimensional systems, the transfer function is just a rational function of a complex variable. This function in other words can be calculated by inspection or algebraic manipulations of differential equations that depict the systems. This function can be used to describe systems of a very high order, even infinite dimensional systems that are governed by partial differential equations.

In engineering, a transfer function is very well used; for instance, a transfer function of an electronic or control system component is a mathematical function modeling hypothetically the device's output for each potential input. In other words, in this field, the function can be viewed as a two-dimensional graph of an independent scalar input against the dependent scalar output, known as a transfer curve.

It is therefore important to note that, all these analyses originated from the results obtained from the Laplace transform, the question, and the fundamental motivation of this section is that, what happens if one uses the results obtained from the Sumudu transform to perform the same analysis? Could we get better results when using Sumudu than Laplace? In this section, we provide a simple yet very important analysis using both Laplace and Sumudu transform.

DEFINITION 2.1 We define the transfer function $H(s)$ for continuous-time input signal $x(t)$ and output $y(t)$ as [4, 7, 34, 35]

$$H(s) = \frac{Y(s)}{X(s)} = \frac{L(y(t))}{L(x(t))}. \tag{2.1}$$

We use the transfer functions in the analysis of systems such as single-input single-output filters in the areas of signal processing, communication theory, and control theory. These functions were implemented in classical control engineering.

In this section, we present the transfer function of the delay differential equation with four different differential operators including classical differentiation, Caputo

DOI: 10.1201/9781003359869-2   **13** fractional differentiation, Caputo-Fabrizio, and the Atangana-Baleanu fractional operators.

Bode gain plot and Bode phase plot is a simple but accurate method for graphing gain and phase-shift plots on circuit theory and control theory. The Bode plot is an experiment of analysis in the frequency domain. The Bode plot for a linear, time-invariant system with transfer function $H(s)$ (s being the complex frequency in the Laplace domain) contains a magnitude plot and a phase plot. The Bode magnitude plot is the graph of the function $|H(s = j\omega)|$ of frequency ω (with j being the imaginary unit). The ω -axis of the magnitude plot is logarithmic and the magnitude is presented in decibels; i.e., a value for the magnitude $|H|$ is plotted on the axis at $20\log|H|$. The Bode phase plot is the graph of the phase, commonly given in degrees, of the transfer function $\arg(H(s = j\omega))$ as a function of ω. The phase is plotted on the same logarithmic ω-axis as the magnitude plot, but the value for the phase is plotted on a linear vertical axis. The phase Bode plot is acquired by plotting the phase angle of the transfer function presented as [36–39]

$$\arg H_{lp}(j\omega) = -\tan^{-1}\frac{\omega}{\omega_c},$$

versus ω, where ω and ω_c are the input and cutoff angular frequencies, respectively. The horizontal frequency axis, in both the magnitude and phase plots, can be replaced by the normalized (nondimensional) frequency ratio $\frac{\omega}{\omega_c}$. In such a case the plot is said to be normalized, and units of the frequencies are no longer utilized since all input frequencies are now presented as multiples of the cutoff frequency ω_c. The Bode plotter is an electronic instrument resembling an oscilloscope, which produces a Bode diagram, or a graph, of a circuit's voltage gain or phase shift plotted against frequency in a feedback control system or a filter. It is extremely useful for analyzing and testing filters and the stability of feedback control systems, through the measurement of corner (cutoff) frequencies and gain and phase margins. This is identical to the function applied by a vector network analyzer, but the network analyzer is typically utilized at much higher frequencies.

A Nyquist plot is a parametric plot of a frequency response applied in automatic control and signal processing. The most common use of Nyquist plots is for obtaining the stability of a system with feedback. In Cartesian coordinates, the real part of the transfer function is plotted on the X axis. The imaginary part is plotted on the Y axis. The frequency is swept as a parameter, resulting in a plot per frequency. The same plot can be defined utilizing polar coordinates, where the gain of the transfer function is the radial coordinate, and the phase of the transfer function is the corresponding angular coordinate. Consideration of the stability of a closed-loop negative feedback system is made by implementing the Nyquist stability criterion to the Nyquist plot of the open-loop system. This technique is easily applicable even for systems with delays and other non-rational transfer functions, which may appear difficult to analyze using other techniques. Stability is defined by investigating the number of encirclements of the point at $(-1,0)$. The range of gains over which the system will be stable can be determined by looking at crossings of the real axis. The Nyquist plot can present some information about the shape of the transfer function.

For instance, the plot presents information on the difference between the number of zeros and poles of the transfer function by the angle at which the curve approaches the origin.

The Nichols plot is a plot utilized in signal processing and control design. We consider the transfer function as [40]

$$G(s) = \frac{Y(s)}{X(s)},$$

with the closed-loop transfer function described by

$$M(s) = \frac{G(s)}{1 + G(s)}.$$

The Nichols plots demonstrates $20\log(|G(s)|)$ versus $\arg(G(s))$. In feedback control design, the plot is useful for obtaining the stability and robustness of a linear system. This implementation of the Nichols plot is central to the quantitative feedback theory (QFT) of Horowitz and Sidi, which is a well-known technique for robust control system design.

3 Analysis of First-order Circuit Model 1

First-order differential equations have found applications in many real-world problems. One of the most used of such equations is the well-known decay equation that explains the decay of a given entity. However, this class of differential equations has also found applications in electrical mechanics, in this field, first-order circuits that contain only one energy element, for example, capacitor or inductor. This is modeled using only the first-order differential equation. In general, there are two types including RC, which contains a resistor and capacitor. They can be placed in parallel or series. It is also known that inductors are best solved by considering the current flowing via the inductor. On the other hand, an RC circuit is a circuit that contains a resistor R and a capacitor C. In this chapter, therefore, we will consider first-order RC where the first-time derivation is classical, fractional in Caputo, Caputo-Fabrizio, and Atangana-Baleanu types. The equations will be solved with both Sumudu and Laplace transforms. The associated transfer functions will be obtained, and their corresponding Bode diagrams will be plotted and compared. Of course, in this book we will not be in the position to say which Bode diagram is more realistic or practical, we will refrain from the obtained results [41, 42].

We consider the first-order circuit model 1 with classical, Caputo, Caputo Fabrizio, and Atangana-Baleanu derivatives.

3.1 ANALYSIS OF FIRST-ORDER CIRCUIT MODEL 1 WITH CLASSICAL DERIVATIVE

We consider the model with classical derivative. We apply Laplace and Sumudu transforms to obtain the solution.

$$V_2(t) + RC\frac{dV_2(t)}{dt} = \frac{1}{2}\left(V_1(t) - RC\frac{dV_1(t)}{dt}\right) \tag{3.1}$$

If we take the Laplace transform of the both sides of Eq. (3.1), we will obtain

$$L(V_2(t)) + RCL\left(\frac{dV_2(t)}{dt}\right) = \frac{1}{2}L(V_1(t)) - \frac{RC}{2}L\left(\frac{dV_1(t)}{dt}\right). \tag{3.2}$$

Then, we get

$$L(V_2(t)) + sRCL(V_2(t)) - RCV_2(0) = \frac{1}{2}L(V_1(t)) - \frac{sRC}{2}L(V_1(t)) + \frac{RC}{2}V_1(0). \tag{3.3}$$

$$L(V_2(t))(1 + sRC) = L(V_1(t))\left(\frac{1 - sRC}{2}\right). \tag{3.4}$$

DOI: 10.1201/9781003359869-3

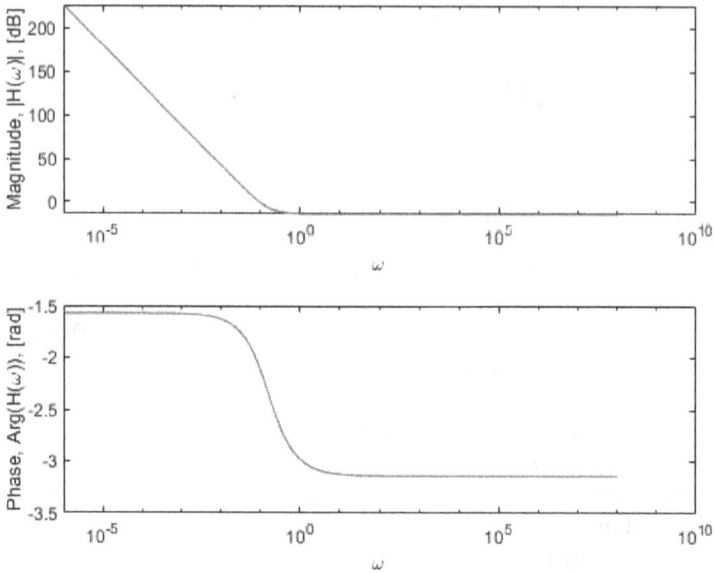

Figure 3.1 Transfer function of the first-order circuit model 1 with the classical derivative using the Laplace transform.

Therefore, we obtain the transfer function as

$$\frac{L(V_2(t))}{L(V_1(t))} = \frac{1 - sRC}{2(1 + sRC)}. \tag{3.5}$$

The graphical representation of the above transfer function is presented in Figure 3.1 as magnitude and phase.

If we take the Sumudu transform of the both sides of Eq. (3.1), we will obtain

$$S(V_2(t)) + RCS\left(\frac{dV_2(t)}{dt}\right) = \frac{1}{2}S(V_1(t)) - \frac{RC}{2}S\left(\frac{dV_1(t)}{dt}\right). \tag{3.6}$$

Then, we get

$$S(V_2(t)) + \frac{RC}{s}(S(V_2(t)) - V_2(0)) = \frac{S(V_1(t))}{2} - \frac{RC}{2s}(S(V_1(t)) - V_1(0)). \tag{3.7}$$

$$S(V_2(t))\left(1 + \frac{RC}{s}\right) = S(V_1(t))\left(\frac{s - RC}{2s}\right). \tag{3.8}$$

Therefore, we obtain the transfer function as

$$\frac{S(V_2(t))}{S(V_1(t))} = \frac{s - RC}{2(s + RC)}. \tag{3.9}$$

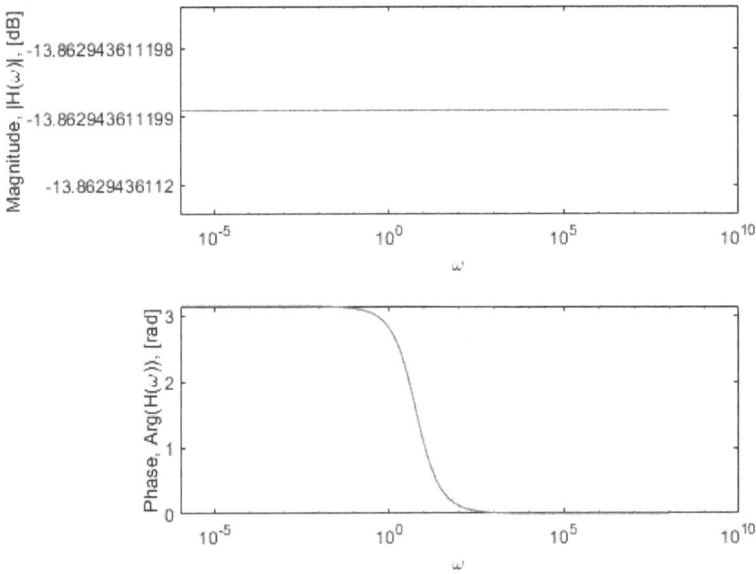

Figure 3.2 Transfer function of the first-order circuit model 1 with the classical derivative using the Sumudu transform.

The graphical representation of the above transfer function is presented in Figure 3.2 as magnitude and phase.

3.2 ANALYSIS OF FIRST-ORDER CIRCUIT MODEL 1 WITH CAPUTO DERIVATIVE

We consider the model with Caputo derivative. We apply Laplace and Sumudu transforms to obtain the solution.

$$V_2(t) + RC \left({}_0^C D_t^\alpha V_2(t) \right) = \frac{1}{2} \left(V_1(t) - RC\frac{dV_1(t)}{dt} \right) \tag{3.10}$$

If we take the Laplace transform of the both sides of Eq. (3.10), we will obtain

$$L(V_2(t)) + RCL \left({}_0^C D_t^\alpha V_2(t) \right) = \frac{1}{2}L(V_1(t)) - \frac{RC}{2}L\left(\frac{dV_1(t)}{dt} \right). \tag{3.11}$$

Then, we get

$$L(V_2(t)) + RCs^{\alpha-1}\left(sL(V_2(t)) - V_2(0) \right) = \frac{1}{2}L(V_1(t)) - \frac{sRC}{2}L(V_1(t)) + \frac{RC}{2}V_1(0). \tag{3.12}$$

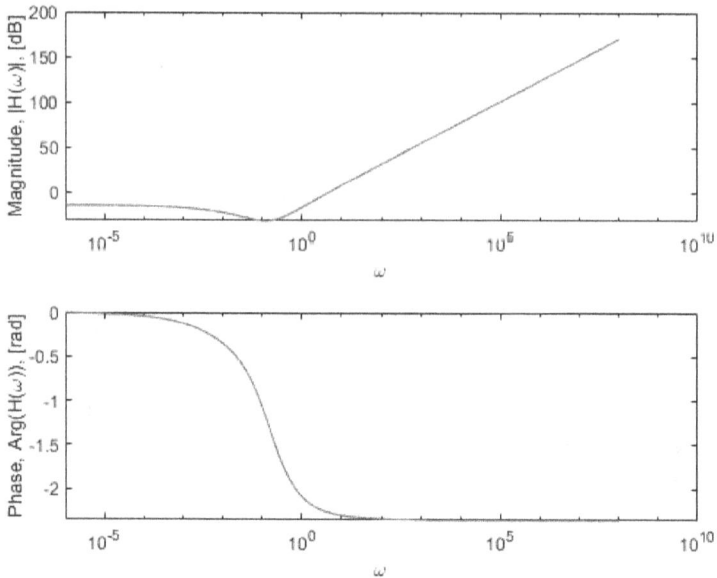

Figure 3.3 Transfer function of the first-order circuit model 1 with the Caputo derivative using the Laplace transform.

$$L(V_2(t))(1+s^\alpha RC) = L(V_1(t))\left(\frac{1-sRC}{2}\right). \tag{3.13}$$

Therefore, we obtain the transfer function as

$$\frac{L(V_2(t))}{L(V_2(t))} = \frac{1-sRC}{2(1+s^\alpha RC)}. \tag{3.14}$$

The graphical representation of the above transfer function is presented in Figure 3.3 as magnitude and phase.

If we take the Sumudu transform of the both sides of Eq. (3.10), we will obtain

$$S(V_2(t)) + RCS\left({}_0^C D_t^\alpha V_2(t)\right) = \frac{1}{2}S(V_1(t)) - \frac{RC}{2}S\left(\frac{dV_1(t)}{dt}\right). \tag{3.15}$$

Then, we get

$$S(V_2(t)) + \frac{RC}{s^\alpha}\left(S(V_2(t)) - V_2(0)\right) = \frac{S(V_1(t))}{2} - \frac{RC}{2s}\left(S(V_1(t)) - V_1(0)\right). \tag{3.16}$$

$$S(V_2(t))\left(1 + \frac{RC}{s^\alpha}\right) = S(V_1(t))\left(\frac{s-RC}{2s}\right). \tag{3.17}$$

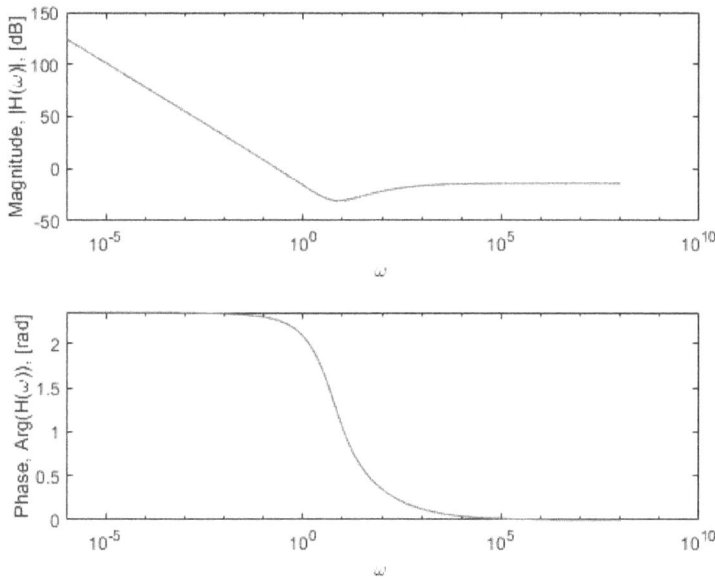

Figure 3.4 Transfer function of the first order circuit model 1 with the Caputo derivative using the Sumudu transform.

Therefore, we obtain the transfer function as

$$\frac{S(V_2(t))}{S(V_1(t))} = \frac{(s - RC)s^{\alpha-1}}{2(s^\alpha + RC)}.$$ (3.18)

The graphical representation of the above transfer function is presented in Figure 3.4 as magnitude and phase.

3.3 ANALYSIS OF FIRST-ORDER CIRCUIT MODEL 1 WITH CAPUTO-FABRIZIO DERIVATIVE

We consider the model with Caputo-Fabrizio derivative. We apply Laplace and Sumudu transform to obtain the solution.

$$V_2(t) + RC \left({}^{CF}_0 D^\alpha_t V_2(t) \right) = \frac{1}{2} \left(V_1(t) - RC \frac{dV_1(t)}{dt} \right)$$ (3.19)

If we take the Laplace transform of the both sides of Eq. (3.19), we will obtain

$$L(V_2(t)) + RCL \left({}^{CF}_0 D^\alpha_t V_2(t) \right) = \frac{1}{2} L(V_1(t)) - \frac{RC}{2} L \left(\frac{dV_1(t)}{dt} \right).$$ (3.20)

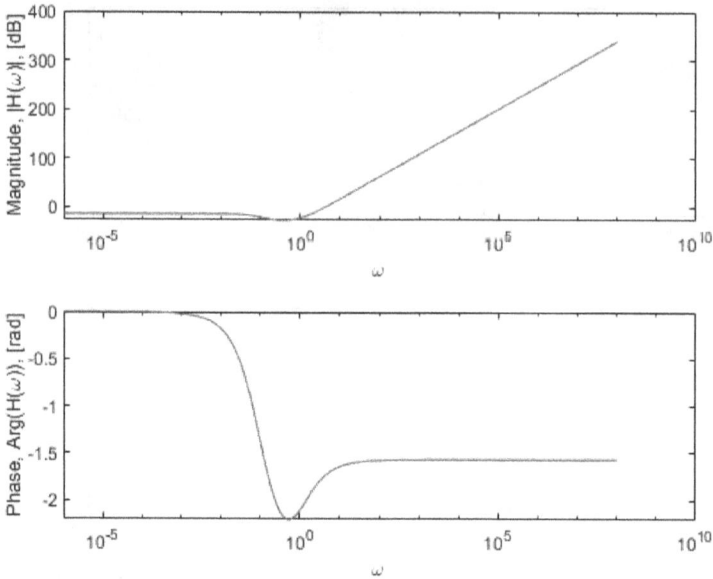

Figure 3.5 Transfer function of the first-order circuit model 1 with the Caputo-Fabrizio derivative using the Laplace transform.

Then, we get

$$L(V_2(t)) - \frac{M(\alpha)}{s\alpha - s - \alpha} RC(sL(V_2(t)) - V_2(0)) = \frac{1}{2}L(V_1(t)) - \frac{sRC}{2}L(V_1(t))$$
$$+ \frac{RC}{2}V_1(0). \tag{3.21}$$

$$L(V_2(t))\left(1 + \frac{sRCM(\alpha)}{s+\alpha - s\alpha}\right) = L(V_1(t))\left(\frac{1-sRC}{2}\right). \tag{3.22}$$

Therefore, we obtain the transfer function as

$$\frac{L(V_2(t))}{L(V_1(t))} = \frac{(s+\alpha - s\alpha)(1-sRC)}{2(s+\alpha - s\alpha + sRCM(\alpha))}. \tag{3.23}$$

The graphical representation of the above transfer function is presented in Figure 3.5 as magnitude and phase.

If we take the Sumudu transform of the both sides of Eq. (3.19), we will obtain

$$S(V_2(t)) + RCS\left({}_0^{CF}D_t^\alpha V_2(t)\right) = \frac{1}{2}S(V_1(t)) - \frac{RC}{2}S\left(\frac{dV_1(t)}{dt}\right). \tag{3.24}$$

Then, we get

$$S(V_2(t)) + \frac{RCM(\alpha)}{s\alpha+1-\alpha}(S(V_2(t))-V_2(0)) = \frac{S(V_1(t))}{2} - \frac{RC}{2s}(S(V_1(t))-V_1(0)).$$
(3.25)

$$S(V_2(t))\left(1+\frac{RCM(\alpha)}{s\alpha+1-\alpha}\right) = S(V_1(t))\left(\frac{s-RC}{2s}\right).$$
(3.26)

Therefore, we obtain the transfer function as

$$\frac{S(V_2(t))}{S(V_1(t))} = \frac{(s-RC)(s\alpha+1-\alpha)}{2s(s\alpha+1-\alpha+RCM(\alpha))}.$$
(3.27)

The graphical representation of the above transfer function is presented in Figure 3.6 as magnitude and phase.

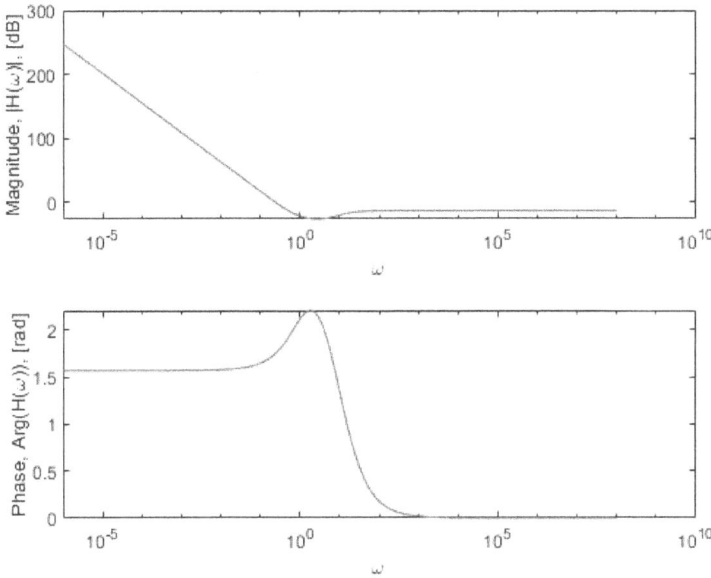

Figure 3.6 Transfer function of the first-order circuit model 1 with the Caputo-Fabrizio derivative using the Sumudu transform.

3.4 ANALYSIS OF FIRST-ORDER CIRCUIT MODEL 1 WITH ATANGANA-BALEANU DERIVATIVE

We consider the model with Atangana-Baleanu derivative. We apply Laplace and Sumudu transforms to obtain the solution.

$$V_2(t) + RC\left(_0^{ABC}D_t^\alpha V_2(t)\right) = \frac{1}{2}\left(V_1(t) - RC\frac{dV_1(t)}{dt}\right)$$
(3.28)

If we take the Laplace transform of the both sides of Eq. (3.28), we will obtain

$$L(V_2(t)) + RCL \left({}_0^{ABC}D_t^\alpha V_2(t) \right) = \frac{1}{2}L(V_1(t)) - \frac{RC}{2}L\left(\frac{dV_1(t)}{dt} \right). \qquad (3.29)$$

Then, we get

$$L(V_2(t)) + \frac{AB(\alpha)s^{\alpha-1}}{s^\alpha(1-\alpha)+\alpha}RC\left(sL(V_2(t)) - V_2(0)\right) = \frac{1}{2}L(V_1(t)) - \frac{sRC}{2}L(V_1(t))$$
$$+ \frac{RC}{2}V_1(0). \qquad (3.30)$$

$$L(V_2(t)) \left(1 + \frac{s^\alpha RCAB(\alpha)}{s^\alpha(1-\alpha)+\alpha} \right) = L(V_1(t)) \left(\frac{1 - sRC}{2} \right). \qquad (3.31)$$

Therefore, we obtain the transfer function as

$$\frac{L(V_2(t))}{L(V_1(t))} = \frac{(1 - sRC)(s^\alpha(1-\alpha)+\alpha)}{2(s^\alpha(1-\alpha)+\alpha+s^\alpha RCAB(\alpha))}. \qquad (3.32)$$

The graphical representation of the above transfer function is presented in Figure 3.7 as magnitude and phase.

If we take the Sumudu transform of the both sides of Eq. (3.28), we will obtain

$$S(V_2(t)) + RCS \left({}_0^{ABC}D_t^\alpha V_2(t) \right) = \frac{1}{2}S(V_1(t)) - \frac{RC}{2}S\left(\frac{dV_1(t)}{dt} \right). \qquad (3.33)$$

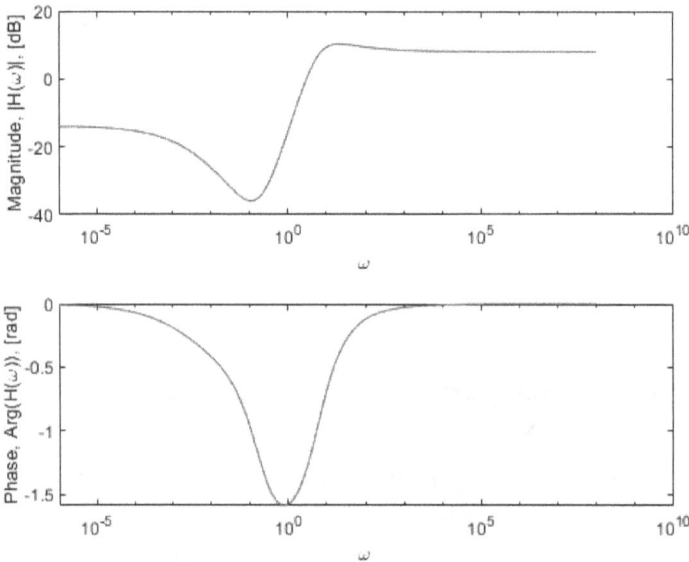

Figure 3.7 Transfer function of the first-order circuit model 1 with the ABC derivative using the Laplace transform.

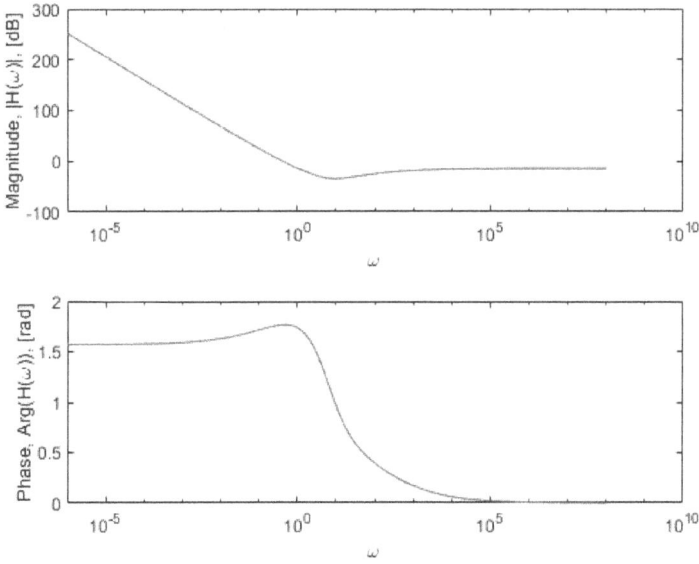

Figure 3.8   Transfer function of the first-order circuit model 1 with the ABC derivative using the Sumudu transform.

Then, we get

$$S(V_2(t)) + \frac{RCAB(\alpha)}{s^\alpha \alpha + 1 - \alpha}(S(V_2(t)) - V_2(0)) = \frac{S(V_1(t))}{2} - \frac{RC}{2s}(S(V_1(t)) - V_1(0)).$$
$$(3.34)$$

$$S(V_2(t))\left(1 + \frac{RCAB(\alpha)}{s^\alpha \alpha + 1 - \alpha}\right) = S(V_1(t))\left(\frac{s - RC}{2s}\right).$$ (3.35)

Therefore, we obtain the transfer function as

$$\frac{S(V_2(t))}{S(V_1(t))} = \frac{(s - RC)(1 - \alpha + \alpha s^\alpha)}{2s(1 - \alpha + \alpha s^\alpha + RCAB(\alpha))}.$$ (3.36)

The graphical representation of the above transfer function is presented in Figure 3.8 as magnitude and phase.

4 Analysis of First-order Circuit Model 2

Many real-world problems can be modeled using first-order systems. They occur often in many physical problems, and a simple example is the decay process. In mathematics and applied mathematics, differential equations with the first order, or first-order system can be viewed as any system that can absorb energy via a storage element and release that stored energy. The available literature proved that a capacitor stores energy in the electric field within its dielectric medium, and an inductor stores energy in the magnetic area induced by the current flowing through its conductors. Therefore, for electric circuits, any circuit that includes a single capacitor or a single inductor in addition to resistors, voltage, and/or current sources can be classified as a first-order circuit. First-order circuits are called RC or RL circuits, respectively, and can be defined by a first-order differential equation. The analysis of first-order circuits contains investigating the behavior of the circuit as a function of time before and after a sudden change in the circuit due to switching actions. There are many approaches applied to research first-order circuits. Electric circuits that include only resistors in addition to current and/or voltage sources are called static circuits and are given by algebraic equations. The circuits that include storage elements (capacitors and inductors) in addition to current and/or voltage sources are called dynamic circuits and are given by differential equations. Capacitors and inductors are called storage elements due to their ability to store energy. Inductors can store magnetic energy in their area, whereas capacitors can store electric energy in their area. The analysis of first-order circuits needs the solution of differential equations. The complete solution contains two parts: the homogeneous solution and the particular solution. The particular solution of a first-order circuit with DC sources and switching action is the steady-state response and called the forced response. The homogenous solution consists of the characteristic mode of the first-order circuit, which decays to zero after a few time constants and is called the transient response [43–45].

We consider the first-order circuit model 2 with classical, Caputo, Caputo Fabrizio, and Atangana-Baleanu derivatives.

4.1 ANALYSIS OF FIRST-ORDER CIRCUIT MODEL 2 WITH CLASSICAL DERIVATIVE

We consider the model with classical derivative. We apply Laplace and Sumudu transforms to obtain the solution.

$$V_2(t) + R_1 C_1 \frac{dV_2(t)}{dt} = -\frac{C_1}{C_2}\left(V_1(t) - R_2 C_2 \frac{dV_1(t)}{dt}\right). \tag{4.1}$$

DOI: 10.1201/9781003359869-4

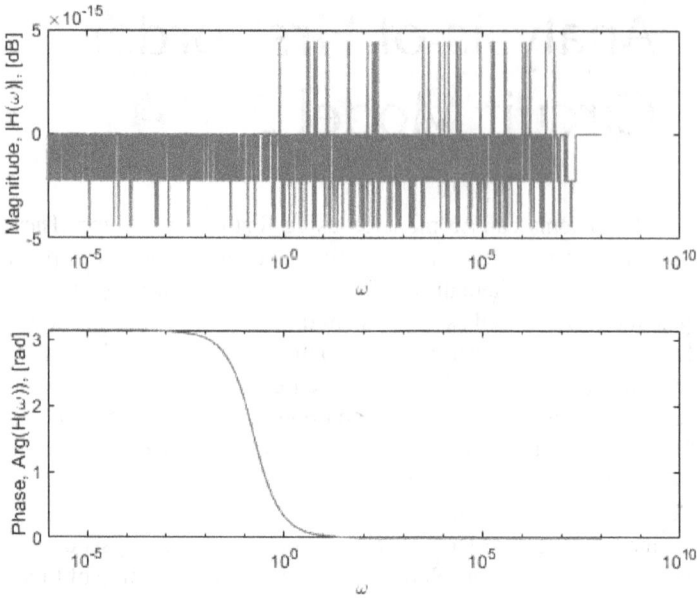

Figure 4.1 Transfer function of the first-order circuit model 2 with the classical derivative using the Laplace transform.

If we take the Laplace transform of the both sides of Eq. (4.1), we will obtain

$$L(V_2(t)) + R_1 C_1 L\left(\frac{dV_2(t)}{dt}\right) = -\frac{C_1}{C_2}\left(L(V_1(t)) - R_2 C_2 L\left(\frac{dV_1(t)}{dt}\right)\right). \quad (4.2)$$

Then, we will get

$$L(V_2(t)) + sR_1 C_1 L(V_2(t)) - R_1 C_1 V_2(0) = -\frac{C_1}{C_2}(L(V_1(t)) - sR_2 C_2 L(V_1(t))$$
$$+ R_2 C_2 V_1(0)). \quad (4.3)$$

$$L(V_2(t))(1 + sR_1 C_1) = L(V_1(t))\left(-\frac{C_1}{C_2} + sC_1 R_2\right). \quad (4.4)$$

Then, we get the transfer function as

$$\frac{L(V_2(t))}{L(V_1(t))} = \frac{-C_1 + sC_1 C_2 R_2}{C_2(1 + sR_1 C_1)}. \quad (4.5)$$

The graphical representation of the above transfer function is presented in Figure 4.1 as magnitude and phase.

If we take the Sumudu transform of the both sides of Eq. (4.1), we will obtain

$$S(V_2(t)) + R_1 C_1 S\left(\frac{dV_2(t)}{dt}\right) = -\frac{C_1}{C_2}\left(S(V_1(t)) - R_2 C_2 S\left(\frac{dV_1(t)}{dt}\right)\right). \quad (4.6)$$

Then, we will get

$$S\left(V_2(t)\right)+R_1C_1\frac{1}{s}S\left(V_2(t)\right)-R_1C_1\frac{f(0)}{s} = -\frac{C_1}{C_2}S\left(V_1(t)\right)+R_2C_2\frac{1}{s}S\left(V_1(t)\right)$$
$$-R_2C_2\frac{f(0)}{s}. \tag{4.7}$$

$$S\left(V_2(t)\right)\left(1+\frac{R_1C_1}{s}\right)=S\left(V_1(t)\right)\left(-\frac{C_1}{C_2}+\frac{C_1R_2}{s}\right). \tag{4.8}$$

Then, we get the Transfer function as

$$\frac{S\left(V_2(t)\right)}{S\left(V_1(t)\right)}=\frac{-C_1s+C_1R_1}{C_2(s+R_1C_1)}. \tag{4.9}$$

The graphical representation of the above transfer function is presented in Figure 4.2 as magnitude and phase.

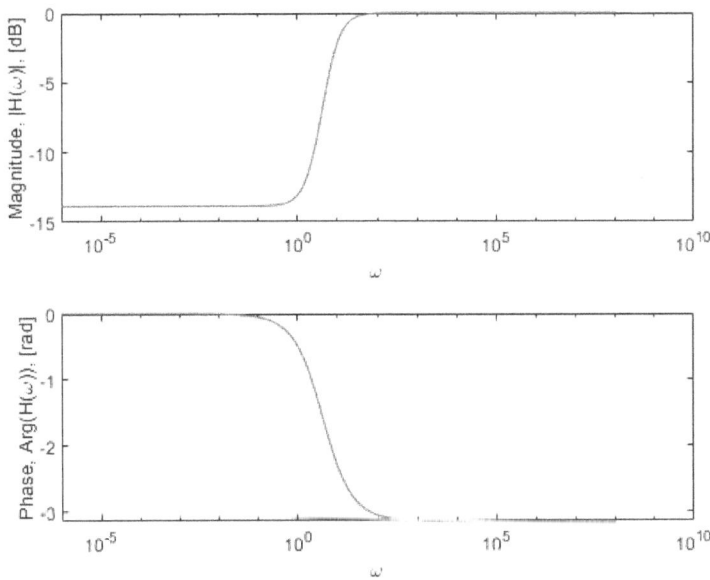

Figure 4.2 Transfer function of the first-order circuit model 2 with the classical derivative using the Sumudu transform.

4.2 ANALYSIS OF FIRST-ORDER CIRCUIT MODEL 2 WITH CAPUTO DERIVATIVE

We consider the model with Caputo derivative. We apply Laplace and Sumudu transforms to obtain the solution.

$$V_2(t) + R_1C_1 \left({}_0^C D_t^\alpha V_2(t) \right) = -\frac{C_1}{C_2} \left(V_1(t) - R_2C_2 \frac{dV_1(t)}{dt} \right). \tag{4.10}$$

If we take the Laplace transform of the both sides of Eq. (4.10), we will obtain

$$L(V_2(t)) + R_1C_1L \left({}_0^C D_t^\alpha V_2(t) \right) = -\frac{C_1}{C_2} \left(L(V_1(t)) - R_2C_2L \left(\frac{dV_1(t)}{dt} \right) \right). \tag{4.11}$$

Then, we will get

$$L(V_2(t)) + s^\alpha R_1C_1L(V_2(t)) - s^{\alpha-1}R_1C_1V_2(0) \tag{4.12}$$

$$= -\frac{C_1}{C_2} \left(L(V_1(t)) - sR_2C_2L(V_1(t)) + R_2C_2V_1(0) \right). \tag{4.13}$$

$$L(V_2(t))(1 + s^\alpha R_1C_1) = L(V_1(t)) \left(-\frac{C_1}{C_2} + sC_1R_2 \right). \tag{4.14}$$

Then, we get the Transfer function as

$$\frac{L(V_2(t))}{L(V_1(t))} = \frac{-C_1 + sC_1C_2R_2}{C_2(1 + s^\alpha R_1C_1)}. \tag{4.15}$$

The graphical representation of the above transfer function is presented in Figure 4.3 as magnitude and phase.

If we take the Sumudu transform of the both sides of Eq. (4.10), we will obtain

$$S(V_2(t)) + R_1C_1S \left({}_0^C D_t^\alpha V_2(t) \right) = -\frac{C_1}{C_2} \left(S(V_1(t)) - R_2C_2S \left(\frac{dV_1(t)}{dt} \right) \right). \tag{4.16}$$

Then, we will get

$$S(V_2(t)) + R_1C_1 \frac{1}{s^\alpha} S(V_2(t)) - R_1C_1 \frac{f(0)}{s^\alpha} = -\frac{C_1}{C_2} S(V_1(t)) + R_2C_2 \frac{1}{s} S(V_1(t))$$

$$- R_2C_2 \frac{f(0)}{s}. \tag{4.17}$$

$$S(V_2(t)) \left(1 + \frac{R_1C_1}{s^\alpha} \right) = S(V_1(t)) \left(-\frac{C_1}{C_2} + \frac{C_1R_2}{s} \right). \tag{4.18}$$

Then, we get the Transfer function as

$$\frac{S(V_2(t))}{S(V_1(t))} = s^{\alpha-1} \frac{-C_1s + C_1R_1}{C_2(s^\alpha + R_1C_1)}. \tag{4.19}$$

The graphical representation of the above transfer function is presented in Figure 4.4 as magnitude and phase.

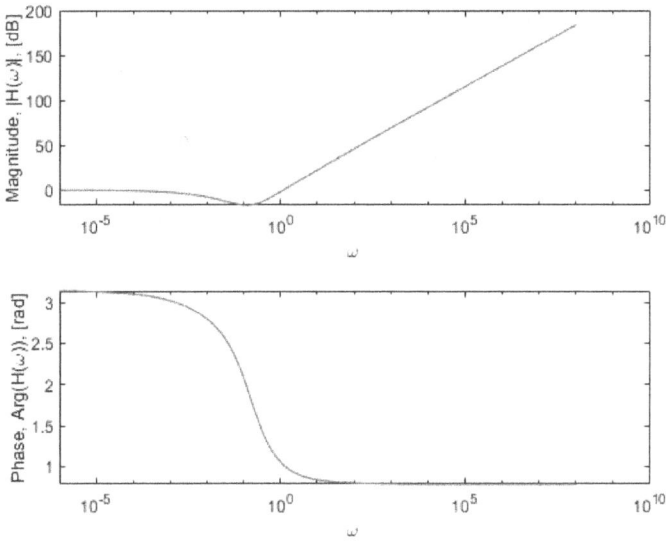

Figure 4.3 Transfer function of the first-order circuit model 2 with the Caputo derivative using the Laplace transform.

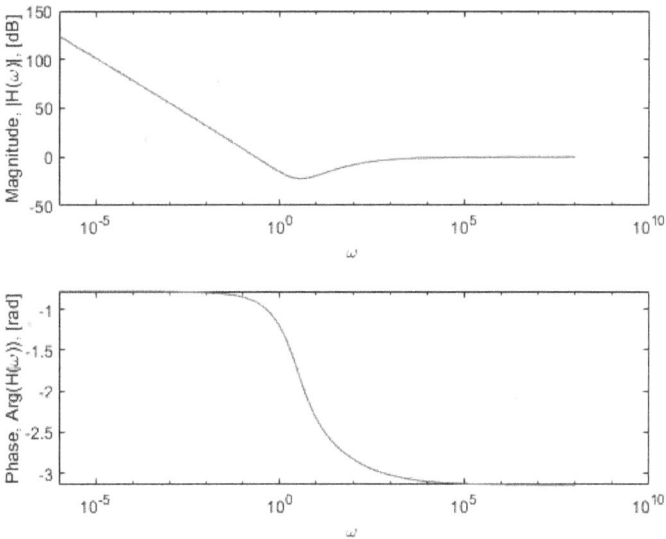

Figure 4.4 Transfer function of the first-order circuit model 2 with the Caputo derivative using the Sumudu transform.

4.3   ANALYSIS OF FIRST-ORDER CIRCUIT MODEL 2 WITH CAPUTO-FABRIZIO DERIVATIVE

We consider the model with Caputo-Fabrizio derivative. We apply Laplace and Sumudu transforms to obtain the solution.

$$V_2(t) + R_1 C_1 \left({}_0^{CF} D_t^\alpha V_2(t) \right) = -\frac{C_1}{C_2} \left(V_1(t) - R_2 C_2 \frac{dV_1(t)}{dt} \right). \tag{4.20}$$

If we take the Laplace transform of the both sides of Eq. (4.20), we will obtain

$$L(V_2(t)) + R_1 C_1 L \left({}_0^{CF} D_t^\alpha V_2(t) \right) = -\frac{C_1}{C_2} \left(L(V_1(t)) - R_2 C_2 L \left(\frac{dV_1(t)}{dt} \right) \right). \tag{4.21}$$

Then, we will get

$$L(V_2(t)) + \frac{sM(\alpha)}{s+\alpha-s\alpha} R_1 C_1 L(V_2(t)) - \frac{sM(\alpha)}{s+\alpha-s\alpha} R_1 C_1 V_2(0)$$
$$= -\frac{C_1}{C_2} \left(L(V_1(t)) - sR_2 C_2 L(V_1(t)) + R_2 C_2 V_1(0) \right).$$

$$L(V_2(t)) \left(1 + \frac{sM(\alpha)}{s+\alpha-s\alpha} R_1 C_1 \right) = L(V_1(t)) \left(-\frac{C_1}{C_2} + sC_1 R_2 \right). \tag{4.22}$$

Then, we get the Transfer function as

$$\frac{L(V_2(t))}{L(V_1(t))} = \frac{(-C_1 + sC_1 C_2 R_2)(s+\alpha-s\alpha)}{C_2(s+\alpha-s\alpha+sM(\alpha)R_1 C_1)}. \tag{4.23}$$

The graphical representation of the above transfer function is presented in Figure 4.5 as magnitude and phase.

If we take the Sumudu transform of the both sides of Eq. (4.20), we will obtain

$$S(V_2(t)) + R_1 C_1 S \left({}_0^{CF} D_t^\alpha V_2(t) \right) = -\frac{C_1}{C_2} \left(S(V_1(t)) - R_2 C_2 S \left(\frac{dV_1(t)}{dt} \right) \right). \tag{4.24}$$

Then, we will get

$$S(V_2(t)) + R_1 C_1 \frac{M(\alpha)}{s\alpha+1-\alpha} S(V_2(t)) - V_2(0)R_1 C_1 \frac{M(\alpha)}{s\alpha+1-\alpha}$$
$$= -\frac{C_1}{C_2} S(V_1(t)) + R_2 C_2 \frac{1}{s} S(V_1(t)) - R_2 C_2 \frac{f(0)}{s}.$$

$$S(V_2(t)) \left(1 + R_1 C_1 \frac{M(\alpha)}{s\alpha+1-\alpha} \right) = S(V_1(t)) \left(-\frac{C_1}{C_2} + \frac{C_1 R_2}{s} \right). \tag{4.25}$$

Then, we get the Transfer function as

$$\frac{S(V_2(t))}{S(V_1(t))} = \frac{(-C_1 s + C_1 C_2 R_2)(s\alpha+1-\alpha)}{C_2 s(s\alpha+1-\alpha+R_1 C_1 M(\alpha))}. \tag{4.26}$$

The graphical representation of the above transfer function is presented in Figure 4.6 as magnitude and phase.

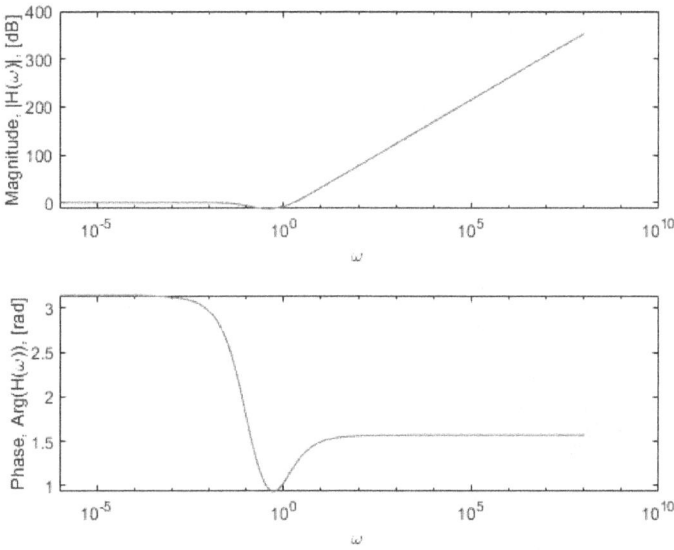

Figure 4.5 Transfer function of the first-order circuit model 2 with the Caputo-Fabrizio derivative using the Laplace transform.

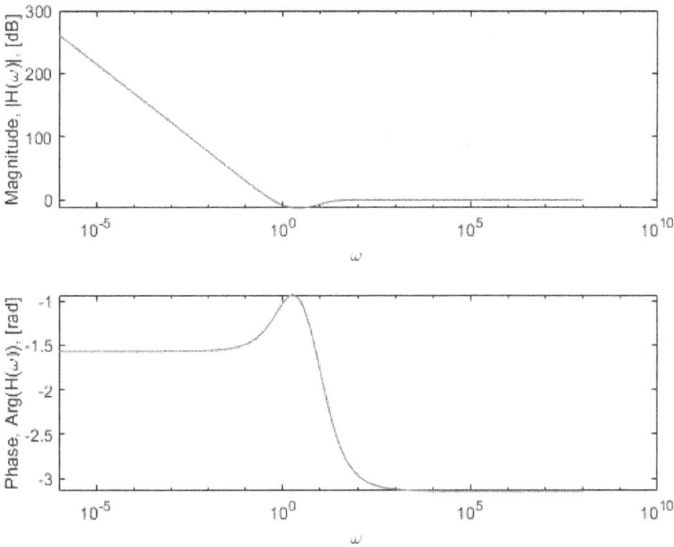

Figure 4.6 Transfer function of the first-order circuit model 2 with the Caputo-Fabrizio derivative using the Sumudu transform.

4.4 ANALYSIS OF FIRST-ORDER CIRCUIT MODEL 2 WITH ATANGANA-BALEANU DERIVATIVE

We consider the model with Atangana-Baleanu derivative. We apply Laplace and Sumudu transforms to obtain the solution.

$$V_2(t) + R_1 C_1 \left({}_0^{ABC} D_t^\alpha V_2(t) \right) = -\frac{C_1}{C_2} \left(V_1(t) - R_2 C_2 \frac{dV_1(t)}{dt} \right). \tag{4.27}$$

If we take the Laplace transform of the both sides of Eq. (4.27), we will obtain

$$L(V_2(t)) + R_1 C_1 L \left({}_0^{ABC} D_t^\alpha V_2(t) \right) = -\frac{C_1}{C_2} \left(L(V_1(t)) - R_2 C_2 L \left(\frac{dV_1(t)}{dt} \right) \right). \tag{4.28}$$

Then, we will get

$$L(V_2(t)) + \frac{AB(\alpha)s^{\alpha-1}}{s^\alpha(1-\alpha)+\alpha} R_1 C_1 L(V_2(t)) - \frac{AB(\alpha)s^{\alpha-1}}{s^\alpha(1-\alpha)+\alpha} R_1 C_1 V_2(0)$$

$$= -\frac{C_1}{C_2} \left(L(V_1(t)) - sR_2 C_2 L(V_1(t)) + R_2 C_2 V_1(0) \right).$$

$$L(V_2(t)) \left(1 + \frac{AB(\alpha)s^{\alpha-1}}{s^\alpha(1-\alpha)+\alpha} R_1 C_1 \right) = L(V_1(t)) \left(-\frac{C_1}{C_2} + sC_1 R_2 \right). \tag{4.29}$$

Then, we get the transfer function as

$$\frac{L(V_2(t))}{L(V_1(t))} = \frac{(-C_1 + sC_1 C_2 R_2)(s^\alpha(1-\alpha)+\alpha)}{C_2(s^\alpha(1-\alpha)+\alpha+AB(\alpha)s^{\alpha-1}R_1 C_1)}. \tag{4.30}$$

The graphical representation of the above transfer function is presented in Figure 4.7 as magnitude and phase.

If we take the Sumudu transform of the both sides of Eq. (4.27), we will obtain

$$S(V_2(t)) + R_1 C_1 S \left({}_0^{ABC} D_t^\alpha V_2(t) \right) = -\frac{C_1}{C_2} \left(S(V_1(t)) - R_2 C_2 S \left(\frac{dV_1(t)}{dt} \right) \right). \tag{4.31}$$

Then, we will get

$$S(V_2(t)) + R_1 C_1 \frac{AB(\alpha)}{1-\alpha+\alpha s^\alpha} S(V_2(t)) - V_2(0) \frac{AB(\alpha)}{1-\alpha+\alpha s^\alpha}$$

$$= -\frac{C_1}{C_2} S(V_1(t)) + R_2 C_2 \frac{1}{s} S(V_1(t)) - R_2 C_2 \frac{f(0)}{s}.$$

$$S(V_2(t)) \left(1 + R_1 C_1 \frac{AB(\alpha)}{1-\alpha+\alpha s^\alpha} \right) = S(V_1(t)) \left(-\frac{C_1}{C_2} + \frac{C_1 R_2}{s} \right). \tag{4.32}$$

Then, we get the transfer function as

$$\frac{S(V_2(t))}{S(V_1(t))} = \frac{(-C_1 s + C_1 C_2 R_2)(\alpha s^\alpha + 1 - \alpha)}{C_2 s(\alpha s^\alpha + 1 - \alpha + AB(\alpha)R_1 C_1)}. \tag{4.33}$$

The graphical representation of the above transfer function is presented in Figure 4.8 as magnitude and phase.

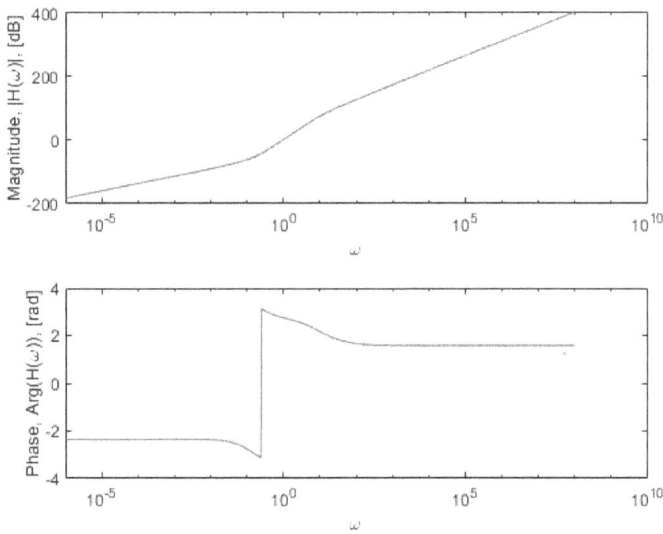

Figure 4.7 Transfer function of the first-order circuit model 2 with the ABC derivative using the Laplace transform.

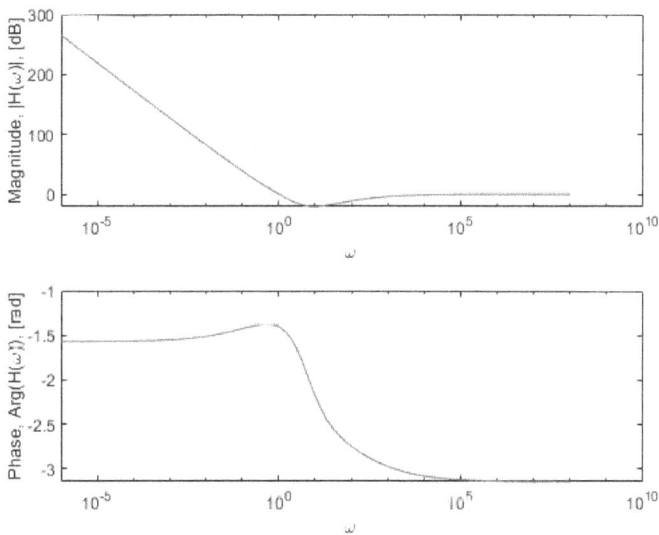

Figure 4.8 Transfer function of the first-order circuit model 2 with the ABC derivative using the Sumudu transform.

5 Analysis of Noninverting Integrators Model 1

Before the transition from the old electronic to the digital world, important technology like control systems, which have been solutions obtained from ordinary differential equations, these solutions were obtained using the so-called analog computation to derive solutions to those equations. Thus, as a result, it can be concluded that this analog computer was moderately common, as almost all solutions to differential equations required the capability to integral signal. On a serious note, control systems have mostly gone digital, thus numerical integration has replaced integration, but one must note that there is still a place for analog integrator circuits for the operation of sensors, signal generation, and filtering. It is worth noting that a rudimentary integrator inverts the integral of the signal, while the second inverting op amp connected in series with the straightforward integrator can reinstate the original phase, thus there is a possibility to construct a non-inverting integrator in a single state. These types of integrators use also differential integrators to keep the result in phase with the input signal. This version has additional properties, for example, the addition of passive components, which should be matched to better the performance. For this version, the connection between the input and the output voltages is the same as the basic integrator except for the sign. In this chapter, the combination of a high-pass filter with the noninverting-integrator circuit system is considered with the aim to improve low-frequency performance. The mathematical models associated with these systems will be considered with classical and fractional differential operators. We aim to obtain solutions using the Sumudu and the Laplace transforms in the complex number space [46, 47].

5.1 ANALYSIS OF NONINVERTING INTEGRATORS MODEL 1 WITH CLASSICAL DERIVATIVE

We consider the model with classical derivative. We apply Laplace and Sumudu transforms to obtain the solution.

$$V_1(t) = \frac{RC}{2} \frac{dV_2(t)}{dt} \tag{5.1}$$

If we take the Laplace transform of the both sides of Eq. (5.1), we will obtain

$$L(V_1(t)) = \frac{RC}{2} L\left(\frac{dV_2(t)}{dt}\right). \tag{5.2}$$

$$L(V_1(t)) = \frac{sRC}{2} L(V_2(t)) - \frac{RC}{2} V_2(0). \tag{5.3}$$

DOI: 10.1201/9781003359869-5

Thus, we get the transfer function as

$$\frac{L(V_2(t))}{L(V_1(t))} = \frac{2}{sRC}.$$ (5.4)

The graphical representation of the above transfer function is presented in Figure 5.1 as magnitude and phase.

If we take the Sumudu transform of the both sides of Eq. (5.1), we will obtain

$$S(V_1(t)) = \frac{RC}{2} S\left(\frac{dV_2(t)}{dt}\right).$$ (5.5)

$$L(V_1(t)) = \frac{RC}{2s} L(V_2(t)) - \frac{RC}{2s} V_2(0).$$ (5.6)

Thus, we get the transfer function as

$$\frac{L(V_2(t))}{L(V_1(t))} = \frac{2s}{RC}.$$ (5.7)

The graphical representation of the above transfer function is presented in Figure 5.2 as magnitude and phase.

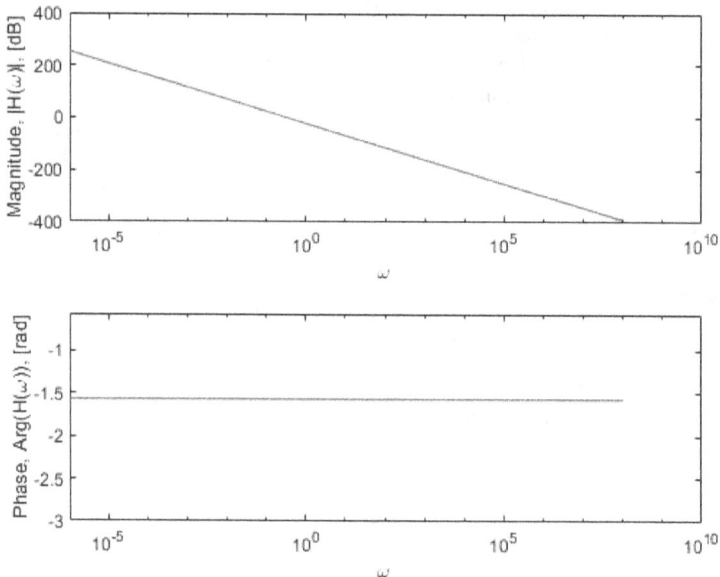

Figure 5.1 Transfer function of the noninverting integrators model 1 with the classical derivative using the Laplace transform.

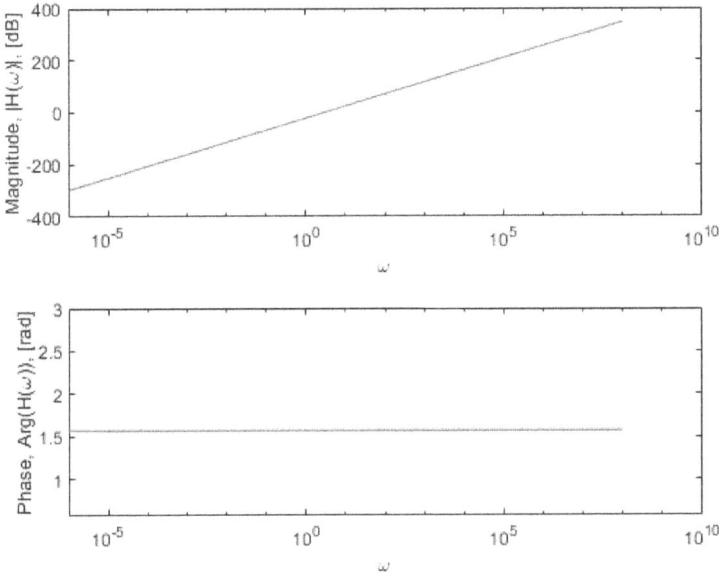

Figure 5.2 Transfer function of the noninverting integrators model 1 with the classical derivative using the Sumudu transform.

5.2 ANALYSIS OF NONINVERTING INTEGRATORS MODEL 1 WITH CAPUTO DERIVATIVE

We consider the model with Caputo derivative. We apply Laplace and Sumudu transforms to obtain the solution.

$$V_1(t) = \frac{RC}{2} \left({}^C_0 D^\alpha_t V_2(t) \right) \tag{5.8}$$

If we take the Laplace transform of the both sides of Eq. (5.8), we will obtain

$$L(V_1(t)) = \frac{RC}{2} L \left({}^C_0 D^\alpha_t V_2(t) \right). \tag{5.9}$$

$$L(V_1(t)) = \frac{s^\alpha RC}{2} L(V_2(t)) - s^{\alpha-1} \frac{RC}{2} V_2(0). \tag{5.10}$$

Thus, we get the transfer function as

$$\frac{L(V_2(t))}{L(V_1(t))} = \frac{2}{s^\alpha RC}. \tag{5.11}$$

The graphical representation of the above transfer function is presented in Figure 5.3 as magnitude and phase.

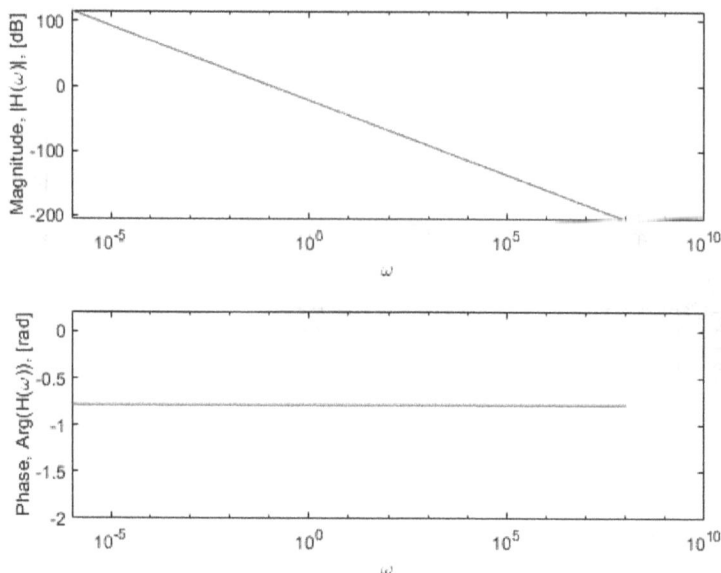

Figure 5.3 Transfer function of the noninverting integrators model 1 with the Caputo derivative using the Laplace transform.

If we take the Sumudu transform of the both sides of Eq. (5.8), we will obtain

$$S(V_1(t)) = \frac{RC}{2} S\left(_0^C D_t^\alpha V_2(t)\right). \tag{5.12}$$

$$S(V_1(t)) = \frac{RC}{2s^\alpha} S(V_2(t)) - \frac{RC}{2s^\alpha} V_2(0). \tag{5.13}$$

Thus, we get the transfer function as

$$\frac{S(V_2(t))}{S(V_1(t))} = \frac{2s^\alpha}{RC}. \tag{5.14}$$

The graphical representation of the above transfer function is presented in Figure 5.4 as magnitude and phase.

5.3 ANALYSIS OF NONINVERTING INTEGRATORS MODEL 1 WITH CAPUTO-FABRIZIO DERIVATIVE

We consider the model with Caputo-Fabrizio derivative. We apply Laplace and Sumudu transforms to obtain the solution.

$$V_1(t) = \frac{RC}{2}\left(_0^{CF} D_t^\alpha V_2(t)\right) \tag{5.15}$$

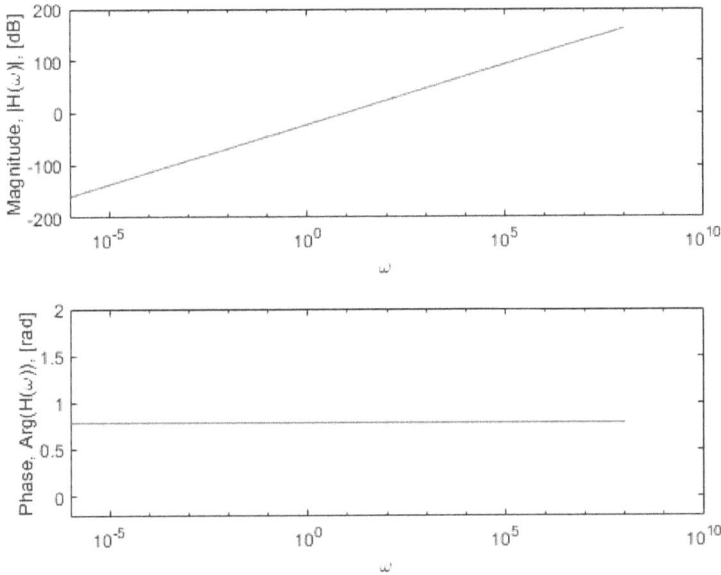

Figure 5.4 Transfer function of the noninverting integrators model 1 with the Caputo derivative using the Sumudu transform.

If we take the Laplace transform of the both sides of Eq. (5.15), we will obtain

$$L(V_1(t)) = \frac{RC}{2}L\left({}_0^{CF}D_t^\alpha V_2(t)\right). \tag{5.16}$$

$$L(V_1(t)) = \frac{sM(\alpha)RC}{2(s+\alpha-s\alpha)}L(V_2(t)) - \frac{RCM(\alpha)}{2(s+\alpha-s\alpha)}V_2(0). \tag{5.17}$$

Thus, we get the transfer function as

$$\frac{L(V_2(t))}{L(V_1(t))} = \frac{2(s+\alpha-\alpha s)}{sM(\alpha)RC}. \tag{5.18}$$

The graphical representation of the above transfer function is presented in Figure 5.5 as magnitude and phase.

If we take the Sumudu transform of the both sides of Eq. (5.15), we will obtain

$$S(V_1(t)) = \frac{RC}{2}S\left({}_0^{CF}D_t^\alpha V_2(t)\right). \tag{5.19}$$

$$L(V_1(t)) = \frac{M(\alpha)RC}{2(\alpha s+1-\alpha)}L(V_2(t)) - \frac{M(\alpha)RC}{2(\alpha s+1-\alpha)}V_2(0). \tag{5.20}$$

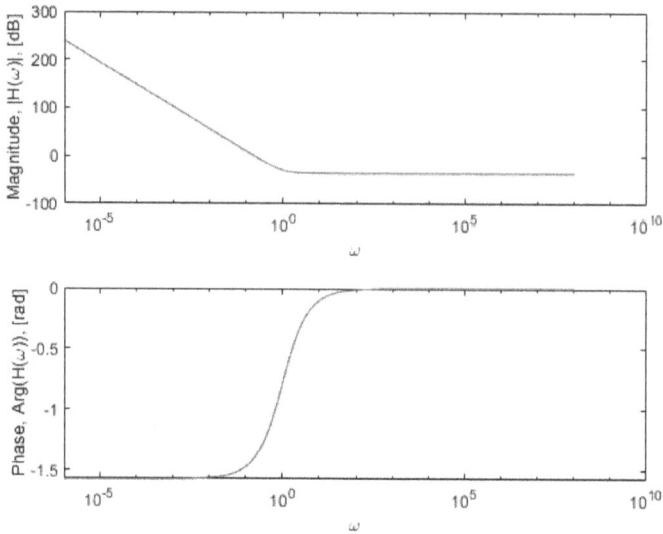

Figure 5.5 Transfer function of the noninverting integrators model 1 with the Caputo-Fabrizio derivative using the Laplace transform.

Thus, we get the transfer function as

$$\frac{L(V_2(t))}{L(V_1(t))} = \frac{2(\alpha s + 1 - \alpha)}{M(\alpha)RC}. \tag{5.21}$$

The graphical representation of the above transfer function is presented in Figure 5.6 as magnitude and phase.

5.4 ANALYSIS OF NONINVERTING INTEGRATORS MODEL 1 WITH ATANGANA-BALEANU DERIVATIVE

We consider the model with Atangana-Baleanu derivative. We apply Laplace and Sumudu transforms to obtain the solution.

$$V_1(t) = \frac{RC}{2} \left({}^{ABC}_0 D_t^\alpha V_2(t) \right) \tag{5.22}$$

If we take the Laplace transform of the both sides of Eq. (5.22), we will obtain

$$L(V_1(t)) = \frac{RC}{2} L \left({}^{ABC}_0 D_t^\alpha V_2(t) \right). \tag{5.23}$$

$$L(V_1(t)) = \frac{s^\alpha AB(\alpha)RC}{2(s^\alpha(1-\alpha)+\alpha)} L(V_2(t)) - \frac{s^{\alpha-1}AB(\alpha)RC}{2(s^\alpha(1-\alpha)+\alpha)} V_2(0). \tag{5.24}$$

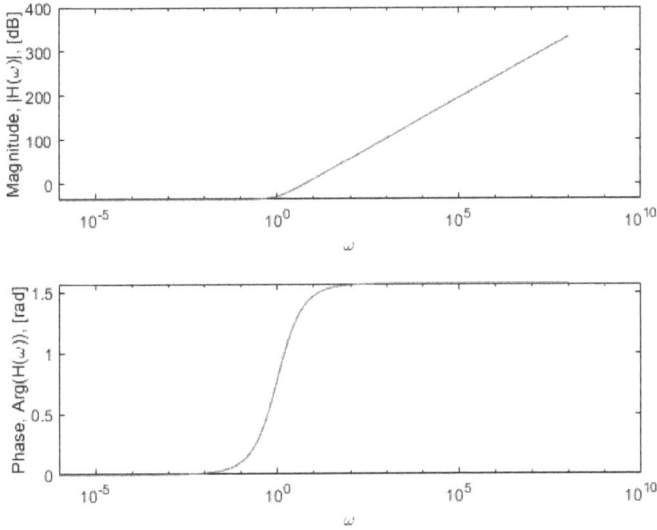

Figure 5.6 Transfer function of the noninverting integrators model 1 with the Caputo-Fabrizio derivative using the Sumudu transform.

Thus, we get the transfer function as

$$\frac{L(V_2(t))}{L(V_1(t))} = \frac{2(s^\alpha(1-\alpha)+\alpha)}{s^\alpha AB(\alpha)RC}. \qquad (5.25)$$

The graphical representation of the above transfer function is presented in Figure 5.7 as magnitude and phase.

If we take the Sumudu transform of the both sides of Eq. (5.22), we will obtain

$$S(V_1(t)) = \frac{RC}{2}S\left({}_0^{ABC}D_t^\alpha V_2(t)\right). \qquad (5.26)$$

$$S(V_1(t)) = \frac{AB(\alpha)RC}{2(\alpha s^\alpha + 1 - \alpha)}S(V_2(t)) - \frac{AB(\alpha)RC}{2(\alpha s^\alpha + 1 - \alpha)}V_2(0). \qquad (5.27)$$

Thus, we get the transfer function as

$$\frac{S(V_2(t))}{S(V_1(t))} = \frac{2(\alpha s^\alpha + 1 - \alpha)}{AB(\alpha)RC}. \qquad (5.28)$$

The graphical representation of the above transfer function is presented in Figure 5.8 as magnitude and phase.

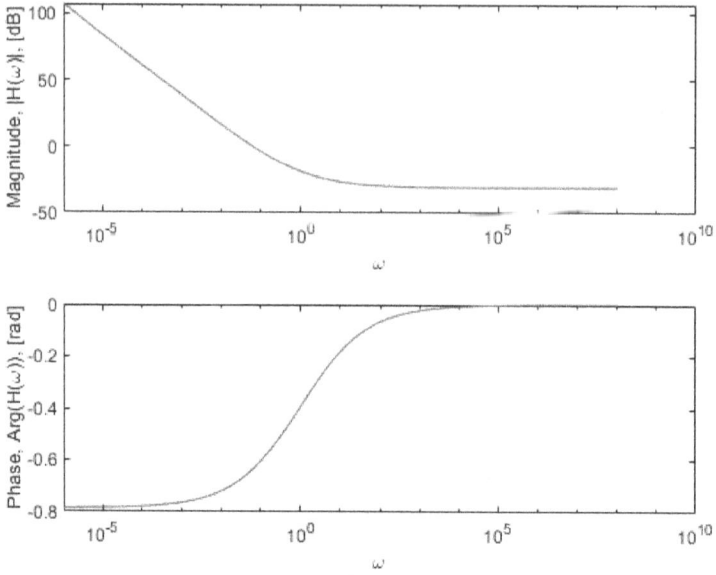

Figure 5.7 Transfer function of the noninverting integrators model 1 with the ABC derivative using the Laplace transform.

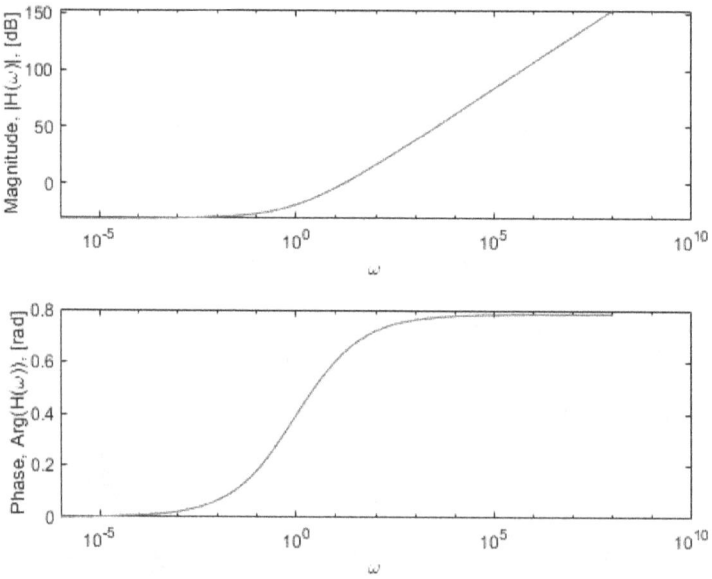

Figure 5.8 Transfer function of the noninverting integrators model 1 with the ABC derivative using the Sumudu transform.

6 Analysis of Noninverting Integrators Model 2

There are several types and models of noninverting-integrator circuit models. In the previous chapter, we gave a brief history of this type and presented an analysis of one model. Note that these classes of circuit problems have given birth to a class of differential equations that need to be solved using either analytical method or numerical methods in the case of nonlinear equations. In this chapter, a different mathematical model able to replicate the dynamic of the noninverting-integrator circuit will be subjected to some analysis. The classical model will be solved using the Laplace and the Sumudu transforms to obtain a transfer function. Both transfer functions will be compared to see the major difference between the Sumudu and the Laplace transforms when both are mapping in complex space. Additionally, to include in the mathematical formulation the effect of nonlocality, the classical model will be extended by replacing the classical time derivative with fractional derivatives. The mathematical model under investigation here is given as follows [48–50]:

6.1 ANALYSIS OF NONINVERTING INTEGRATORS MODEL 2 WITH CLASSICAL DERIVATIVE

We consider the model with classical derivative. We apply Laplace and Sumudu transforms to obtain the solution.

$$V_1(t) = 2RC\frac{dV_2(t)}{dt} \qquad (6.1)$$

If we take the Laplace transform of the both sides of Eq. (6.1), we will obtain

$$L(V_1(t)) = 2RCL\left(\frac{dV_2(t)}{dt}\right). \qquad (6.2)$$

$$L(V_1(t)) = 2sRCL(V_2(t)) - 2RCV_2(0). \qquad (6.3)$$

Thus, we get the transfer function as

$$\frac{L(V_2(t))}{L(V_1(t))} = \frac{1}{2sRC}. \qquad (6.4)$$

The graphical representation of the above transfer function is presented in Figure 6.1 as magnitude and phase.

DOI: 10.1201/9781003359869-6

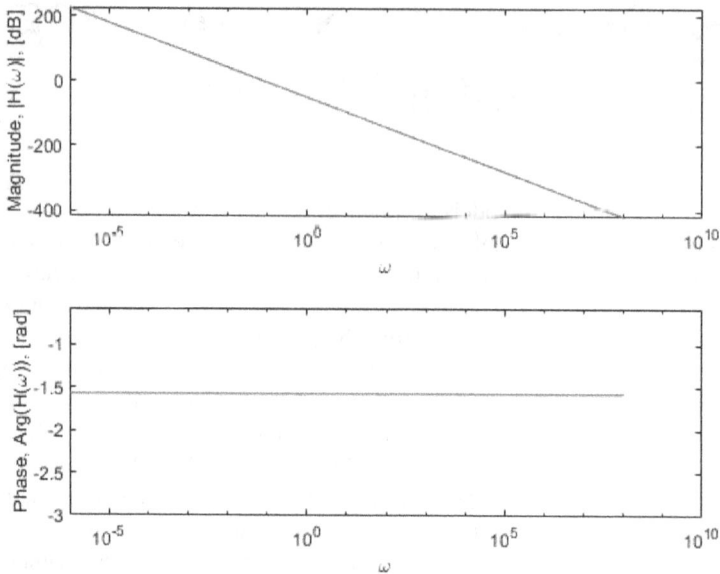

Figure 6.1 Transfer function of the noninverting integrators model 2 with the classical derivative using the Laplace transform.

If we take the Sumudu transform of the both sides of Eq. (6.1), we will obtain

$$S(V_1(t)) = 2RCS\left(\frac{dV_2(t)}{dt}\right). \tag{6.5}$$

$$S(V_1(t)) = \frac{2RC}{s}S(V_2(t)) - \frac{2RC}{s}V_2(0). \tag{6.6}$$

Thus, we get the transfer function as

$$\frac{S(V_2(t))}{S(V_1(t))} = \frac{s}{2RC}. \tag{6.7}$$

The graphical representation of the above transfer function is presented in Figure 6.2 as magnitude and phase.

6.2 ANALYSIS OF NONINVERTING INTEGRATORS MODEL 2 WITH CAPUTO DERIVATIVE

We consider the model with Caputo derivative. We apply Laplace and Sumudu transforms to obtain the solution.

$$V_1(t) = 2RC\left({}_0^C D_t^\alpha V_2(t)\right) \tag{6.8}$$

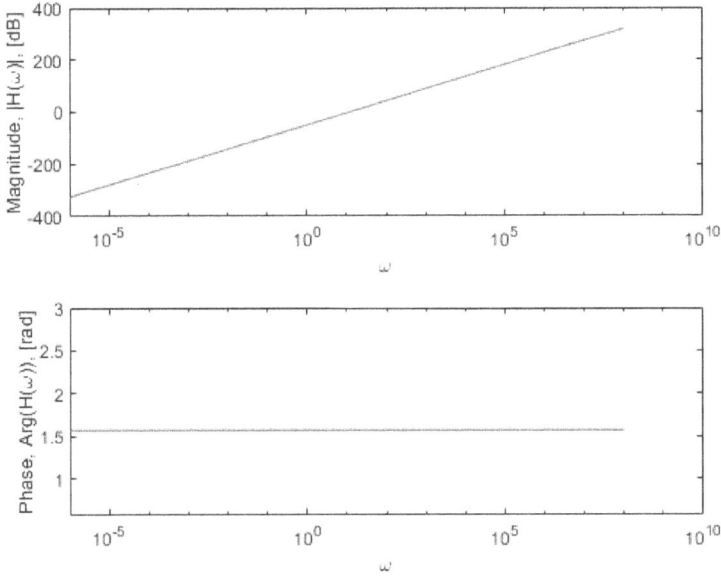

Figure 6.2 Transfer function of the noninverting integrators model 2 with the classical derivative using the Sumudu transform.

If we take the Laplace transform of the both sides of Eq. (6.8), we will obtain

$$L(V_1(t)) = 2RCL\left({}_0^C D_t^\alpha V_2(t)\right).$$ (6.9)

$$L(V_1(t)) = 2s^\alpha RCL(V_2(t)) - s^{\alpha-1}2RCV_2(0).$$ (6.10)

Thus, we get the Transfer function as

$$\frac{L(V_2(t))}{L(V_1(t))} = \frac{1}{2s^\alpha RC}.$$ (6.11)

The graphical representation of the above transfer function is presented in Figure 6.3 as magnitude and phase.

If we take the Sumudu transform of the both sides of Eq. (6.8), we will obtain

$$S(V_1(t)) = 2RCS\left({}_0^C D_t^\alpha V_2(t)\right).$$ (6.12)

$$L(V_1(t)) = \frac{2RC}{s^\alpha}L(V_2(t)) - \frac{2RC}{s^\alpha}V_2(0).$$ (6.13)

Thus, we get the transfer function as

$$\frac{L(V_2(t))}{L(V_1(t))} = \frac{s^\alpha}{2RC}.$$ (6.14)

The graphical representation of the above transfer function is presented in Figure 6.4 as magnitude and phase.

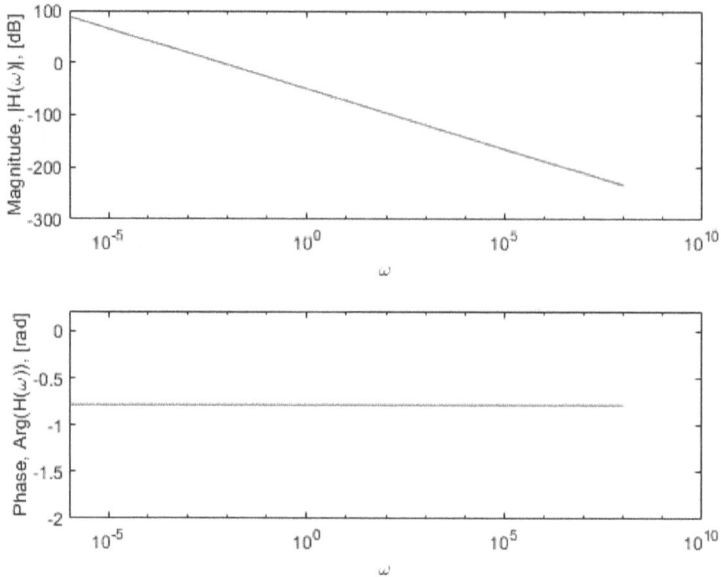

Figure 6.3   Transfer function of the noninverting integrators model 2 with the Caputo derivative using the Laplace transform.

6.3   ANALYSIS OF NONINVERTING INTEGRATORS MODEL 2 WITH CAPUTO-FABRIZIO DERIVATIVE

We consider the model with Caputo-Fabrizio derivative. We apply Laplace and Sumudu transforms to obtain the solution.

$$V_1(t) = 2RC \left({}_0^{CF} D_t^\alpha V_2(t) \right) \tag{6.15}$$

If we take the Laplace transform of the both sides of Eq. (6.15), we will obtain

$$L(V_1(t)) = 2RCL \left({}_0^{CF} D_t^\alpha V_2(t) \right). \tag{6.16}$$

$$L(V_1(t)) = \frac{2sM(\alpha)RC}{s + \alpha - s\alpha} L(V_2(t)) - \frac{2RCM(\alpha)}{s + \alpha - s\alpha} V_2(0). \tag{6.17}$$

Thus, we get the transfer function as

$$\frac{L(V_2(t))}{L(V_1(t))} = \frac{s + \alpha - \alpha s}{2sM(\alpha)RC}. \tag{6.18}$$

The graphical representation of the above transfer function is presented in Figure 6.5 as magnitude and phase.

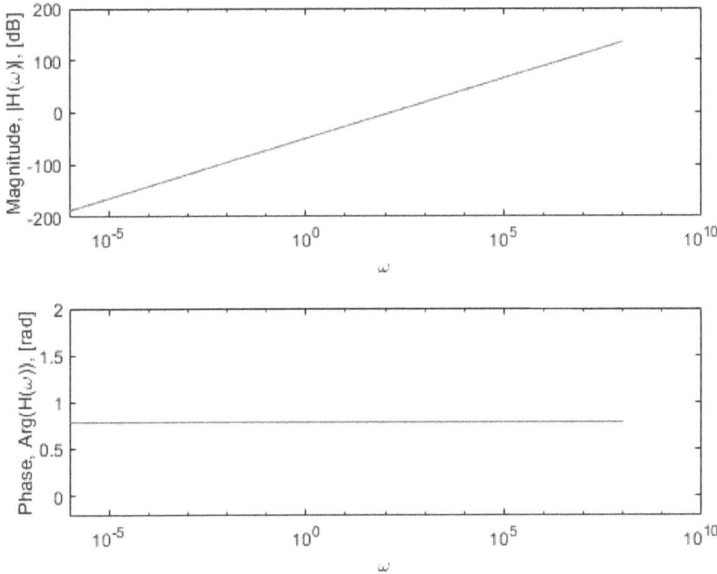

Figure 6.4 Transfer function of the noninverting integrators model 2 with the Caputo derivative using the Sumudu transform.

If we take the Sumudu transform of the both sides of Eq. (6.15), we will obtain

$$S(V_1(t)) = 2RCS \left({}^{CF}_0 D_t^\alpha V_2(t) \right). \tag{6.19}$$

$$L(V_1(t)) = \frac{2M(\alpha)RC}{\alpha s + 1 - \alpha} L(V_2(t)) - \frac{2M(\alpha)RC}{\alpha s + 1 - \alpha} V_2(0). \tag{6.20}$$

Thus, we get the transfer function as

$$\frac{L(V_2(t))}{L(V_1(t))} = \frac{\alpha s + 1 - \alpha}{2M(\alpha)RC}. \tag{6.21}$$

The graphical representation of the above transfer function is presented in Figure 6.6 as magnitude and phase.

6.4 ANALYSIS OF NONINVERTING INTEGRATORS MODEL 2 WITH ATANGANA-BALEANU DERIVATIVE

We consider the model with Atangana-Baleanu derivative. We apply Laplace and Sumudu transforms to obtain the solution.

$$V_1(t) = 2RC \left({}^{ABC}_0 D_t^\alpha V_2(t) \right) \tag{6.22}$$

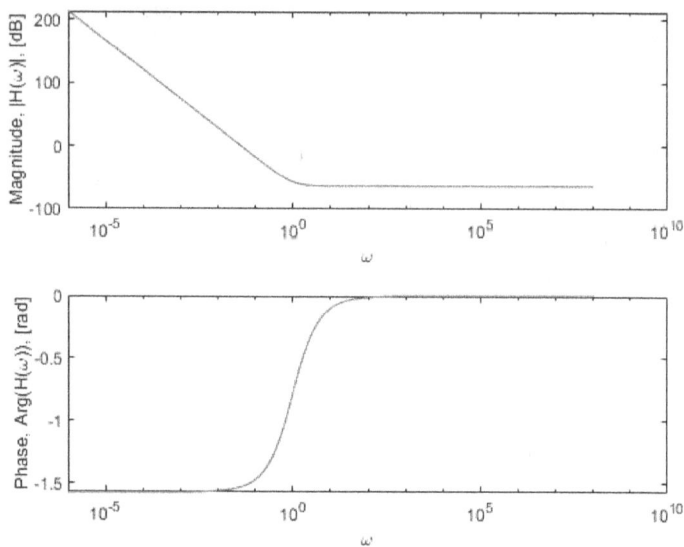

Figure 6.5 Transfer function of the noninverting integrators model 2 with the Caputo-Fabrizio derivative using the Laplace transform.

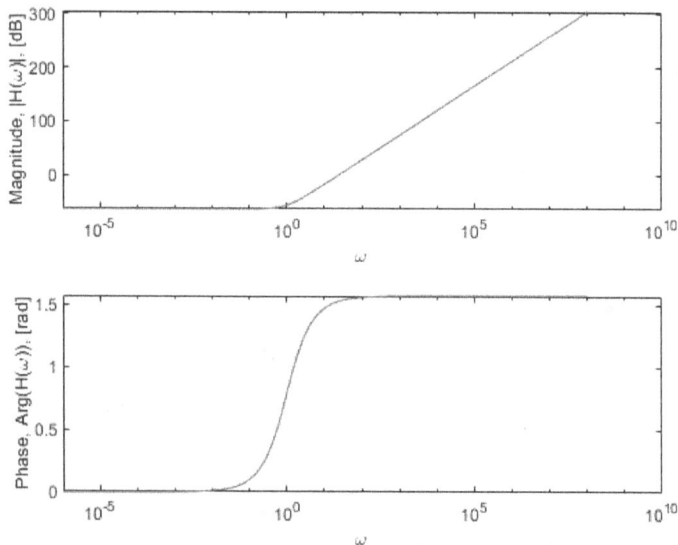

Figure 6.6 Transfer function of the Noninverting integrators model 2 with the Caputo-Fabrizio derivative using the Sumudu transform.

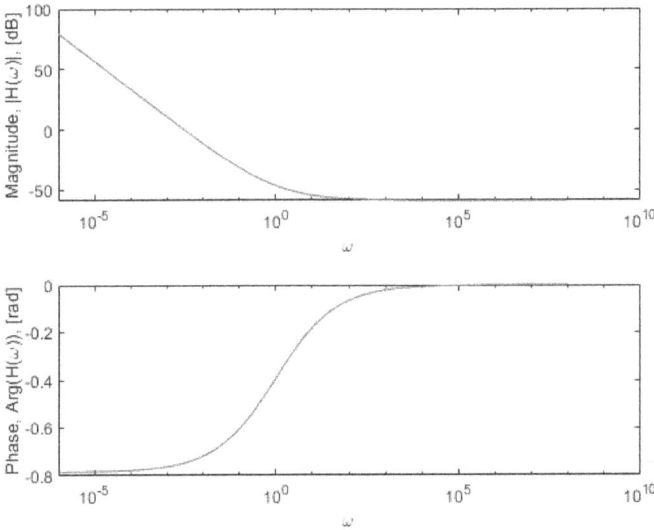

Figure 6.7   Transfer function of the noninverting integrators model 2 with the ABC derivative using the Laplace transform.

If we take the Laplace transform of the both sides of Eq. (6.22), we will obtain

$$L(V_1(t)) = 2RCL\left({}_0^{ABC}D_t^{\alpha}V_2(t)\right).\tag{6.23}$$

$$L(V_1(t)) = \frac{2s^{\alpha}AB(\alpha)RC}{s^{\alpha}(1-\alpha)+\alpha}L(V_2(t)) - \frac{s^{\alpha-1}AB(\alpha)2RC}{s^{\alpha}(1-\alpha)+\alpha}V_2(0).\tag{6.24}$$

Thus, we get the transfer function as

$$\frac{L(V_2(t))}{L(V_1(t))} = \frac{s^{\alpha}(1-\alpha)+\alpha}{2s^{\alpha}AB(\alpha)RC}.\tag{6.25}$$

The graphical representation of the above transfer function is presented in Figure 6.7 as magnitude and phase.

If we take the Sumudu transform of the both sides of Eq. (6.22), we will obtain

$$S(V_1(t)) = 2RCS\left({}_0^{ABC}D_t^{\alpha}V_2(t)\right).\tag{6.26}$$

$$S(V_1(t)) = \frac{2AB(\alpha)RC}{\alpha s^{\alpha}+1-\alpha}S(V_2(t)) - \frac{2AB(\alpha)RC}{\alpha s^{\alpha}+1-\alpha}V_2(0).\tag{6.27}$$

Thus, we get the transfer function as

$$\frac{S(V_2(t))}{S(V_1(t))} = \frac{\alpha s^{\alpha}+1-\alpha}{2AB(\alpha)RC}.\tag{6.28}$$

The graphical representation of the above transfer function is presented in Figure 6.8 as magnitude and phase.

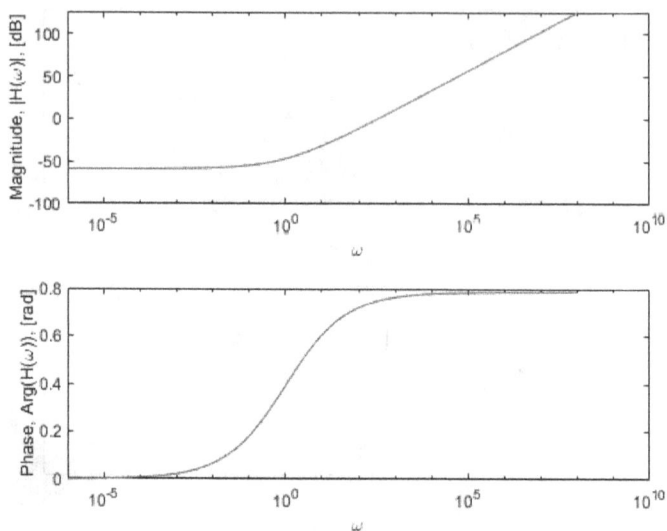

Figure 6.8   Transfer function of the noninverting integrators model 2 with the ABC derivative using the Sumudu transform.

7 Analysis of Lag Network Model

To obtain a better frequency response in the feedback and control system, engineers have designed a type of circuit known as the lag compensator, which is known to be an electrical network that provides sinusoidal results having the phase lag when a sinusoidal output is instigated. These models have found application in several sub-fields of engineering, for example, they have been applied in robotics, automobile diagnostics, laser frequency stabilization, satellite control, liquid-crystal display, and many others. They play a significant role as a building block in analog control systems. More importantly, they can be used in digital control. However, it is noted that both analog and digital control systems are employed lead-lag compensators. Nevertheless, the skill used for the construction of this class is different in each case; nonetheless, the underlying principles are identical. In this case, the transfer function is organized such that the output is in terms of sum of terms involving the contributing parameters. In general, the construction of the lead-lag compensator required that the system needed correction can be categorized within a lead network, a lag network, or even a combination of both. In this case, electrical feedback of the constructed network to an input signal is obtained by the mean of the Laplace, transform, within the Laplace domain. Of course, the obtained complex mathematic function could be represented in two ways accordingly to how the current-gain ratio of the transfer function or the voltage-gain ratio transfer function. In this chapter we shall focus on the lag-compensator, as said before, the lag-compensator is an electrical network for which the outcome is sinusoidal having the phase lag when this sinusoidal input is applied. Thus, in this chapter analysis of the mathematical model will be undertaken by first obtaining the Laplace and the Sumudu transforms of this mathematical model and then converting the classical derivative to fractional types [51–53].

7.1 ANALYSIS OF LAG NETWORK MODEL WITH CLASSICAL DERIVATIVE

We consider the model with classical derivative. We apply Laplace and Sumudu transforms to obtain the solution.

$$V_2(t) + (R_1 + R_2)C \frac{dV_2(t)}{dt} = V_1 + R_2 C \frac{dV_1(t)}{dt} \tag{7.1}$$

If we take the Laplace transform of the both sides of Eq. (7.1), we will obtain

$$L(V_2(t)) + (R_1 + R_2)CL\left(\frac{dV_2(t)}{dt}\right) = L(V_1(t)) + R_2 CL\left(\frac{dV_1(t)}{dt}\right) \tag{7.2}$$

DOI: 10.1201/9781003359869-7

53

Then, we will get

$$L(V_2(t)) + (R_1 + R_2)CsL(V_2(t)) - (R_1 + R_2)CV_2(0)$$
$$= L(V_1(t)) + R_2CsL(V_1(t)) - R_2CV_1(0)$$

Thus, we will obtain

$$L(V_2(t))(1 + (R_1 + R_2)Cs) = L(V_1(t))(1 + R_2Cs). \qquad (7.3)$$

Then, we get the transfer function as

$$\frac{L(V_2(t))}{L(V_1(t))} = \frac{1 + R_2Cs}{1 + (R_1 + R_2)Cs}. \qquad (7.4)$$

The graphical representation of the above transfer function is presented in Figure 7.1 as magnitude and phase.

If we take the Sumudu transform of the both sides of Eq. (7.1), we will obtain

$$S(V_2(t)) + (R_1 + R_2)CS\left(\frac{dV_2(t)}{dt}\right) = S(V_1(t)) + R_2CS\left(\frac{dV_1(t)}{dt}\right) \qquad (7.5)$$

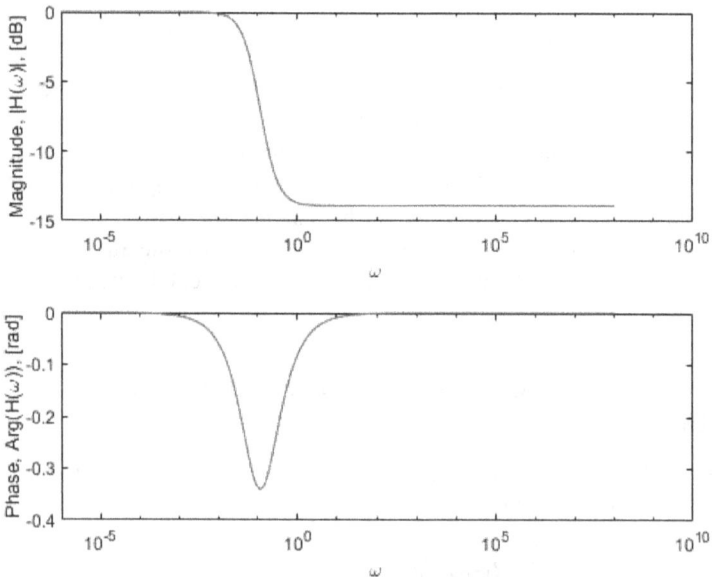

Figure 7.1 Transfer function of the lag network model with the classical derivative using the Laplace transform.

Then, we will get

$$S(V_2(t)) + (R_1+R_2)C\frac{1}{s}S(V_2(t)) - \frac{1}{s}(R_1+R_2)CV_2(0)$$
$$= S(V_1(t)) + R_2C\frac{1}{s}S(V_1(t)) - \frac{1}{s}R_2CV_1(0)$$

Thus, we will obtain

$$S(V_2(t))\left(1+(R_1+R_2)C\frac{1}{s}\right) = S(V_1(t))\left(1+R_2C\frac{1}{s}\right). \tag{7.6}$$

Then, we get the transfer function as

$$\frac{S(V_2(t))}{S(V_1(t))} = \frac{s+R_2C}{s+(R_1+R_2)C}. \tag{7.7}$$

The graphical representation of the above transfer function is presented in Figure 7.2 as magnitude and phase.

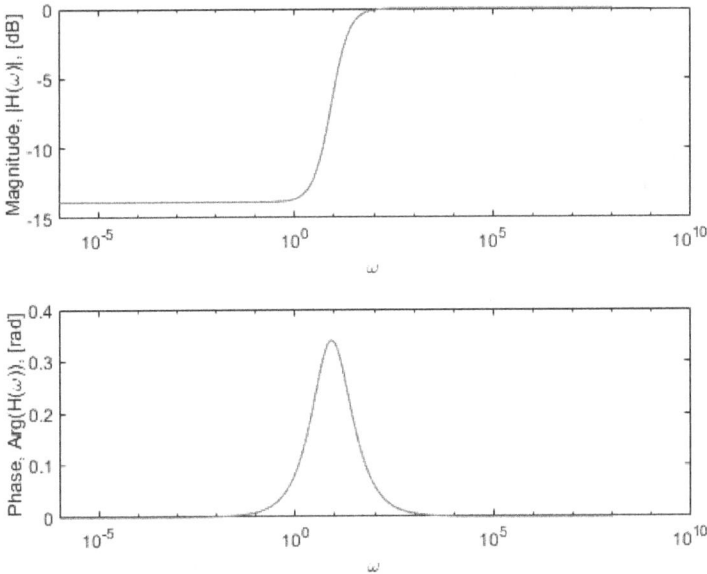

Figure 7.2   Transfer function of the lag network model with the classical derivative using the Sumudu transform.

7.2 ANALYSIS OF LAG NETWORK MODEL WITH CAPUTO DERIVATIVE

We consider the model with Caputo derivative. We apply Laplace and Sumudu transforms to obtain the solution.

$$V_2(t) + (R_1 + R_2)C\left({}_0^C D_t^\alpha V_2(t)\right) = V_1 + R_2 C \frac{dV_1(t)}{dt} \tag{7.8}$$

If we take the Laplace transform of the both sides of Eq. (7.8), we will obtain

$$L(V_2(t)) + (R_1 + R_2)CL\left({}_0^C D_t^\alpha V_2(t)\right) = L(V_1(t)) + R_2 CL\left(\frac{dV_1(t)}{dt}\right) \tag{7.9}$$

Then, we will get

$$L(V_2(t)) + (R_1 + R_2)Cs^\alpha L(V_2(t)) - s^{\alpha-1}(R_1 + R_2)CV_2(0)$$
$$= L(V_1(t)) + R_2 CsL(V_1(t)) - R_2 CV_1(0)$$

Thus, we will obtain

$$L(V_2(t))\left(1 + (R_1 + R_2)Cs^\alpha\right) = L(V_1(t))\left(1 + R_2 Cs\right). \tag{7.10}$$

Then, we get the transfer function as

$$\frac{L(V_2(t))}{L(V_1(t))} = \frac{1 + R_2 Cs}{1 + (R_1 + R_2)Cs^\alpha}. \tag{7.11}$$

The graphical representation of the above transfer function is presented in Figure 7.3 as magnitude and phase.

If we take the Sumudu transform of the both sides of Eq. (7.8), we will obtain

$$S(V_2(t)) + (R_1 + R_2)CS\left({}_0^C D_t^\alpha V_2(t)\right) = S(V_1(t)) + R_2 CS\left(\frac{dV_1(t)}{dt}\right) \tag{7.12}$$

Then, we will get

$$S(V_2(t)) + (R_1 + R_2)C\frac{1}{s^\alpha}S(V_2(t)) - \frac{1}{s^\alpha}(R_1 + R_2)CV_2(0)$$
$$= S(V_1(t)) + R_2 C\frac{1}{s}S(V_1(t)) - \frac{1}{s}R_2 CV_1(0)$$

Thus, we will obtain

$$S(V_2(t))\left(1 + (R_1 + R_2)C\frac{1}{s^\alpha}\right) = S(V_1(t))\left(1 + R_2 C\frac{1}{s}\right). \tag{7.13}$$

Then, we get the transfer function as

$$\frac{S(V_2(t))}{S(V_1(t))} = s^{\alpha-1}\frac{s + R_2 C}{s^\alpha + (R_1 + R_2)C}. \tag{7.14}$$

The graphical representation of the above transfer function is presented in Figure 7.4 as magnitude and phase.

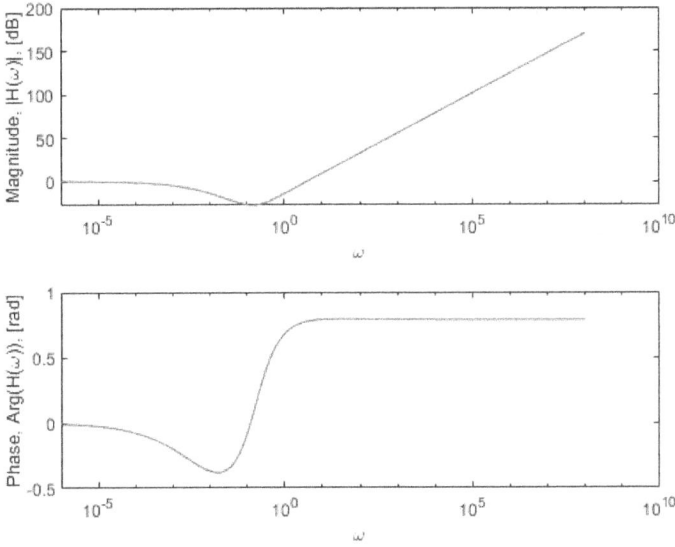

Figure 7.3 Transfer function of the lag network model with the Caputo derivative using the Laplace transform.

7.3 ANALYSIS OF LAG NETWORK MODEL WITH CAPUTO-FABRIZIO DERIVATIVE

We consider the model with Caputo-Fabrizio derivative. We apply Laplace and Sumudu transforms to obtain the solution.

$$V_2(t) + (R_1 + R_2)C\left({}_0^{CF}D_t^{\alpha}V_2(t)\right) = V_1 + R_2C\frac{dV_1(t)}{dt} \qquad (7.15)$$

If we take the Laplace transform of the both sides of Eq. (7.15), we will obtain

$$L(V_2(t)) + (R_1 + R_2)CL\left({}_0^{CF}D_t^{\alpha}V_2(t)\right) = L(V_1(t)) + R_2CL\left(\frac{dV_1(t)}{dt}\right) \qquad (7.16)$$

Then, we will get

$$L(V_2(t)) + (R_1 + R_2)C\frac{sM(\alpha)}{s + \alpha - s\alpha}L(V_2(t)) - \frac{M(\alpha)}{s + \alpha - s\alpha}(R_1 + R_2)CV_2(0)$$
$$= L(V_1(t)) + R_2CsL(V_1(t)) - R_2CV_1(0)$$

Thus, we will obtain

$$L(V_2(t))\left(1 + (R_1 + R_2)C\frac{sM(\alpha)}{s + \alpha - s\alpha}\right) = L(V_1(t))(1 + R_2Cs). \qquad (7.17)$$

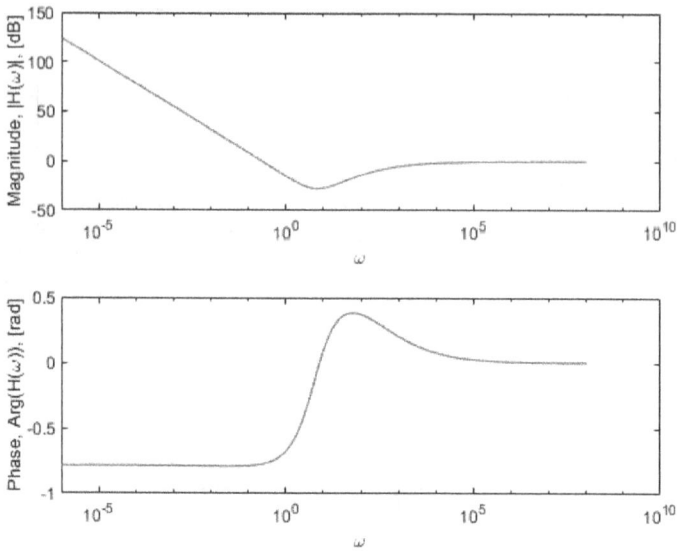

Figure 7.4 Transfer function of the lag network model with the Caputo derivative using the Sumudu transform.

Then, we get the transfer function as

$$\frac{L(V_2(t))}{L(V_1(t))} = \frac{(1+R_2Cs)(s+\alpha-s\alpha)}{s+\alpha-s\alpha+sM(\alpha)C(R_1+R_2)} \quad (7.18)$$

The graphical representation of the above transfer function is presented in Figure 7.5 as magnitude and phase.

If we take the Sumudu transform of the both sides of Eq. (7.15), we will obtain

$$S(V_2(t)) + (R_1+R_2)CS\left(_0^{CF}D_t^{\alpha}V_2(t)\right) = S(V_1(t)) + R_2CS\left(\frac{dV_1(t)}{dt}\right) \quad (7.19)$$

Then, we will get

$$S(V_2(t)) + (R_1+R_2)C\frac{M(\alpha)}{s\alpha+1-\alpha}S(V_2(t)) - \frac{M(\alpha)}{s\alpha+1-\alpha}(R_1+R_2)CV_2(0)$$
$$=S(V_1(t)) + R_2C\frac{1}{s}S(V_1(t)) - \frac{1}{s}R_2CV_1(0)$$

Thus, we will obtain

$$S(V_2(t))\left(1+(R_1+R_2)C\frac{M(\alpha)}{s\alpha+1-\alpha}\right) = S(V_1(t))\left(1+R_2C\frac{1}{s}\right). \quad (7.20)$$

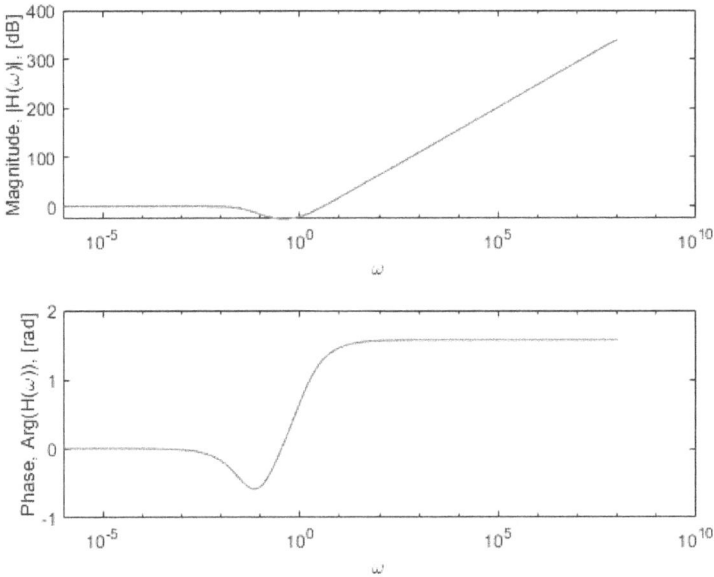

Figure 7.5 Transfer function of the lag network model with the Caputo-Fabrizio derivative using the Laplace transform.

Then, we get the transfer function as

$$\frac{S(V_2(t))}{S(V_1(t))} = \frac{(s+R_2C)(s\alpha+1-\alpha)}{s(s\alpha+1-\alpha+(R_1+R_2)CM(\alpha))}. \tag{7.21}$$

The graphical representation of the above transfer function is presented in Figure 7.6 as magnitude and phase.

7.4 ANALYSIS OF LAG NETWORK MODEL WITH ATANGANA-BALEANU DERIVATIVE

We consider the model with Atangana-Baleanu derivative. We apply Laplace and Sumudu transforms to obtain the solution.

$$V_2(t) + (R_1+R_2)C \left({}^{ABC}_0 D_t^\alpha V_2(t) \right) = V_1 + R_2C\frac{dV_1(t)}{dt} \tag{7.22}$$

If we take the Laplace transform of the both sides of Eq. (7.22), we will obtain

$$L(V_2(t)) + (R_1+R_2)CL \left({}^{ABC}_0 D_t^\alpha V_2(t) \right) = L(V_1(t)) + R_2CL \left(\frac{dV_1(t)}{dt} \right) \tag{7.23}$$

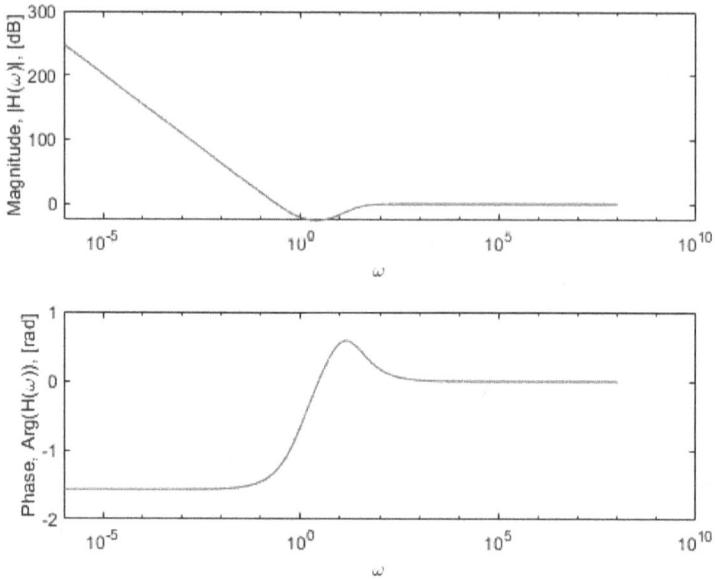

Figure 7.6 Transfer function of the lag network model with the Caputo-Fabrizio derivative using the Sumudu transform.

Then, we will get

$$L(V_2(t)) + (R_1 + R_2)C \frac{s^\alpha AB(\alpha)}{s^\alpha(1-\alpha)+\alpha} L(V_2(t)) - \frac{s^{\alpha-1}AB(\alpha)}{s^\alpha(1-\alpha)+\alpha}(R_1+R_2)CV_2(0)$$
$$= L(V_1(t)) + R_2CsL(V_1(t)) - R_2CV_1(0)$$

Thus, we will obtain

$$L(V_2(t)) \left(1 + (R_1+R_2)C \frac{s^\alpha AB(\alpha)}{s^\alpha(1-\alpha)+\alpha}\right) = L(V_1(t))(1+R_2Cs). \qquad (7.24)$$

Then, we get the transfer function as

$$\frac{L(V_2(t))}{L(V_1(t))} = \frac{(1+R_2Cs)(s^\alpha(1-\alpha)+\alpha)}{s^\alpha(1-\alpha)+\alpha+(R_1+R_2)Cs^\alpha AB(\alpha)}. \qquad (7.25)$$

The graphical representation of the above transfer function is presented in Figure 7.7 as magnitude and phase.

If we take the Sumudu transform of the both sides of Eq. (7.22), we will obtain

$$S(V_2(t)) + (R_1+R_2)CS \left({}_0^{ABC}D_t^\alpha V_2(t)\right) = S(V_1(t)) + R_2CS \left(\frac{dV_1(t)}{dt}\right) \qquad (7.26)$$

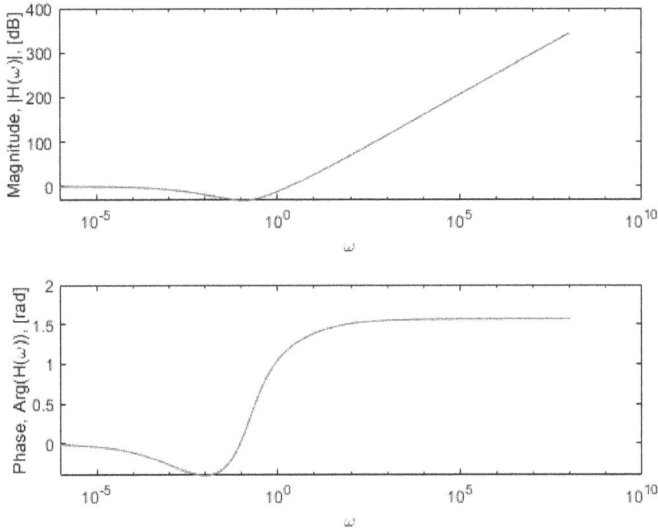

Figure 7.7 Transfer function of the lag network model with the ABC derivative using the Laplace transform.

Then, we will get

$$S(V_2(t)) + (R_1 + R_2)C \frac{AB(\alpha)}{\alpha s^\alpha + 1 - \alpha} S(V_2(t)) - \frac{AB(\alpha)}{\alpha s^\alpha + 1 - \alpha}(R_1 + R_2)CV_2(0)$$

$$= S(V_1(t)) + R_2 C \frac{1}{s} S(V_1(t)) - \frac{1}{s} R_2 CV_1(0)$$

Thus, we will obtain

$$S(V_2(t)) \left(1 + (R_1 + R_2)C \frac{AB(\alpha)}{\alpha s^\alpha + 1 - \alpha} \right) = S(V_1(t)) \left(1 + R_2 C \frac{1}{s} \right). \qquad (7.27)$$

Then, we get the transfer function as

$$\frac{S(V_2(t))}{S(V_1(t))} = \frac{(s + R_2 C)(\alpha s^\alpha + 1 - \alpha)}{s(\alpha s^\alpha + 1 - \alpha + (R_1 R_2)CAB(\alpha))}. \qquad (7.28)$$

The graphical representation of the above transfer function is presented in Figure 7.8 as magnitude and phase.

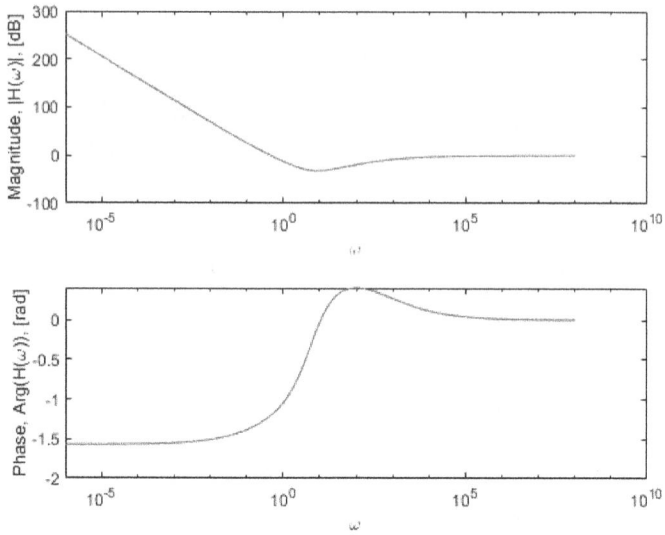

Figure 7.8 Transfer function of the lag network model with the ABC derivative using the Sumudu transform.

8 Analysis of Lead Network Model

The current electrical technology and its different applications showed that lead compensators are almost present everywhere in control. It is also known that a lead compensator can optimize the stability or the speed of response of a system. However, a lag compensator can be used to get rid of the steady-state error. The available literature indicates that based on the need of the engineer, one or more lead compensators can be used in various combinations. As indicated in the previous chapter, lead, lag, and lead/lag compensators are commonly constructed for a system in the transfer function form. The major disparity between the lag and lead compensator is that the lag compensator increases the negative phase of the system over the quantified frequency collection. While on the other hand, a lead compensator augments the positive phase over the specified frequency. In this chapter a mathematical model depicting the dynamic of the lead network is considered. The classical model will be subjected to the Laplace transform and the Sumudu transform. Note that the main aim is not to derive the exact solutions using these integral operators but to obtain the associate transfer functions for comparison. The model will later be extended within the framework of fractional differentiation [51–53].

8.1 ANALYSIS OF ANALYSIS OF LEAD NETWORK MODEL WITH CLASSICAL DERIVATIVE

We consider the model with classical derivative. We apply Laplace and Sumudu transforms to obtain the solution.

$$\frac{dV_2(t)}{dt} + \left(\frac{R_1+R_2}{R_1R_2C}\right)V_2(t) = \frac{dV_1(t)}{dt} + \frac{1}{R_1C}V_1(t) \tag{8.1}$$

If we take the Laplace transform of the both sides of Eq. (8.1), we will obtain

$$L\left(\frac{dV_2(t)}{dt}\right) + \left(\frac{R_1+R_2}{R_1R_2C}\right)L(V_2(t)) = L\left(\frac{dV_1(t)}{dt}\right) + \frac{1}{R_1C}L(V_1(t)). \tag{8.2}$$

Then, we can get

$$sL(V_2(t)) - V_2(0) + \left(\frac{R_1+R_2}{R_1R_2C}\right)L(V_2(t)) = sL(V_1(t)) - V_1(0)$$

$$+ \frac{1}{R_1C}L(V_1(t)). \tag{8.3}$$

$$L(V_2(t))\left(s + \left(\frac{R_1+R_2}{R_1R_2C}\right)\right) = L(V_1(t))\left(s + \frac{1}{R_1C}\right). \tag{8.4}$$

DOI: 10.1201/9781003359869-8

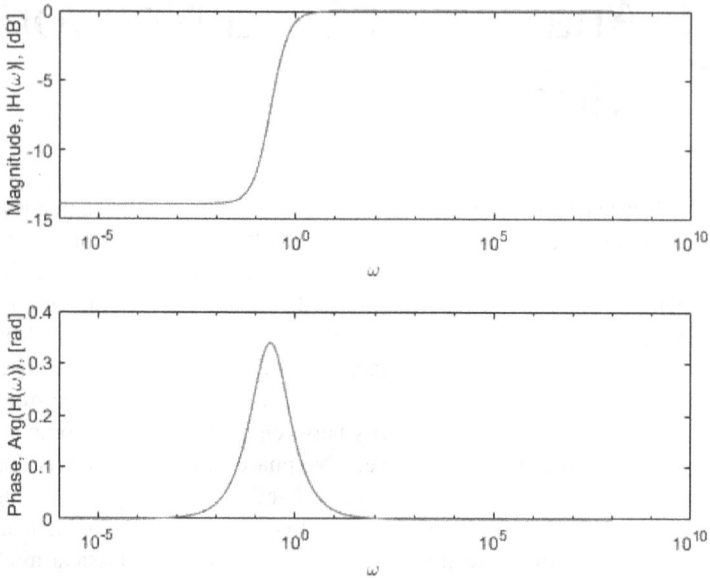

Figure 8.1 Transfer function of the lead network model with the classical derivative using the Laplace transform.

Therefore, we obtain the transfer function as

$$\frac{L(V_2(t))}{L(V_1(t))} = \frac{R_2(1+sR_1C)}{sR_1R_2C+R_1+R_2}.$$ (8.5)

The graphical representation of the above transfer function is presented in Figure 8.1 as magnitude and phase.

If we take the Sumudu transform of the both sides of Eq. (8.1), we will obtain

$$S\left(\frac{dV_2(t)}{dt}\right) + \left(\frac{R_1+R_2}{R_1R_2C}\right)S(V_2(t)) = S\left(\frac{dV_1(t)}{dt}\right) + \frac{1}{R_1C}S(V_1(t)).$$ (8.6)

Then, we can get

$$\frac{1}{s}L(V_2(t)) - \frac{1}{s}V_2(0) + \left(\frac{R_1+R_2}{R_1R_2C}\right)S(V_2(t)) = \frac{1}{s}S(V_1(t)) - \frac{1}{s}V_1(0)$$

$$+ \frac{1}{R_1C}S(V_1(t)).$$ (8.7)

$$S(V_2(t))\left(\frac{1}{s} + \left(\frac{R_1+R_2}{R_1R_2C}\right)\right) = S(V_1(t))\left(\frac{1}{s} + \frac{1}{R_1C}\right).$$ (8.8)

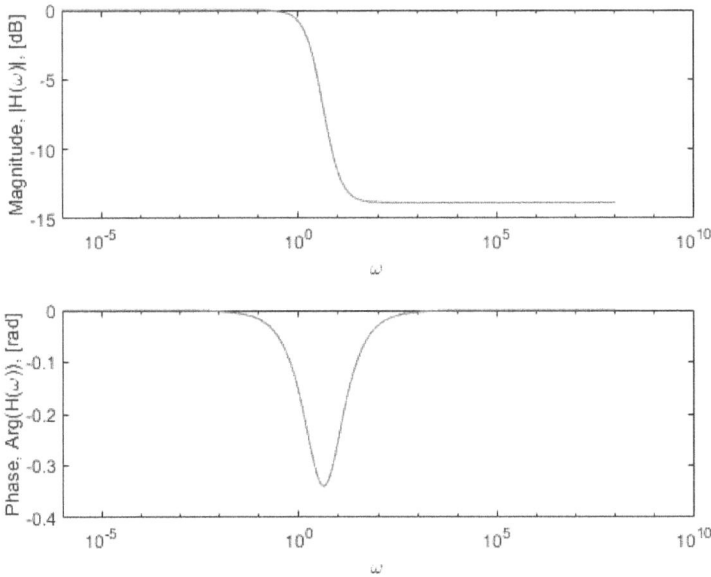

Figure 8.2 Transfer function of the lead network model with the classical derivative using the Sumudu transform.

Therefore, we obtain the transfer function as

$$\frac{S(V_2(t))}{S(V_1(t))} = \frac{R_2(R_1C+s)}{R_1R_2C+sR_1+sR_2}.\qquad(8.9)$$

The graphical representation of the above transfer function is presented in Figure 8.2 as magnitude and phase.

8.2 ANALYSIS OF LEAD NETWORK MODEL WITH CAPUTO DERIVATIVE

We consider the model with Caputo derivative. We apply Laplace and Sumudu transforms to obtain the solution.

$$\left({}_0^C D_t^\alpha V_2(t)\right) + \left(\frac{R_1+R_2}{R_1R_2C}\right)V_2(t) = \frac{dV_1(t)}{dt} + \frac{1}{R_1C}V_1(t)\qquad(8.10)$$

If we take the Laplace transform of the both sides of Eq. (8.10), we will obtain

$$L\left({}_0^C D_t^\alpha V_2(t)\right) + \left(\frac{R_1+R_2}{R_1R_2C}\right)L(V_2(t)) = L\left(\frac{dV_1(t)}{dt}\right) + \frac{1}{R_1C}L(V_1(t)).\qquad(8.11)$$

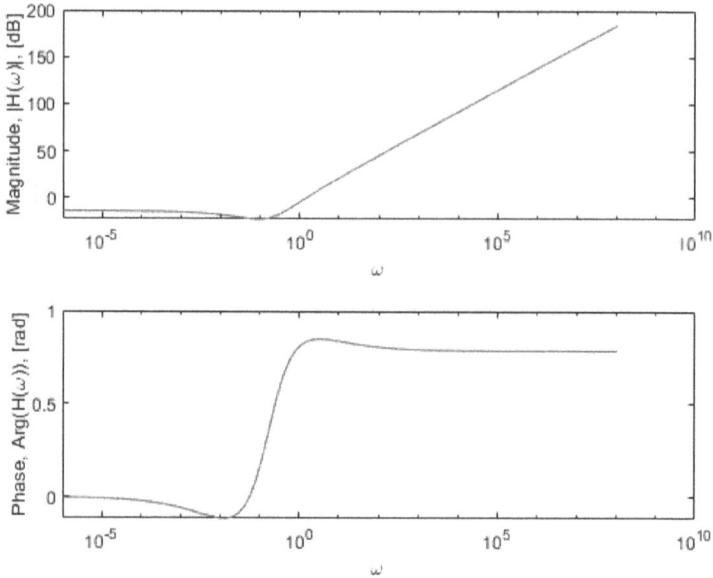

Figure 8.3 Transfer function of the lead network model with the Caputo derivative using the Laplace transform.

Then, we can get

$$s^{\alpha} L\left(V_2(t)\right) - s^{\alpha-1} V_2(0) + \left(\frac{R_1 + R_2}{R_1 R_2 C}\right) L\left(V_2(t)\right) = sL\left(V_1(t)\right) - V_1(0)$$

$$+ \frac{1}{R_1 C} L\left(V_1(t)\right). \qquad (8.12)$$

$$L\left(V_2(t)\right)\left(s^{\alpha} + \left(\frac{R_1 + R_2}{R_1 R_2 C}\right)\right) = L\left(V_1(t)\right)\left(s + \frac{1}{R_1 C}\right). \qquad (8.13)$$

Therefore, we obtain the transfer function as

$$\frac{L\left(V_2(t)\right)}{L\left(V_1(t)\right)} = \frac{R_2(1 + sR_1 C)}{s^{\alpha} R_1 R_2 C + R_1 + R_2}. \qquad (8.14)$$

The graphical representation of the above transfer function is presented in Figure 8.3 as magnitude and phase.

If we take the Sumudu transform of the both sides of Eq. (8.10), we will obtain

$$S\left({}_0^C D_t^{\alpha} V_2(t)\right) + \left(\frac{R_1 + R_2}{R_1 R_2 C}\right) S\left(V_2(t)\right) = S\left(\frac{dV_1(t)}{dt}\right) + \frac{1}{R_1 C} S\left(V_1(t)\right). \qquad (8.15)$$

Then, we can get

$$\frac{1}{s^{\alpha}}L\left(V_2(t)\right) - \frac{1}{s^{\alpha}}V_2(0) + \left(\frac{R_1 + R_2}{R_1 R_2 C}\right) S\left(V_2(t)\right) = \frac{1}{s}S\left(V_1(t)\right) - \frac{1}{s}V_1(0)$$

$$+ \frac{1}{R_1 C}S\left(V_1(t)\right). \qquad (8.16)$$

$$S\left(V_2(t)\right)\left(\frac{1}{s^{\alpha}} + \left(\frac{R_1 + R_2}{R_1 R_2 C}\right)\right) = S\left(V_1(t)\right)\left(\frac{1}{s} + \frac{1}{R_1 C}\right). \qquad (8.17)$$

Therefore, we obtain the transfer function as

$$\frac{S\left(V_2(t)\right)}{S\left(V_1(t)\right)} = \frac{s^{\alpha-1}R_2(R_1 C + s)}{R_1 R_2 C + s^{\alpha}(R_1 + R_2)}. \qquad (8.18)$$

The graphical representation of the above transfer function is presented in Figure 8.4 as magnitude and phase.

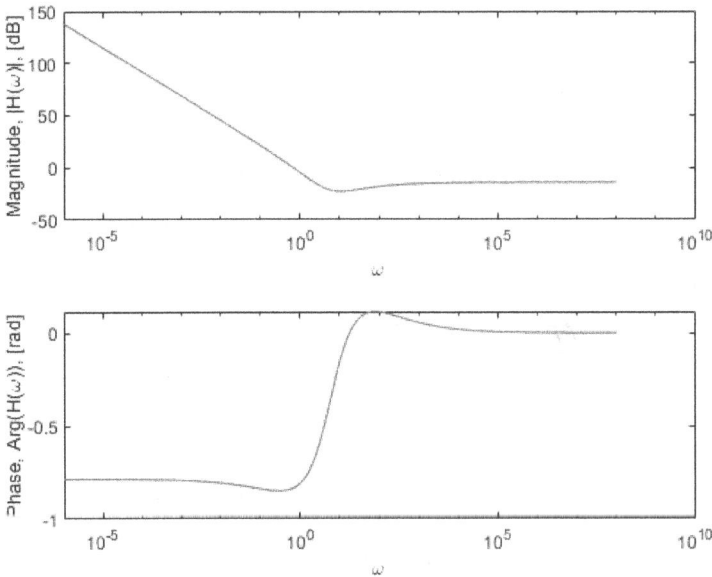

Figure 8.4 Transfer function of the lead network model with the Caputo derivative using the Sumudu transform.

8.3   ANALYSIS OF LEAD NETWORK MODEL WITH CAPUTO-FABRIZIO DERIVATIVE

We consider the model with Caputo-Fabrizio derivative. We apply Laplace and Sumudu transform to obtain the solution.

$$\left({}_0^{CF}D_t^{\alpha}V_2(t)\right) + \left(\frac{R_1+R_2}{R_1R_2C}\right)V_2(t) = \frac{dV_1(t)}{dt} + \frac{1}{R_1C}V_1(t) \tag{8.19}$$

If we take the Laplace transform of the both sides of Eq. (8.19), we will obtain:

$$L\left({}_0^{CF}D_t^{\alpha}V_2(t)\right) + \left(\frac{R_1+R_2}{R_1R_2C}\right)L\left(V_2(t)\right) = L\left(\frac{dV_1(t)}{dt}\right) + \frac{1}{R_1C}L\left(V_1(t)\right). \tag{8.20}$$

Then, we can get

$$\frac{sM(\alpha)}{s+\alpha-s\alpha}L\left(V_2(t)\right) - \frac{M(\alpha)}{s+\alpha-s\alpha}V_2(0) + \left(\frac{R_1+R_2}{R_1R_2C}\right)L\left(V_2(t)\right) \tag{8.21}$$

$$=sL\left(V_1(t)\right) - V_1(0) + \frac{1}{R_1C}L\left(V_1(t)\right). \tag{8.22}$$

$$L\left(V_2(t)\right)\left(\frac{sM(\alpha)}{s+\alpha-s\alpha} + \left(\frac{R_1+R_2}{R_1R_2C}\right)\right) = L\left(V_1(t)\right)\left(s+\frac{1}{R_1C}\right). \tag{8.23}$$

Therefore, we obtain the transfer function as

$$\frac{L\left(V_2(t)\right)}{L\left(V_1(t)\right)} = \frac{R_2(sR_1C+1)(s+\alpha-s\alpha)}{sM(\alpha)R_1R_2C + (R_1+R_2)(s+\alpha-s\alpha)}. \tag{8.24}$$

The graphical representation of the above transfer function is presented in Figure 8.5 as magnitude and phase.

If we take the Sumudu transform of the both sides of Eq. (8.19), we will obtain

$$S\left({}_0^{CF}D_t^{\alpha}V_2(t)\right) + \left(\frac{R_1+R_2}{R_1R_2C}\right)S\left(V_2(t)\right) = S\left(\frac{dV_1(t)}{dt}\right) + \frac{1}{R_1C}S\left(V_1(t)\right). \tag{8.25}$$

Then, we can get

$$\frac{M(\alpha)}{\alpha s+1-\alpha}L\left(V_2(t)\right) - \frac{M(\alpha)}{\alpha s+1-\alpha}V_2(0) + \left(\frac{R_1+R_2}{R_1R_2C}\right)S\left(V_2(t)\right) \tag{8.26}$$

$$=\frac{1}{s}S\left(V_1(t)\right) - \frac{1}{s}V_1(0) + \frac{1}{R_1C}S\left(V_1(t)\right). \tag{8.27}$$

$$S\left(V_2(t)\right)\left(\frac{M(\alpha)}{\alpha s+1-\alpha} + \left(\frac{R_1+R_2}{R_1R_2C}\right)\right) = S\left(V_1(t)\right)\left(\frac{1}{s} + \frac{1}{R_1C}\right). \tag{8.28}$$

Therefore, we obtain the transfer function as

$$\frac{S\left(V_2(t)\right)}{S\left(V_1(t)\right)} = \frac{R_2(R_1C+s)(\alpha s+1-\alpha)}{s(R_1R_2CM(\alpha) + (R_1+R_2)(\alpha s+1-\alpha))}. \tag{8.29}$$

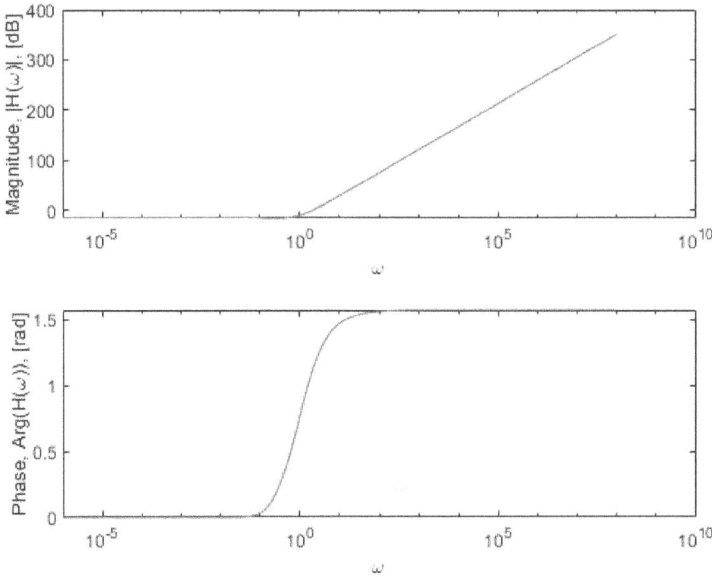

Figure 8.5 Transfer function of the lead network model with the Caputo-Fabrizio derivative using the Laplace transform.

The graphical representation of the above transfer function is presented in Figure 8.6 as magnitude and phase.

8.4 ANALYSIS OF LEAD NETWORK MODEL WITH ATANGANA-BALEANU DERIVATIVE

We consider the model with Atangana-Baleanu derivative. We apply Laplace and Sumudu transforms to obtain the solution.

$$\left(_0^{ABC}D_t^\alpha V_2(t)\right) + \left(\frac{R_1 + R_2}{R_1 R_2 C}\right) V_2(t) = \frac{dV_1(t)}{dt} + \frac{1}{R_1 C} V_1(t) \qquad (8.30)$$

If we take the Laplace transform of the both sides of Eq. (8.30), we will obtain

$$L\left(_0^{ABC}D_t^\alpha V_2(t)\right) + \left(\frac{R_1 + R_2}{R_1 R_2 C}\right) L\left(V_2(t)\right) = L\left(\frac{dV_1(t)}{dt}\right) + \frac{1}{R_1 C} L\left(V_1(t)\right). \qquad (8.31)$$

Then, we can get

$$\frac{s^\alpha AB(\alpha)}{s^\alpha(1-\alpha)+\alpha} L\left(V_2(t)\right) - \frac{s^{\alpha-1}AB(\alpha)}{s^\alpha(1-\alpha)+\alpha} V_2(0) + \left(\frac{R_1 + R_2}{R_1 R_2 C}\right) L\left(V_2(t)\right)$$

$$= sL\left(V_1(t)\right) - V_1(0) + \frac{1}{R_1 C} L\left(V_1(t)\right).$$

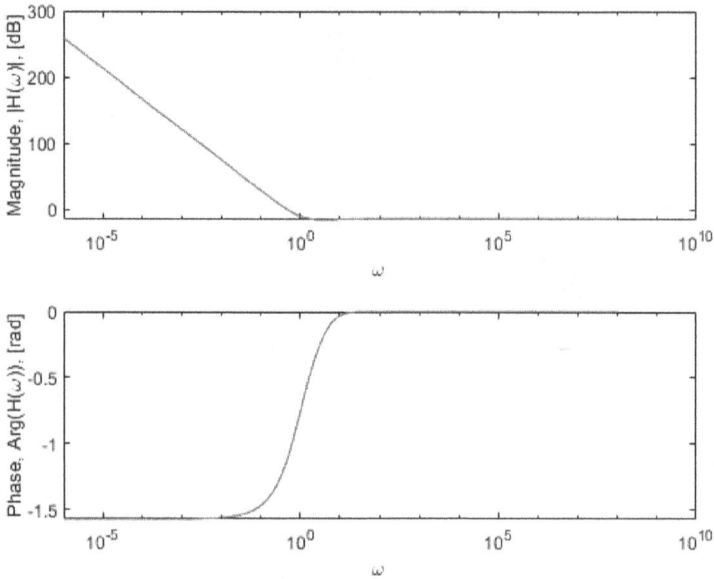

Figure 8.6   Transfer function of the lead network model with the Caputo-Fabrizio derivative using the Sumudu transform.

$$L(V_2(t))\left(\frac{s^\alpha AB(\alpha)}{s^\alpha(1-\alpha)+\alpha}+\left(\frac{R_1+R_2}{R_1R_2C}\right)\right)=L(V_1(t))\left(s+\frac{1}{R_1C}\right). \qquad (8.32)$$

Therefore, we obtain the transfer function as

$$\frac{L(V_2(t))}{L(V_1(t))}=\frac{R_2(1+sR_1C)(s^\alpha(1-\alpha)+\alpha)}{s^\alpha AB(\alpha)R_1R_2C+(R_1+R_2)(s^\alpha(1-\alpha)+\alpha)}. \qquad (8.33)$$

The graphical representation of the above transfer function is presented in Figure 8.7 as magnitude and phase.

If we take the Sumudu transform of the both sides of Eq. (8.30), we will obtain

$$S\left({}_0^{ABC}D_t^\alpha V_2(t)\right)+\left(\frac{R_1+R_2}{R_1R_2C}\right)S(V_2(t))=S\left(\frac{dV_1(t)}{dt}\right)+\frac{1}{R_1C}S(V_1(t)). \qquad (8.34)$$

Then, we can get

$$\frac{AB(\alpha)}{\alpha s^\alpha+1-\alpha}S(V_2(t))-\frac{AB(\alpha)}{\alpha s^\alpha+1-\alpha}V_2(0)+\left(\frac{R_1+R_2}{R_1R_2C}\right)S(V_2(t)) \qquad (8.35)$$

$$=\frac{1}{s}S(V_1(t))-\frac{1}{s}V_1(0)+\frac{1}{R_1C}S(V_1(t)). \qquad (8.36)$$

$$S(V_2(t))\left(\frac{AB(\alpha)}{\alpha s^\alpha+1-\alpha}+\left(\frac{R_1+R_2}{R_1R_2C}\right)\right)=S(V_1(t))\left(\frac{1}{s}+\frac{1}{R_1C}\right). \qquad (8.37)$$

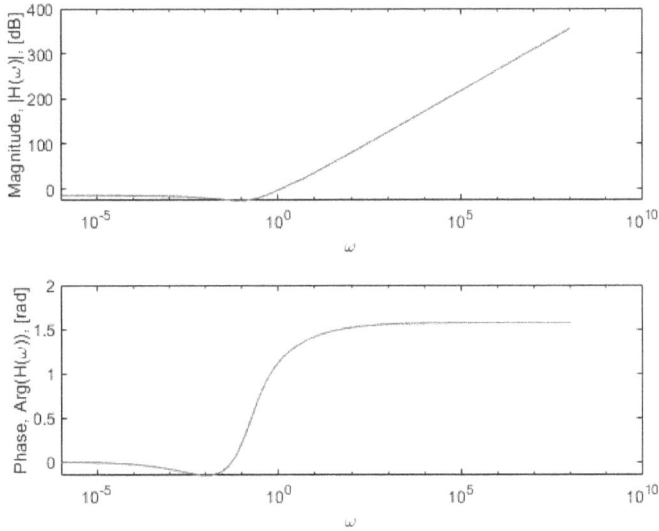

Figure 8.7 Transfer function of the lead network model with the ABC derivative using the Laplace transform.

Therefore, we obtain the transfer function as

$$\frac{S(V_2(t))}{S(V_1(t))} = \frac{R_2(R_1C+s)(\alpha s^\alpha + 1 - \alpha)}{s(R_1R_2CAB(\alpha) + (R_1+R_2)(\alpha s^\alpha + 1 - \alpha))}. \quad (8.38)$$

The graphical representation of the above transfer function is presented in Figure 8.8 as magnitude and phase.

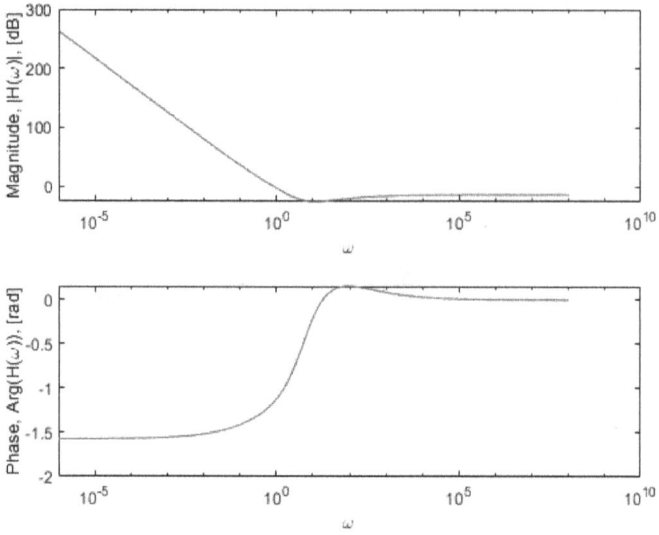

Figure 8.8 Transfer function of the lead network model with the ABC derivative using the Sumudu transform.

9 Analysis of First-order Circuit Model 3

First-order differential equations have attracted the attention of many researchers from different backgrounds as they have been recognized as powerful mathematics tools to model some basic behaviors found in many real-world problems. In particular they are used to model behaviors of the first-order circuits, which are known to include one energy storage element, for example, capacitor or an inductor. In this chapter we shall consider the following linear differential equation will be subject to some analysis including the application of the Sumudu and the Laplace transforms to obtain the transfer functions. Latter the model will be extended within the framework of fractional differential equations and then Laplace and Sumudu transforms will be applied to obtain transfer functions. These transfer functions will be compared to see the major difference between Sumudu and Laplace transforms on the one hand and to also see the effect of fractional differential operators on the other hand [54, 55].

9.1 ANALYSIS OF FIRST-ORDER CIRCUIT MODEL 3 WITH CLASSICAL DERIVATIVE

We consider the model with classical derivative. We apply Laplace and Sumudu transforms to obtain the solution.

$$\frac{dV_2(t)}{dt} + 4(10^6)V_2(t) = -4(10^7)V_1(t) \tag{9.1}$$

If we take the Laplace transform of the both sides of Eq. (9.1), we will obtain

$$L\left(\frac{dV_2(t)}{dt}\right) + 4(10^6)L(V_2(t)) = -4(10^7)L(V_1(t)) \tag{9.2}$$

$$sL(V_2(t)) - V_2(0) + 4(10^6)L(V_2(t)) = -4(10^7)L(V_1(t)) \tag{9.3}$$

$$\left(s + 4(10^6)\right)L(V_2(t)) = -4(10^7)L(V_1(t)) \tag{9.4}$$

Therefore, we get the transfer function as

$$\frac{L(V_2(t))}{L(V_1(t))} = \frac{-4(10^7)}{s + 4(10^6)}. \tag{9.5}$$

The graphical representation of the above transfer function is presented in Figure 9.1 as magnitude and phase.

DOI: 10.1201/9781003359869-9

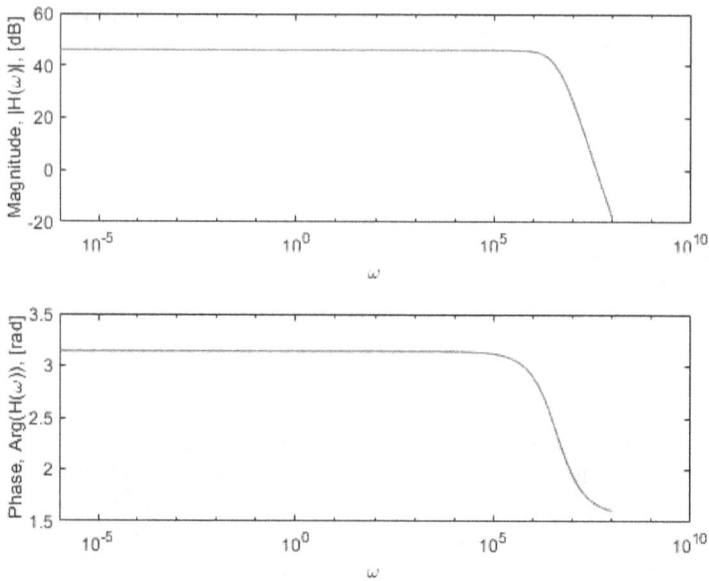

Figure 9.1 Transfer function of the first-order circuit model 3 with the classical derivative using the Laplace transform.

If we take the Sumudu transform of the both sides of Eq. (9.1), we will obtain

$$S\left(\frac{dV_2(t)}{dt}\right) + 4(10^6)S(V_2(t)) = -4(10^7)S(V_1(t)) \tag{9.6}$$

$$\frac{1}{s}S(V_2(t)) - \frac{1}{s}V_2(0) + 4(10^6)S(V_2(t)) = -4(10^7)S(V_1(t)) \tag{9.7}$$

$$\left(\frac{1}{s} + 4(10^6)\right)S(V_2(t)) = -4(10^7)S(V_1(t)) \tag{9.8}$$

Therefore, we get the transfer function as

$$\frac{S(V_2(t))}{S(V_1(t))} = \frac{-4(10^7)s}{1 + s4(10^6)}. \tag{9.9}$$

The graphical representation of the above transfer function is presented in Figure 9.2 as magnitude and phase.

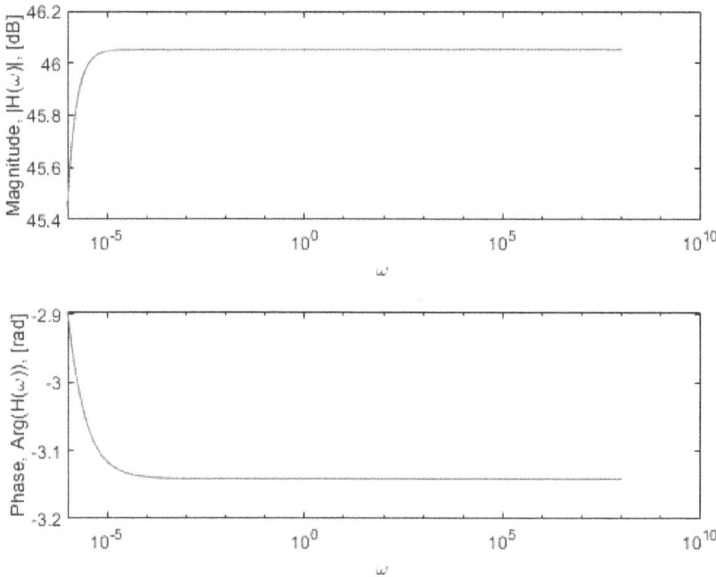

Figure 9.2 Transfer function of the first-order circuit model 3 with the classical derivative using the Sumudu transform.

9.2 ANALYSIS OF FIRST-ORDER CIRCUIT MODEL 3 WITH CAPUTO DERIVATIVE

We consider the model with Caputo derivative. We apply Laplace and Sumudu transforms to obtain the solution.

$$\left({}_{0}^{C}D_{t}^{\alpha}V_2(t)\right) + 4(10^6)V_2(t) = -4(10^7)V_1(t) \tag{9.10}$$

If we take the Laplace transform of the both sides of Eq. (9.10), we will obtain

$$L\left({}_{0}^{C}D_{t}^{\alpha}V_2(t)\right) + 4(10^6)L\left(V_2(t)\right) = -4(10^7)L\left(V_1(t)\right) \tag{9.11}$$

$$s^{\alpha}L\left(V_2(t)\right) - s^{\alpha-1}V_2(0) + 4(10^6)L\left(V_2(t)\right) = -4(10^7)L\left(V_1(t)\right) \tag{9.12}$$

$$\left(s^{\alpha} + 4(10^6)\right)L\left(V_2(t)\right) = -4(10^7)L\left(V_1(t)\right) \tag{9.13}$$

Therefore, we get the transfer function as

$$\frac{L\left(V_2(t)\right)}{L\left(V_1(t)\right)} = \frac{-4(10^7)}{s^{\alpha} + 4(10^6)}. \tag{9.14}$$

The graphical representation of the above transfer function is presented in Figure 9.3 as magnitude and phase.

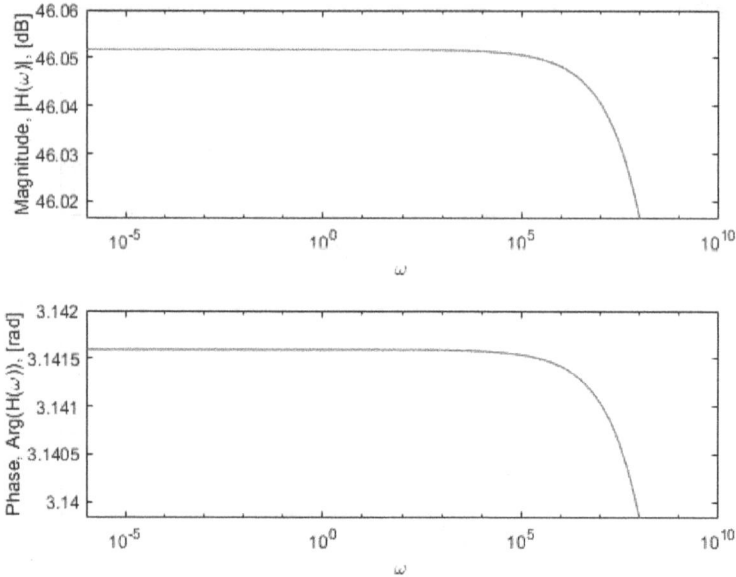

Figure 9.3 Transfer function of the first-order circuit model 3 with the Caputo derivative using the Laplace transform.

If we take the Sumudu transform of the both sides of Eq. (9.10), we will obtain

$$S\left({}^{C}_{0}D^{\alpha}_{t}V_2(t)\right) + 4(10^6)S(V_2(t)) = -4(10^7)S(V_1(t)) \qquad (9.15)$$

$$\frac{1}{s^{\alpha}}S(V_2(t)) - \frac{1}{s^{\alpha}}V_2(0) + 4(10^6)S(V_2(t)) = -4(10^7)S(V_1(t)) \qquad (9.16)$$

$$\left(\frac{1}{s^{\alpha}} + 4(10^6)\right)S(V_2(t)) = -4(10^7)S(V_1(t)) \qquad (9.17)$$

Therefore, we get the transfer function as

$$\frac{S(V_2(t))}{S(V_1(t))} = \frac{-4(10^7)s^{\alpha}}{1 + s^{\alpha}4(10^6)}. \qquad (9.18)$$

The graphical representation of the above transfer function is presented in Figure 9.4 as magnitude and phase.

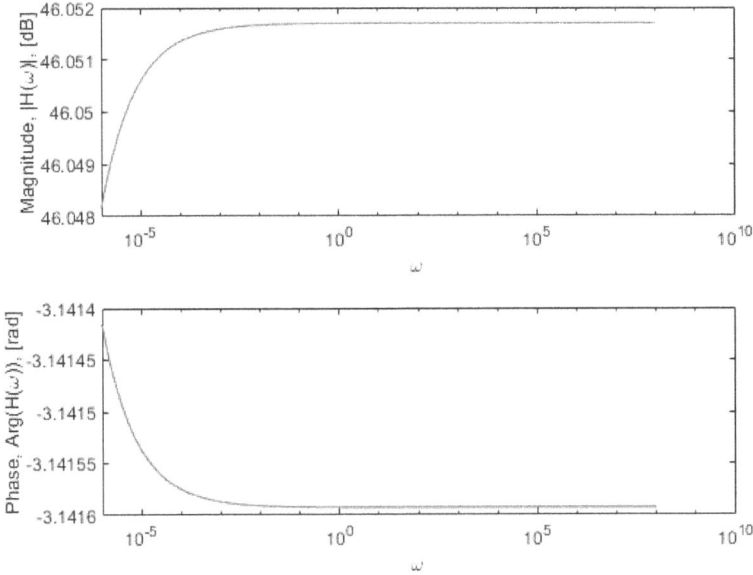

Figure 9.4 Transfer function of the first-order circuit model 3 with the Caputo derivative using the Sumudu transform.

9.3 ANALYSIS OF FIRST-ORDER CIRCUIT MODEL 3 WITH CAPUTO-FABRIZIO DERIVATIVE

We consider the model with Caputo-Fabrizio derivative. We apply Laplace and Sumudu transforms to obtain the solution.

$$\left({}^{CF}_0 D^\alpha_t V_2(t) \right) + 4(10^6) V_2(t) = -4(10^7) V_1(t) \tag{9.19}$$

If we take the Laplace transform of the both sides of Eq. (9.19), we will obtain

$$L\left({}^{CF}_0 D^\alpha_t V_2(t) \right) + 4(10^6) L\left(V_2(t) \right) = -4(10^7) L\left(V_1(t) \right) \tag{9.20}$$

$$\frac{sM(\alpha)}{s+\alpha-s\alpha} L\left(V_2(t) \right) - \frac{M(\alpha)}{s+\alpha-s\alpha} V_2(0) + 4(10^6) L\left(V_2(t) \right) = -4(10^7) L\left(V_1(t) \right) \tag{9.21}$$

$$\left(\frac{sM(\alpha)}{s+\alpha-s\alpha} + 4(10^6) \right) L\left(V_2(t) \right) = -4(10^7) L\left(V_1(t) \right) \tag{9.22}$$

Therefore, we get the transfer function as

$$\frac{L\left(V_2(t) \right)}{L\left(V_1(t) \right)} = \frac{-4(10^7)(s+\alpha-s\alpha)}{sM(\alpha) + 4(10^6)(s+\alpha-s\alpha)}. \tag{9.23}$$

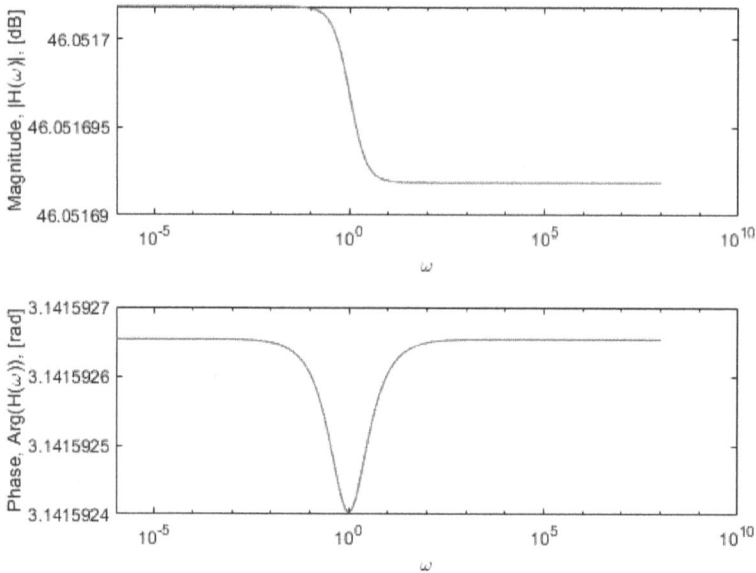

Figure 9.5 Transfer function of the first-order circuit model 3 with the Caputo-Fabrizio derivative using the Laplace transform.

The graphical representation of the above transfer function is presented in Figure 9.5 as magnitude and phase. If we take the Sumudu transform of the both sides of Eq. (9.19), we will obtain

$$S\left({}_0^{CF}D_t^\alpha V_2(t)\right) + 4(10^6)S\left(V_2(t)\right) = -4(10^7)S\left(V_1(t)\right) \tag{9.24}$$

$$\frac{M(\alpha)}{\alpha s + 1 - \alpha}S\left(V_2(t)\right) - \frac{M(\alpha)}{\alpha s + 1 - \alpha}V_2(0) + 4(10^6)S\left(V_2(t)\right) = -4(10^7)S\left(V_1(t)\right) \tag{9.25}$$

$$\left(\frac{M(\alpha)}{\alpha s + 1 - \alpha} + 4(10^6)\right)S\left(V_2(t)\right) = -4(10^7)S\left(V_1(t)\right) \tag{9.26}$$

Therefore, we get the transfer function as

$$\frac{S\left(V_2(t)\right)}{S\left(V_1(t)\right)} = \frac{-4(10^7)(\alpha s + 1 - \alpha)}{M(\alpha) + 4(10^6)(\alpha s + 1 - \alpha)}. \tag{9.27}$$

The graphical representation of the above transfer function is presented in Figure 9.6 as magnitude and phase.

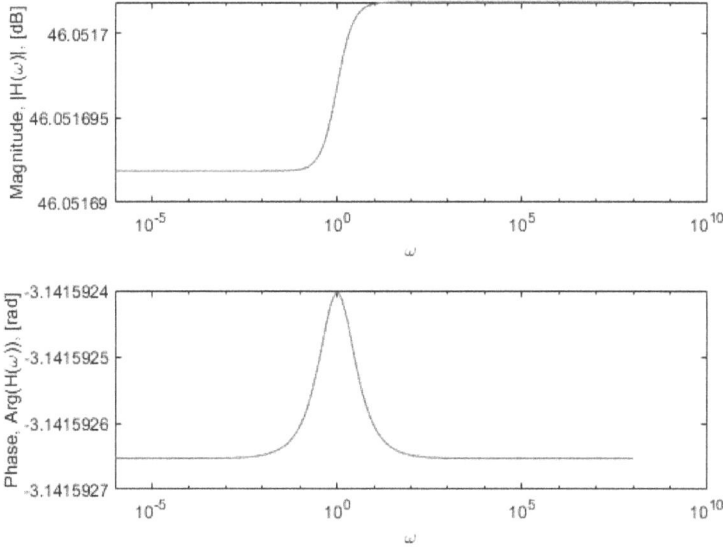

Figure 9.6   Transfer function of the first-order circuit model 3 with the Caputo-Fabrizio derivative using the Sumudu transform.

9.4   ANALYSIS OF FIRST-ORDER CIRCUIT MODEL 3 WITH ATANGANA-BALEANU DERIVATIVE

We consider the model with Atangana-Baleanu derivative. We apply Laplace and Sumudu transforms to obtain the solution.

$$\left({}_0^{ABC}D_t^\alpha V_2(t) \right) + 4(10^6)V_2(t) = -4(10^7)V_1(t) \tag{9.28}$$

If we take the Laplace transform of the both sides of Eq. (9.28), we will obtain

$$L\left({}_0^{ABC}D_t^\alpha V_2(t) \right) + 4(10^6)L\left(V_2(t)\right) = -4(10^7)L\left(V_1(t)\right) \tag{9.29}$$

$$\frac{s^\alpha AB(\alpha)}{s^\alpha(1-\alpha)+\alpha}L\left(V_2(t)\right) - \frac{s^{\alpha-1}AB(\alpha)}{s^\alpha(1-\alpha)+\alpha}V_2(0) + 4(10^6)L\left(V_2(t)\right)$$
$$= -4(10^7)L\left(V_1(t)\right) \tag{9.30}$$

$$\left(\frac{s^\alpha AB(\alpha)}{s^\alpha(1-\alpha)+\alpha} + 4(10^6) \right)L\left(V_2(t)\right) = -4(10^7)L\left(V_1(t)\right) \tag{9.31}$$

Therefore, we get the transfer function as

$$\frac{L\left(V_2(t)\right)}{L\left(V_1(t)\right)} = \frac{-4(10^7)(s^\alpha(1-\alpha)+\alpha)}{s^\alpha AB(\alpha)+4(10^6)(s^\alpha(1-\alpha)+\alpha)}. \tag{9.32}$$

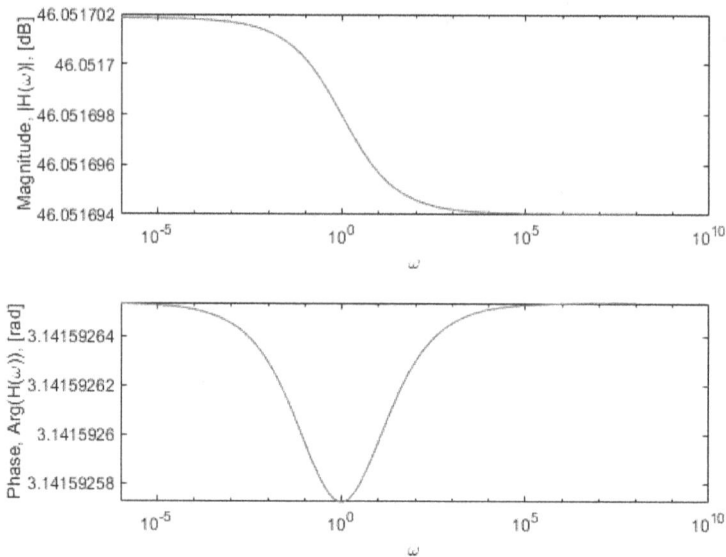

Figure 9.7   Transfer function of the first-order circuit model 3 with the ABC derivative using the Laplace transform.

The graphical representation of the above transfer function is presented in Figure 9.7 as magnitude and phase.

If we take the Sumudu transform of the both sides of Eq. (9.28), we will obtain

$$S\left({}_{0}^{ABC}D_{t}^{\alpha}V_2(t)\right) + 4(10^6)S\left(V_2(t)\right) = -4(10^7)S\left(V_1(t)\right) \tag{9.33}$$

$$\frac{AB(\alpha)}{\alpha s^\alpha + 1 - \alpha}S\left(V_2(t)\right) - \frac{AB(\alpha)}{\alpha s^\alpha + 1 - \alpha}V_2(0) + 4(10^6)S\left(V_2(t)\right) = -4(10^7)S\left(V_1(t)\right) \tag{9.34}$$

$$\left(\frac{AB(\alpha)}{\alpha s^\alpha + 1 - \alpha} + 4(10^6)\right)S\left(V_2(t)\right) = -4(10^7)S\left(V_1(t)\right) \tag{9.35}$$

Therefore, we get the transfer function as

$$\frac{S\left(V_2(t)\right)}{S\left(V_1(t)\right)} = \frac{-4(10^7)(\alpha s^\alpha + 1 - \alpha)}{AB(\alpha) + 4(10^6)(\alpha s^\alpha + 1 - \alpha)}. \tag{9.36}$$

The graphical representation of the above transfer function is presented in Figure 9.8 as magnitude and phase.

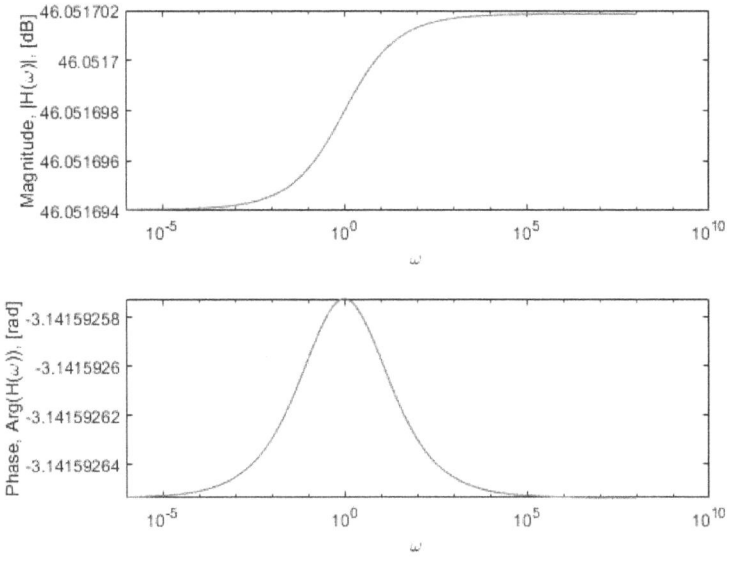

Figure 9.8   Transfer function of the first-order circuit model 3 with the ABC derivative using the Sumudu transform.

10 Analysis of First-order Circuit Model 4

Electronic elements such as inductors and capacitors are seen as nonlinear components. When they are introduced into a circuit system, the behaviors are not instantaneous as they would be when other components like resistors are introduced. For instance, changing the state will disturb the circuit system, and the nonlinear elements necessitate time to reply to the change. It was observed in several cases that some feedback could lead to jumps in the voltage and current, which may eventually cause damage to the circuit system. Thus to solve this problem, transient feedback with circuit design can help to reduce the risk of ill behaviors. In this chapter, therefore, a mathematical model of a circuit that explores the complete response of inductors and capacitors to a state variation, this model will take into account the transient response. The classical model will be solved using the Laplace and the Sumudu transforms, and the solutions will be obtained in complex space to enable us to evaluate the transfer functions of both. Fractional differential operators with different kernels including power law, exponential decay, and the generalized Mittag-Leffler function will be used to include in the mathematical model the effect of nonlocalities associated with a power law, exponential decay, and the generalized Mittag-Leffler function. The classical differential equation under investigation here is given as follows [54, 55]:

10.1 ANALYSIS OF FIRST-ORDER CIRCUIT MODEL 4 WITH CLASSICAL DERIVATIVE

We consider the model with classical derivative. We apply Laplace and Sumudu transforms to obtain the solution.

$$R_1R_2(C_1+C_2)\frac{dV_2(t)}{dt} + (R_1+R_2)V_2(t) = R_1R_2C_1\frac{dV_1(t)}{dt} + R_2V_1(t) \qquad (10.1)$$

If we take the Laplace transform of the both sides of Eq. (10.1), we will obtain

$$R_1R_2(C_1+C_2)L\left(\frac{dV_2(t)}{dt}\right) + (R_1+R_2)L(V_2(t)) = R_1R_2C_1L\left(\frac{dV_1(t)}{dt}\right)$$
$$+ R_2L(V_1(t)). \qquad (10.2)$$

Then, we will reach

$$sR_1R_2(C_1+C_2)L(V_2(t)) - R_1R_2(C_1+C_2)V_2(0)$$
$$= (R_1+R_2)L(V_2(t)) + sR_1R_2C_1L(V_1(t)) - R_1R_2C_1V_1(0) + R_2L(V_1(t)).$$

$$L(V_2(t))(sR_1R_2(C_1+C_2) + (R_1+R_2)) = L(V_1(t))(sR_1R_2C_1 + R_2). \qquad (10.3)$$

DOI: 10.1201/9781003359869-10

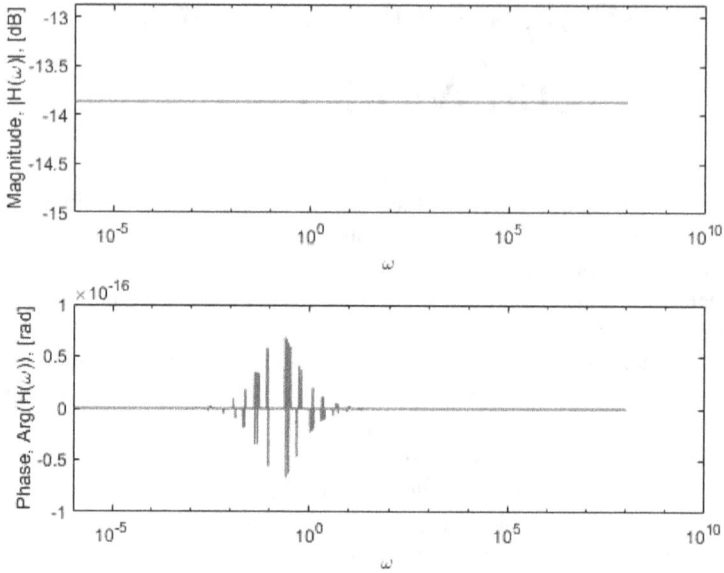

Figure 10.1 Transfer function of the first-order circuit model 4 with the classical derivative using the Laplace transform.

Then, we obtain the transfer function as

$$\frac{L(V_2(t))}{L(V_1(t))} = \frac{sR_1R_2C_1 + R_2}{sR_1R_2(C_1 + C_2) + (R_1 + R_2)}. \tag{10.4}$$

The graphical representation of the above transfer function is presented in Figure 10.1 as magnitude and phase.

If we take the Sumudu transform of the both sides of Eq. (10.1), we will obtain

$$R_1R_2(C_1 + C_2)S\left(\frac{dV_2(t)}{dt}\right) + (R_1 + R_2)S(V_2(t)) = R_1R_2C_1S\left(\frac{dV_1(t)}{dt}\right)$$
$$+ R_2S(V_1(t)). \tag{10.5}$$

Then, we will reach

$$\frac{1}{s}R_1R_2(C_1 + C_2)S(V_2(t)) - R_1R_2(C_1 + C_2)\frac{1}{s}V_2(0) + (R_1 + R_2)S(V_2(t))$$
$$= \frac{1}{s}R_1R_2C_1S(V_1(t)) - \frac{1}{s}R_1R_2C_1V_1(0) + R_2S(V_1(t)).$$

$$S(V_2(t))\left(\frac{1}{s}R_1R_2(C_1 + C_2) + (R_1 + R_2)\right) = S(V_1(t))\left(\frac{1}{s}R_1R_2C_1 + R_2\right). \tag{10.6}$$

Analysis of First-order Circuit Model 4   85

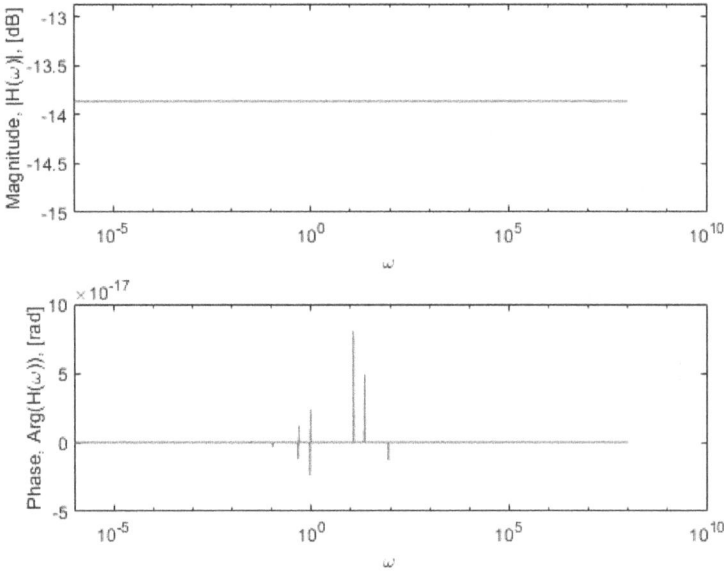

Figure 10.2 Transfer function of the first-order circuit model 4 with the classical derivative using the Sumudu transform.

Then, we obtain the transfer function as

$$\frac{S(V_2(t))}{S(V_1(t))} = \frac{s^{\alpha-1}(R_1R_2C_1 + sR_2)}{R_1R_2(C_1 + C_2) + s^{\alpha}(R_1 + R_2)}. \tag{10.7}$$

The graphical representation of the above transfer function is presented in Figure 10.2 as magnitude and phase.

10.2 ANALYSIS OF FIRST-ORDER CIRCUIT MODEL 4 WITH CAPUTO DERIVATIVE

We consider the model with Caputo derivative. We apply Laplace and Sumudu transforms to obtain the solution.

$$R_1R_2(C_1 + C_2)\left({}_0^C D_t^{\alpha}V_2(t)\right) + (R_1 + R_2)V_2(t) = R_1R_2C_1\frac{dV_1(t)}{dt} + R_2V_1(t) \tag{10.8}$$

If we take the Laplace transform of the both sides of Eq. (10.8), we will obtain

$$R_1R_2(C_1 + C_2)L\left({}_0^C D_t^{\alpha}V_2(t)\right) + (R_1 + R_2)L\left(V_2(t)\right) = R_1R_2C_1L\left(\frac{dV_1(t)}{dt}\right)$$
$$+ R_2L\left(V_1(t)\right). \tag{10.9}$$

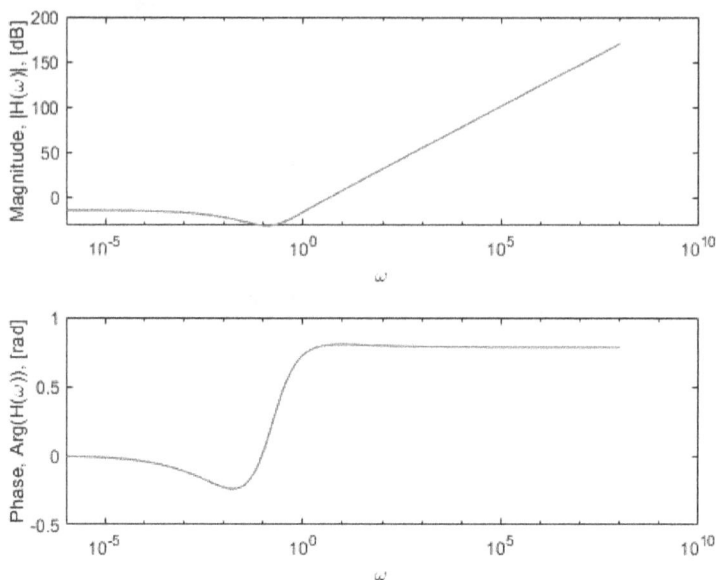

Figure 10.3 Transfer function of the first-order circuit model 4 with the Caputo derivative using the Laplace transform.

Then, we will reach

$$s^{\alpha} R_1 R_2 (C_1 + C_2) L(V_2(t)) - s^{\alpha-1} R_1 R_2 (C_1 + C_2) V_2(0)$$

$$= (R_1 + R_2) L(V_2(t)) + s R_1 R_2 C_1 L(V_1(t)) - R_1 R_2 C_1 V_1(0) + R_2 L(V_1(t)).$$

$$L(V_2(t))(s^{\alpha} R_1 R_2 (C_1 + C_2) + (R_1 + R_2)) = L(V_1(t))(s R_1 R_2 C_1 + R_2). \quad (10.10)$$

Then, we obtain the transfer function as

$$\frac{L(V_2(t))}{L(V_1(t))} = \frac{s R_1 R_2 C_1 + R_2}{s^{\alpha} R_1 R_2 (C_1 + C_2) + (R_1 + R_2)}. \quad (10.11)$$

The graphical representation of the above transfer function is presented in Figure 10.3 as magnitude and phase.

If we take the Sumudu transform of the both sides of Eq. (10.8), we will obtain

$$R_1 R_2 (C_1 + C_2) S\left({}_0^C D_t^{\alpha} V_2(t)\right) + (R_1 + R_2) S(V_2(t)) = R_1 R_2 C_1 S\left(\frac{dV_1(t)}{dt}\right)$$

$$+ R_2 S(V_1(t)). \quad (10.12)$$

Then, we will reach

$$\frac{1}{s^\alpha}R_1R_2(C_1+C_2)S(V_2(t)) - R_1R_2(C_1+C_2)\frac{1}{s^\alpha}V_2(0) + (R_1+R_2)S(V_2(t))$$

$$= \frac{1}{s}R_1R_2C_1S(V_1(t)) - \frac{1}{s}R_1R_2C_1V_1(0) + R_2S(V_1(t)).$$

$$S(V_2(t))\left(\frac{1}{s^\alpha}R_1R_2(C_1+C_2) + (R_1+R_2)\right) = S(V_1(t))\left(\frac{1}{s}R_1R_2C_1 + R_2\right). \quad (10.13)$$

Then, we obtain the transfer function as

$$\frac{S(V_2(t))}{S(V_1(t))} = \frac{R_1R_2C_1 + sR_2}{R_1R_2(C_1+C_2) + s(R_1+R_2)}. \quad (10.14)$$

The graphical representation of the above transfer function is presented in Figure 10.4 as magnitude and phase.

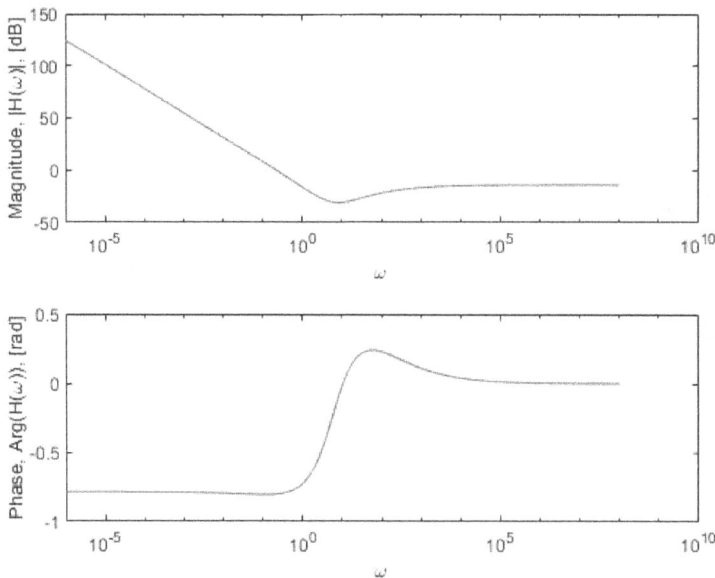

Figure 10.4   Transfer function of the first-order circuit model 4 with the Caputo derivative using the Sumudu transform.

10.3   ANALYSIS OF FIRST-ORDER CIRCUIT MODEL 4 WITH CAPUTO-FABRIZIO DERIVATIVE

We consider the model with Caputo-Fabrizio derivative. We apply Laplace and Sumudu transforms to obtain the solution.

$$R_1R_2(C_1+C_2)\left({}_0^{CF}D_t^\alpha V_2(t)\right)+(R_1+R_2)V_2(t)=R_1R_2C_1\frac{dV_1(t)}{dt}+R_2V_1(t) \quad (10.15)$$

If we take the Laplace transform of the both sides of Eq. (10.15), we will obtain

$$R_1R_2(C_1+C_2)L\left({}_0^{CF}D_t^\alpha V_2(t)\right)+(R_1+R_2)L\left(V_2(t)\right)=R_1R_2C_1L\left(\frac{dV_1(t)}{dt}\right)$$
$$+R_2L\left(V_1(t)\right). \quad (10.16)$$

Then, we will reach

$$\frac{sM(\alpha)}{s+\alpha-s\alpha}R_1R_2(C_1+C_2)L\left(V_2(t)\right)-\frac{M(\alpha)}{s+\alpha-s\alpha}R_1R_2(C_1+C_2)V_2(0)$$
$$=\ (R_1+R_2)L\left(V_2(t)\right)+sR_1R_2C_1L\left(V_1(t)\right)-R_1R_2C_1V_1(0)+R_2L\left(V_1(t)\right).$$

$$L\left(V_2(t)\right)\left(\frac{sM(\alpha)}{s+\alpha-s\alpha}R_1R_2(C_1+C_2)+(R_1+R_2)\right)=L\left(V_1(t)\right)\left(sR_1R_2C_1+R_2\right).$$
$$(10.17)$$

Then, we obtain the transfer function as

$$\frac{L\left(V_2(t)\right)}{L\left(V_1(t)\right)}=\frac{(sR_1R_2C_1+R_2)(s+\alpha-s\alpha)}{sM(\alpha)R_1R_2(C_1+C_2)+(s+\alpha-s\alpha)(R_1+R_2)}. \quad (10.18)$$

The graphical representation of the above transfer function is presented in Figure 10.5 as magnitude and phase.

If we take the Sumudu transform of the both sides of Eq. (10.15), we will obtain

$$R_1R_2(C_1+C_2)S\left({}_0^{CF}D_t^\alpha V_2(t)\right)+(R_1+R_2)S\left(V_2(t)\right)=R_1R_2C_1S\left(\frac{dV_1(t)}{dt}\right)$$
$$+R_2S\left(V_1(t)\right). \quad (10.19)$$

Then, we will reach

$$\frac{M(\alpha)}{s\alpha+1-\alpha}R_1R_2(C_1+C_2)S\left(V_2(t)\right)-R_1R_2(C_1+C_2)\frac{M(\alpha)}{s\alpha+1-\alpha}V_2(0)$$
$$=\ (R_1+R_2)S\left(V_2(t)\right)+\frac{1}{s}R_1R_2C_1S\left(V_1(t)\right)-\frac{1}{s}R_1R_2C_1V_1(0)+R_2S\left(V_1(t)\right).$$

$$S\left(V_2(t)\right)\left(\frac{M(\alpha)}{s\alpha+1-\alpha}R_1R_2(C_1+C_2)+(R_1+R_2)\right)$$
$$=S\left(V_1(t)\right)\left(\frac{1}{s}R_1R_2C_1+R_2\right). \quad (10.20)$$

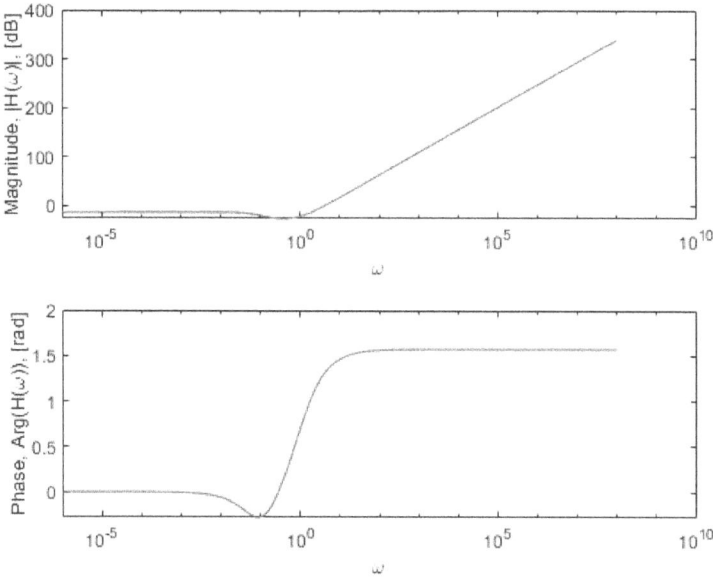

Figure 10.5 Transfer function of the first-order circuit model 4 with the Caputo-Fabrizio derivative using the Laplace transform.

Then, we obtain the transfer function as

$$\frac{S(V_2(t))}{S(V_1(t))} = \frac{(R_1 R_2 C_1 + sR_2)(s\alpha + 1 - \alpha)}{M(\alpha)R_1 R_2(C_1 + C_2) + (s\alpha + 1 - \alpha)(R_1 + R_2)}. \tag{10.21}$$

The graphical representation of the above transfer function is presented in Figure 10.6 as magnitude and phase.

10.4   ANALYSIS OF FIRST-ORDER CIRCUIT MODEL 4 WITH ATANGANA-BALEANU DERIVATIVE

We consider the model with Atangana-Baleanu derivative. We apply Laplace and Sumudu transforms to obtain the solution.

$$R_1 R_2(C_1 + C_2)\left({}^{ABC}_{0}D^{\alpha}_t V_2(t)\right) + (R_1 + R_2)V_2(t) = R_1 R_2 C_1 \frac{dV_1(t)}{dt} + R_2 V_1(t) \tag{10.22}$$

If we take the Laplace transform of the both sides of Eq. (10.22), we will obtain

$$R_1 R_2(C_1 + C_2)L\left({}^{ABC}_{0}D^{\alpha}_t V_2(t)\right) + (R_1 + R_2)L(V_2(t)) = R_1 R_2 C_1 L\left(\frac{dV_1(t)}{dt}\right)$$
$$+ R_2 L(V_1(t)). \tag{10.23}$$

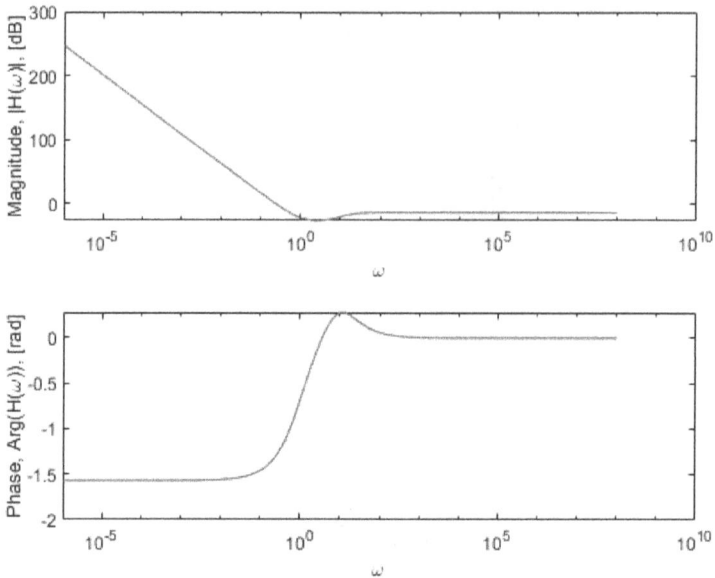

Figure 10.6 Transfer function of the first-order circuit model 4 with the Caputo-Fabrizio derivative using the Sumudu transform.

Then, we will reach

$$\frac{s^\alpha AB(\alpha)}{s^\alpha(1-\alpha)+\alpha}R_1R_2(C_1+C_2)L(V_2(t)) - \frac{s^{\alpha-1}AB(\alpha)}{s^\alpha(1-\alpha)+\alpha}R_1R_2(C_1+C_2)V_2(0)$$
$$+(R_1+R_2)L(V_2(t)) = sR_1R_2C_1L(V_1(t)) - R_1R_2C_1V_1(0) + R_2L(V_1(t)).$$

$$L(V_2(t))\left(\frac{s^\alpha AB(\alpha)}{s^\alpha(1-\alpha)+\alpha}R_1R_2(C_1+C_2)+(R_1+R_2)\right) = L(V_1(t))(sR_1R_2C_1+R_2).$$

(10.24)

Then, we obtain the transfer function as

$$\frac{L(V_2(t))}{L(V_1(t))} = \frac{(sR_1R_2C_1+R_2)(s^\alpha(1-\alpha)+\alpha)}{s^\alpha AB(\alpha)R_1R_2(C_1+C_2)+(s^\alpha(1-\alpha)+\alpha)(R_1+R_2)}.$$

(10.25)

The graphical representation of the above transfer function is presented in Figure 10.7 as magnitude and phase.

If we take the Sumudu transform of the both sides of Eq. (10.22), we will obtain

$$R_1R_2(C_1+C_2)S\left(^{ABC}_0D_t^\alpha V_2(t)\right) + (R_1+R_2)S(V_2(t)) = R_1R_2C_1S\left(\frac{dV_1(t)}{dt}\right)$$
$$+ R_2S(V_1(t)). \quad (10.26)$$

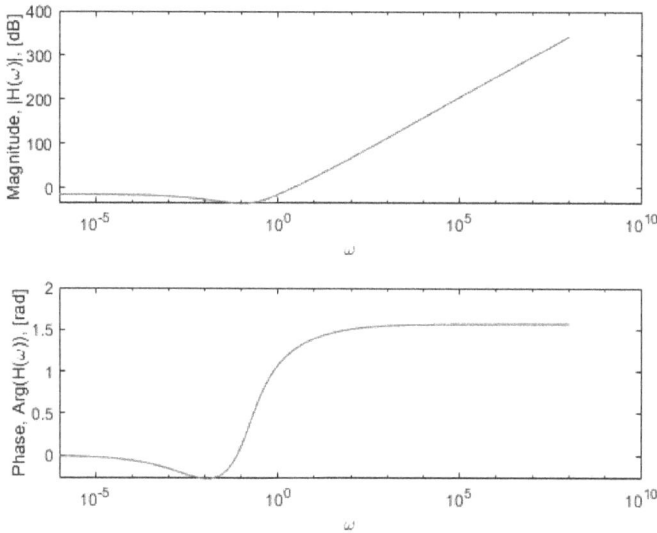

Figure 10.7 Transfer function of the first-order circuit model 4 with the ABC derivative using the Laplace transform.

Then, we will reach

$$\frac{AB(\alpha)}{\alpha s^\alpha + 1 - \alpha} R_1 R_2 (C_1 + C_2) S(V_2(t)) - R_1 R_2 (C_1 + C_2) \frac{AB(\alpha)}{\alpha s^\alpha + 1 - \alpha} V_2(0)$$

$$+ (R_1 + R_2) S(V_2(t)) = \frac{1}{s} R_1 R_2 C_1 S(V_1(t)) - \frac{1}{s} R_1 R_2 C_1 V_1(0) + R_2 S(V_1(t)).$$

$$S(V_2(t)) \left(\frac{AB(\alpha)}{\alpha s^\alpha + 1 - \alpha} R_1 R_2 (C_1 + C_2) + (R_1 + R_2) \right) = S(V_1(t)) \left(\frac{1}{s} R_1 R_2 C_1 + R_2 \right).$$

$$(10.27)$$

Then, we obtain the transfer function as

$$\frac{S(V_2(t))}{S(V_1(t))} = \frac{(R_1 R_2 C_1 + s R_2)(\alpha s^\alpha + 1 - \alpha)}{s(AB(\alpha)R_1 R_2(C_1 + C_2) + (\alpha s^\alpha + 1 - \alpha)(R_1 + R_2))}. \qquad (10.28)$$

The graphical representation of the above transfer function is presented in Figure 10.8 as magnitude and phase.

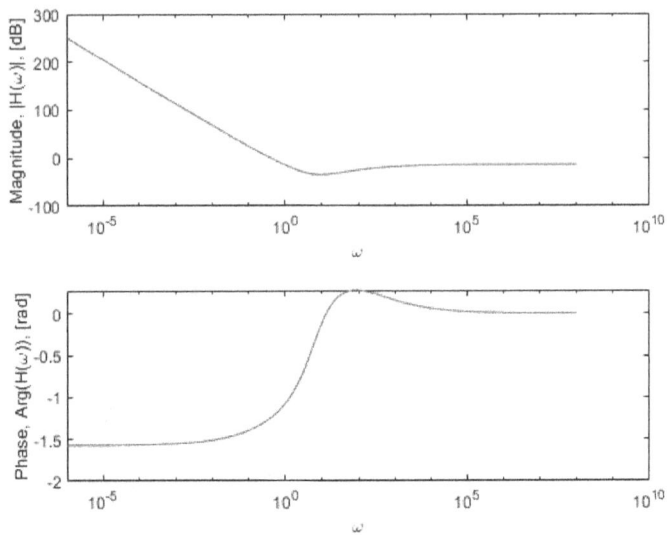

Figure 10.8 Transfer function of the first-order circuit model 4 with the ABC derivative using the Sumudu transform.

11 Analysis of First-order Circuit Model 5

Mathematical equations based on the first derivative have also been found suitable for modeling some complex real-world problems in different fields of science, technology, and engineering. Within the field of electrical engineering, a resistor-inductor circuit model can be efficiently modeled using first-order differential equations. It is worth noting that resistor-inductor circuits have been labeled RL circuits, or RL networks, and have been known to be comprised of resistors and inductors driven by a voltage or current source. Particularly, a first-order RL circuit contains one resistor and one inductor which makes it a fundamental type of RL circuit. Additionally, one should note that this class is one of the main equivalent infinite impulse response electric filters, since it contains a resistor and an inductor, either in series driving by a voltage source or in parallel driving by a current source. In this chapter we consider the following linear ordinary differential equation, different differential operators with integer and non-integer orders will be applied. We consider the first order circuit model 5 with classical, Caputo, Caputo Fabrizio, and Atangana-Baleanu derivatives [54, 55].

11.1 ANALYSIS OF FIRST-ORDER CIRCUIT MODEL 5 WITH CLASSICAL DERIVATIVE

We consider the model with classical derivative. We apply Laplace and Sumudu transform to obtain the solution.

$$Ri(t) + K\frac{di(t)}{dt} = V(t) \tag{11.1}$$

If we take the Laplace transform of the both sides of Eq. (11.1), we will obtain

$$RL(i(t)) + KL\left(\frac{di(t)}{dt}\right) = L(V(t)). \tag{11.2}$$

$$RL(i(t)) + KsL(i(t)) - Li(0) = L(V(t)). \tag{11.3}$$

$$L(i(t))(R + sK) = L(V(t)) \tag{11.4}$$

Then, we get the transfer function as

$$\frac{L(V(t))}{L(i(t))} = R + sK. \tag{11.5}$$

The graphical representation of the above transfer function is presented in Figure 11.1 as magnitude and phase.

DOI: 10.1201/9781003359869-11

If we take the Sumudu transform of the both sides of Eq. (11.1), we will obtain

$$RS(i(t)) + KS\left(\frac{di(t)}{dt}\right) = S(V(t)).$$ (11.6)

$$RS(i(t)) + K\frac{1}{s}S(i(t)) - \frac{1}{s}Li(0) = S(V(t)).$$ (11.7)

$$S(i(t))\left(R + \frac{L}{s}\right) = S(V(t))$$ (11.8)

Then, we get the transfer function as

$$\frac{S(V(t))}{S(i(t))} = R + \frac{K}{s}.$$ (11.9)

The graphical representation of the above transfer function is presented in Figure 11.2 as magnitude and phase.

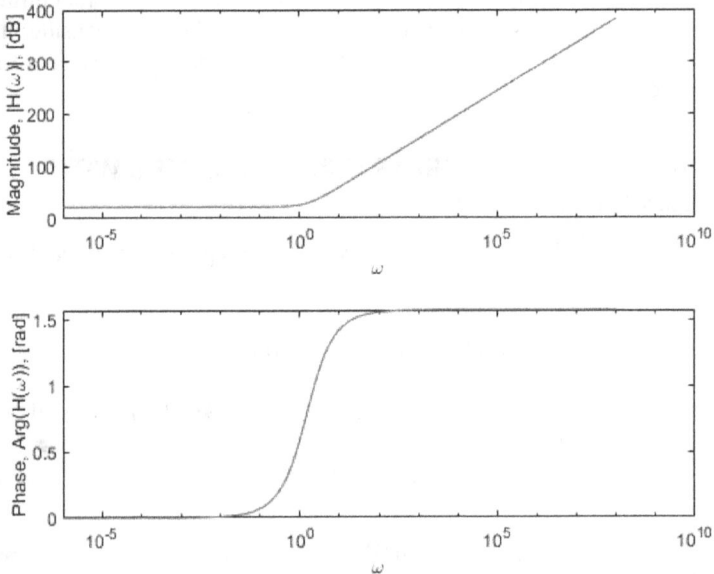

Figure 11.1 Transfer function of the first-order circuit model 5 with the classical derivative using the Laplace transform.

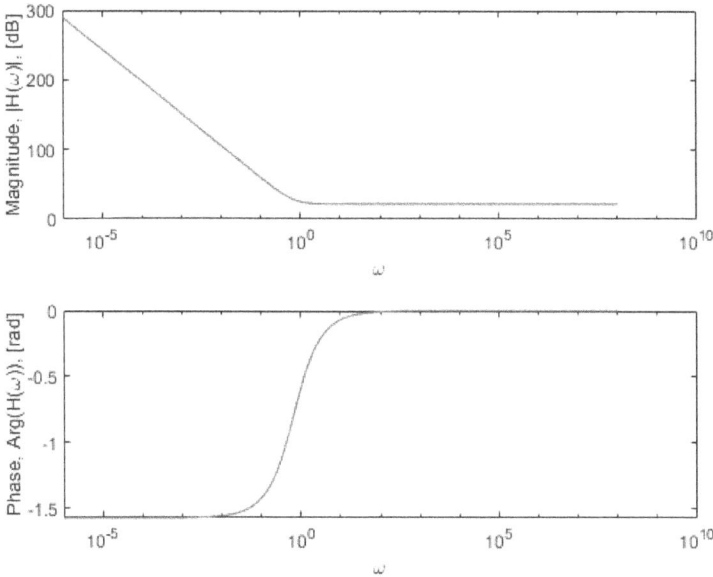

Figure 11.2 Transfer function of the first-order circuit model 5 with the classical derivative using the Sumudu transform.

11.2   ANALYSIS OF FIRST-ORDER CIRCUIT MODEL 5 WITH CAPUTO DERIVATIVE

We consider the model with Caputo derivative. We apply Laplace and Sumudu transforms to obtain the solution.

$$Ri(t) + K \left({}_0^C D_t^\alpha i(t) \right) = V(t) \tag{11.10}$$

If we take the Laplace transform of the both sides of Eq. (11.10), we will obtain

$$RL(i(t)) + KL \left({}_0^C D_t^\alpha i(t) \right) = L(V(t)). \tag{11.11}$$

$$RL(i(t)) + Ks^\alpha L(i(t)) - s^{\alpha-1} Li(0) = L(V(t)). \tag{11.12}$$

$$L(i(t)) (R + s^\alpha K) = L(V(t)) \tag{11.13}$$

Then, we get the transfer function as

$$\frac{L(V(t))}{L(i(t))} = R + s^\alpha K. \tag{11.14}$$

The graphical representation of the above transfer function is presented in Figure 11.3 as magnitude and phase.

If we take the Sumudu transform of the both sides of Eq. (11.10), we will obtain

$$RS\left(i(t)\right) + KS\left({}_{0}^{C}D_{t}^{\alpha}i(t)\right) = S\left(V(t)\right). \tag{11.15}$$

$$RS\left(i(t)\right) + K\frac{1}{s^{\alpha}}S\left(i(t)\right) - \frac{1}{s^{\alpha}}Li(0) = S\left(V(t)\right). \tag{11.16}$$

$$S\left(i(t)\right)\left(R + \frac{K}{s^{\alpha}}\right) = S\left(V(t)\right) \tag{11.17}$$

Then, we get the transfer function as

$$\frac{S\left(V(t)\right)}{S\left(i(t)\right)} = R + \frac{K}{s^{\alpha}}. \tag{11.18}$$

The graphical representation of the above transfer function is presented in Figure 11.4 as magnitude and phase.

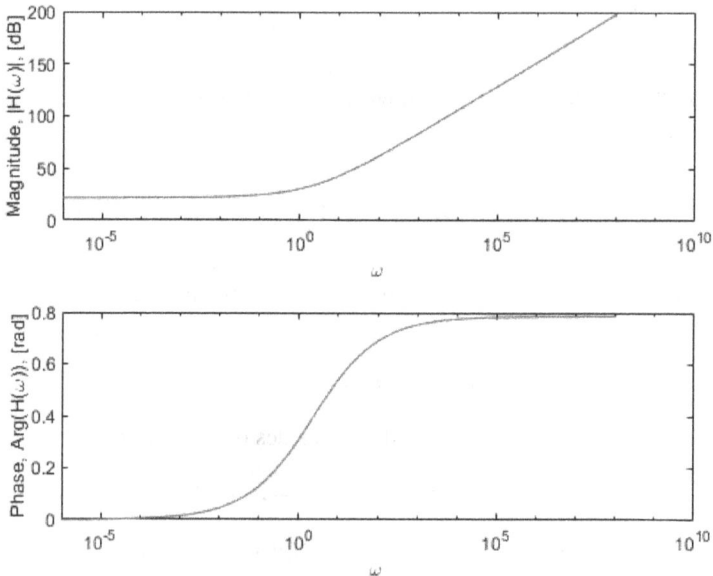

Figure 11.3 Transfer function of the first order circuit model 5 with the Caputo derivative using the Laplace transform.

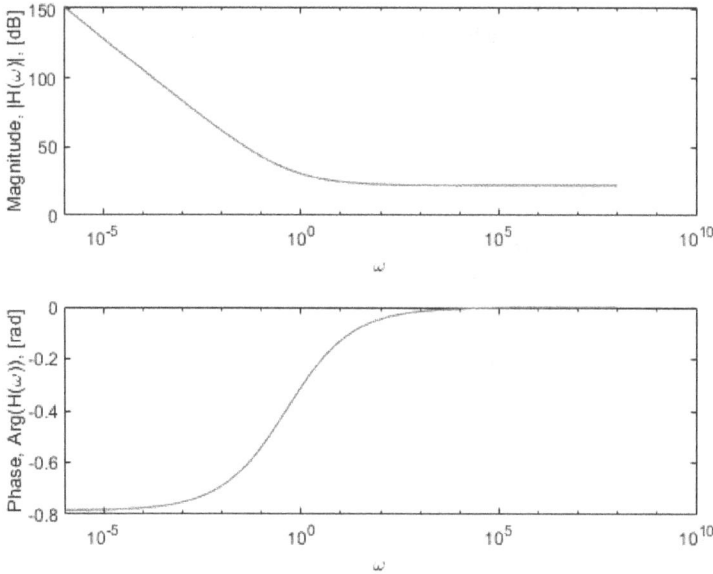

Figure 11.4 Transfer function of the first-order circuit model 5 with the Caputo derivative using the Sumudu transform.

11.3 ANALYSIS OF FIRST-ORDER CIRCUIT MODEL 5 WITH CAPUTO-FABRIZIO DERIVATIVE

We consider the model with Caputo-Fabrizio derivative. We apply Laplace and Sumudu transforms to obtain the solution [92].

$$Ri(t) + K \left({}_0^{CF} D_t^\alpha i(t) \right) = V(t) \tag{11.19}$$

If we take the Laplace transform of the both sides of Eq. (11.19), we will obtain

$$RL(i(t)) + KL \left({}_0^{CF} D_t^\alpha i(t) \right) = L(V(t)). \tag{11.20}$$

$$RL(i(t)) + K \frac{sM(\alpha)}{s + \alpha - s\alpha} L(i(t)) - \frac{M(\alpha)}{s + \alpha - s\alpha} Li(0) = L(V(t)). \tag{11.21}$$

$$L(i(t)) \left(R + \frac{sKM(\alpha)}{s + \alpha - s\alpha} \right) = L(V(t)) \tag{11.22}$$

Then, we get the transfer function as

$$\frac{L(V(t))}{L(i(t))} = R + \frac{sKM(\alpha)}{s + \alpha - s\alpha}. \tag{11.23}$$

The graphical representation of the above transfer function is presented in Figure 11.5 as magnitude and phase.

If we take the Sumudu transform of the both sides of Eq. (11.19), we will obtain

$$RS(i(t)) + KS\left({}^{CF}_{0}D^{\alpha}_t i(t)\right) = S(V(t)). \tag{11.24}$$

$$RS(i(t)) + K\frac{M(\alpha)}{s\alpha + 1 - \alpha}S(i(t)) - \frac{M(\alpha)}{s\alpha + 1 - \alpha}Li(0) = S(V(t)). \tag{11.25}$$

$$S(i(t))\left(R \mid \frac{KM(\alpha)}{s\alpha + 1 - \alpha}\right) = S(V(t)) \tag{11.26}$$

Then, we get the transfer function as

$$\frac{S(V(t))}{S(i(t))} = R + \frac{LM(\alpha)}{s\alpha + 1 - \alpha}. \tag{11.27}$$

The graphical representation of the above transfer function is presented in Figure 11.6 as magnitude and phase.

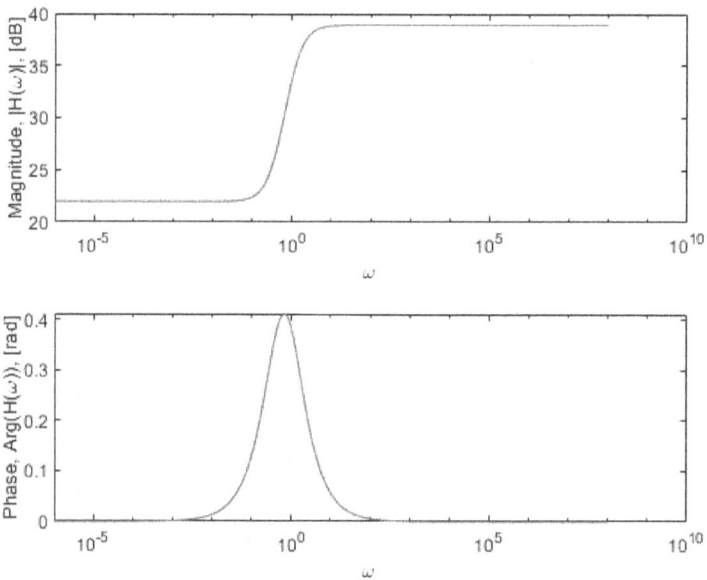

Figure 11.5 Transfer function of the first-order circuit model 5 with the Caputo-Fabrizio derivative using the Laplace transform.

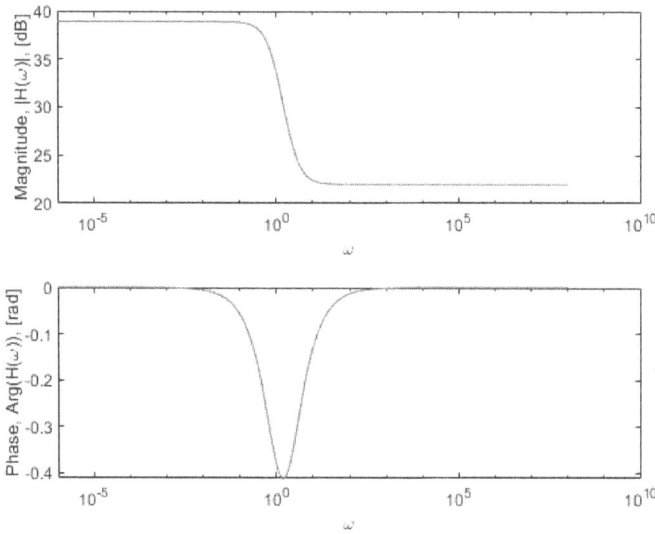

Figure 11.6 Transfer function of the first-order circuit model 5 with the Caputo-Fabrizio derivative using the Sumudu transform.

11.4   ANALYSIS OF FIRST-ORDER CIRCUIT MODEL 5 WITH ATANGANA-BALEANU DERIVATIVE

We consider the model with Atangana-Baleanu derivative. We apply Laplace and Sumudu transforms to obtain the solution.

$$Ri(t) + K\left(_0^{ABC}D_t^{\alpha}i(t)\right) = V(t) \tag{11.28}$$

If we take the Laplace transform of the both sides of Eq. (11.28), we will obtain

$$RL(i(t)) + KL\left(_0^{ABC}D_t^{\alpha}i(t)\right) = L(V(t)). \tag{11.29}$$

$$RL(i(t)) + K\frac{s^{\alpha}AB(\alpha)}{s^{\alpha}(1-\alpha) \mid \alpha}L(i(t)) - \frac{s^{\alpha-1}AB(\alpha)}{s^{\alpha}(1-\alpha)+\alpha}Li(0) - L(V(t)). \tag{11.30}$$

$$L(i(t))\left(R + \frac{Ks^{\alpha}AB(\alpha)}{s^{\alpha}(1-\alpha)+\alpha}\right) = L(V(t)) \tag{11.31}$$

Then, we get the transfer function as

$$\frac{L(V(t))}{L(i(t))} = R + \frac{Ks^{\alpha}AB(\alpha)}{s^{\alpha}(1-\alpha)+\alpha}. \tag{11.32}$$

The graphical representation of the above transfer function is presented in Figure 11.7 as magnitude and phase.

If we take the Sumudu transform of the both sides of Eq. (11.28), we will obtain

$$RS(i(t)) + KS\left({}_0^{ABC}D_t^\alpha i(t)\right) = S(V(t)). \tag{11.33}$$

$$RS(i(t)) + K\frac{AB(\alpha)}{\alpha s^\alpha + 1 - \alpha}S(i(t)) - \frac{AB(\alpha)}{\alpha s^\alpha + 1 - \alpha}Li(0) = S(V(t)). \tag{11.34}$$

$$S(i(t))\left(R + \frac{AB(\alpha)K}{\alpha s^\alpha + 1 - \alpha}\right) = S(V(t)) \tag{11.35}$$

Then, we get the transfer function as

$$\frac{S(V(t))}{S(i(t))} = R + \frac{AB(\alpha)K}{\alpha s^\alpha + 1 - \alpha}. \tag{11.36}$$

The graphical representation of the above transfer function is presented in Figure 11.8 as magnitude and phase.

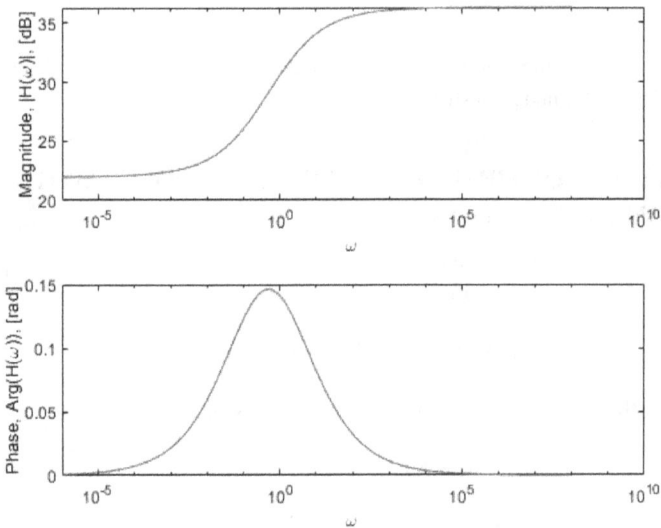

Figure 11.7 Transfer function of the first-order circuit model 5 with the ABC derivative using the Laplace transform.

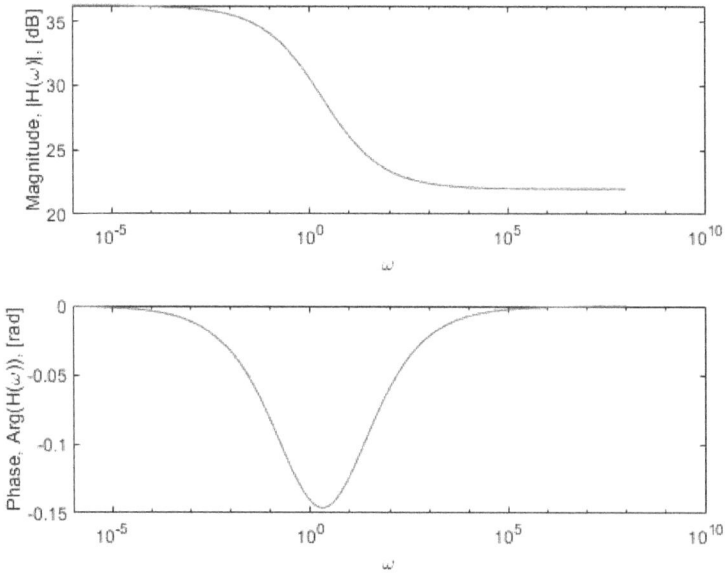

Figure 11.8 Transfer function of the first-order circuit model 5 with the ABC derivative using the Sumudu transform.

12 Analysis of a Series RLC Circuit Model

Besides the well-established theory of RL circuits, there is another more complex type called RLC circuits. In this case, R represents the resistor, L stands for the inductor, and C indicates the capacitor. These electronic devices can be connected in series or parallel accordingly. Generality speaking the nomenclature comes from the letters that are used to define the constituent components of the system. It is worth noting that, for this system, some resistance is unavoidable especially when a resistor is not included as a component. Thus, in an ideal situation, one will expect a pure LC circuit to exist only in the domain of superconductivity. Indeed, in this case, a physical effect plays a significant role at temperatures beneath and/or pressures far above what is found in nature on the Earth's surface. As an application, RLC circuits [52,53] have got several implementations as oscillator circuits. In other applications, the radio receivers, and television systems use them for tuning to select a narrow frequency range from ambient radio waves. Thus, this circuit is frequently named tuned circuit, and they are also known to apply to band-pass filters, band-stop filters, low-pass filters, or high-pass filters. These circuits are classified under second-order circuits, and this implies any voltage or current in the system can be modeled using ordinary differential equations with second order. To obtain several different topologies, one can combine in different ways the R, L, and C. In this chapter, we shall analyze differential equations describing the dynamics of the RLC circuit with different differential operators. We consider a series RLC circuit model with classical, Caputo, Caputo Fabrizio, and Atangana-Baleanu derivatives.

12.1 ANALYSIS OF A SERIES RLC CIRCUIT MODEL WITH CLASSICAL DERIVATIVE

We consider the model with classical derivative. We apply Laplace and Sumudu transforms to obtain the solution.

$$\frac{d^2V(t)}{dt^2} + \frac{R}{L}\frac{dV}{dt} = \frac{Ri(t)}{CL} + \frac{1}{C}\frac{di(t)}{dt} \tag{12.1}$$

If we take the Laplace transform of the both sides of Eq. (12.1), we will obtain

$$L\left(\frac{d^2V(t)}{dt^2}\right) + \frac{R}{L}L\left(\frac{dV}{dt}\right) = L\left(\frac{Ri(t)}{CL}\right) + \frac{1}{C}L\left(\frac{di(t)}{dt}\right). \tag{12.2}$$

DOI: 10.1201/9781003359869-12

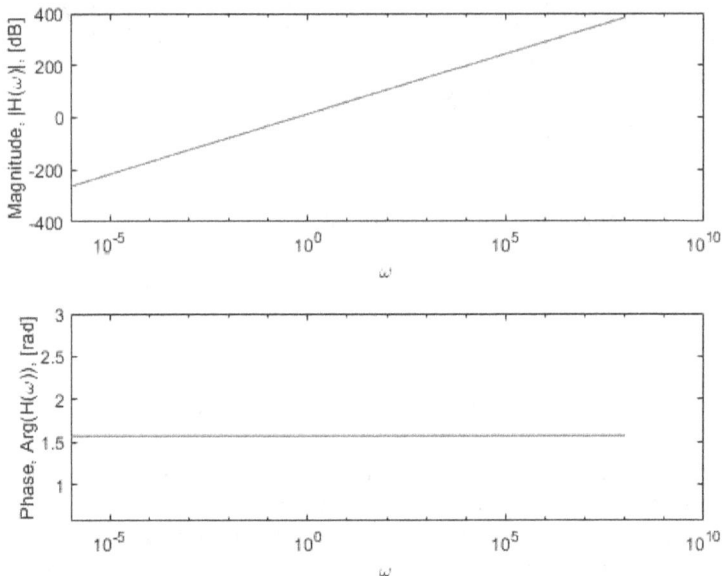

Figure 12.1 Transfer function of the series RLC circuit model with the classical derivative using the Laplace transform.

Then, we obtain

$$s^2 L(V(t)) - sV(0) - V'(0) + \frac{sR}{L}L(V(t)) - \frac{R}{L}V(0)$$
$$= \frac{R}{CL}L(i(t)) + \frac{s}{C}L(i(t)) - \frac{1}{C}i(0)$$

Thus, we have

$$L(V(t))\left(s^2 + \frac{sR}{L}\right) = L(i(t))\left(\frac{R}{CL} + \frac{s}{C}\right) \tag{12.3}$$

Therefore, we find the transfer functions as

$$\frac{L(i(t))}{L(V(t))} = Cs. \tag{12.4}$$

The graphical representation of the above transfer function is presented in Figure 12.1 as magnitude and phase.

If we take the Sumudu transform of the both sides of Eq. (12.1), we will obtain

$$S\left(\frac{d^2V(t)}{dt^2}\right) + \frac{R}{L}S\left(\frac{dV}{dt}\right) = S\left(\frac{Ri(t)}{CL}\right) + \frac{1}{C}S\left(\frac{di(t)}{dt}\right). \tag{12.5}$$

Then, we obtain

$$\frac{1}{s^2}S(V(t)) - \frac{1}{s^2}V(0) - \frac{1}{s}V'(0) + \frac{R}{sL}S(V(t)) - \frac{R}{Ls}V(0)$$

$$= \frac{R}{CL}S(i(t)) + \frac{1}{Cs}S(i(t)) - \frac{1}{Cs}i(0)$$

Thus, we have

$$S(V(t)) \left(\frac{1}{s^2} + \frac{R}{sL} \right) = S(i(t)) \left(\frac{R}{CL} + \frac{1}{Cs} \right) \tag{12.6}$$

Therefore, we find the transfer functions as

$$\frac{S(i(t))}{S(V(t))} = \frac{C}{s}. \tag{12.7}$$

The graphical representation of the above transfer function is presented in Figure 12.2 as magnitude and phase.

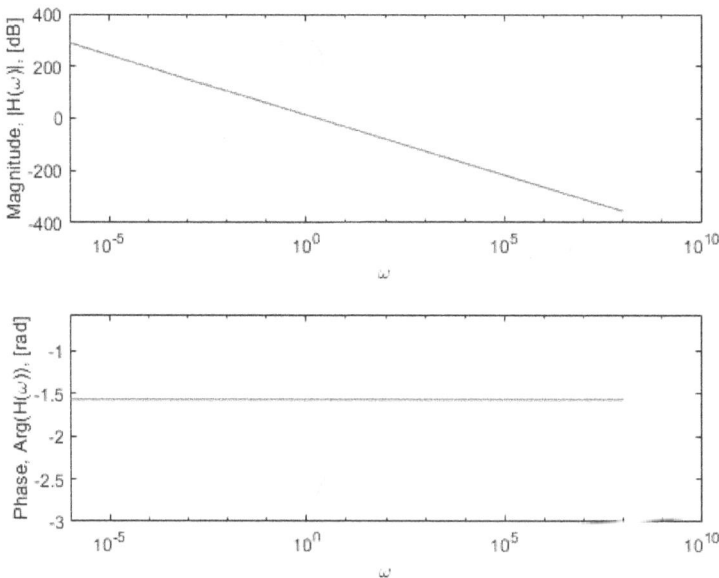

Figure 12.2 Transfer function of the series RLC circuit model with the classical derivative using the Sumudu transform.

12.2 ANALYSIS OF A SERIES RLC CIRCUIT MODEL WITH CAPUTO DERIVATIVE

We consider the model with Caputo derivative. We apply Laplace and Sumudu transforms to obtain the solution.

$$\left({}_0^C D_t^\alpha V(t)\right) + \frac{R}{L}\frac{dV}{dt} = \frac{Ri(t)}{CL} + \frac{1}{C}\frac{di(t)}{dt} \tag{12.8}$$

If we take the Laplace transform of the both sides of Eq. (12.8), we will obtain

$$L\left({}_0^C D_t^\alpha V(t)\right) + \frac{R}{L}L\left(\frac{dV}{dt}\right) = L\left(\frac{Ri(t)}{CL}\right) + \frac{1}{C}L\left(\frac{di(t)}{dt}\right). \tag{12.9}$$

Then, we obtain

$$\left(s^2 L(V(t)) - sV(0) - V'(0)\right)s^{\alpha-2} + \frac{sR}{L}L(V(t)) - \frac{R}{L}V(0)$$
$$= \frac{R}{CL}L(i(t)) + \frac{s}{C}L(i(t)) - \frac{1}{C}i(0)$$

Thus, we have

$$L(V(t))\left(s^\alpha + \frac{sR}{L}\right) = L(i(t))\left(\frac{R}{CL} + \frac{s}{C}\right) \tag{12.10}$$

Therefore, we find the transfer functions as

$$\frac{L(i(t))}{L(V(t))} = \frac{s^\alpha L + sR}{R + sL}. \tag{12.11}$$

The graphical representation of the above transfer function is presented in Figure 12.3 as magnitude and phase.

If we take the Sumudu transform of the both sides of Eq. (12.8), we will obtain

$$S\left({}_0^C D_t^\alpha V(t)\right) + \frac{R}{L}S\left(\frac{dV}{dt}\right) = S\left(\frac{Ri(t)}{CL}\right) + \frac{1}{C}S\left(\frac{di(t)}{dt}\right). \tag{12.12}$$

Then, we obtain

$$s^{2-\alpha}\left(\frac{S[V] - V(0)}{s^2} - \frac{V'(0)}{s}\right) + \frac{R}{sL}S(V(t)) - \frac{R}{Ls}V(0)$$
$$= \frac{R}{CL}S(i(t)) + \frac{1}{Cs}S(i(t)) - \frac{1}{Cs}i(0)$$

Thus, we have

$$S(V(t))\left(\frac{1}{s^\alpha} + \frac{R}{sL}\right) = S(i(t))\left(\frac{R}{CL} + \frac{1}{Cs}\right) \tag{12.13}$$

Therefore, we find the transfer functions as

$$\frac{S(i(t))}{S(V(t))} = \frac{C(sL + Rs^\alpha)}{s^{\alpha+1}(sR + L)}. \tag{12.14}$$

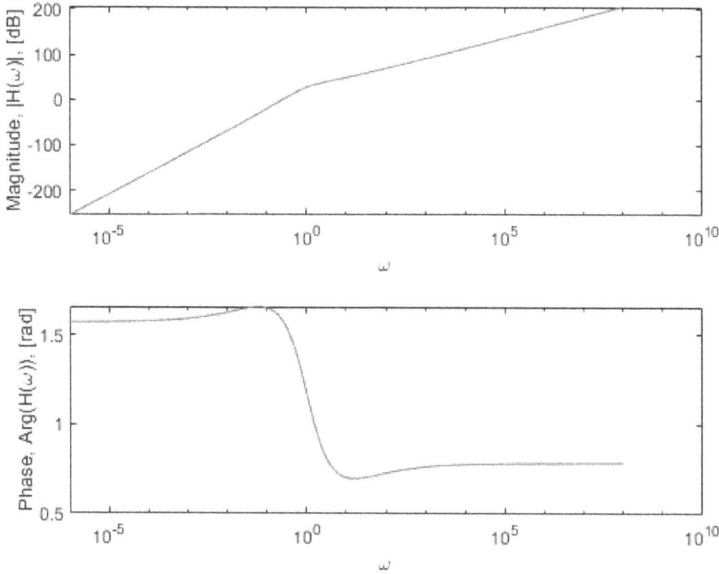

Figure 12.3 Transfer function of the series RLC circuit model with the Caputo derivative using the Laplace transform.

The graphical representation of the above transfer function is presented in Figure 12.4 as magnitude and phase.

12.3 ANALYSIS OF A SERIES RLC CIRCUIT MODEL WITH CAPUTO-FABRIZIO DERIVATIVE

We consider the model with Caputo-Fabrizio derivative. We apply Laplace and Sumudu transforms to obtain the solution.

$$\left(_0^{CF}D_t^\alpha V(t)\right) + \frac{R}{L}\frac{dV}{dt} = \frac{Ri(t)}{CL} + \frac{1}{C}\frac{di(t)}{dt} \tag{12.15}$$

If we take the Laplace transform of the both sides of Eq. (12.15), we will obtain

$$L\left(_0^{CF}D_t^\alpha V(t)\right) + \frac{R}{L}L\left(\frac{dV}{dt}\right) = L\left(\frac{Ri(t)}{CL}\right) + \frac{1}{C}L\left(\frac{di(t)}{dt}\right). \tag{12.16}$$

Then, we obtain

$$\frac{M(\alpha)}{\alpha + s(2-\alpha)}\left(s^2 L(V(t)) - sV(0) - V'(0)\right) + \frac{sR}{L}L(V(t)) - \frac{R}{L}V(0)$$
$$= \frac{R}{CL}L(i(t)) + \frac{s}{C}L(i(t)) - \frac{1}{C}i(0)$$

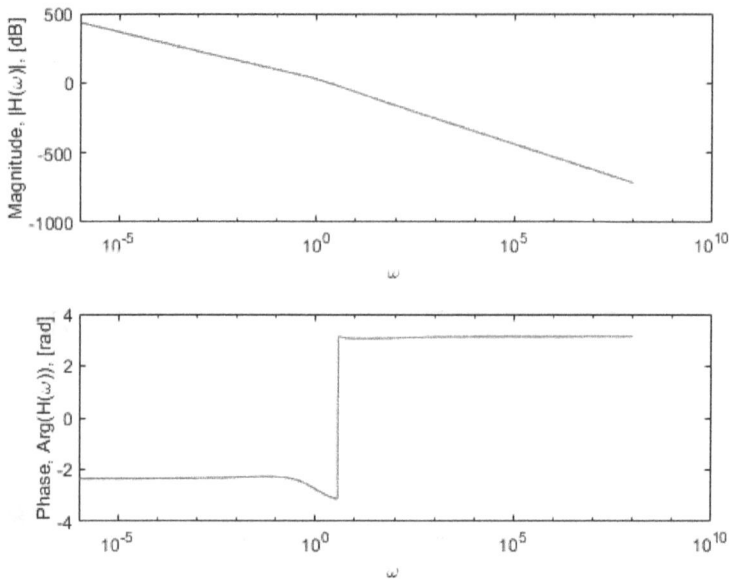

Figure 12.4 Transfer function of the series RLC circuit model with the Caputo derivative using the Sumudu transform.

Thus, we have

$$L(V(t))\left(\frac{M(\alpha)s^2}{\alpha+s(2-\alpha)}+\frac{sR}{L}\right)=L(i(t))\left(\frac{R}{CL}+\frac{s}{C}\right) \qquad (12.17)$$

Therefore, we find the transfer functions as

$$\frac{L(i(t))}{L(V(t))}=\frac{C\left(LM(\alpha)s^2+sR(\alpha+s(2-\alpha))\right)}{(R+sL)(\alpha+s(2-\alpha))}. \qquad (12.18)$$

The graphical representation of the above transfer function is presented in Figure 12.5 as magnitude and phase.

If we take the Sumudu transform of the both sides of Eq. (12.15), we will obtain

$$S\left({}_0^{CF}D_t^\alpha V(t)\right)+\frac{R}{L}S\left(\frac{dV}{dt}\right)=S\left(\frac{Ri(t)}{CL}\right)+\frac{1}{C}S\left(\frac{di(t)}{dt}\right). \qquad (12.19)$$

Then, we obtain

$$\frac{sM(\alpha)}{2-\alpha+\alpha s}\left(\frac{S[V]-V(0)}{s^2}-\frac{V'(0)}{s}\right)+\frac{R}{sL}S(V(t))-\frac{R}{Ls}V(0)$$

$$=\frac{R}{CL}S(i(t))+\frac{1}{Cs}S(i(t))-\frac{1}{Cs}i(0)$$

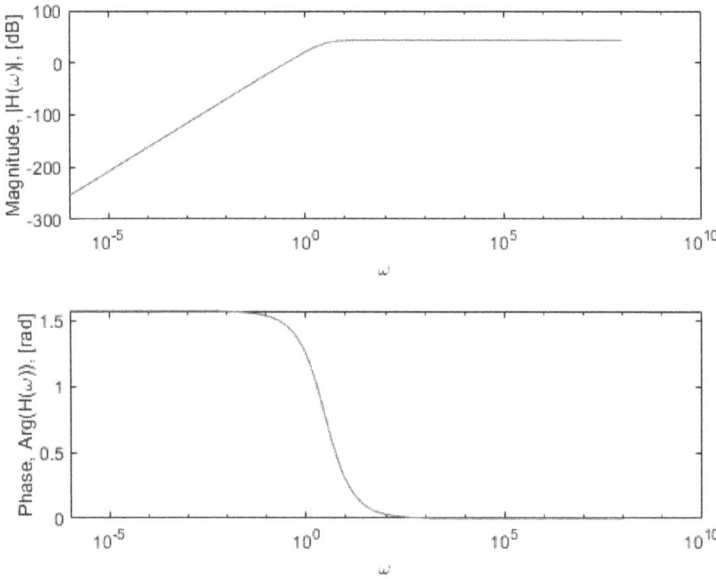

Figure 12.5 Transfer function of the series RLC circuit model with the Caputo-Fabrizio derivative using the Laplace transform.

Thus, we have

$$S(V(t))\left(\frac{M(\alpha)}{s(2-\alpha+\alpha s)} + \frac{R}{sL}\right) = S(i(t))\left(\frac{R}{CL} + \frac{1}{Cs}\right) \qquad (12.20)$$

Therefore, we find the transfer functions as

$$\frac{S(i(t))}{S(V(t))} = \frac{C[LM(\alpha) + R(2-\alpha+\alpha s)]}{(L+sR)(2-\alpha+\alpha s)}. \qquad (12.21)$$

The graphical representation of the above transfer function is presented in Figure 12.6 as magnitude and phase.

12.4 ANALYSIS OF A SERIES RLC CIRCUIT MODEL WITH ATANGANA-BALEANU DERIVATIVE

We consider the model with Atangana-Baleanu derivative. We apply Laplace and Sumudu transforms to obtain the solution.

$$\left(^{ABC}_{0}D^\alpha_t V(t)\right) + \frac{R}{L}\frac{dV}{dt} = \frac{Ri(t)}{CL} + \frac{1}{C}\frac{di(t)}{dt} \qquad (12.22)$$

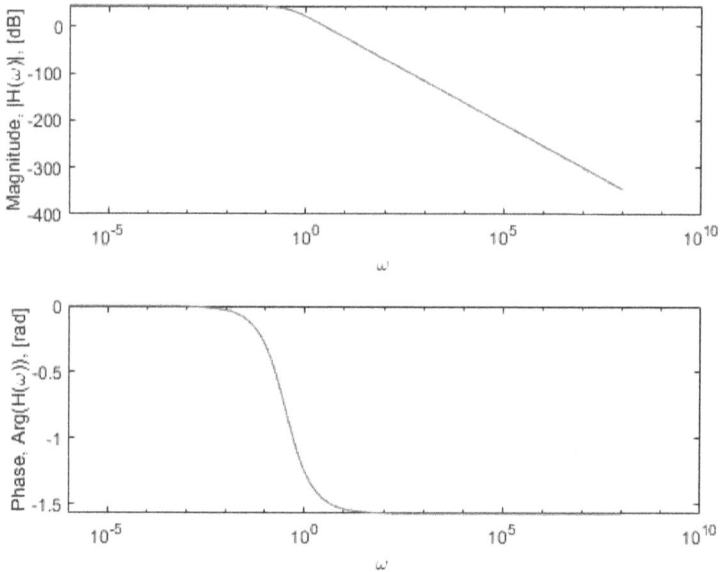

Figure 12.6 Transfer function of the series RLC circuit model with the Caputo-Fabrizio derivative using the Sumudu transform.

If we take the Laplace transform of the both sides of Eq. (12.22), we will obtain

$$L\left(^{ABC}_0 D^\alpha_t V(t)\right) + \frac{R}{L}L\left(\frac{dV}{dt}\right) = L\left(\frac{Ri(t)}{CL}\right) + \frac{1}{C}L\left(\frac{di(t)}{dt}\right). \qquad (12.23)$$

Then, we obtain

$$\frac{AB(\alpha)s^{\alpha-1}}{\alpha+(2-\alpha)s^\alpha}\left(s^2 L(V(t)) - sV(0) - V'(0)\right) + \frac{sR}{L}L(V(t)) - \frac{R}{L}V(0)$$

$$= \frac{R}{CL}L(i(t)) + \frac{s}{C}L(i(t)) - \frac{1}{C}i(0)$$

Thus, we have

$$L(V(t))\left(\frac{AB(\alpha)s^{\alpha+1}}{\alpha+s^\alpha(2-\alpha)} + \frac{sR}{L}\right) = L(i(t))\left(\frac{R}{CL} + \frac{s}{C}\right) \qquad (12.24)$$

Therefore, we find the transfer functions as

$$\frac{L(i(t))}{L(V(t))} = \frac{C\left[AB(\alpha)Ls^{\alpha+1} + sR(\alpha+(2-\alpha)s^\alpha)\right]}{(R+sL)(\alpha+(2-\alpha)s^\alpha)}. \qquad (12.25)$$

The graphical representation of the above transfer function is presented in Figure 12.7 as magnitude and phase.

If we take the Sumudu transform of the both sides of Eq. (12.22), we will obtain

$$S\left(_0^{ABC}D_t^{\alpha}V(t)\right) + \frac{R}{L}S\left(\frac{dV}{dt}\right) = S\left(\frac{Ri(t)}{CL}\right) + \frac{1}{C}S\left(\frac{di(t)}{dt}\right). \tag{12.26}$$

Then, we obtain

$$\frac{sAB(\alpha)}{2-\alpha+\alpha s^{\alpha}}\left(\frac{S[V]-V(0)}{s^2} - \frac{V'(0)}{s}\right) + \frac{R}{sL}S(V(t)) - \frac{R}{Ls}V(0)$$

$$= \frac{R}{CL}S(i(t)) + \frac{1}{Cs}S(i(t)) - \frac{1}{Cs}i(0)$$

Thus, we have

$$S(V(t))\left(\frac{AB(\alpha)}{s(2-\alpha+\alpha s^{\alpha})} + \frac{R}{sL}\right) = S(i(t))\left(\frac{R}{CL} + \frac{1}{Cs}\right) \tag{12.27}$$

Therefore, we find the transfer functions as

$$\frac{S(i(t))}{S(V(t))} = \frac{C[LAB(\alpha)+R(2-\alpha+\alpha s^{\alpha})]}{(L+sR)(2-\alpha+\alpha s^{\alpha})}. \tag{12.28}$$

The graphical representation of the above transfer function is presented in Figure 12.8 as magnitude and phase.

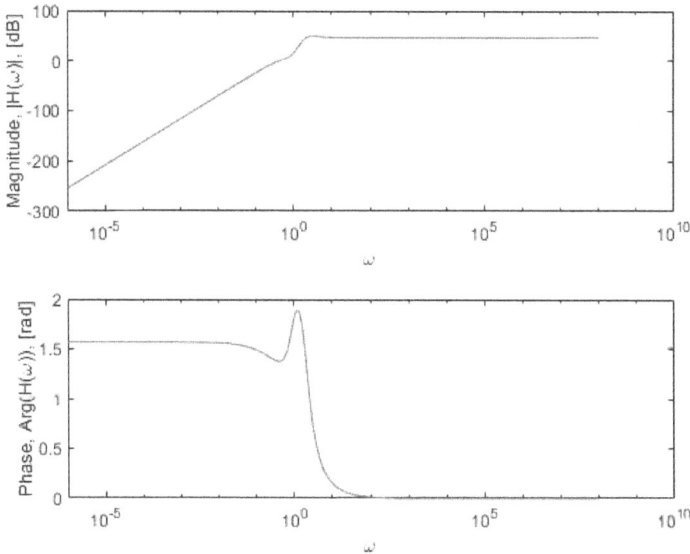

Figure 12.7 Transfer function of the series RLC circuit model with the ABC derivative using the Laplace transform.

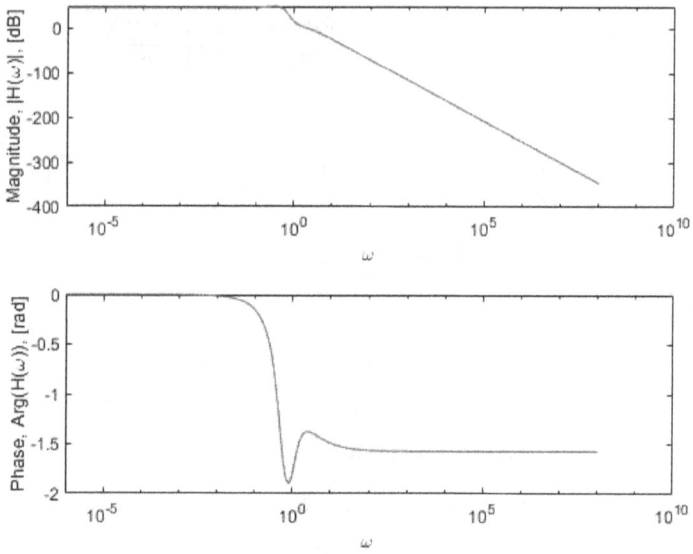

Figure 12.8 Transfer function of the series RLC circuit model with the ABC derivative using the Sumudu transform.

13 Analysis of a Parallel RLC Circuit Model

There are several ways to build a system that includes R, L, and C as three circuit elements. One way is to place them in series, where the three elements share the same current. For this first simple case, two ways can be recognized. The second way is to place them in parallel, noting the system is in parallel if all three components share the same pair of nodes. These are the simplest in notion and most straightforward for investigations. Nonetheless, several arrangements could be made, with some being very important for practical purposes in real circuits. A well-known problem is usually found in the need to account for inductor resistance. This is due uniquely because inductors are classically built from coils of wire, the resistance of which is not frequently necessary. However, it has an important effect on the circuit system. The investigation of a parallel RLC circuit can be more complex when converted into mathematical terms than the series version. Because, as an alternative to the current being common to the circuit components, the applied voltage is in this case common to all, thus there is a need to find the individual branch currents via each element. Now the overall impedance, known as Z of a parallel RLC system [56–58], is computed by inducing the current of the system similar to that of a DC parallel system; thus, the alteration is that admittance is applied instead of impedance. These systems have been used in many industrial settlements, cities, towns, and transport. In this chapter we shall therefore consider the mathematical model describing the RLC circuit problem in parallel mode. Different differential operators will be considered, for each the Laplace and the Sumudu transforms will be applied with the aim of obtaining the transfer function under the condition that the u-domain of the Sumudu transform is the complex number as in the Laplace transform case.

13.1 ANALYSIS OF A PARALLEL RLC CIRCUIT MODEL WITH CLASSICAL DERIVATIVE

We consider the model with classical derivative. We apply Laplace and Sumudu transforms to obtain the solution.

$$\frac{d^2 i(t)}{dt^2} + \frac{1}{RC}\frac{di}{dt} = \frac{V(t)}{RCL} + \frac{1}{L}\frac{dV(t)}{dt} \tag{13.1}$$

If we take the Laplace transform of the both sides of Eq. (13.1), we will obtain

$$L\left(\frac{d^2 i(t)}{dt^2}\right) + \frac{1}{RC}L\left(\frac{di}{dt}\right) = L\left(\frac{V(t)}{RCL}\right) + L\left(\frac{1}{L}\frac{dV(t)}{dt}\right)$$

DOI: 10.1201/9781003359869-13

113

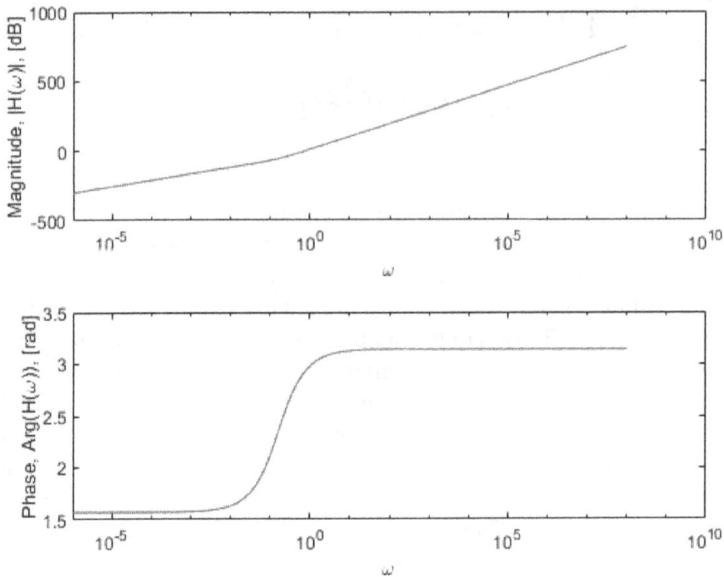

Figure 13.1 Transfer function of the parallel RLC circuit model with the classical derivative using the Laplace transform.

Then, we obtain

$$\left(s^2 L(i(t)) - si(0) - i'(0)\right) + \frac{1}{RC}\left(sL(i(t)) - i(0)\right) = \left(\frac{L(V(t))}{RCL}\right)$$
$$+ \frac{1}{L}\left(L(V(t)) - V(0)\right)$$

If we simplify the above equation, we will obtain

$$L(i(t))\left(s^2 + \frac{s}{RC}\right) = L(V(t))\left(\frac{1}{RCL} + \frac{1}{L}\right)$$

Then, we obtain the transfer function as

$$\frac{L(V(t))}{L(i(t))} = \frac{Ls(1 + sRC)}{1 + RC}.$$

The graphical representation of the above transfer function is presented in Figure 13.1 as magnitude and phase.

If we take the Sumudu transform of the both sides of Eq. (13.1), we will obtain

$$S\left(\frac{d^2 i(t)}{dt^2}\right) + \frac{1}{RC} S\left(\frac{di}{dt}\right) = S\left(\frac{V(t)}{RCL}\right) + S\left(\frac{1}{L}\frac{dV(t)}{dt}\right)$$

Then, we obtain

$$\left(\frac{S[i(t)] - i(0)}{s^2} - \frac{i'(0)}{s}\right) + \frac{1}{sRC}\left(S(i(t)) - i(0)\right) = \left(\frac{S(V(t))}{RCL}\right)$$
$$+ \frac{1}{sL}\left(S(V(t)) - V(0)\right)$$

If we simplify the above equation, we will obtain

$$S(i(t))\left(\frac{1}{s^2} + \frac{1}{sRC}\right) = S(V(t))\left(\frac{1}{RCL} + \frac{1}{sL}\right)$$

Then, we obtain the transfer function as

$$\frac{S(V(t))}{S(i(t))} = \frac{L}{s}.$$

The graphical representation of the above transfer function is presented in Figure 13.2 as magnitude and phase.

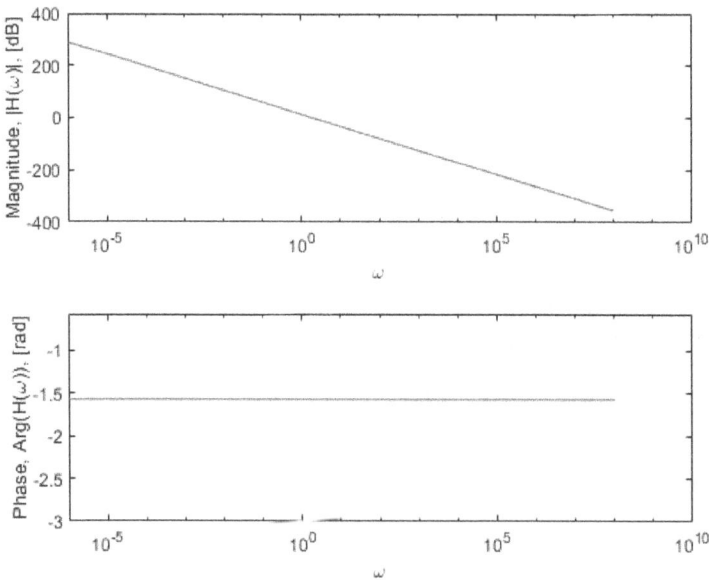

Figure 13.2 Transfer function of the parallel RLC circuit model with the classical derivative using the Sumudu transform.

13.2 ANALYSIS OF A PARALLEL RLC CIRCUIT MODEL WITH CAPUTO DERIVATIVE

We consider the model with Caputo derivative. We apply Laplace and Sumudu transforms to obtain the solution.

$$\left({}_{0}^{C}D_{t}^{\alpha}i(t)\right) + \frac{1}{RC}\frac{di}{dt} = \frac{V(t)}{RCL} + \frac{1}{L}\frac{dV(t)}{dt} \tag{13.2}$$

If we take the Laplace transform of the both sides of Eq. (13.2), we will obtain

$$L\left({}_{0}^{C}D_{t}^{\alpha}i(t)\right) + \frac{1}{RC}L\left(\frac{di}{dt}\right) = L\left(\frac{V(t)}{RCL}\right) + L\left(\frac{1}{L}\frac{dV(t)}{dt}\right)$$

Then, we obtain

$$s^{\alpha-2}\left(s^2 L(i(t)) - si(0) - i'(0)\right) + \frac{1}{RC}\left(sL(i(t)) - i(0)\right)$$
$$= \left(\frac{L(V(t))}{RCL}\right) + \frac{1}{L}\left(L(V(t)) - V(0)\right)$$

If we simplify the above equation, we will obtain

$$L(i(t))\left(s^{\alpha} + \frac{s}{RC}\right) = L(V(t))\left(\frac{1}{RCL} + \frac{1}{L}\right)$$

Then, we obtain the transfer function as

$$\frac{L(V(t))}{L(i(t))} = \frac{L(s + s^{\alpha}RC)}{1 + RC}.$$

The graphical representation of the above transfer function is presented in Figure 13.3 as magnitude and phase.

If we take the Sumudu transform of the both sides of Eq. (13.2), we will obtain

$$S\left({}_{0}^{C}D_{t}^{\alpha}i(t)\right) + \frac{1}{RC}S\left(\frac{di}{dt}\right) = S\left(\frac{V(t)}{RCL}\right) + S\left(\frac{1}{L}\frac{dV(t)}{dt}\right)$$

Then, we obtain

$$s^{2-\alpha}\left(\frac{S[i(t)] - i(0)}{s^2} - \frac{i'(0)}{s}\right) + \frac{1}{sRC}\left(S(i(t)) - i(0)\right)$$
$$= \left(\frac{S(V(t))}{RCL}\right) + \frac{1}{sL}\left(S(V(t)) - V(0)\right)$$

If we simplify the above equation, we will obtain

$$S(i(t))\left(\frac{1}{s^{\alpha}} + \frac{1}{sRC}\right) = S(V(t))\left(\frac{1}{RCL} + \frac{1}{sL}\right)$$

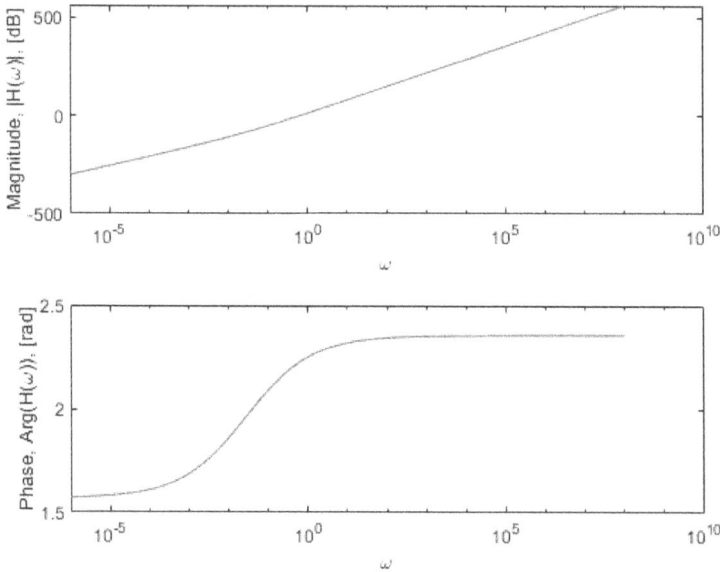

Figure 13.3 Transfer function of the parallel RLC circuit model with the Caputo derivative using the Laplace transform.

Then, we obtain the transfer function as

$$\frac{S(V(t))}{S(i(t))} = \frac{(sRC + s^\alpha)L}{s^\alpha(s + RC)}.$$

The graphical representation of the above transfer function is presented in Figure 13.4 as magnitude and phase.

13.3 ANALYSIS OF A PARALLEL RLC CIRCUIT MODEL WITH CAPUTO-FABRIZIO DERIVATIVE

We consider the model with Caputo-Fabrizio derivative. We apply Laplace and Sumudu transforms to obtain the solution.

$$\left(_0^{CF}D_t^\alpha i(t)\right) + \frac{1}{RC}\frac{di}{dt} = \frac{V(t)}{RCL} + \frac{1}{L}\frac{dV(t)}{dt} \tag{13.3}$$

If we take the Laplace transform of the both sides of Eq. (13.3), we will obtain

$$L\left(_0^{CF}D_t^\alpha i(t)\right) + \frac{1}{RC}L\left(\frac{di}{dt}\right) = L\left(\frac{V(t)}{RCL}\right) + L\left(\frac{1}{L}\frac{dV(t)}{dt}\right)$$

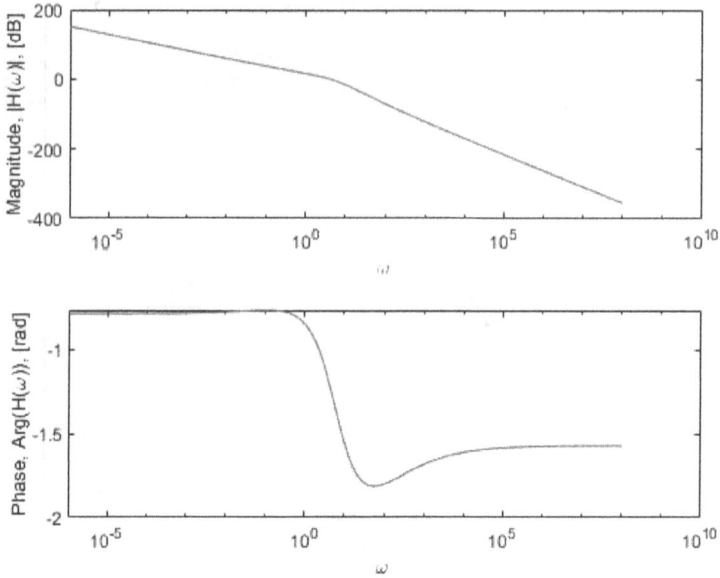

Figure 13.4   Transfer function of the parallel RLC circuit model with the Caputo derivative using the Sumudu transform.

Then, we obtain

$$\frac{M(\alpha)}{\alpha + s(2-\alpha)} \left(s^2 L(i(t)) - si(0) - i'(0) \right) + \frac{1}{RC} \left(sL(i(t)) - i(0) \right)$$
$$= \left(\frac{L(V(t))}{RCL} \right) + \frac{1}{L} \left(L(V(t)) - V(0) \right)$$

If we simplify the above equation, we will obtain

$$L(i(t)) \left(\frac{s^2 M(\alpha)}{\alpha + s(2-\alpha)} + \frac{s}{RC} \right) = L(V(t)) \left(\frac{1}{RCL} + \frac{1}{L} \right)$$

Then, we obtain the transfer function as

$$\frac{L(V(t))}{L(i(t))} = \frac{L\left[RCs^2 M(\alpha) + s(\alpha + s(2-\alpha)) \right]}{(RC+1)(\alpha + s(2-\alpha))}.$$

The graphical representation of the above transfer function is presented in Figure 13.5 as magnitude and phase.

If we take the Sumudu transform of the both sides of Eq. (13.3), we will obtain

$$S\left({}_0^{CF} D_t^\alpha i(t) \right) + \frac{1}{RC} S\left(\frac{di}{dt} \right) = S\left(\frac{V(t)}{RCL} \right) + S\left(\frac{1}{L} \frac{dV(t)}{dt} \right)$$

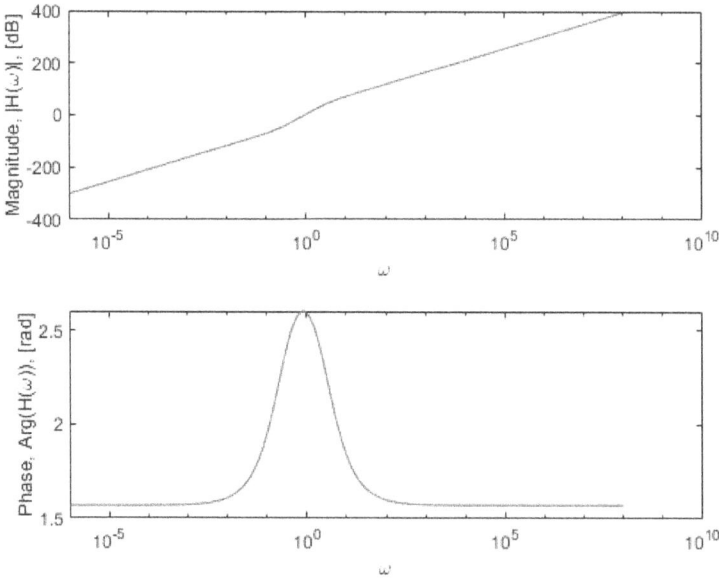

Figure 13.5 Transfer function of the parallel RLC circuit model with the Caputo-Fabrizio derivative using the Laplace transform.

Then, we obtain

$$\frac{sM(\alpha)}{2-\alpha+\alpha s}\left(\frac{S[i(t)]-i(0)}{s^2}-\frac{i'(0)}{s}\right)+\frac{1}{sRC}(S(i(t))-i(0))$$
$$=\left(\frac{S(V(t))}{RCL}\right)+\frac{1}{sL}(S(V(t))-V(0))$$

If we simplify the above equation, we will obtain

$$S(i(t))\left(\frac{M(\alpha)}{s(2-\alpha+s\alpha)}+\frac{1}{sRC}\right)=S(V(t))\left(\frac{1}{RCL}+\frac{1}{sL}\right)$$

Then, we obtain the transfer function as

$$\frac{S(V(t))}{S(i(t))}=\frac{L(RCM(\alpha)+2-\alpha+\alpha s)}{(RC+s)(2-\alpha+\alpha s)}.$$

The graphical representation of the above transfer function is presented in Figure 13.6 as magnitude and phase.

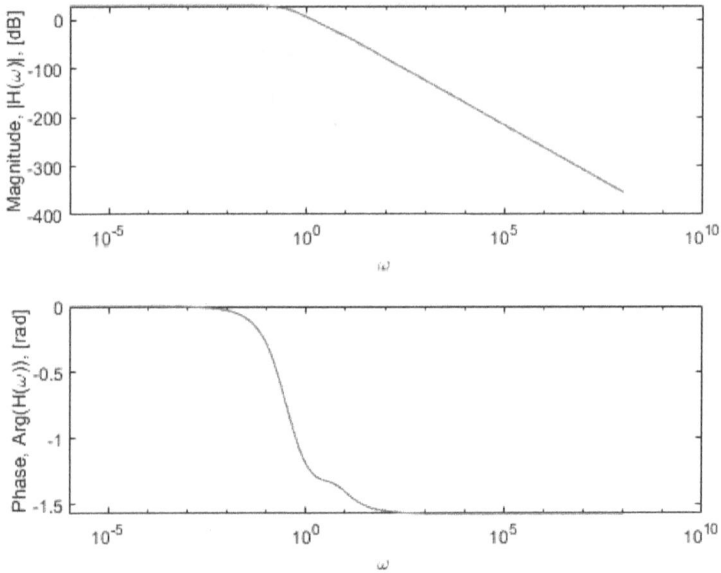

Figure 13.6   Transfer function of the parallel RLC circuit model with the Caputo-Fabrizio derivative using the Sumudu transform.

13.4   ANALYSIS OF A PARALLEL RLC CIRCUIT MODEL WITH ATANGANA-BALEANU DERIVATIVE

We consider the model with Atangana-Baleanu derivative. We apply Laplace and Sumudu transforms to obtain the solution.

$$\left(_0^{ABC}D_t^\alpha i(t)\right) + \frac{1}{RC}\frac{di}{dt} = \frac{V(t)}{RCL} + \frac{1}{L}\frac{dV(t)}{dt} \tag{13.4}$$

If we take the Laplace transform of the both sides of Eq. (13.4), we will obtain

$$L\left(_0^{ABC}D_t^\alpha i(t)\right) + \frac{1}{RC}L\left(\frac{di}{dt}\right) = L\left(\frac{V(t)}{RCL}\right) + L\left(\frac{1}{L}\frac{dV(t)}{dt}\right)$$

Then, we obtain

$$\frac{AB(\alpha)s^{\alpha-1}}{\alpha+(2-\alpha)s^\alpha}\left(s^2 L(i(t)) - si(0) - i'(0)\right) + \frac{1}{RC}\left(sL(i(t)) - i(0)\right)$$

$$= \left(\frac{L(V(t))}{RCL}\right) + \frac{1}{L}\left(L(V(t)) - V(0)\right)$$

If we simplify the above equation, we will obtain

$$L(i(t))\left(\frac{s^{\alpha+1}AB(\alpha)}{\alpha+s^\alpha(2-\alpha)} + \frac{s}{RC}\right) = L(V(t))\left(\frac{1}{RCL} + \frac{1}{L}\right)$$

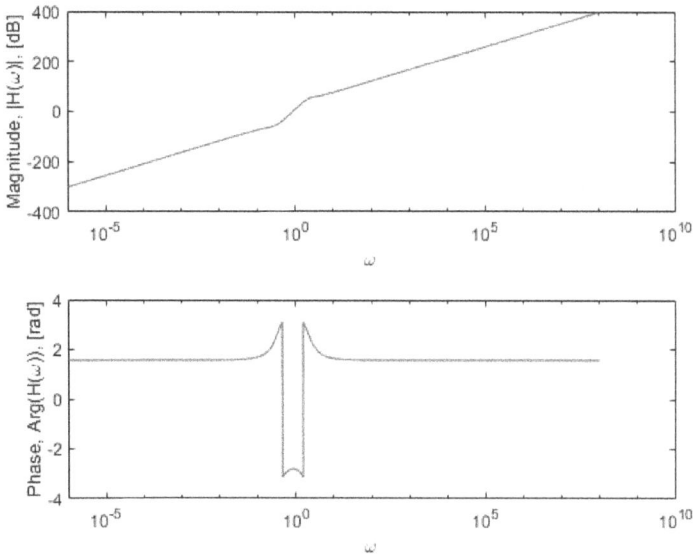

Figure 13.7 Transfer function of the parallel RLC circuit model with the ABC derivative using the Laplace transform.

Then, we obtain the transfer function as

$$\frac{L(V(t))}{L(i(t))} = \frac{L\left[RCs^{\alpha+1}AB(\alpha) + s(\alpha + s^{\alpha}(2-\alpha))\right]}{(RC+1)(\alpha + s^{\alpha}(2-\alpha))}.$$

The graphical representation of the above transfer function is presented in Figure 13.7 as magnitude and phase.

If we take the Sumudu transform of the both sides of Eq. (13.4), we will obtain

$$S\left(_{0}^{ABC}D_{t}^{\alpha}i(t)\right) + \frac{1}{RC}S\left(\frac{di}{dt}\right) = S\left(\frac{V(t)}{RCL}\right) + S\left(\frac{1}{L}\frac{dV(t)}{dt}\right)$$

Then, we obtain

$$\frac{sAB(\alpha)}{2 - \alpha + \alpha s^{\alpha}}\left(\frac{S[i(t)] - i(0)}{s^2} - \frac{i'(0)}{s}\right) + \frac{1}{sRC}\left(S(i(t)) - i(0)\right)$$

$$= \left(\frac{S(V(t))}{RCL}\right) + \frac{1}{sL}\left(S(V(t)) - V(0)\right)$$

If we simplify the above equation, we will obtain

$$S(i(t))\left(\frac{AB(\alpha)}{s(2 - \alpha + s^{\alpha}\alpha)} + \frac{1}{sRC}\right) = S(V(t))\left(\frac{1}{RCL} + \frac{1}{sL}\right)$$

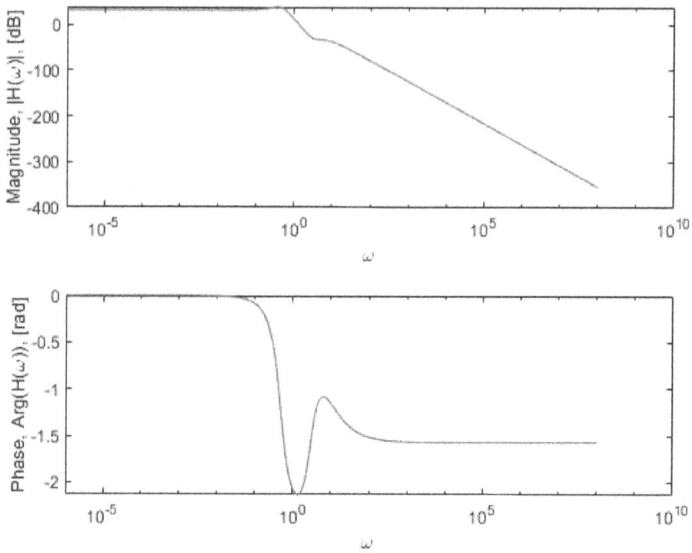

Figure 13.8 Transfer function of the parallel RLC circuit model with the ABC derivative using the Sumudu transform.

Then, we obtain the transfer function as

$$\frac{S(V(t))}{S(i(t))} = \frac{L(RCAB(\alpha) + 2 - \alpha + \alpha s^{\alpha})}{(RC + s)(2 - \alpha + \alpha s^{\alpha})}.$$

The graphical representation of the above transfer function is presented in Figure 13.8 as magnitude and phase.

14 Analysis of Higher Order Circuit Model 1

To capture processes exhibiting second-order variation, for example, problems displaying processes involving accelerative behaviors, or change in concerning time or space, the concept of second differentiation has been introduced and gave birth to an important class of differential equations known as second order. This class of ordinary or partial differential equations has been found powerful mathematical tools to replicate behavior with accelerative or dispersive properties. In electrical engineering, these equations are used to model the transfer of electricity. In this field, it is known that the order of a circuit is equivalent to the number of energy storage elements resulting from all the possible possibilities that include series and parallel combinations of inductors and capacitors [59–62]. In this section, we shall devote our attention to applying the Laplace and the Sumudu transforms to solving second-order circuit problems. Our aim is not to obtain exact solutions to these equations in real space, but to obtain the solutions in the complex space for us to determine the Transfer function.

We consider the higher order circuit model 1 with classical, Caputo, Caputo Fabrizio, and Atangana-Baleanu derivatives.

14.1 ANALYSIS OF HIGHER ORDER CIRCUIT MODEL 1 WITH CLASSICAL DERIVATIVE

We consider the model with classical derivative. We apply Laplace and Sumudu transforms to obtain the solution.

$$\frac{R_1}{CL}i(t) + \frac{1}{C}\frac{di(t)}{dt} + \frac{V(t)}{CL} = \frac{d^2V(t)}{dt^2} + \left(\frac{R_1}{L} + \frac{1}{R_2C}\right)\frac{dV(t)}{dt} + \frac{R_1+R_2}{R_2CL}V(t) \quad (14.1)$$

If we take the Laplace transform of the both sides of Eq. (14.1), we will obtain

$$\frac{R_1}{CL}L(i(t)) + \frac{1}{C}L\left(\frac{di(t)}{dt}\right) + \frac{L(V(t))}{CL} = L\left(\frac{d^2V(t)}{dt^2}\right)$$
$$+ \left(\frac{R_1}{L} + \frac{1}{R_2C}\right)L\left(\frac{dV(t)}{dt}\right) + \frac{R_1+R_2}{R_2CL}L(V(t))$$

DOI: 10.1201/9781003359869-14

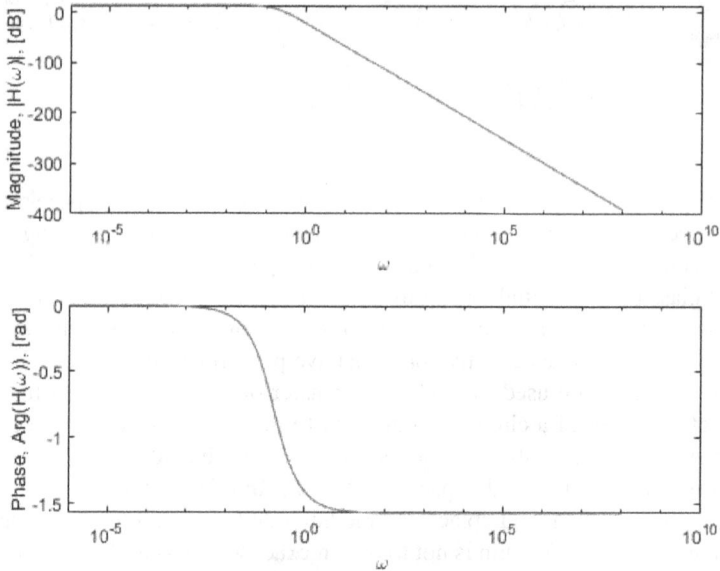

Figure 14.1 Transfer function of the higher order circuit model 1 with the classical derivative using the Laplace transform.

Then, we obtain

$$\frac{R_1}{CL}L(i(t)) + \frac{1}{C}(sL(i(t)) - i(0)) + \frac{L(V(t))}{CL} = (s^2L(V(t)) - sV(0) - V'(0))$$
$$+ \left(\frac{R_1}{L} + \frac{1}{R_2C}\right)(sL(V(t)) - V(0)) + \frac{R_1 + R_2}{R_2CL}L(V(t))$$

If we simplify the above equation, we will get

$$L(i(t))\left(\frac{R_1}{CL} + \frac{s}{C}\right) = L(V(t))\left(s^2 - \frac{1}{CL} + \frac{sR_1}{L} + \frac{s}{R_2C} + \frac{R_1 + R_2}{R_2CL}\right)$$

Therefore, we obtain the transfer function as

$$\frac{L(V(t))}{L(i(t))} = \frac{R_2(R_1 + sL)}{s^2R_2CL + sR_1R_2C + sL + R_1}. \quad (14.2)$$

The graphical representation of the above transfer function is presented in Figure 14.1 as magnitude and phase.

If we take the Sumudu transform of the both sides of Eq. (14.1), we will obtain

$$\frac{R_1}{CL}S(i(t)) + \frac{1}{C}S\left(\frac{di(t)}{dt}\right) + \frac{S(V(t))}{CL} = S\left(\frac{d^2V(t)}{dt^2}\right)$$

$$+ \left(\frac{R_1}{L} + \frac{1}{R_2C}\right)S\left(\frac{dV(t)}{dt}\right) + \frac{R_1+R_2}{R_2CL}S(V(t))$$

Then, we obtain

$$\frac{R_1}{CL}S(i(t)) + \frac{1}{sC}\left(S[i(t)] - i(0)\right) + \frac{S(V(t))}{CL} = \left(\frac{S[V(t)] - V(0)}{s^2} - \frac{V'(0)}{s}\right)$$

$$+ \left(\frac{R_1}{sL} + \frac{1}{sR_2C}\right)\left(S[V(t)] - V(0)\right) + \frac{R_1+R_2}{R_2CL}S(V(t))$$

If we simplify the above equation, we will get

$$S(i(t))\left(\frac{R_1}{CL} + \frac{1}{sC}\right) = S(V(t))\left(-\frac{1}{CL} + \frac{1}{s^2} + \frac{R_1}{SL} + \frac{1}{SR_2C} + \frac{R_1+R_2}{R_2CL}\right)$$

Therefore, we obtain the transfer function as

$$\frac{S(V(t))}{S(i(t))} = \frac{sR_2(sR_1 + L)}{CLR_2 + R_1CSR_2 + sL + s^2R_1}. \tag{14.3}$$

The graphical representation of the above transfer function is presented in Figure 14.2 as magnitude and phase.

14.2   ANALYSIS OF HIGHER ORDER CIRCUIT MODEL 1 WITH CAPUTO DERIVATIVE

We consider the model with Caputo derivative. We apply Laplace and Sumudu transforms to obtain the solution.

$$\frac{R_1}{CL}i(t) + \frac{1}{C}\left({}_0^C D_t^\alpha i(t)\right) + \frac{V(t)}{CL} = \frac{d^2V(t)}{dt^2} + \left(\frac{R_1}{L} + \frac{1}{R_2C}\right)\frac{dV(t)}{dt} + \frac{R_1+R_2}{R_2CL}V(t) \tag{14.4}$$

If we take the Laplace transform of the both sides of Eq. (14.4), we will obtain

$$\frac{R_1}{CL}L(i(t)) + \frac{1}{C}L\left({}_0^C D_t^\alpha i(t)\right) + \frac{L.(V(t))}{CL} = L\left(\frac{d^2V(t)}{dt^2}\right)$$

$$+ \left(\frac{R_1}{L} + \frac{1}{R_2C}\right)L\left(\frac{dV(t)}{dt}\right) + \frac{R_1+R_2}{R_2CL}L(V(t))$$

Then, we obtain

$$\frac{R_1}{CL}L(i(t)) + \frac{1}{C}s^{\alpha-1}\left(sL(i(t)) - i(0)\right) + \frac{L(V(t))}{CL} = \left(s^2L(V(t)) - sV(0) - V'(0)\right)$$

$$\left(\frac{R_1}{L} + \frac{1}{R_2C}\right)\left(sL(V(t)) - V(0)\right) + \frac{R_1+R_2}{R_2CL}L(V(t))$$

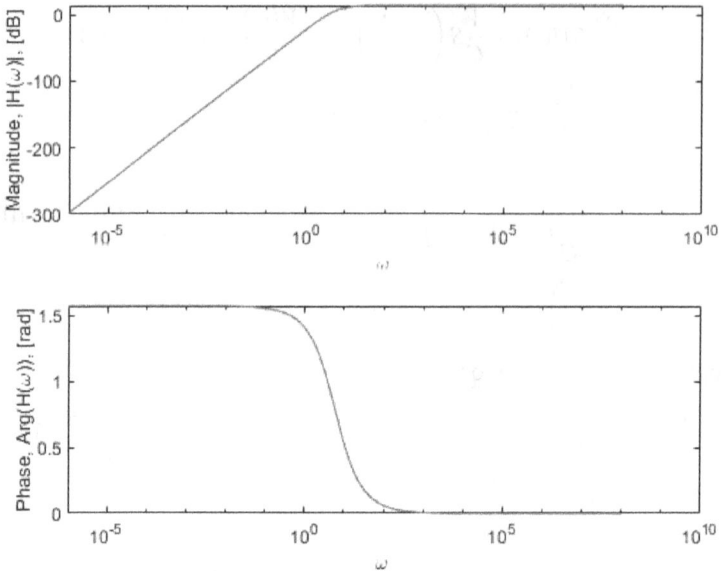

Figure 14.2   Transfer function of the higher order circuit model 1 with the classical derivative using the Sumudu transform.

If we simplify the above equation, we will get

$$L(i(t)) \left(\frac{R_1}{CL} + \frac{s^\alpha}{C} \right) = L(V(t)) \left(s^2 - \frac{1}{CL} + \frac{sR_1}{L} + \frac{s}{R_2 C} + \frac{R_1 + R_2}{R_2 CL} \right)$$

$$L(i(t)) \left(\frac{R_1 + Ls^\alpha}{CL} \right) = L(V(t)) \left(\frac{s^2 CLR_2 + CsR_1 R2 + Ls + R_1}{CLR_2} \right)$$

Therefore, we obtain the transfer function as

$$\frac{L(V(t))}{L(i(t))} = \frac{R_2(R_1 + s^\alpha L)}{s^2 CLR_2 + CsR_1 R_2 + Ls + R_1}. \tag{14.5}$$

The graphical representation of the above transfer function is presented in Figure 14.3 as magnitude and phase.

If we take the Sumudu transform of the both sides of Eq. (14.4), we will obtain

$$\frac{R_1}{CL} S(i(t)) + \frac{1}{C} S \left({}_0^C D_t^\alpha i(t) \right) + \frac{S(V(t))}{CL} = S \left(\frac{d^2 V(t)}{dt^2} \right)$$

$$+ \left(\frac{R_1}{L} + \frac{1}{R_2 C} \right) S \left(\frac{dV(t)}{dt} \right) + \frac{R_1 + R_2}{R_2 CL} S(V(t))$$

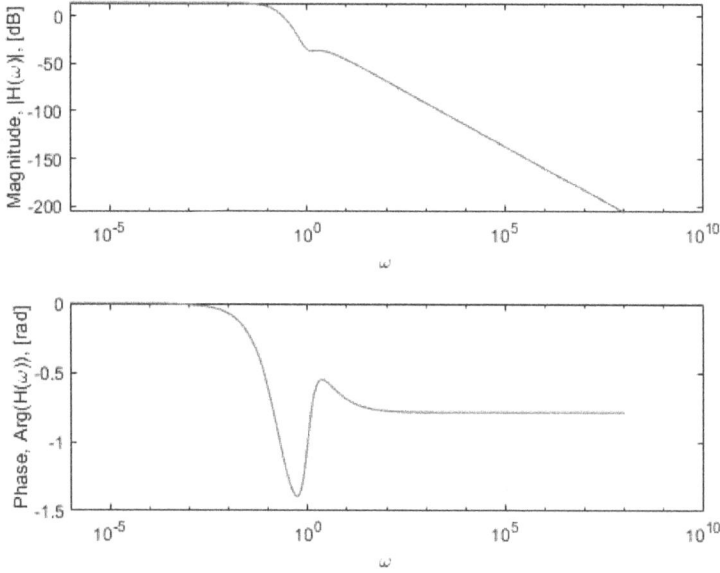

Figure 14.3   Transfer function of the higher order circuit model 1 with the Caputo derivative using the Laplace transform.

Then, we obtain

$$\frac{R_1}{CL}S(i(t)) + \frac{1}{C}\frac{S(i(t)) - i(0)}{s^\alpha} + \frac{S(V(t))}{CL} = \left(\frac{S[V(t)] - V(0)}{s^2} - \frac{V'(0)}{s}\right)$$
$$+ \left(\frac{R_1}{sL} + \frac{1}{sR_2C}\right)(S[V(t)] - V(0)) + \frac{R_1 + R_2}{R_2CL}S(V(t))$$

If we simplify the above equation, we will get

$$S(i(t))\left(\frac{R_1}{CL} + \frac{1}{Cs^\alpha}\right) = S(V(t))\left(-\frac{1}{CL} + \frac{1}{s^2} + \frac{R_1}{SL} + \frac{1}{SR_2C} + \frac{R_1 + R_2}{R_2CL}\right)$$

$$S(i(t))\left(\frac{R_1 s^\alpha + L}{CLs^\alpha}\right) = S(V(t))\left(\frac{CLR_2 + sCR_2R_1 + Ls + s^2R_1}{s^2CLR_2}\right)$$

Therefore, we obtain the transfer function as

$$\frac{S(V(t))}{S(i(t))} = \frac{(R_2 s^{2-\alpha})(L + R_1 s^\alpha)}{CLR_2 + sCR_2R_1 + Ls + s^2R_1}. \tag{14.6}$$

The graphical representation of the above transfer function is presented in Figure 14.4 as magnitude and phase.

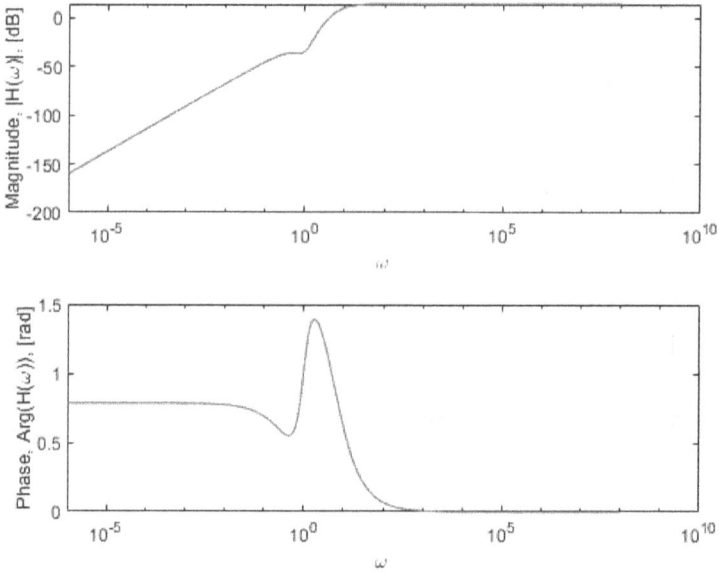

Figure 14.4 Transfer function of the higher order circuit model 1 with the Caputo derivative using the Sumudu transform.

14.3 ANALYSIS OF HIGHER ORDER CIRCUIT MODEL 1 WITH CAPUTO-FABRIZIO DERIVATIVE

We consider the model with Caputo-Fabrizio derivative. We apply Laplace and Sumudu transforms to obtain the solution.

$$\frac{R_1}{CL}i(t) + \frac{1}{C}\left(_0^{CF}D_t^\alpha i(t)\right) + \frac{V(t)}{CL} = \frac{d^2V(t)}{dt^2} + \left(\frac{R_1}{L} + \frac{1}{R_2C}\right)\frac{dV(t)}{dt} + \frac{R_1+R_2}{R_2CL}V(t)$$

(14.7)

If we take the Laplace transform of the both sides of Eq. (14.7), we will obtain

$$\frac{R_1}{CL}L(i(t)) + \frac{1}{C}L\left(_0^{CF}D_t^\alpha i(t)\right) + \frac{L(V(t))}{CL} = L\left(\frac{d^2V(t)}{dt^2}\right)$$

$$+ \left(\frac{R_1}{L} + \frac{1}{R_2C}\right)L\left(\frac{dV(t)}{dt}\right) + \frac{R_1+R_2}{R_2CL}L(V(t))$$

Then, we obtain

$$\frac{R_1}{CL}L(i(t)) + \frac{1}{C}\frac{M(\alpha)}{s+\alpha-s\alpha}(sL(i(t))-i(0)) + \frac{L(V(t))}{CL}$$
$$= \left(s^2 L(V(t)) - sV(0) - V'(0)\right) + \left(\frac{R_1}{L} + \frac{1}{R_2 C}\right)(sL(V(t)) - V(0))$$
$$+ \frac{R_1+R_2}{R_2 CL}L(V(t))$$

If we simplify the above equation, we will get

$$L(i(t))\left(\frac{R_1}{CL} + \frac{s}{C}\frac{M(\alpha)}{s+\alpha-s\alpha}\right) = L(V(t))\left(s^2 - \frac{1}{CL} + \frac{sR_1}{L} + \frac{s}{R_2 C} + \frac{R_1+R_2}{R_2 CL}\right)$$

$$L(i(t))\left(\frac{R_1(s+\alpha-s\alpha)+LsM(\alpha)}{CL(s+\alpha-s\alpha)}\right) = L(V(t))\left(\frac{s^2 CLR_2 + CsR_1 R2 + Ls + R_1}{CLR_2}\right)$$

Therefore, we obtain the transfer function as

$$\frac{L(V(t))}{L(i(t))} = \frac{R_2\left(R_1(s+\alpha-s\alpha)+LsM(\alpha)\right)}{(s+\alpha-s\alpha)(s^2 CLR_2 + CsR_1 R_2 + Ls + R_1)}. \tag{14.8}$$

The graphical representation of the above transfer function is presented in Figure 14.5 as magnitude and phase.

If we take the Sumudu transform of the both sides of Eq. (14.7), we will obtain

$$\frac{R_1}{CL}S(i(t)) + \frac{1}{C}S\left({}_0^{CF}D_t^\alpha i(t)\right) + \frac{S(V(t))}{CL} = S\left(\frac{d^2 V(t)}{dt^2}\right)$$
$$+ \left(\frac{R_1}{L} + \frac{1}{R_2 C}\right)S\left(\frac{dV(t)}{dt}\right) + \frac{R_1+R_2}{R_2 CL}S(V(t))$$

Then, we obtain

$$\frac{R_1}{CL}S(i(t)) + \frac{1}{C}\frac{M(\alpha)}{s\alpha+1-\alpha}(S[i(t)]-i(0)) + \frac{S(V(t))}{CL} = \left(\frac{S[V(t)]-V(0)}{s^2} - \frac{V'(0)}{s}\right)$$
$$+ \left(\frac{R_1}{sL} + \frac{1}{sR_2 C}\right)(S[V(t)]-V(0)) + \frac{R_1 \mid R_2}{R_2 CL}S(V(t))$$

If we simplify the above equation, we will get

$$S(i(t))\left(\frac{R_1}{CL} + \frac{1}{C}\frac{M(\alpha)}{s\alpha+1-\alpha}\right) = S(V(t))\left(-\frac{1}{CL} + \frac{1}{s^2} + \frac{R_1}{SL} + \frac{1}{SR_2 C} + \frac{R_1 \mid R_2}{R_2 CL}\right)$$

$$S(i(t))\left(\frac{R_1(s\alpha+1-\alpha)+LM(\alpha)}{CL(s\alpha+1-\alpha)}\right) = S(V(t))\left(\frac{CLR_2 + sCR_2 R_1 + Ls + s^2 R_1}{s^2 CLR_2}\right)$$

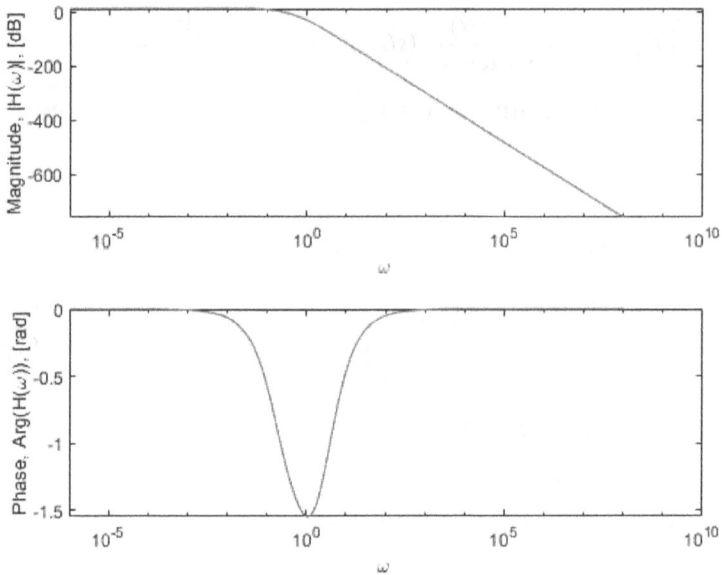

Figure 14.5 Transfer function of the higher order circuit model 1 with the Caputo-Fabrizio derivative using the Laplace transform.

Therefore, we obtain the transfer function as

$$\frac{S(V(t))}{S(i(t))} = \frac{((s\alpha + 1 - \alpha)R_1 + M(\alpha)L)s^2 R_2}{(CLR_2 + sCR_2 R_1 + Ls + s^2 R_1)(s\alpha + 1 - \alpha)}. \tag{14.9}$$

The graphical representation of the above transfer function is presented in Figure 14.6 as magnitude and phase.

14.4 ANALYSIS OF HIGHER ORDER CIRCUIT MODEL 1 WITH ATANGANA-BALEANU DERIVATIVE

We consider the model with Atangana-Baleanu derivative. We apply Laplace and Sumudu transforms to obtain the solution.

$$\frac{R_1}{CL}i(t) + \frac{1}{C}\left({}_0^{ABC}D_t^\alpha i(t)\right) + \frac{V(t)}{CL} = \frac{d^2 V(t)}{dt^2} + \left(\frac{R_1}{L} + \frac{1}{R_2 C}\right)\frac{dV(t)}{dt}$$

$$+ \frac{R_1 + R_2}{R_2 CL}V(t) \tag{14.10}$$

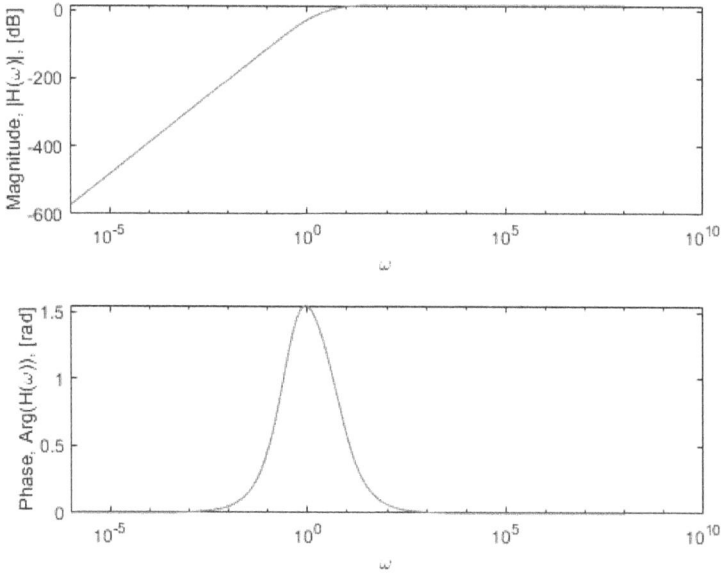

Figure 14.6 Transfer function of the higher order circuit model 1 with the Caputo-Fabrizio derivative using the Sumudu transform.

If we take the Laplace transform of the both sides of Eq. (14.10), we will obtain

$$\frac{R_1}{CL}L(i(t)) + \frac{1}{C}L\left(^{ABC}_{0}D^{\alpha}_t i(t)\right) + \frac{L(V(t))}{CL} = L\left(\frac{d^2V(t)}{dt^2}\right)$$
$$+ \left(\frac{R_1}{L} + \frac{1}{R_2C}\right)L\left(\frac{dV(t)}{dt}\right) + \frac{R_1+R_2}{R_2CL}L(V(t))$$

Then, we obtain

$$\frac{R_1}{CL}L(i(t)) + \frac{1}{C}\frac{s^{\alpha-1}AB(\alpha)}{s^{\alpha}(1-\alpha)+\alpha}(sL(i(t)) - i(0)) + \frac{L(V(t))}{CL}$$
$$= \left(s^2L(V(t)) - sV(0) - V'(0)\right) + \left(\frac{R_1}{L} + \frac{1}{R_2C}\right)(sL(V(t)) - V(0))$$
$$+ \frac{R_1+R_2}{R_2CL}L(V(t))$$

If we simplify the above equation, we will get

$$L(i(t))\left(\frac{R_1}{CL} + \frac{1}{C}\frac{s^{\alpha}AB(\alpha)}{s^{\alpha}(1-\alpha)+\alpha}\right) = L(V(t))\left(s^2 - \frac{1}{CL} + \frac{sR_1}{L} + \frac{s}{R_2C} + \frac{R_1+R_2}{R_2CL}\right)$$

$$L(i(t)) \left(\frac{R_1 (s^\alpha (1-\alpha) + \alpha)) + Ls^\alpha AB(\alpha)}{CL(s^\alpha (1-\alpha) + \alpha)} \right)$$

$$= L(V(t)) \left(\frac{s^2 CLR_2 + CsR_1 R2 + Ls + R_1}{CLR_2} \right)$$

Therefore, we obtain the transfer function as

$$\frac{L(V(t))}{L(i(t))} = \frac{R_2 \left(R_1 (s^\alpha (1-\alpha) + \alpha) + Ls^\alpha AB(\alpha) \right)}{(s^\alpha (1-\alpha) + \alpha)(s^2 CLR_2 + CsR_1 R_2 + Ls + R_1)}. \tag{14.11}$$

The graphical representation of the above transfer function is presented in Figure 14.7 as magnitude and phase.

If we take the Sumudu transform of the both sides of Eq. (14.10), we will obtain

$$\frac{R_1}{CL} S(i(t)) + \frac{1}{C} S \left({}^{ABC}_0 D^\alpha_t i(t) \right) + \frac{S(V(t))}{CL} = S \left(\frac{d^2 V(t)}{dt^2} \right)$$

$$+ \left(\frac{R_1}{L} + \frac{1}{R_2 C} \right) S \left(\frac{dV(t)}{dt} \right) + \frac{R_1 + R_2}{R_2 CL} S(V(t))$$

Then, we obtain

$$\frac{R_1}{CL} S(i(t)) + \frac{1}{C} \frac{AB(\alpha)}{\alpha s^\alpha + 1 - \alpha} (S[i(t)] - i(0)) + \frac{S(V(t))}{CL}$$

$$= \left(\frac{S[V(t)] - V(0)}{s^2} - \frac{V'(0)}{s} \right) + \left(\frac{R_1}{sL} + \frac{1}{sR_2 C} \right) (S[V(t)] - V(0))$$

$$+ \frac{R_1 + R_2}{R_2 CL} S(V(t))$$

If we simplify the above equation, we will get

$$S(i(t)) \left(\frac{R_1}{CL} + \frac{1}{C} \frac{AB(\alpha)}{\alpha s^\alpha + 1 - \alpha} \right) = S(V(t)) \left(-\frac{1}{CL} + \frac{1}{s^2} + \frac{R_1}{SL} + \frac{1}{SR_2 C} + \frac{R_1 + R_2}{R_2 CL} \right)$$

$$S(i(t)) \left(\frac{R_1 (\alpha s^\alpha + 1 - \alpha) + LAB(\alpha)}{CL(\alpha s^\alpha + 1 - \alpha)} \right) = S(V(t)) \left(\frac{CLR_2 + sCR_2 R_1 + Ls + s^2 R_1}{s^2 CLR_2} \right)$$

Therefore, we obtain the transfer function as

$$\frac{S(V(t))}{S(i(t))} = \frac{((\alpha s^\alpha + 1 - \alpha) R_1 + AB(\alpha)L) s^2 R_2}{(CLR_2 + sCR_2 R_1 + Ls + s^2 R_1)(\alpha s^\alpha + 1 - \alpha)}. \tag{14.12}$$

The graphical representation of the above transfer function is presented in Figure 14.8 as magnitude and phase.

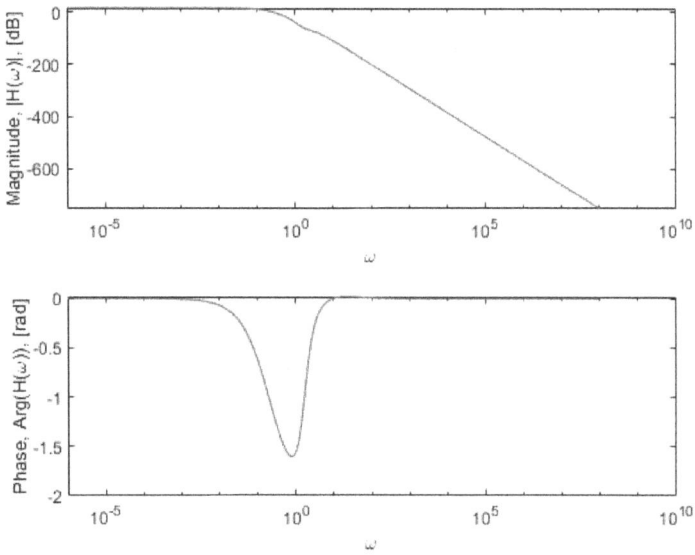

Figure 14.7 Transfer function of the higher order circuit model 1 with the ABC derivative using the Laplace transform.

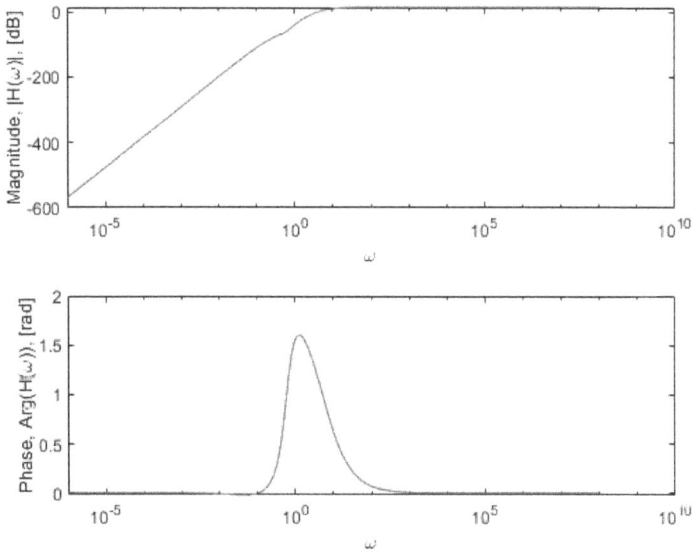

Figure 14.8 Transfer function of the higher order circuit model 1 with the ABC derivative using the Sumudu transform.

15 Analysis of Higher Order Circuit Model 2

Besides the use of first- and second-order differentiation to construct linear ordinary differential equations, mathematicians, and engineers have also made use of higher-order differential equations to capture the dynamics of circuit systems with more complex components. They are referred to as higher orders because they are constructed using orders of more than two. In general, it is possible to form arbitrary orders of differential equations if the function under investigation is several times differentiable. Within the framework of an electrical circuit, having in mind that they are called filters, an order of a different operator will then correspond to the order of the filter, these orders describe its response. For a specific energy store element in the circuit involving an inductor or capacitor, it is possible to add one order to the circuit as long as the elements are not connected in a manner that leads to degeneracy, for example, two capacitors in parallel [63–66]. In this chapter higher order circuits will be subjected to some analysis, in particular, a Laplace transform and Sumudu transforms will be used to obtain the transfer functions that will also be compared. The classical model will be extended using fractional differential operators with different kernels. These models will also be solved via Sumudu and Laplace transforms, and their transfer functions will be compared for different fractional orders.

15.1 ANALYSIS OF HIGHER ORDER CIRCUIT MODEL 2 WITH CLASSICAL DERIVATIVE

We consider the model with classical derivative. We apply Laplace and Sumudu transforms to obtain the solution.

$$\frac{1}{CL}i(t) + \frac{1}{L}\frac{dV(t)}{dt} + \frac{V(t)}{R_1 CL} = \frac{d^2 i(t)}{dt^2} + \left(\frac{R_2}{L} + \frac{1}{R_1 C}\right)\frac{di(t)}{dt} + \frac{R_1 + R_2}{R_1 CL}i(t) \quad (15.1)$$

If we take the Laplace transform of the both sides of Eq. (15.1), we will obtain

$$\frac{1}{CL}L(i(t)) + \frac{1}{L}L\left(\frac{dV(t)}{dt}\right) + \frac{L(V(t))}{R_1 CL}$$

$$= L\left(\frac{d^2 i(t)}{dt^2}\right) + \left(\frac{R_2}{L} + \frac{1}{R_1 C}\right)L\left(\frac{di(t)}{dt}\right) + \frac{R_1 + R_2}{R_1 CL}L(i(t))$$

Then, we have

$$\frac{1}{CL}L(i(t)) + \frac{1}{L}(sL(V(t) - V(0))) + \frac{L(V(t))}{R_1 CL}$$

$$= s^2 L(i(t)) - si(0) - i'(0) + \left(\frac{R_2}{L} + \frac{1}{R_1 C}\right)(sL(i(t)) - i(0)) + \frac{R_1 + R_2}{R_1 CL}L(i(t))$$

DOI: 10.1201/9781003359869-15

135

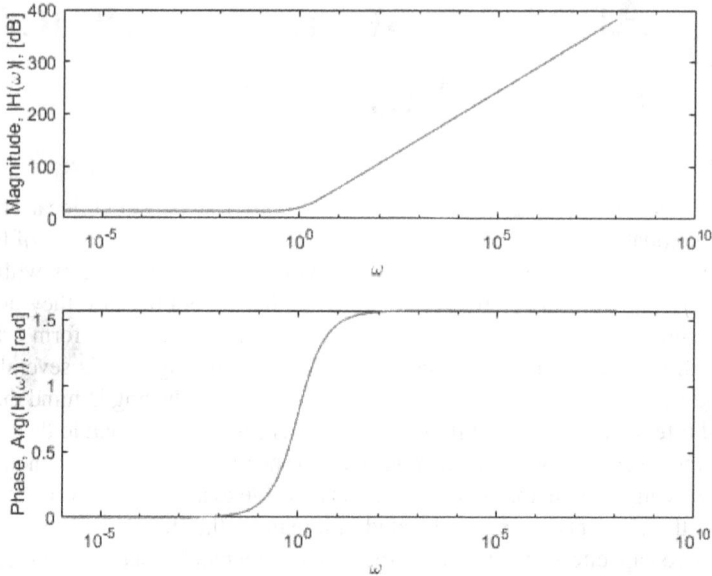

Figure 15.1 Transfer function of the higher order circuit model 2 with the classical derivative using the Laplace transform.

If we simplify the above equation, we will get

$$L(V(t)) \left(\frac{s}{L} + \frac{1}{R_1CL} \right) = L(i(t)) \left(s^2 + \frac{sR_2}{L} + \frac{s}{R_1C} + \frac{R_2}{R_1CL} \right)$$

Then, we reach

$$L(V(t)) \left(\frac{sR_1C+1}{LR_1C} \right) = L(i(t)) \left(\frac{s^2R_1CL + sR_2R_1C + Ls + R_2}{R_1CL} \right)$$

Therefore, we will obtain the transfer function as

$$\frac{L(V(t))}{L(i(t))} = \frac{s^2R_1CL + sR_2R_1C + Ls + R_2}{sR_1C+1}.$$

The graphical representation of the above transfer function is presented in Figure 15.1 as magnitude and phase.

If we take the Sumudu transform of the both sides of Eq. (15.1), we will obtain

$$\frac{1}{CL}S(i(t)) + \frac{1}{L}S\left(\frac{dV(t)}{dt} \right) + \frac{S(V(t))}{R_1CL}$$

$$= S\left(\frac{d^2i(t)}{dt^2} \right) + \left(\frac{R_2}{L} + \frac{1}{R_1C} \right)S\left(\frac{di(t)}{dt} \right) + \frac{R_1+R_2}{R_1CL}S(i(t))$$

Then, we have

$$
\frac{1}{CL}S(i(t)) + \frac{1}{sL}(S(V(t) - V(0))) + \frac{S(V(t))}{R_1CL}
$$
$$
= \left(\frac{S[i(t)] - i(0)}{s^2} - \frac{i'(0)}{s}\right) + \left(\frac{R_2}{sL} + \frac{1}{sR_1C}\right)(S(i(t)) - i(0)) + \frac{R_1 + R_2}{R_1CL}S(i(t))
$$

If we simplify the above equation, we will get

$$
S(V(t))\left(\frac{1}{sL} + \frac{1}{R_1CL}\right) = S(i(t))\left(\frac{1}{s^2} + \frac{R_2}{sL} + \frac{1}{sR_1C} + \frac{R_2}{R_1CL}\right)
$$

Then, we reach

$$
S(V(t))\left(\frac{s + R_1C}{sLR_1C}\right) = S(i(t))\left(\frac{R_1CL + sR_2R_1C + Ls + s^2R_2}{s^2R_1CL}\right)
$$

Therefore, we will obtain the transfer function as

$$
\frac{S(V(t))}{S(i(t))} = \frac{R_1CL + sR_2R_1C + sL + s^2R_2}{s(R_1C + s)}.
$$

The graphical representation of the above transfer function is presented in Figure 15.2 as magnitude and phase.

15.2 ANALYSIS OF HIGHER ORDER CIRCUIT MODEL 2 WITH CAPUTO DERIVATIVE

We consider the model with Caputo derivative. We apply Laplace and Sumudu transforms to obtain the solution.

$$
\frac{1}{CL}i(t) + \frac{1}{L}\left({}_0^C D_t^\alpha V(t)\right) + \frac{V(t)}{R_1CL} = \frac{d^2i(t)}{dt^2} + \left(\frac{R_2}{L} + \frac{1}{R_1C}\right)\frac{di(t)}{dt} + \frac{R_1 + R_2}{R_1CL}i(t)
$$
(15.2)

If we take the Laplace transform of the both sides of Eq. (15.2), we will obtain

$$
\frac{1}{CL}L(i(t)) + \frac{1}{L}L\left({}_0^C D_t^\alpha V(t)\right) + \frac{L(V(t))}{R_1CL}
$$
$$
= L\left(\frac{d^2i(t)}{dt^2}\right) + \left(\frac{R_2}{L} + \frac{1}{R_1C}\right)L\left(\frac{di(t)}{dt}\right) + \frac{R_1 + R_2}{R_1CL}L(i(t))
$$

Then, we have

$$
\frac{1}{CL}L(i(t)) + \frac{1}{L}s^{\alpha-1}(sL(V(t)) - V(0)) + \frac{L(V(t))}{R_1CL}
$$
$$
= s^2L(i(t)) - si(0) - i'(0) + \left(\frac{R_2}{L} + \frac{1}{R_1C}\right)(sL(i(t)) - i(0)) + \frac{R_1 + R_2}{R_1CL}L(i(t))
$$

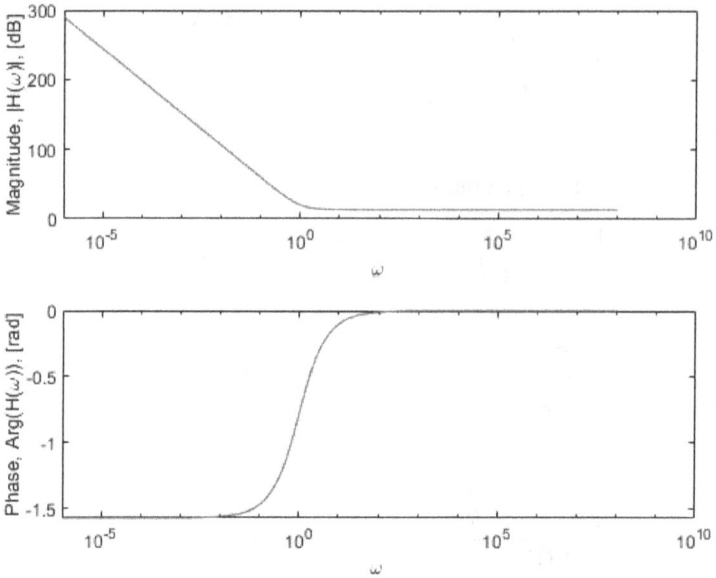

Figure 15.2   Transfer function of the higher order circuit model 2 with the classical derivative using the Sumudu transform.

If we simplify the above equation, we will get

$$L(V(t))\left(\frac{s^\alpha}{L}+\frac{1}{R_1CL}\right)=L(i(t))\left(s^2+\frac{sR_2}{L}+\frac{s}{R_1C}+\frac{R_2}{R_1CL}\right)$$

Then, we reach

$$L(V(t))\left(\frac{s^\alpha R_1C+1}{LR_1C}\right)=L(i(t))\left(\frac{s^2R_1CL+sR_2R_1C+Ls+R_2}{R_1CL}\right)$$

Therefore, we will obtain the transfer function as

$$\frac{L(V(t))}{L(i(t))}=\frac{s^2R_1CL+sR_2R_1C+Ls+R_2}{s^\alpha R_1C+1}.$$

The graphical representation of the above transfer function is presented in Figure 15.3 as magnitude and phase.

If we take the Sumudu transform of the both sides of Eq. (15.2), we will obtain

$$\frac{1}{CL}S(i(t))+\frac{1}{L}S\left({}_0^CD_t^\alpha V(t)\right)+\frac{S(V(t))}{R_1CL}$$

$$=S\left(\frac{d^2i(t)}{dt^2}\right)+\left(\frac{R_2}{L}+\frac{1}{R_1C}\right)S\left(\frac{di(t)}{dt}\right)+\frac{R_1+R_2}{R_1CL}S(i(t))$$

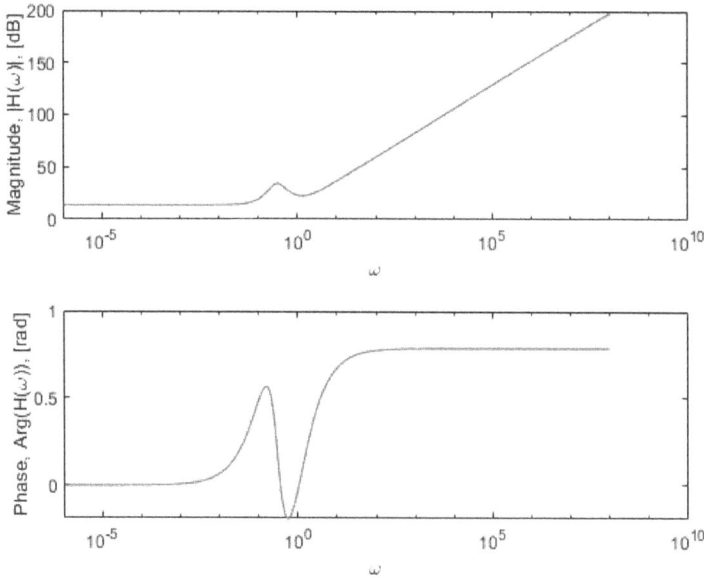

Figure 15.3 Transfer function of the higher order circuit model 2 with the Caputo derivative using the Laplace transform.

Then, we have

$$\frac{1}{CL}S(i(t)) + \frac{1}{L}\frac{S(i(t)) - i(0)}{s^\alpha} + \frac{S(V(t))}{R_1 CL}$$

$$= \left(\frac{S[i(t)] - i(0)}{s^2} - \frac{i'(0)}{s}\right) + \left(\frac{R_2}{sL} + \frac{1}{sR_1 C}\right)(S(i(t)) - i(0)) + \frac{R_1 + R_2}{R_1 CL}S(i(t))$$

If we simplify the above equation, we will get

$$S(V(t))\left(\frac{1}{s^\alpha L} + \frac{1}{R_1 CL}\right) = S(i(t))\left(\frac{1}{s^2} + \frac{R_2}{sL} + \frac{1}{sR_1 C} + \frac{R_2}{R_1 CL}\right)$$

Then, we reach

$$S(V(t))\left(\frac{s^\alpha + R_1 C}{s^\alpha LR_1 C}\right) = S(i(t))\left(\frac{R_1 CL + sR_2 R_1 C + Ls + s^2 R_2}{s^2 R_1 CL}\right)$$

Therefore, we will obtain the transfer function as

$$\frac{S(V(t))}{S(i(t))} = \frac{R_1 CL + sR_2 R_1 C + sL + s^2 R_2}{s^{2-\alpha}(R_1 C + s^\alpha)}.$$

The graphical representation of the above transfer function is presented in Figure 15.4 as magnitude and phase.

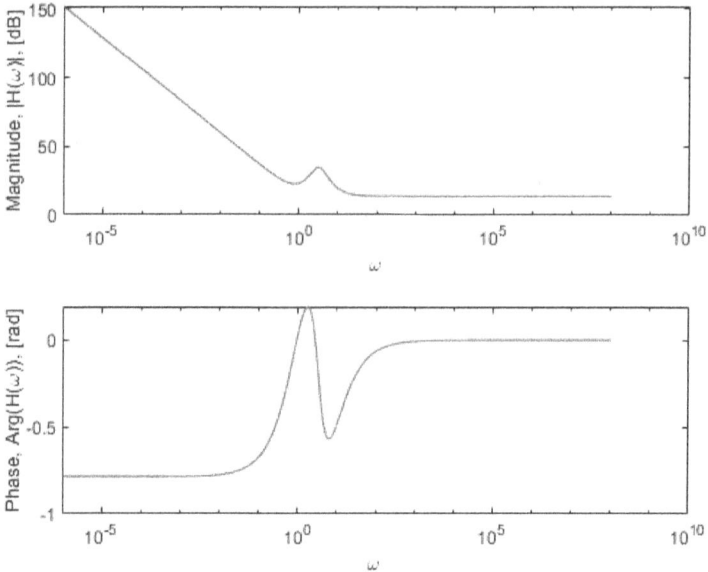

Figure 15.4 Transfer function of the higher order circuit model 2 with the Caputo derivative using the Sumudu transform.

15.3 ANALYSIS OF HIGHER ORDER CIRCUIT MODEL 2 WITH CAPUTO-FABRIZIO DERIVATIVE

We consider the model with Caputo-Fabrizio derivative. We apply Laplace and Sumudu transforms to obtain the solution.

$$\frac{1}{CL}i(t) + \frac{1}{L}\left(_0^{CF}D_t^\alpha V(t)\right) + \frac{V(t)}{R_1CL} = \frac{d^2i(t)}{dt^2} + \left(\frac{R_2}{L} + \frac{1}{R_1C}\right)\frac{di(t)}{dt} + \frac{R_1+R_2}{R_1CL}i(t)$$
(15.3)

If we take the Laplace transform of the both sides of Eq. (15.3), we will obtain

$$\frac{1}{CL}L(i(t)) + \frac{1}{L}L\left(_0^{CF}D_t^\alpha V(t)\right) + \frac{L(V(t))}{R_1CL}$$
$$= L\left(\frac{d^2i(t)}{dt^2}\right) + \left(\frac{R_2}{L} + \frac{1}{R_1C}\right)L\left(\frac{di(t)}{dt}\right) + \frac{R_1+R_2}{R_1CL}L(i(t))$$

Then, we have

$$\frac{1}{CL}L(i(t)) + \frac{1}{L}\frac{M(\alpha)}{s+\alpha-s\alpha}(sL(V(t))-V(0)) + \frac{L(V(t))}{R_1CL}$$
$$= s^2L(i(t)) - si(0) - i'(0) + \left(\frac{R_2}{L} + \frac{1}{R_1C}\right)(sL(i(t))-i(0)) + \frac{R_1+R_2}{R_1CL}L(i(t))$$

If we simplify the above equation, we will get

$$L(V(t))\left(\frac{sM(\alpha)}{L(s+\alpha-s\alpha)}+\frac{1}{R_1CL}\right)=L(i(t))\left(s^2+\frac{sR_2}{L}+\frac{s}{R_1C}+\frac{R_2}{R_1CL}\right)$$

Then, we reach

$$L(V(t))\left(\frac{sM(\alpha)R_1C+s+\alpha-s\alpha}{LR_1C(s+\alpha-s\alpha)}\right)=L(i(t))\left(\frac{s^2R_1CL+sR_2R_1C+Ls+R_2}{R_1CL}\right)$$

Therefore, we will obtain the transfer function as

$$\frac{L(V(t))}{L(i(t))}=\frac{(s+\alpha-s\alpha)(s^2R_1CL+sR_1R_2C+Ls+R_2)}{sM(\alpha)R_1C+s+\alpha-s\alpha}.$$

The graphical representation of the above transfer function is presented in Figure 15.5 as magnitude and phase.

If we take the Sumudu transform of the both sides of Eq. (15.3), we will obtain

$$\frac{1}{CL}S(i(t))+\frac{1}{L}S\left({}_0^{CF}D_t^\alpha V(t)\right)+\frac{S(V(t))}{R_1CL}$$
$$=S\left(\frac{d^2i(t)}{dt^2}\right)+\left(\frac{R_2}{L}+\frac{1}{R_1C}\right)S\left(\frac{di(t)}{dt}\right)+\frac{R_1+R_2}{R_1CL}S(i(t))$$

Then, we have

$$\frac{1}{CL}S(i(t))+\frac{1}{L}\frac{M(\alpha)}{s\alpha+1-\alpha}(S[V(t)]-V(0))+\frac{S(V(t))}{R_1CL}$$
$$=\left(\frac{S[i(t)]-i(0)}{s^2}-\frac{i'(0)}{s}\right)+\left(\frac{R_2}{sL}+\frac{1}{sR_1C}\right)(S(i(t))-i(0))+\frac{R_1+R_2}{R_1CL}S(i(t))$$

If we simplify the above equation, we will get

$$S(V(t))\left(\frac{M(\alpha)}{L(s\alpha+1-\alpha)}+\frac{1}{R_1CL}\right)=S(i(t))\left(\frac{1}{s^2}+\frac{R_2}{sL}+\frac{1}{sR_1C}+\frac{R_2}{R_1CL}\right)$$

Then, we reach

$$S(V(t))\left(\frac{M(\alpha)R_1C+1-\alpha+s\alpha}{(1-\alpha+s\alpha)LR_1C}\right)=S(i(t))\left(\frac{R_1CL+sR_2R_1C+Ls+s^2R_2}{s^2R_1CL}\right)$$

Therefore, we will obtain the transfer function as

$$\frac{S(V(t))}{S(i(t))}=\frac{(R_1CL+sR_2R_1C+sL+s^2R_2)(1-\alpha-s\alpha)}{s^2(M(\alpha)R_1C+1-\alpha+s\alpha)}.$$

The graphical representation of the above transfer function is presented in Figure 15.6 as magnitude and phase.

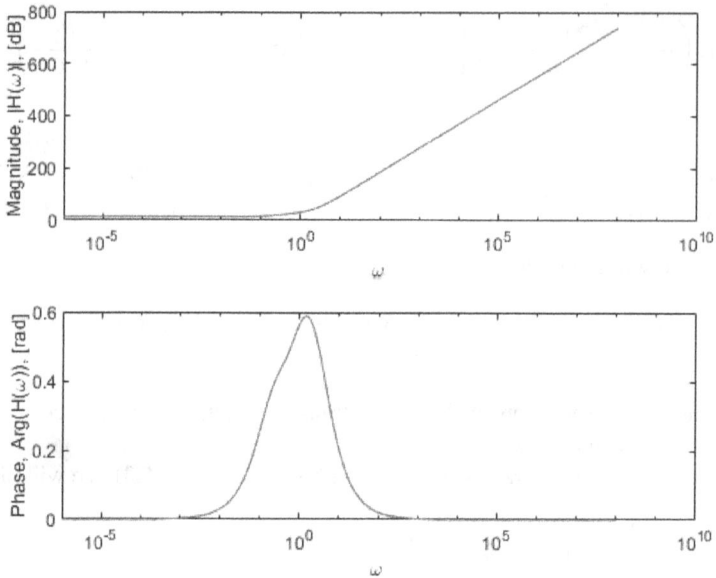

Figure 15.5 Transfer function of the higher order circuit model 2 with the Caputo-Fabrizio derivative using the Laplace transform.

15.4 ANALYSIS OF HIGHER ORDER CIRCUIT MODEL 2 WITH ATANGANA-BALEANU DERIVATIVE

We consider the model with Atangana-Baleanu derivative. We apply Laplace and Sumudu transforms to obtain the solution.

$$\frac{1}{CL}i(t) + \frac{1}{L}\left({}_0^{ABC}D_t^\alpha V(t)\right) + \frac{V(t)}{R_1CL} = \frac{d^2i(t)}{dt^2} + \left(\frac{R_2}{L} + \frac{1}{R_1C}\right)\frac{di(t)}{dt} + \frac{R_1+R_2}{R_1CL}i(t)$$
(15.4)

If we take the Laplace transform of the both sides of Eq. (15.4), we will obtain

$$\frac{1}{CL}L(i(t)) + \frac{1}{L}L\left({}_0^{ABC}D_t^\alpha V(t)\right) + \frac{L(V(t))}{R_1CL}$$

$$= L\left(\frac{d^2i(t)}{dt^2}\right) + \left(\frac{R_2}{L} + \frac{1}{R_1C}\right)L\left(\frac{di(t)}{dt}\right) + \frac{R_1+R_2}{R_1CL}L(i(t))$$

Then, we have

$$\frac{1}{CL}L(i(t)) + \frac{1}{L}\frac{s^{\alpha-1}AB(\alpha)}{s^\alpha(1-\alpha)+\alpha}\left(sL(V(t)) - V(0)\right) + \frac{L(V(t))}{R_1CL}$$

$$= s^2L(i(t)) - si(0) - i'(0) + \left(\frac{R_2}{L} + \frac{1}{R_1C}\right)\left(sL(i(t)) - i(0)\right) + \frac{R_1+R_2}{R_1CL}L(i(t))$$

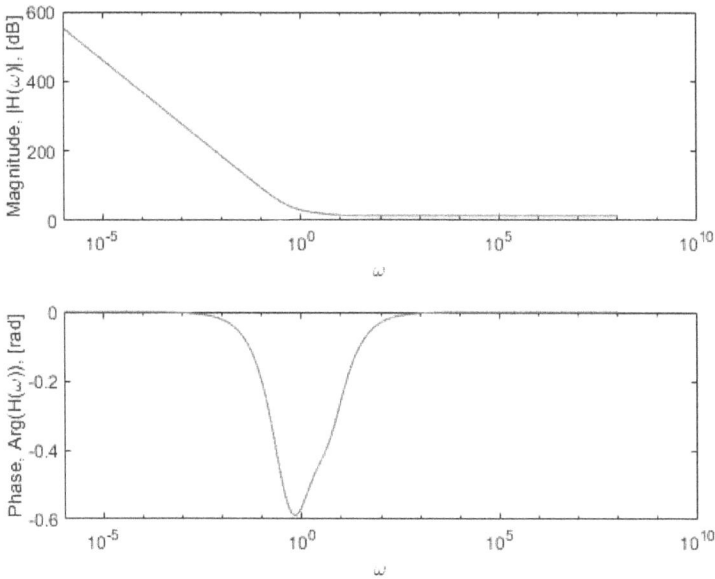

Figure 15.6 Transfer function of the higher order circuit model 2 with the Caputo-Fabrizio derivative using the Sumudu transform.

If we simplify the above equation, we will get

$$L(V(t))\left(\frac{1}{L}\frac{s^\alpha AB(\alpha)}{s^\alpha(1-\alpha)+\alpha} + \frac{1}{R_1CL}\right) = L(i(t))\left(s^2 + \frac{sR_2}{L} + \frac{s}{R_1C} + \frac{R_2}{R_1CL}\right)$$

Then, we reach

$$L(V(t))\left(\frac{s^\alpha AB(\alpha)R_1C + s^\alpha(1-\alpha)+\alpha}{R_1CL(s^\alpha(1-\alpha)+\alpha)}\right) = L(i(t))\left(\frac{s^2R_1CL + sR_2R_1C + Ls + R_2}{R_1CL}\right)$$

Therefore, we will obtain the transfer function as

$$\frac{L(V(t))}{L(i(t))} = \frac{(s^\alpha(1-\alpha)+\alpha)(s^2R_1CL + sR_1R_2C + Ls + R_2)}{s^\alpha AB(\alpha)R_1C + s^\alpha(1-\alpha)+\alpha}.$$

The graphical representation of the above transfer function is presented in Figure 15.7 as magnitude and phase.

If we take the Sumudu transform of the both sides of Eq. (15.4), we will obtain

$$\frac{1}{CL}S(i(t)) + \frac{1}{L}S\left(^{ABC}_{0}D^\alpha_t V(t)\right) + \frac{S(V(t))}{R_1CL}$$

$$= S\left(\frac{d^2i(t)}{dt^2}\right) + \left(\frac{R_2}{L} + \frac{1}{R_1C}\right)S\left(\frac{di(t)}{dt}\right) + \frac{R_1+R_2}{R_1CL}S(i(t))$$

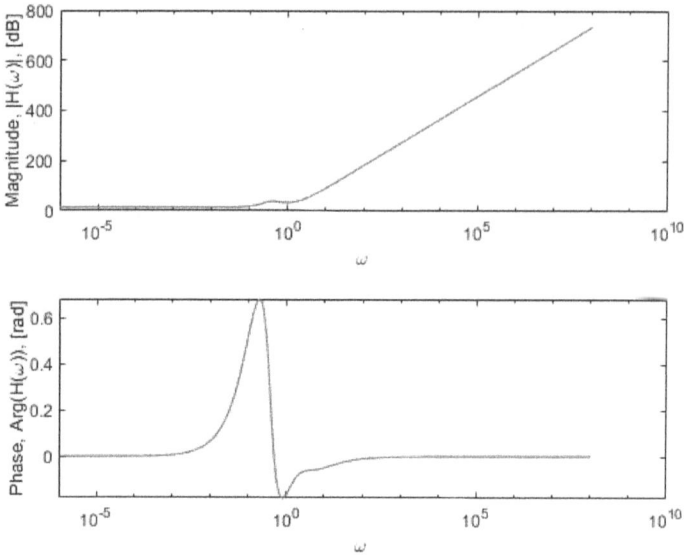

Figure 15.7 Transfer function of the higher order circuit model 2 with the ABC derivative using the Laplace transform.

Then, we have

$$\frac{1}{CL}S\left(i(t)\right)+\frac{1}{L}\frac{AB(\alpha)}{\alpha s^{\alpha}+1-\alpha}\left(S[V(t)]-V(0)\right)+\frac{S\left(V(t)\right)}{R_1CL}$$
$$=\left(\frac{S[i(t)]-i(0)}{s^2}-\frac{i'(0)}{s}\right)+\left(\frac{R_2}{sL}+\frac{1}{sR_1C}\right)\left(S(i(t))-i(0)\right)+\frac{R_1+R_2}{R_1CL}S\left(i(t)\right)$$

If we simplify the above equation, we will get

$$S(V(t))\left(\frac{1}{L}\frac{AB(\alpha)}{\alpha s^{\alpha}+1-\alpha}+\frac{1}{R_1CL}\right)=S(i(t))\left(\frac{1}{s^2}+\frac{R_2}{sL}+\frac{1}{sR_1C}+\frac{R_2}{R_1CL}\right)$$

Then, we reach

$$S(V(t))\left(\frac{AB(\alpha)R_1C+\alpha s^{\alpha}+1-\alpha}{R_1CL(\alpha s^{\alpha}+1-\alpha)}\right)=S(i(t))\left(\frac{R_1CL+sR_2R_1C+Ls+s^2R_2}{s^2R_1CL}\right)$$

Therefore, we will obtain the transfer function as

$$\frac{S(V(t))}{S(i(t))}=\frac{(R_1CL+sR_2R_1C+sL+s^2R_2)(\alpha s^{\alpha}+1-\alpha)}{s^2(AB(\alpha)R_1C+1-\alpha+\alpha s^{\alpha})}.$$

The graphical representation of the above transfer function is presented in Figure 15.8 as magnitude and phase.

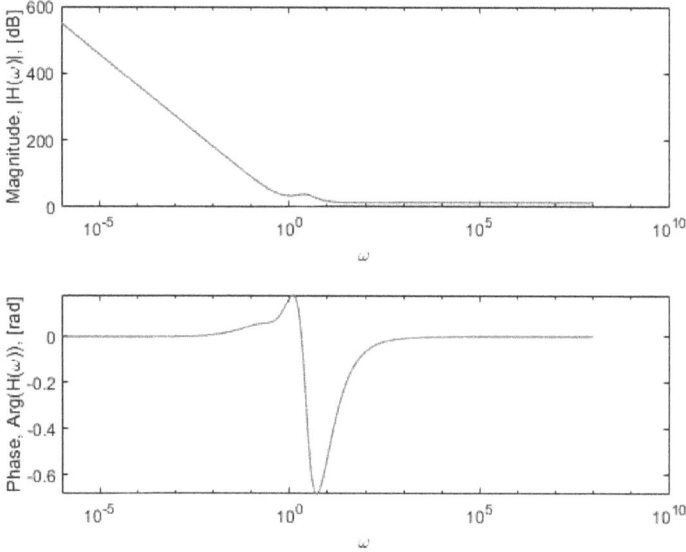

Figure 15.8 Transfer function of the higher order circuit model 2 with the ABC derivative using the Sumudu transform.

16 Analysis of Higher Order Circuit Model 3

A different linear ordinary differential equation with two orders is under investigation in this chapter. The two orders correspond to two energy-saving components including two capacitors, two inductors, or a combination of one capacitor and one resistor where the system depends and does not depend on voltage sources and dependent and independent current sources [67–70]. Since this mathematical model is a linear equation, therefore can be solved using an integral transform like Sumudu and Laplace transforms. However, we do not aim to obtain the exact solution to this problem but to obtain the associate transfer functions. Besides this exercise, we also aim to extend the model by replacing the time classical derivative with different fractional differential operators including the Caputo derivative, the Caputo-Fabrizio derivative, and finally the Atangana-Baleanu fractional derivative. For each model, the Laplace transform and Sumudu transform are applied to obtain transfer functions. These transfer functions are compared to access the effect of Sumudu and each fractional derivative.

16.1 ANALYSIS OF HIGHER ORDER CIRCUIT MODEL 3 WITH CLASSICAL DERIVATIVE

We consider the model with classical derivative. We apply Laplace and Sumudu transforms to obtain the solution.

$$\frac{1}{R_2C_2}\frac{dV_1(t)}{dt} + \frac{V_1(t)}{R_1R_2C_1C_2} = \frac{d^2V_2(t)}{dt^2} + \left(\frac{1}{R_1C_1} + \frac{1}{R_2C_2}\right)V_2(t) + \frac{1}{R_1R_2C_1C_2}V_2(t)$$

(16.1)

If we take the Laplace transform of the both sides of Eq. (16.1), we will obtain

$$\frac{1}{R_2C_2}L\left(\frac{dV_1(t)}{dt}\right) + \frac{L(V_1(t))}{R_1R_2C_1C_2}$$

$$= L\left(\frac{d^2V_2(t)}{dt^2}\right) + \left(\frac{1}{R_1C_1} + \frac{1}{R_2C_2}\right)L(V_2(t)) + \frac{1}{R_1R_2C_1C_2}L(V_2(t))$$

Then, we obtain

$$\frac{1}{R_2C_2}(sL(V_1(t)) - V_1(0)) + \frac{L(V_1(t))}{R_1R_2C_1C_2}$$

$$= (s^2L(V_2(t)) - sV_2(0) - V_2'(0)) + \left(\frac{1}{R_1C_1} + \frac{1}{R_2C_2}\right)L(V_2(t))$$

DOI: 10.1201/9781003359869-16

If we simplify the above equation, we will get

$$L(V_1(t)) \left(\frac{s}{R_2C_2} + \frac{1}{R_1R_2C_1C_2} \right) = L(V_2(t)) \left(s^2 + \frac{1}{R_1C_1} + \frac{1}{R_2C_2} + \frac{1}{R_1R_2C_1C_2} \right)$$

Thus, we have

$$L(V_1(t)) \left(\frac{sR_1C_1 + 1}{R_1R_2C_1C_2} \right) = L(V_2(t)) \left(\frac{s^2R_1R_2C_1C_2 + R_2C_2 + R_1C_1 + 1}{R_1R_2C_1C_2} \right)$$

Therefore, we obtain the transfer function as

$$\frac{L(V_2(t))}{L(V_1(t))} = \frac{sR_1C_1 + 1}{s^2R_1R_2C_1C_2 + R_2C_2 + R_1C_1 + 1}. \tag{16.2}$$

The graphical representation of the above transfer function is presented in Figure 16.1 as magnitude and phase.

If we take the Sumudu transform of the both sides of Eq. (16.1), we will obtain

$$\frac{1}{R_2C_2} S \left(\frac{dV_1(t)}{dt} \right) + \frac{S(V_1(t))}{R_1R_2C_1C_2}$$

$$= S \left(\frac{d^2V_2(t)}{dt^2} \right) + \left(\frac{1}{R_1C_1} + \frac{1}{R_2C_2} \right) S(V_2(t)) + \frac{1}{R_1R_2C_1C_2} S(V_2(t))$$

Then, we obtain

$$\frac{1}{sR_2C_2} (S(V_1(t)) - V_1(0)) + \frac{S(V_1(t))}{R_1R_2C_1C_2}$$

$$= \left(\frac{S[V_2(t)] - V_2(0)}{s^2} - \frac{V_2'(0)}{s} \right) + \left(\frac{1}{R_1C_1} + \frac{1}{R_2C_2} \right) S(V_2(t)) + \frac{1}{R_1R_2C_1C_2} S(V_2(t))$$

If we simplify the above equation, we will get

$$S(V_1(t)) \left(\frac{1}{sR_2C_2} + \frac{1}{R_1R_2C_1C_2} \right) = S(V_2(t)) \left(\frac{1}{s^2} + \frac{1}{R_1C_1} + \frac{1}{R_2C_2} + \frac{1}{R_1R_2C_1C_2} \right)$$

Thus, we have

$$S(V_1(t)) \left(\frac{R_1C_1 + s}{sR_1R_2C_1C_2} \right) = S(V_2(t)) \left(\frac{R_1R_2C_1C_2 + s^2R_2C_2 + s^2R_1C_1 + s^2}{s^2R_1R_2C_1C_2} \right)$$

Therefore, we obtain the transfer function as

$$\frac{S(V_2(t))}{S(V_1(t))} = \frac{s(R_1C_1 + s)}{R_1R_2C_1C_2 + s^2R_2C_2 + s^2R_1C_1 + s^2}. \tag{16.3}$$

The graphical representation of the above transfer function is presented in Figure 16.2 as magnitude and phase.

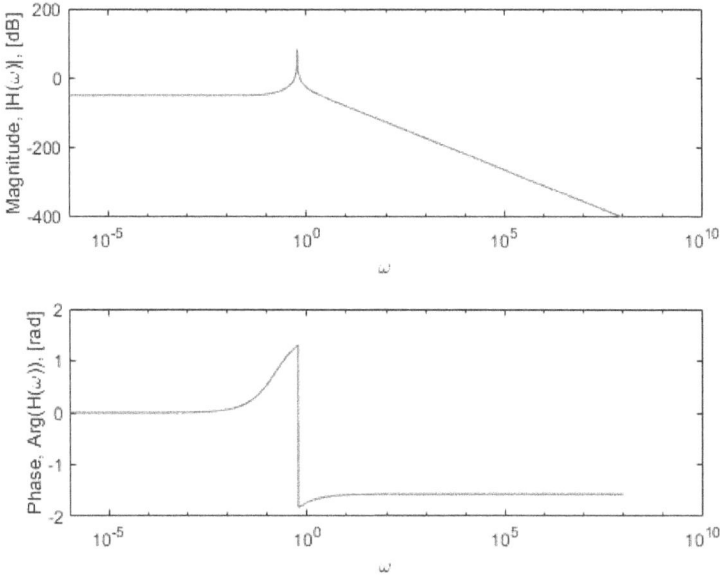

Figure 16.1 Transfer function of the higher order circuit model 3 with the classical derivative using the Laplace transform.

16.2 ANALYSIS OF HIGHER ORDER CIRCUIT MODEL 3 WITH CAPUTO DERIVATIVE

We consider the model with Caputo derivative. We apply Laplace and Sumudu transforms to obtain the solution.

$$\frac{1}{R_2C_2}\left({}_0^CD_t^\alpha V_1(t)\right) + \frac{V_1(t)}{R_1R_2C_1C_2} = \frac{d^2V_2(t)}{dt^2} + \left(\frac{1}{R_1C_1} + \frac{1}{R_2C_2}\right)V_2(t)$$

$$+ \frac{1}{R_1R_2C_1C_2}V_2(t) \qquad (16.4)$$

If we take the Laplace transform of the both sides of Eq. (16.4), we will obtain

$$\frac{1}{R_2C_2}L\left({}_0^CD_t^\alpha V_1(t)\right) + \frac{L(V_1(t))}{R_1R_2C_1C_2}$$

$$= L\left(\frac{d^2V_2(t)}{dt^2}\right) + \left(\frac{1}{R_1C_1} + \frac{1}{R_2C_2}\right)L(V_2(t)) + \frac{1}{R_1R_2C_1C_2}L(V_2(t))$$

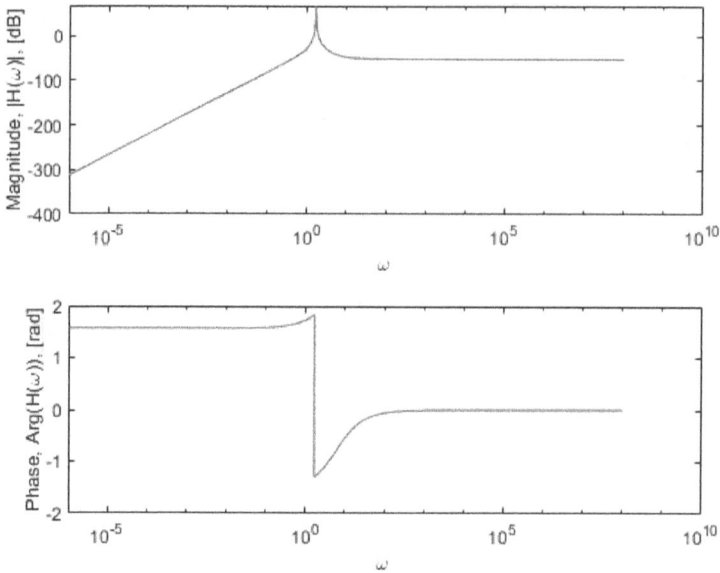

Figure 16.2 Transfer function of the higher order circuit model 3 with the classical derivative using the Sumudu transform.

Then, we obtain

$$\frac{1}{R_2 C_2} s^{\alpha-1} \left(sL(V_1(t)) - V_1(0) \right) + \frac{L(V_1(t))}{R_1 R_2 C_1 C_2}$$

$$= \left(s^2 L(V_2(t)) - sV_2(0) - V_2'(0) \right) + \left(\frac{1}{R_1 C_1} + \frac{1}{R_2 C_2} \right) L(V_2(t)) + \frac{1}{R_1 R_2 C_1 C_2} L(V_2(t))$$

If we simplify the above equation, we will get

$$L(V_1(t)) \left(\frac{s^{\alpha}}{R_2 C_2} + \frac{1}{R_1 R_2 C_1 C_2} \right) = L(V_2(t)) \left(s^2 + \frac{1}{R_1 C_1} + \frac{1}{R_2 C_2} + \frac{1}{R_1 R_2 C_1 C_2} \right)$$

Thus, we have

$$L(V_1(t)) \left(\frac{s^{\alpha} R_1 C_1 + 1}{R_1 R_2 C_1 C_2} \right) = L(V_2(t)) \left(\frac{s^2 R_1 R_2 C_1 C_2 + R_2 C_2 + R_1 C_1 + 1}{R_1 R_2 C_1 C_2} \right)$$

Therefore, we obtain the transfer function as

$$\frac{L(V_2(t))}{L(V_1(t))} = \frac{s^{\alpha} R_1 C_1 + 1}{s^2 R_1 R_2 C_1 C_2 + R_2 C_2 + R_1 C_1 + 1}. \tag{16.5}$$

The graphical representation of the above transfer function is presented in Figure 16.3 as magnitude and phase.

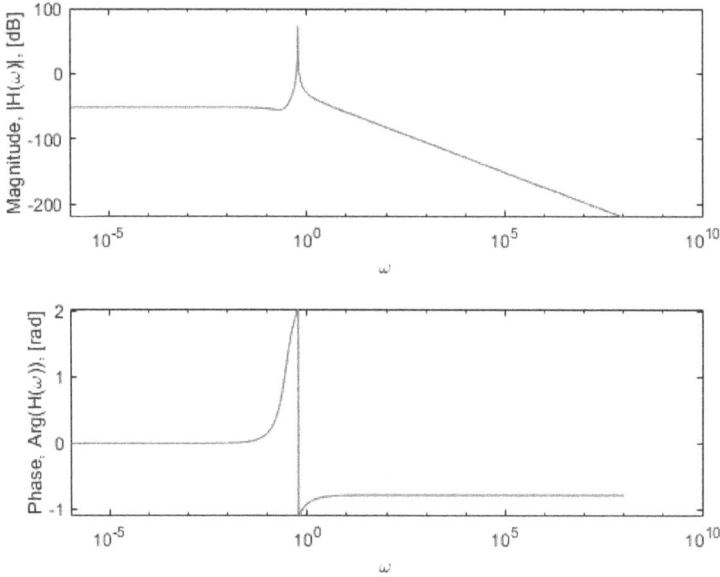

Figure 16.3 Transfer function of the higher order circuit model 3 with the Caputo derivative using the Laplace transform.

If we take the Sumudu transform of the both sides of Eq. (16.4), we will obtain

$$\frac{1}{R_2 C_2} S \left({}_0^C D_t^\alpha V_1(t) \right) + \frac{S(V_1(t))}{R_1 R_2 C_1 C_2}$$

$$= S \left(\frac{d^2 V_2(t)}{dt^2} \right) + \left(\frac{1}{R_1 C_1} + \frac{1}{R_2 C_2} \right) S(V_2(t)) + \frac{1}{R_1 R_2 C_1 C_2} S(V_2(t))$$

Then, we obtain

$$\frac{1}{R_2 C_2} \frac{S(V_1(t)) - V_1(0)}{s^\alpha} + \frac{S(V_1(t))}{R_1 R_2 C_1 C_2}$$

$$= \left(\frac{S[V_2(t)] - V_2(0)}{s^2} - \frac{V_2'(0)}{s} \right) + \left(\frac{1}{R_1 C_1} + \frac{1}{R_2 C_2} \right) S(V_2(t)) + \frac{1}{R_1 R_2 C_1 C_2} S(V_2(t))$$

If we simplify the above equation, we will get

$$S(V_1(t)) \left(\frac{1}{s^\alpha R_2 C_2} + \frac{1}{R_1 R_2 C_1 C_2} \right) = S(V_2(t)) \left(\frac{1}{s^2} + \frac{1}{R_1 C_1} + \frac{1}{R_2 C_2} + \frac{1}{R_1 R_2 C_1 C_2} \right)$$

Thus, we have

$$S(V_1(t)) \left(\frac{R_1 C_1 + s^\alpha}{s^\alpha R_1 R_2 C_1 C_2} \right) = S(V_2(t)) \left(\frac{R_1 R_2 C_1 C_2 + s^2 R_2 C_2 + s^2 R_1 C_1 + s^2}{s^2 R_1 R_2 C_1 C_2} \right)$$

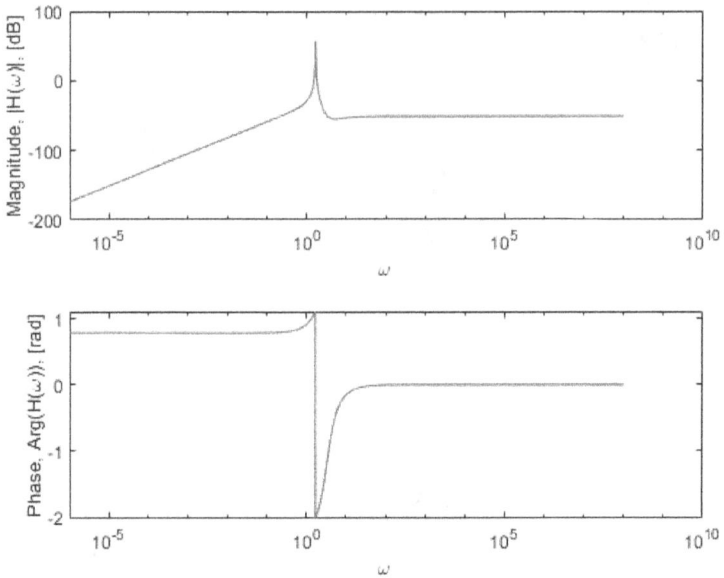

Figure 16.4 Transfer function of the higher order circuit model 3 with the Caputo derivative using the Sumudu transform.

Therefore, we obtain the transfer function as

$$\frac{S(V_2(t))}{S(V_1(t))} = \frac{s^{2-\alpha}(R_1C_1 + s^\alpha)}{R_1R_2C_1C_2 + s^2R_2C_2 + s^2R_1C_1 + s^2}. \tag{16.6}$$

The graphical representation of the above transfer function is presented in Figure 16.4 as magnitude and phase.

16.3 ANALYSIS OF HIGHER ORDER CIRCUIT MODEL 3 WITH CPUTO-FABRIZIO DERIVATIVE

We consider the model with Caputo-Fabrizio derivative. We apply Laplace and Sumudu transforms to obtain the solution.

$$\frac{1}{R_2C_2}\left({}^{CF}_0D_t^\alpha V_1(t)\right) + \frac{V_1(t)}{R_1R_2C_1C_2} = \frac{d^2V_2(t)}{dt^2} + \left(\frac{1}{R_1C_1} + \frac{1}{R_2C_2}\right)V_2(t)$$

$$+ \frac{1}{R_1R_2C_1C_2}V_2(t) \tag{16.7}$$

If we take the Laplace transform of the both sides of Eq. (16.7), we will obtain

$$\frac{1}{R_2C_2}L\left({}^{CF}_{0}D^\alpha_t V_1(t)\right) + \frac{L(V_1(t))}{R_1R_2C_1C_2}$$

$$= L\left(\frac{d^2V_2(t)}{dt^2}\right) + \left(\frac{1}{R_1C_1} + \frac{1}{R_2C_2}\right)L(V_2(t)) + \frac{1}{R_1R_2C_1C_2}L(V_2(t))$$

Then, we obtain

$$\frac{1}{R_2C_2}\frac{M(\alpha)}{s+\alpha-s\alpha}(sL(V_1(t)) - V_1(0)) + \frac{L(V_1(t))}{R_1R_2C_1C_2}$$

$$= \left(s^2L(V_2(t)) - sV_2(0) - V_2'(0)\right) + \left(\frac{1}{R_1C_1} + \frac{1}{R_2C_2}\right)L(V_2(t)) + \frac{1}{R_1R_2C_1C_2}L(V_2(t))$$

If we simplify the above equation, we will get

$$L(V_1(t))\left(\frac{1}{R_2C_2}\frac{sM(\alpha)}{s+\alpha-s\alpha} + \frac{1}{R_1R_2C_1C_2}\right)$$

$$= L(V_2(t))\left(s^2 + \frac{1}{R_1C_1} + \frac{1}{R_2C_2} + \frac{1}{R_1R_2C_1C_2}\right)$$

Thus, we have

$$L(V_1(t))\left(\frac{sR_1C_1M(\alpha) + s + \alpha - s\alpha}{(R_1R_2C_1C_2)(s+\alpha-s\alpha)}\right)$$

$$= L(V_2(t))\left(\frac{s^2R_1R_2C_1C_2 + R_2C_2 + R_1C_1 + 1}{R_1R_2C_1C_2}\right)$$

Therefore, we obtain the transfer function as

$$\frac{L(V_2(t))}{L(V_1(t))} = \frac{sM(\alpha)R_1C_1 + s + \alpha - s\alpha}{(s^2R_1R_2C_1C_2 + R_2C_2 + R_1C_1 + 1)(s+\alpha-s\alpha)}. \tag{16.8}$$

The graphical representation of the above transfer function is presented in Figure 16.5 as magnitude and phase.

If we take the Sumudu transform of the both sides of Eq. (16.7), we will obtain

$$\frac{1}{R_2C_2}S\left({}^{CF}_{0}D^\alpha_t V_1(t)\right) + \frac{S(V_1(t))}{R_1R_2C_1C_2}$$

$$= S\left(\frac{d^2V_2(t)}{dt^2}\right) + \left(\frac{1}{R_1C_1} + \frac{1}{R_2C_2}\right)S(V_2(t)) + \frac{1}{R_1R_2C_1C_2}S(V_2(t))$$

Then, we obtain

$$\frac{1}{R_2C_2}\frac{M(\alpha)}{s\alpha+1-\alpha}(S[i(t)] - i(0)) + \frac{S(V_1(t))}{R_1R_2C_1C_2}$$

$$= \left(\frac{S[V_2(t)] - V_2(0)}{s^2} - \frac{V_2'(0)}{s}\right) + \left(\frac{1}{R_1C_1} + \frac{1}{R_2C_2}\right)S(V_2(t)) + \frac{1}{R_1R_2C_1C_2}S(V_2(t))$$

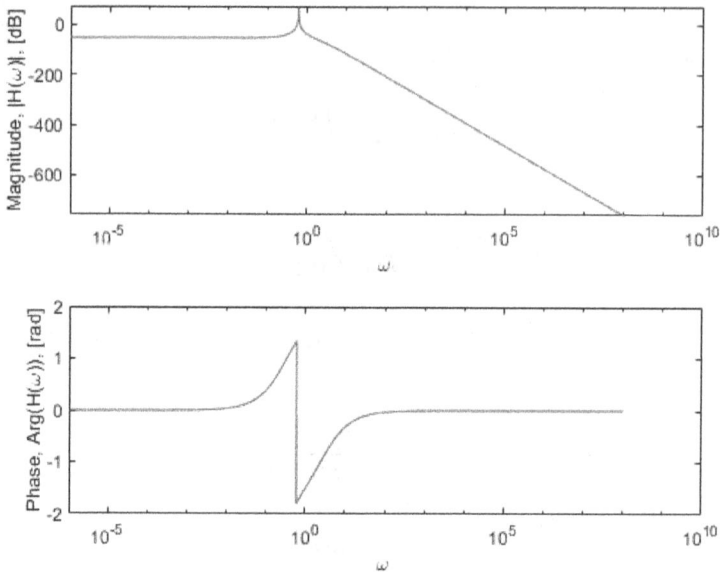

Figure 16.5 Transfer function of the higher order circuit model 3 with the Caputo-Fabrizio derivative using the Laplace transform.

If we simplify the above equation, we will get

$$S(V_1(t)) \left(\frac{1}{R_2 C_2} \frac{M(\alpha)}{s\alpha + 1 - \alpha} + \frac{1}{R_1 R_2 C_1 C_2} \right)$$
$$= S(V_2(t)) \left(\frac{1}{s^2} + \frac{1}{R_1 C_1} + \frac{1}{R_2 C_2} + \frac{1}{R_1 R_2 C_1 C_2} \right)$$

Thus, we have

$$S(V_1(t)) \left(\frac{R_1 C_1 M(\alpha) + s\alpha + 1 - \alpha}{(s\alpha + 1 - \alpha) R_1 R_2 C_1 C_2} \right)$$
$$= S(V_2(t)) \left(\frac{R_1 R_2 C_1 C_2 + s^2 R_2 C_2 + s^2 R_1 C_1 + s^2}{s^2 R_1 R_2 C_1 C_2} \right)$$

Therefore, we obtain the transfer function as

$$\frac{S(V_2(t))}{S(V_1(t))} = \frac{s^2 (R_1 C_1 M(\alpha) + s\alpha + 1 - \alpha)}{(R_1 R_2 C_1 C_2 + s^2 R_2 C_2 + s^2 R_1 C_1 + s^2)(s\alpha + 1 - \alpha)}. \quad (16.9)$$

The graphical representation of the above transfer function is presented in Figure 16.6 as magnitude and phase.

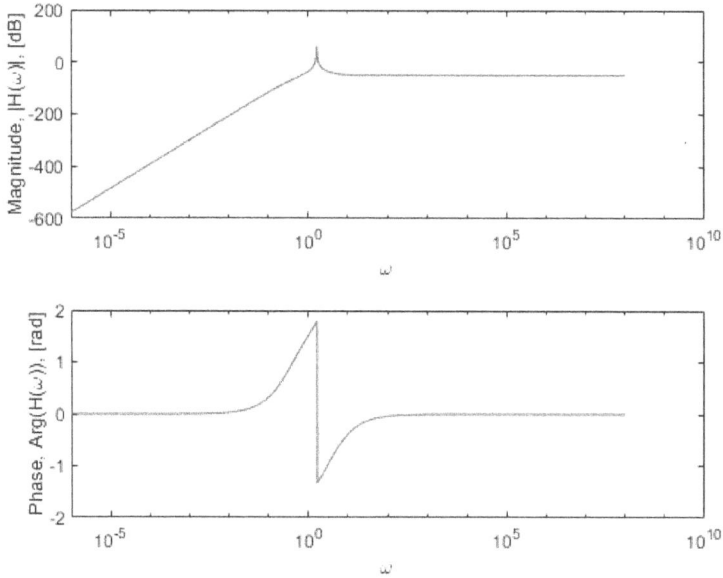

Figure 16.6 Transfer function of the higher order circuit model 3 with the Caputo-Fabrizio derivative using the Sumudu transform.

16.4   ANALYSIS OF HIGHER ORDER CIRCUIT MODEL 3 WITH ATANGANA-BALEANU DERIVATIVE

We consider the model with Atangana-Baleanu derivative. We apply Laplace and Sumudu transforms to obtain the solution.

$$\frac{1}{R_2C_2}\left({}_0^{ABC}D_t^\alpha V_1(t)\right) + \frac{V_1(t)}{R_1R_2C_1C_2} = \frac{d^2V_2(t)}{dt^2} + \left(\frac{1}{R_1C_1} + \frac{1}{R_2C_2}\right)V_2(t)$$

$$+ \frac{1}{R_1R_2C_1C_2}V_2(t) \qquad (16.10)$$

If we take the Laplace transform of the both sides of Eq. (16.10), we will obtain

$$\frac{1}{R_2C_2}L\left({}_0^{ABC}D_t^\alpha V_1(t)\right) + \frac{L(V_1(t))}{R_1R_2C_1C_2}$$

$$= L\left(\frac{d^2V_2(t)}{dt^2}\right) + \left(\frac{1}{R_1C_1} + \frac{1}{R_2C_2}\right)L(V_2(t)) + \frac{1}{R_1R_2C_1C_2}L(V_2(t))$$

Then, we obtain

$$\frac{1}{R_2 C_2} \frac{s^{\alpha-1} AB(\alpha)}{s^\alpha (1-\alpha) + \alpha} (sL(V_1(t)) - V_1(0)) + \frac{L(V_1(t))}{R_1 R_2 C_1 C_2}$$

$$= (s^2 L(V_2(t)) - s V_2(0) - V_2'(0)) + \left(\frac{1}{R_1 C_1} + \frac{1}{R_2 C_2} \right) L(V_2(t))$$

$$+ \frac{1}{R_1 R_2 C_1 C_2} L(V_2(t))$$

If we simplify the above equation, we will get

$$L(V_1(t)) \left(\frac{1}{R_2 C_2} \frac{s^\alpha AB(\alpha)}{s^\alpha (1-\alpha) + \alpha} + \frac{1}{R_1 R_2 C_1 C_2} \right)$$

$$= L(V_2(t)) \left(s^2 + \frac{1}{R_1 C_1} + \frac{1}{R_2 C_2} + \frac{1}{R_1 R_2 C_1 C_2} \right)$$

Thus, we have

$$L(V_1(t)) \left(\frac{s^\alpha R_1 C_1 AB(\alpha) + s^\alpha (1-\alpha) + \alpha}{(R_1 R_2 C_1 C_2)(s^\alpha (1-\alpha) + \alpha)} \right)$$

$$= L(V_2(t)) \left(\frac{s^2 R_1 R_2 C_1 C_2 + R_2 C_2 + R_1 C_1 + 1}{R_1 R_2 C_1 C_2} \right)$$

Therefore, we obtain the transfer function as

$$\frac{L(V_2(t))}{L(V_1(t))} = \frac{s^\alpha R_1 C_1 AB(\alpha) + s^\alpha (1-\alpha) + \alpha}{(s^2 R_1 R_2 C_1 C_2 + R_2 C_2 + R_1 C_1 + 1)(s^\alpha (1-\alpha) + \alpha)}. \qquad (16.11)$$

The graphical representation of the above transfer function is presented in Figure 16.7 as magnitude and phase.

If we take the Sumudu transform of the both sides of Eq. (16.10), we will obtain

$$\frac{1}{R_2 C_2} S \left({}_0^{ABC} D_t^\alpha V_1(t) \right) + \frac{S(V_1(t))}{R_1 R_2 C_1 C_2}$$

$$= S \left(\frac{d^2 V_2(t)}{dt^2} \right) + \left(\frac{1}{R_1 C_1} + \frac{1}{R_2 C_2} \right) S(V_2(t)) + \frac{1}{R_1 R_2 C_1 C_2} S(V_2(t))$$

Then, we obtain

$$\frac{1}{R_2 C_2} \frac{AB(\alpha)}{\alpha s^\alpha + 1 - \alpha} (S[V_1(t)] - V_1(0)) + \frac{S(V_1(t))}{R_1 R_2 C_1 C_2}$$

$$= \left(\frac{S[V_2(t)] - V_2(0)}{s^2} - \frac{V_2'(0)}{s} \right) + \left(\frac{1}{R_1 C_1} + \frac{1}{R_2 C_2} \right) S(V_2(t))$$

$$+ \frac{1}{R_1 R_2 C_1 C_2} S(V_2(t))$$

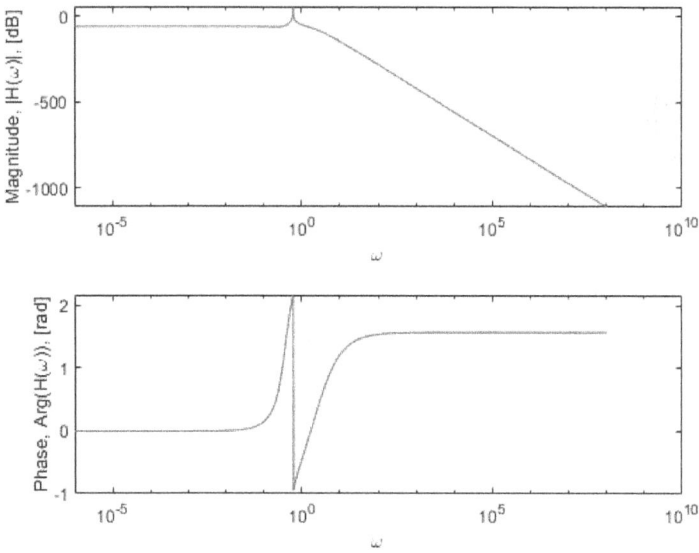

Figure 16.7 Transfer function of the higher order circuit model 3 with the ABC derivative using the Laplace transform.

If we simplify the above equation, we will get

$$S(V_1(t)) \left(\frac{1}{R_2 C_2} \frac{AB(\alpha)}{\alpha s^\alpha + 1 - \alpha} + \frac{1}{R_1 R_2 C_1 C_2} \right)$$

$$= S(V_2(t)) \left(\frac{1}{s^2} + \frac{1}{R_1 C_1} + \frac{1}{R_2 C_2} + \frac{1}{R_1 R_2 C_1 C_2} \right)$$

Thus, we have

$$S(V_1(t)) \left(\frac{R_1 C_1 AB(\alpha) + \alpha s^\alpha + 1 - \alpha}{(\alpha s^\alpha + 1 - \alpha) R_1 R_2 C_1 C_2} \right)$$

$$= S(V_2(t)) \left(\frac{R_1 R_2 C_1 C_2 + s^2 R_2 C_2 + s^2 R_1 C_1 + s^2}{s^2 R_1 R_2 C_1 C_2} \right)$$

Therefore, we obtain the transfer function as

$$\frac{S(V_2(t))}{S(V_1(t))} = \frac{s^2 (R_1 C_1 AB(\alpha) + \alpha s^\alpha + 1 - \alpha)}{(R_1 R_2 C_1 C_2 + s^2 R_2 C_2 + s^2 R_1 C_1 + s^2)(\alpha s^\alpha + 1 - \alpha)}. \qquad (16.12)$$

The graphical representation of the above transfer function is presented in Figure 16.8 as magnitude and phase.

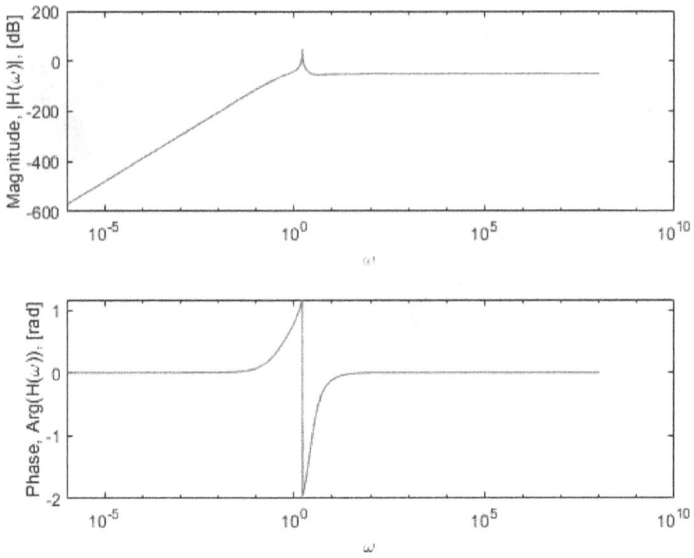

Figure 16.8 Transfer function of the higher order circuit model 3 with the ABC derivative using the Sumudu transform.

17 Nonlinear Model 1

Within the framework of mathematics, applied mathematics, and other fields of science, a nonlinear equation can be viewed as an equation in which the change of the output is not proportional to the change of input. The nonlinearity is an additional force that has a significant impact on the system behavior. These differential operators have been found fundamental and powerful tools for describing processes with nonlinear behaviors; thus they have been used to model real-world problems in engineering, biology, physics, chemistry, and many other fields [71]. These equations have been applied in modeling electric circuits. In particular, a nonlinear circuit can be viewed as an electric circuit for which parameters changed concerning current and voltage. This implies that an electric circuit for which parameters including resistance, inductance, capacitance, waveform, frequency, and many others is not constant. Due to the wider applicability of these models, there is a need to obtain their exact solutions to predict future behaviors of the system. However, due to nonlinearity, it is sometimes impossible to derive an exact solution; therefore, numerical schemes are used to provide numerical solutions to these equations. In this chapter we shall consider a nonlinear differential equation able to depict the dynamic behavior of the nonlinear circuit. The model will be extended to the concept of fractional nonlinear differential equations. Three different fractional differential operators will be employed for this purpose. For each model, a numerical scheme based on the Lagrange interpolation will be used to provide numerical solutions. Different theoretical investigations will be performed to insure the well-posedness of the model.

We consider the following system with the Caputo derivative

$$\begin{cases} {}_0^C D_t^\alpha x(t) = x(t)\,(y(t) - z(t)x(t)) = F(x,y,z,t) \\ {}_0^C D_t^\alpha y(t) = z(t)\,(x(t) - y(t)) = H(x,y,z,t) \\ {}_0^C D_t^\alpha z(t) = x(t) - z(t)y(t) = G(x,y,z,t) \end{cases}$$

We transform the above system to the following system as

$$x(t) - x(0) - \frac{1}{\Gamma(\alpha)} \int_0^t F(x,y,z,\tau)(t-\tau)^{\alpha-1} d\tau$$

$$y(t) - y(0) = \frac{1}{\Gamma(\alpha)} \int_0^t H(x,y,z,\tau)(t-\tau)^{\alpha-1} d\tau$$

$$z(t) - z(0) = \frac{1}{\Gamma(\alpha)} \int_0^t G(x,y,z,\tau)(t-\tau)^{\alpha-1} d\tau$$

DOI: 10.1201/9781003359869-17

Then, we consider the system at $t = t_{n+1}$:

$$x(t_{n+1}) = x(0) + \frac{1}{\Gamma(\alpha)} \int_0^{t_{n+1}} F(x,y,z,\tau)(t_{n+1} - \tau)^{\alpha-1} d\tau$$

$$y(t_{n+1}) = y(0) + \frac{1}{\Gamma(\alpha)} \int_0^{t_{n+1}} H(x,y,z,\tau)(t_{n+1} - \tau)^{\alpha-1} d\tau$$

$$z(t_{n+1}) = z(0) + \frac{1}{\Gamma(\alpha)} \int_0^{t_{n+1}} G(x,y,z,\tau)(t_{n+1} - \tau)^{\alpha-1} d\tau$$

Then, we apply the Atangana Toufik method and obtain

$$x^{n+1} = x^0 + \sum_{k=0}^{n} \left(\frac{(\Delta t)^\alpha}{\Gamma(\alpha+2)} F(x^k,y^k,z^k,t_k) \left((n+1-k)^\alpha(n-k+2+\alpha) \right.\right.$$
$$-(n-k)^\alpha(n-k+2+2\alpha))$$
$$\left.\left. -\frac{(\Delta t)^\alpha}{\Gamma(\alpha+2)} F(x^{k-1},y^{k-1},z^{k-1},t_{k-1}) \left((n+1-k)^{\alpha+1} - (n-k)^\alpha(n-k+1+\alpha)\right) \right) \right.$$

$$y^{n+1} = y^0 + \sum_{k=0}^{n} \left(\frac{(\Delta t)^\alpha}{\Gamma(\alpha+2)} H(x^k,y^k,z^k,t_k) \left((n+1-k)^\alpha(n-k+2+\alpha) \right.\right.$$
$$-(n-k)^\alpha(n-k+2+2\alpha))$$
$$\left.\left. -\frac{(\Delta t)^\alpha}{\Gamma(\alpha+2)} H(x^{k-1},y^{k-1},z^{k-1},t_{k-1}) \left((n+1-k)^{\alpha+1} - (n-k)^\alpha(n-k+1+\alpha)\right) \right) \right.$$

$$z^{n+1} = z^0 + \sum_{k=0}^{n} \left(\frac{(\Delta t)^\alpha}{\Gamma(\alpha+2)} G(x^k,y^k,z^k,t_k) \left((n+1-k)^\alpha(n-k+2+\alpha) \right.\right.$$
$$-(n-k)^\alpha(n-k+2+2\alpha))$$
$$\left.\left. -\frac{(\Delta t)^\alpha}{\Gamma(\alpha+2)} G(x^{k-1},y^{k-1},z^{k-1},t_{k-1}) \left((n+1-k)^{\alpha+1} - (n-k)^\alpha(n-k+1+\alpha)\right) \right) \right.$$

We consider the following system with the ABC derivative:

$$\begin{cases} {}^{ABC}_0 D_t^\alpha x(t) = x(t)\,(y(t) - z(t)x(t)) = F(x,y,z,t) \\ {}^{ABC}_0 D_t^\alpha y(t) = z(t)\,(x(t) - y(t)) = H(x,y,z,t) \\ {}^{ABC}_0 D_t^\alpha z(t) = x(t) - z(t)y(t) = G(x,y,z,t) \end{cases}$$

We transform the above system to the following system as

$$x(t) - x(0) = \frac{1-\alpha}{ABC(\alpha)} F(x,y,z,t) + \frac{\alpha}{ABC(\alpha)\Gamma(\alpha)} \int_0^t F(x,y,z,\tau)(t-\tau)^{\alpha-1} d\tau$$

$$y(t) - y(0) = \frac{1-\alpha}{ABC(\alpha)} H(x,y,z,t) + \frac{\alpha}{ABC(\alpha)\Gamma(\alpha)} \int_0^t H(x,y,z,\tau)(t-\tau)^{\alpha-1} d\tau$$

$$z(t) - z(0) = \frac{1-\alpha}{ABC(\alpha)} G(x,y,z,t) + \frac{\alpha}{ABC(\alpha)\Gamma(\alpha)} \int_0^t G(x,y,z,\tau)(t-\tau)^{\alpha-1} d\tau$$

Then, we consider the system at $t = t_{n+1}$:

$$x(t_{n+1}) = x(0) + \frac{1-\alpha}{ABC(\alpha)} F(x, y, z, t_n)$$

$$+ \frac{\alpha}{ABC(\alpha)\Gamma(\alpha)} \int_0^{t_{n+1}} F(x, y, z, \tau)(t_{n+1} - \tau)^{\alpha-1} d\tau$$

$$y(t_{n+1}) = y(0) + \frac{1-\alpha}{ABC(\alpha)} H(x, y, z, t_n)$$

$$+ \frac{\alpha}{ABC(\alpha)\Gamma(\alpha)} \int_0^{t_{n+1}} H(x, y, z, \tau)(t_{n+1} - \tau)^{\alpha-1} d\tau$$

$$z(t_{n+1}) = z(0) + \frac{1-\alpha}{ABC(\alpha)} G(x, y, z, t_n)$$

$$+ \frac{\alpha}{ABC(\alpha)\Gamma(\alpha)} \int_0^{t_{n+1}} G(x, y, z, \tau)(t_{n+1} - \tau)^{\alpha-1} d\tau$$

Then, we apply the Atangana-Toufik method and obtain

$$x^{n+1} = x^0 + \frac{1-\alpha}{ABC(\alpha)} F(x^n, y^n, z^n, t_n)$$

$$+ \frac{\alpha}{ABC(\alpha)} \sum_{k=0}^{n} \left(\frac{(\Delta t)^\alpha}{\Gamma(\alpha+2)} F(x^k, y^k, z^k, t_k) \left((n+1-k)^\alpha (n-k+2+\alpha) \right. \right.$$

$$-(n-k)^\alpha (n-k+2+2\alpha))$$

$$\left. \left. - \frac{(\Delta t)^\alpha}{\Gamma(\alpha+2)} F(x^{k-1}, y^{k-1}, z^{k-1}, t_{k-1}) \left((n+1-k)^{\alpha+1} - (n-k)^\alpha (n-k+1+\alpha) \right) \right) \right.$$

$$y^{n+1} = y^0 + \frac{1-\alpha}{ABC(\alpha)} H(x^n, y^n, z^n, t_n)$$

$$+ \frac{\alpha}{ABC(\alpha)} \sum_{k=0}^{n} \left(\frac{(\Delta t)^\alpha}{\Gamma(\alpha+2)} H(x^k, y^k, z^k, t_k) \left((n+1-k)^\alpha (n-k+2+\alpha) \right. \right.$$

$$-(n-k)^\alpha (n-k+2+2\alpha))$$

$$\left. \left. - \frac{(\Delta t)^\alpha}{\Gamma(\alpha+2)} H(x^{k-1}, y^{k-1}, z^{k-1}, t_{k-1}) \left((n+1-k)^{\alpha+1} - (n-k)^\alpha (n-k+1+\alpha) \right) \right) \right.$$

$$z^{n+1} = z^0 + \frac{1-\alpha}{ABC(\alpha)} G(x^n, y^n, z^n, t_n)$$

$$+ \frac{\alpha}{ABC(\alpha)} \sum_{k=0}^{n} \left(\frac{(\Delta t)^\alpha}{\Gamma(\alpha+2)} G(x^k, y^k, z^k, t_k) \left((n+1-k)^\alpha (n-k+2+\alpha) \right. \right.$$

$$-(n-k)^\alpha (n-k+2+2\alpha))$$

$$\left. \left. - \frac{(\Delta t)^\alpha}{\Gamma(\alpha+2)} G(x^{k-1}, y^{k-1}, z^{k-1}, t_{k-1}) \left((n+1-k)^{\alpha+1} - (n-k)^\alpha (n-k+1+\alpha) \right) \right) \right.$$

We consider the following system with the Caputo-Fabrizio derivative:

$$\begin{cases} {}_{0}^{CF}D_t^{\alpha}x(t) = x(t)\,(y(t) - z(t)x(t)) = F(x,y,z,t) \\ {}_{0}^{CF}D_t^{\alpha}y(t) = z(t)\,(x(t) - y(t)) = H(x,y,z,t) \\ {}_{0}^{CF}D_t^{\alpha}z(t) = x(t) - z(t)y(t) = G(x,y,z,t) \end{cases}$$

We transform the above system to the following system as

$$x(t) - x(0) = \frac{1-\alpha}{M(\alpha)}F(x,y,z,t) + \frac{\alpha}{M(\alpha)}\int_0^t F(x,y,z,\tau)d\tau$$

$$y(t) - y(0) = \frac{1-\alpha}{M(\alpha)}H(x,y,z,t) + \frac{\alpha}{M(\alpha)}\int_0^t H(x,y,z,\tau)d\tau$$

$$z(t) - z(0) = \frac{1-\alpha}{M(\alpha)}G(x,y,z,t) + \frac{\alpha}{M(\alpha)}\int_0^t G(x,y,z,\tau)d\tau$$

Then, we consider the system at $t = t_{n+1}$ and $t = t_n$ as

$$x(t_{n+1}) = x(0) + \frac{1-\alpha}{M(\alpha)}F(x,y,z,t_n) + \frac{\alpha}{M(\alpha)}\int_0^{t_{n+1}} F(x,y,z,\tau)d\tau$$

$$y(t_{n+1}) = y(0) + \frac{1-\alpha}{M(\alpha)}H(x,y,z,t_n) + \frac{\alpha}{M(\alpha)}\int_0^{t_{n+1}} H(x,y,z,\tau)d\tau$$

$$z(t_{n+1}) = z(0) + \frac{1-\alpha}{M(\alpha)}G(x,y,z,t_n) + \frac{\alpha}{M(\alpha)}\int_0^{t_{n+1}} G(x,y,z,\tau)d\tau$$

$$x(t_n) = x(0) + \frac{1-\alpha}{M(\alpha)}F(x,y,z,t_{n-1}) + \frac{\alpha}{M(\alpha)}\int_0^{t_n} F(x,y,z,\tau)d\tau$$

$$y(t_n) = y(0) + \frac{1-\alpha}{M(\alpha)}H(x,y,z,t_{n-1}) + \frac{\alpha}{M(\alpha)}\int_0^{t_n} H(x,y,z,\tau)d\tau$$

$$z(t_n) = z(0) + \frac{1-\alpha}{M(\alpha)}G(x,y,z,t_{n-1}) + \frac{\alpha}{M(\alpha)}\int_0^{t_n} G(x,y,z,\tau)d\tau$$

Then, we obtain

$$\begin{aligned} x(t_{n+1}) &= x(t_n) + \frac{1-\alpha}{M(\alpha)}\left(F(x,y,z,t_n) - F(x,y,z,t_{n-1})\right) \\ &+ \frac{\alpha}{M(\alpha)}\left(\frac{3\Delta t}{2}F(x,y,z,t_n) - \frac{\Delta t}{2}F(x,y,z,t_{n-1})\right) \end{aligned}$$

$$\begin{aligned} y(t_{n+1}) &= y(t_n) + \frac{1-\alpha}{M(\alpha)}\left(H(x,y,z,t_n) - H(x,y,z,t_{n-1})\right) \\ &+ \frac{\alpha}{M(\alpha)}\left(\frac{3\Delta t}{2}H(x,y,z,t_n) - \frac{\Delta t}{2}H(x,y,z,t_{n-1})\right) \end{aligned}$$

$$\begin{aligned} z(t_{n+1}) &= z(t_n) + \frac{1-\alpha}{M(\alpha)}\left(G(x,y,z,t_n) - G(x,y,z,t_{n-1})\right) \\ &+ \frac{\alpha}{M(\alpha)}\left(\frac{3\Delta t}{2}G(x,y,z,t_n) - \frac{\Delta t}{2}G(x,y,z,t_{n-1})\right) \end{aligned}$$

We consider the following system with the fractal fractional derivative using power law kernel:

$$\begin{cases} {}^{FFP}_0D^\alpha_t x(t) = x(t)\,(y(t) - z(t)x(t)) = F(x,y,z,t) \\ {}^{FFP}_0D^\alpha_t y(t) = z(t)\,(x(t) - y(t)) = H(x,y,z,t) \\ {}^{FFP}_0D^\alpha_t z(t) = x(t) - z(t)y(t) = G(x,y,z,t) \end{cases}$$

We can write the above system as

$$\begin{cases} {}^{RL}_0D^\alpha_t x(t) = \beta t^{\beta-1}F(x,y,z,t) = K(x,y,z,t) \\ {}^{RL}_0D^\alpha_t y(t) = \beta t^{\beta-1}H(x,y,z,t) = L(x,y,z,t) \\ {}^{RL}_0D^\alpha_t z(t) = \beta t^{\beta-1}G(x,y,z,t) = M(x,y,z,t) \end{cases}$$

We transform the above system to the following system as

$$x(t) - x(0) = \frac{1}{\Gamma(\alpha)} \int_0^t K(x,y,z,\tau)(t-\tau)^{\alpha-1}d\tau$$

$$y(t) - y(0) = \frac{1}{\Gamma(\alpha)} \int_0^t L(x,y,z,\tau)(t-\tau)^{\alpha-1}d\tau$$

$$z(t) - z(0) = \frac{1}{\Gamma(\alpha)} \int_0^t M(x,y,z,\tau)(t-\tau)^{\alpha-1}d\tau$$

Then, we consider the system at $t = t_{n+1}$:

$$x(t_{n+1}) = x(0) + \frac{1}{\Gamma(\alpha)} \int_0^{t_{n+1}} K(x,y,z,\tau)(t_{n+1}-\tau)^{\alpha-1}d\tau$$

$$y(t_{n+1}) = y(0) + \frac{1}{\Gamma(\alpha)} \int_0^{t_{n+1}} L(x,y,z,\tau)(t_{n+1}-\tau)^{\alpha-1}d\tau$$

$$z(t_{n+1}) = z(0) + \frac{1}{\Gamma(\alpha)} \int_0^{t_{n+1}} M(x,y,z,\tau)(t_{n+1}-\tau)^{\alpha-1}d\tau$$

Then, we apply the Atangana-Toufik method and obtain

$$x^{n+1} = x^0 + \sum_{k=0}^n \left(\frac{(\Delta t)^\alpha}{\Gamma(\alpha+2)} K(x^k,y^k,z^k,t_k)\,((n+1-k)^\alpha(n-k+2+\alpha) \right.$$

$$-(n-k)^\alpha(n-k+2+2\alpha))$$

$$\left. -\frac{(\Delta t)^\alpha}{\Gamma(\alpha+2)} K(x^{k-1},y^{k-1},z^{k-1},t_{k-1})\,((n+1-k)^{\alpha+1} - (n-k)^\alpha(n-k+1+\alpha)) \right)$$

$$y^{n+1} = y^0 + \sum_{k=0}^n \left(\frac{(\Delta t)^\alpha}{\Gamma(\alpha+2)} L(x^k,y^k,z^k,t_k)\,((n+1\quad k)^\alpha(n-k+2+\alpha) \right.$$

$$-(n-k)^\alpha(n-k+2+2\alpha))$$

$$\left. -\frac{(\Delta t)^\alpha}{\Gamma(\alpha+2)} L(x^{k-1},y^{k-1},z^{k-1},t_{k-1})\,((n+1-k)^{\alpha+1} - (n-k)^\alpha(n-k+1+\alpha)) \right)$$

$$z^{n+1} = z^0 + \sum_{k=0}^{n} \left(\frac{(\Delta t)^{\alpha}}{\Gamma(\alpha+2)} M(x^k, y^k, z^k, t_k) \left((n+1-k)^{\alpha}(n-k+2+\alpha) \right) \right.$$

$$-(n-k)^{\alpha}(n-k+2+2\alpha))$$

$$\left. - \frac{(\Delta t)^{\alpha}}{\Gamma(\alpha+2)} M(x^{k-1}, y^{k-1}, z^{k-1}, t_{k-1}) \left((n+1-k)^{\alpha+1} - (n-k)^{\alpha}(n-k+1+\alpha) \right) \right)$$

We consider the following system with the fractal fractional derivative using the generalized Mittag-Leffler function as

$$\begin{cases} {}_0^{FFM}D_t^{\alpha}x(t) = x(t)\,(y(t)-z(t)x(t)) = F(x,y,z,t) \\ {}_0^{FFM}D_t^{\alpha}y(t) = z(t)\,(x(t)-y(t)) = H(x,y,z,t) \\ {}_0^{FFM}D_t^{\alpha}z(t) = x(t) - z(t)y(t) = G(x,y,z,t) \end{cases}$$

We can write the above system as

$$\begin{cases} {}_0^{ABR}D_t^{\alpha}x(t) = \beta t^{\beta-1}F(x,y,z,t) = K(x,y,z,t) \\ {}_0^{ABR}D_t^{\alpha}y(t) = \beta t^{\beta-1}H(x,y,z,t) = L(x,y,z,t) \\ {}_0^{ABR}D_t^{\alpha}z(t) = \beta t^{\beta-1}G(x,y,z,t) = M(x,y,z,t) \end{cases}$$

We transform the above system to the following system as

$$x(t) - x(0) = \frac{1-\alpha}{ABC(\alpha)}K(x,y,z,t) + \frac{\alpha}{ABC(\alpha)\Gamma(\alpha)}\int_0^t K(x,y,z,\tau)(t-\tau)^{\alpha-1}d\tau$$

$$y(t) - y(0) = \frac{1-\alpha}{ABC(\alpha)}L(x,y,z,t) + \frac{\alpha}{ABC(\alpha)\Gamma(\alpha)}\int_0^t L(x,y,z,\tau)(t-\tau)^{\alpha-1}d\tau$$

$$z(t) - z(0) = \frac{1-\alpha}{ABC(\alpha)}M(x,y,z,t) + \frac{\alpha}{ABC(\alpha)\Gamma(\alpha)}\int_0^t M(x,y,z,\tau)(t-\tau)^{\alpha-1}d\tau$$

Then, we consider the system at $t = t_{n+1}$:

$$x(t_{n+1}) = x(0) + \frac{1-\alpha}{ABC(\alpha)}K(x,y,z,t_n)$$

$$+ \frac{\alpha}{ABC(\alpha)\Gamma(\alpha)}\int_0^{t_{n+1}} K(x,y,z,\tau)(t_{n+1}-\tau)^{\alpha-1}d\tau$$

$$y(t_{n+1}) = y(0) + \frac{1-\alpha}{ABC(\alpha)}L(x,y,z,t_n)$$

$$+ \frac{\alpha}{ABC(\alpha)\Gamma(\alpha)}\int_0^{t_{n+1}} L(x,y,z,\tau)(t_{n+1}-\tau)^{\alpha-1}d\tau$$

$$z(t_{n+1}) = z(0) + \frac{1-\alpha}{ABC(\alpha)}M(x,y,z,t_n)$$

$$+ \frac{\alpha}{ABC(\alpha)\Gamma(\alpha)}\int_0^{t_{n+1}} M(x,y,z,\tau)(t_{n+1}-\tau)^{\alpha-1}d\tau$$

Then, we apply the Atangana-Toufik method and obtain

$$x^{n+1} = x^0 + \frac{1-\alpha}{ABC(\alpha)} K(x^n, y^n, z^n, t_n)$$

$$+ \frac{\alpha}{ABC(\alpha)} \sum_{k=0}^{n} \left(\frac{(\Delta t)^\alpha}{\Gamma(\alpha+2)} K(x^k, y^k, z^k, t_k) \left((n+1-k)^\alpha (n-k+2+\alpha) \right. \right.$$

$$-(n-k)^\alpha (n-k+2+2\alpha))$$

$$\left. - \frac{(\Delta t)^\alpha}{\Gamma(\alpha+2)} K(x^{k-1}, y^{k-1}, z^{k-1}, t_{k-1}) \left((n+1-k)^{\alpha+1} - (n-k)^\alpha (n-k+1+\alpha) \right) \right)$$

$$y^{n+1} = y^0 + \frac{1-\alpha}{ABC(\alpha)} L(x^n, y^n, z^n, t_n)$$

$$+ \frac{\alpha}{ABC(\alpha)} \sum_{k=0}^{n} \left(\frac{(\Delta t)^\alpha}{\Gamma(\alpha+2)} L(x^k, y^k, z^k, t_k) \left((n+1-k)^\alpha (n-k+2+\alpha) \right. \right.$$

$$-(n-k)^\alpha (n-k+2+2\alpha))$$

$$\left. - \frac{(\Delta t)^\alpha}{\Gamma(\alpha+2)} L(x^{k-1}, y^{k-1}, z^{k-1}, t_{k-1}) \left((n+1-k)^{\alpha+1} - (n-k)^\alpha (n-k+1+\alpha) \right) \right)$$

$$z^{n+1} = z^0 + \frac{1-\alpha}{ABC(\alpha)} M(x^n, y^n, z^n, t_n)$$

$$+ \frac{\alpha}{ABC(\alpha)} \sum_{k=0}^{n} \left(\frac{(\Delta t)^\alpha}{\Gamma(\alpha+2)} M(x^k, y^k, z^k, t_k) \left((n+1-k)^\alpha (n-k+2+\alpha) \right. \right.$$

$$-(n-k)^\alpha (n-k+2+2\alpha))$$

$$\left. - \frac{(\Delta t)^\alpha}{\Gamma(\alpha+2)} M(x^{k-1}, y^{k-1}, z^{k-1}, t_{k-1}) \left((n+1-k)^{\alpha+1} - (n-k)^\alpha (n-k+1+\alpha) \right) \right)$$

We consider the following system with the fractal fractional derivative using the exponential decay kernel:

$$\begin{cases} {}^{FFE}_0 D_t^\alpha x(t) = x(t) \left(y(t) - z(t)x(t) \right) = F(x, y, z, t) \\ {}^{FFE}_0 D_t^\alpha y(t) = z(t) \left(x(t) - y(t) \right) = H(x, y, z, t) \\ {}^{FFE}_0 D_t^\alpha z(t) = x(t) - z(t)y(t) = G(x, y, z, t) \end{cases}$$

We can write the above system as

$$\begin{cases} {}^{CF}_0 D_t^\alpha x(t) = \beta t^{\beta-1} F(x, y, z, t) = K(x, y, z, t) \\ {}^{CF}_0 D_t^\alpha y(t) = \beta t^{\beta-1} H(x, y, z, t) = L(x, y, z, t) \\ {}^{CF}_0 D_t^\alpha z(t) = \beta t^{\beta-1} G(x, y, z, t) = M(x, y, z, t) \end{cases}$$

We transform the above system to the following system as

$$x(t) - x(0) = \frac{1-\alpha}{M(\alpha)} K(x,y,z,t) + \frac{\alpha}{M(\alpha)} \int_0^t K(x,y,z,\tau) d\tau$$

$$y(t) - y(0) = \frac{1-\alpha}{M(\alpha)} L(x,y,z,t) + \frac{\alpha}{M(\alpha)} \int_0^t L(x,y,z,\tau) d\tau$$

$$z(t) - z(0) = \frac{1-\alpha}{M(\alpha)} M(x,y,z,t) + \frac{\alpha}{M(\alpha)} \int_0^t M(x,y,z,\tau) d\tau$$

Then, we consider the system at $t = t_{n+1}$ and $t = t_n$ as

$$x(t_{n+1}) = x(0) + \frac{1-\alpha}{M(\alpha)} K(x,y,z,t_n) + \frac{\alpha}{M(\alpha)} \int_0^{t_{n+1}} K(x,y,z,\tau) d\tau$$

$$y(t_{n+1}) = y(0) + \frac{1-\alpha}{M(\alpha)} L(x,y,z,t_n) + \frac{\alpha}{M(\alpha)} \int_0^{t_{n+1}} L(x,y,z,\tau) d\tau$$

$$z(t_{n+1}) = z(0) + \frac{1-\alpha}{M(\alpha)} M(x,y,z,t_n) + \frac{\alpha}{M(\alpha)} \int_0^{t_{n+1}} M(x,y,z,\tau) d\tau$$

$$x(t_n) = x(0) + \frac{1-\alpha}{M(\alpha)} K(x,y,z,t_{n-1}) + \frac{\alpha}{M(\alpha)} \int_0^{t_n} K(x,y,z,\tau) d\tau$$

$$y(t_n) = y(0) + \frac{1-\alpha}{M(\alpha)} L(x,y,z,t_{n-1}) + \frac{\alpha}{M(\alpha)} \int_0^{t_n} L(x,y,z,\tau) d\tau$$

$$z(t_n) = z(0) + \frac{1-\alpha}{M(\alpha)} M(x,y,z,t_{n-1}) + \frac{\alpha}{M(\alpha)} \int_0^{t_n} M(x,y,z,\tau) d\tau$$

Then, we obtain

$$
\begin{aligned}
x(t_{n+1}) &= x(t_n) + \frac{1-\alpha}{M(\alpha)} \left(K(x,y,z,t_n) - K(x,y,z,t_{n-1}) \right) \\
&+ \frac{\alpha}{M(\alpha)} \left(\frac{3\Delta t}{2} K(x,y,z,t_n) - \frac{\Delta t}{2} K(x,y,z,t_{n-1}) \right)
\end{aligned}
$$

$$
\begin{aligned}
y(t_{n+1}) &= y(t_n) + \frac{1-\alpha}{M(\alpha)} \left(L(x,y,z,t_n) - L(x,y,z,t_{n-1}) \right) \\
&+ \frac{\alpha}{M(\alpha)} \left(\frac{3\Delta t}{2} L(x,y,z,t_n) - \frac{\Delta t}{2} L(x,y,z,t_{n-1}) \right)
\end{aligned}
$$

$$
\begin{aligned}
z(t_{n+1}) &= z(t_n) + \frac{1-\alpha}{M(\alpha)} \left(M(x,y,z,t_n) - M(x,y,z,t_{n-1}) \right) \\
&+ \frac{\alpha}{M(\alpha)} \left(\frac{3\Delta t}{2} M(x,y,z,t_n) - \frac{\Delta t}{2} - M(x,y,z,t_{n-1}) \right)
\end{aligned}
$$

18 Chua Circuit Model

In this section we consider the Chua circuit model with the Caputo, Caputo-Fabrizio, and Atangana-Baleanu derivatives. We also take into consideration the fractal fractional derivative using power law, exponential decay law, and the generalized Mittag-Leffler law kernels. We discretize the problem with all derivatives. We use the two-step Lagrange polynomial to get the desired results. We apply the Atangana-Toufik method to the model for all derivatives. We obtain very effective results for this model [16, 17, 72–74].

We consider the following system with the Caputo derivative:

$$\begin{cases} {}_0^C D_t^\alpha x(t) = a\left(y(t) - \phi(x(t))\right) = F(x,y,t) \\ {}_0^C D_t^\alpha y(t) = x(t) - y(t) + z(t) = H(x,y,z,t) \\ {}_0^C D_t^\alpha z(t) = -\lambda y(t) = G(y,t) \end{cases}$$

We transform the above system to the following system as

$$x(t) - x(0) = \frac{1}{\Gamma(\alpha)} \int_0^t F(x,y,\tau)(t-\tau)^{\alpha-1} d\tau$$

$$y(t) - y(0) = \frac{1}{\Gamma(\alpha)} \int_0^t H(x,y,z,\tau)(t-\tau)^{\alpha-1} d\tau$$

$$z(t) - z(0) = \frac{1}{\Gamma(\alpha)} \int_0^t G(y,\tau)(t-\tau)^{\alpha-1} d\tau$$

Then, we consider the system at $t = t_{n+1}$:

$$x(t_{n+1}) = x(0) + \frac{1}{\Gamma(\alpha)} \int_0^{t_{n+1}} F(x,y,\tau)(t_{n+1}-\tau)^{\alpha-1} d\tau$$

$$y(t_{n+1}) = y(0) + \frac{1}{\Gamma(\alpha)} \int_0^{t_{n+1}} H(x,y,z,\tau)(t_{n+1}-\tau)^{\alpha-1} d\tau$$

$$z(t_{n+1}) = z(0) + \frac{1}{\Gamma(\alpha)} \int_0^{t_{n+1}} G(y,\tau)(t_{n+1}-\tau)^{\alpha-1} d\tau$$

Then, we apply the Atangana-Toufik method and obtain

$$\begin{aligned} x^{n+1} = x^0 + \sum_{k=0}^{n} & \left(\frac{(\Delta t)^\alpha}{\Gamma(\alpha+2)} F(x^k,y^k,t_k) \left((n+1-k)^\alpha(n-k+2+\alpha) \right. \right. \\ & - (n-k)^\alpha(n-k+2+2\alpha)) \\ & \left. - \frac{(\Delta t)^\alpha}{\Gamma(\alpha+2)} F(x^{k-1},y^{k-1},t_{k-1}) \left((n+1-k)^{\alpha+1} - (n-k)^\alpha(n-k+1+\alpha)\right) \right) \end{aligned}$$

$$y^{n+1} = y^0 + \sum_{k=0}^{n} \left(\frac{(\Delta t)^\alpha}{\Gamma(\alpha+2)} H(x^k, y^k, z^k, t_k) \left((n+1-k)^\alpha (n-k+2+\alpha)\right.\right.$$

$$-(n-k)^\alpha (n-k+2+2\alpha))$$

$$\left.-\frac{(\Delta t)^\alpha}{\Gamma(\alpha+2)} H(x^{k-1}, y^{k-1}, z^{k-1}, t_{k-1}) \left((n+1-k)^{\alpha+1} - (n-k)^\alpha (n-k+1+\alpha)\right)\right)$$

$$z^{n+1} = z^0 + \sum_{k=0}^{n} \left(\frac{(\Delta t)^\alpha}{\Gamma(\alpha+2)} G(y^k, t_k) \left((n+1-k)^\alpha (n-k+2+\alpha)\right.\right.$$

$$-(n-k)^\alpha (n-k+2+2\alpha))$$

$$\left.-\frac{(\Delta t)^\alpha}{\Gamma(\alpha+2)} G(y^{k-1}, t_{k-1}) \left((n+1-k)^{\alpha+1} - (n-k)^\alpha (n-k+1+\alpha)\right)\right)$$

We consider the following system with the ABC derivative:

$$\begin{cases} {}^{ABC}_{\ 0}D_t^\alpha x(t) = a\left(y(t) - \phi(x(t))\right) = F(x,y,t) \\ {}^{ABC}_{\ 0}D_t^\alpha y(t) = x(t) - y(t) + z(t) = H(x,y,z,t) \\ {}^{ABC}_{\ 0}D_t^\alpha z(t) = -\lambda y(t) = G(y,t) \end{cases}$$

We transform the above system to the following system as

$$x(t) - x(0) = \frac{1-\alpha}{ABC(\alpha)} F(x,y,t) + \frac{\alpha}{ABC(\alpha)\Gamma(\alpha)} \int_0^t F(x,y,\tau)(t-\tau)^{\alpha-1} d\tau$$

$$y(t) - y(0) = \frac{1-\alpha}{ABC(\alpha)} H(x,y,z,t) + \frac{\alpha}{ABC(\alpha)\Gamma(\alpha)} \int_0^t H(x,y,z,\tau)(t-\tau)^{\alpha-1} d\tau$$

$$z(t) - z(0) = \frac{1-\alpha}{ABC(\alpha)} G(y,t) + \frac{\alpha}{ABC(\alpha)\Gamma(\alpha)} \int_0^t G(y,\tau)(t-\tau)^{\alpha-1} d\tau$$

Then, we consider the system at $t = t_{n+1}$:

$$x(t_{n+1}) = x(0) + \frac{1-\alpha}{ABC(\alpha)} F(x,y,t_n)$$

$$+ \frac{\alpha}{ABC(\alpha)\Gamma(\alpha)} \int_0^{t_{n+1}} F(x,y,\tau)(t_{n+1} - \tau)^{\alpha-1} d\tau$$

$$y(t_{n+1}) = y(0) + \frac{1-\alpha}{ABC(\alpha)} H(x,y,z,t_n)$$

$$+ \frac{\alpha}{ABC(\alpha)\Gamma(\alpha)} \int_0^{t_{n+1}} H(x,y,z,\tau)(t_{n+1} - \tau)^{\alpha-1} d\tau$$

$$z(t_{n+1}) = z(0) + \frac{1-\alpha}{ABC(\alpha)} G(y,t_n)$$

$$+ \frac{\alpha}{ABC(\alpha)\Gamma(\alpha)} \int_0^{t_{n+1}} G(y,\tau)(t_{n+1} - \tau)^{\alpha-1} d\tau$$

Then, we apply the Atangana-Toufik method and obtain

$$x^{n+1} = x^0 + \frac{1-\alpha}{ABC(\alpha)} F(x^n, y^n, t_n)$$

$$+ \frac{\alpha}{ABC(\alpha)} \sum_{k=0}^{n} \left(\frac{(\Delta t)^\alpha}{\Gamma(\alpha+2)} F(x^k, y^k, t_k) \left((n+1-k)^\alpha (n-k+2+\alpha) \right. \right.$$

$$-(n-k)^\alpha (n-k+2+2\alpha))$$

$$\left. - \frac{(\Delta t)^\alpha}{\Gamma(\alpha+2)} F(x^{k-1}, y^{k-1}, t_{k-1}) \left((n+1-k)^{\alpha+1} - (n-k)^\alpha (n-k+1+\alpha) \right) \right)$$

$$y^{n+1} = y^0 + \frac{1-\alpha}{ABC(\alpha)} H(x^n, y^n, z^n, t_n)$$

$$+ \frac{\alpha}{ABC(\alpha)} \sum_{k=0}^{n} \left(\frac{(\Delta t)^\alpha}{\Gamma(\alpha+2)} H(x^k, y^k, z^k, t_k) \left((n+1-k)^\alpha (n-k+2+\alpha) \right. \right.$$

$$- (n-k)^\alpha (n-k+2+2\alpha))$$

$$- \frac{(\Delta t)^\alpha}{\Gamma(\alpha+2)} H(x^{k-1}, y^{k-1}, z^{k-1}, t_{k-1})$$

$$\left. \times \left((n+1-k)^{\alpha+1} - (n-k)^\alpha (n-k+1+\alpha) \right) \right)$$

$$z^{n+1} = z^0 + \frac{1-\alpha}{ABC(\alpha)} G(y^n, t_n)$$

$$+ \frac{\alpha}{ABC(\alpha)} \sum_{k=0}^{n} \left(\frac{(\Delta t)^\alpha}{\Gamma(\alpha+2)} G(y^k, t_k) \left((n+1-k)^\alpha (n-k+2+\alpha) \right. \right.$$

$$-(n-k)^\alpha (n-k+2+2\alpha))$$

$$\left. - \frac{(\Delta t)^\alpha}{\Gamma(\alpha+2)} G(y^{k-1}, t_{k-1}) \left((n+1-k)^{\alpha+1} - (n-k)^\alpha (n-k+1+\alpha) \right) \right)$$

We consider the following system with the Caputo-Fabrizio derivative:

$$\begin{cases} {}^{CF}_0 D_t^\alpha x(t) = a\left(y(t) - \phi(x(t))\right) = F(x, y, t) \\ {}^{CF}_0 D_t^\alpha y(t) = x(t) - y(t) + z(t) = H(x, y, z, t) \\ {}^{CF}_0 D_t^\alpha z(t) = -\lambda y(t) = G(y, t) \end{cases}$$

We transform the above system to the following system as

$$x(t) - x(0) = \frac{1-\alpha}{M(\alpha)} F(x, y, t) + \frac{\alpha}{M(\alpha)} \int_0^t F(x, y, \tau) d\tau$$

$$y(t) - y(0) = \frac{1-\alpha}{M(\alpha)} H(x, y, z, t) + \frac{\alpha}{M(\alpha)} \int_0^t H(x, y, z, \tau) d\tau$$

$$z(t) - z(0) = \frac{1-\alpha}{M(\alpha)} G(y, t) + \frac{\alpha}{M(\alpha)} \int_0^t G(y, \tau) d\tau$$

Then, we consider the system at $t = t_{n+1}$ and $t = t_n$ as

$$x(t_{n+1}) = x(0) + \frac{1-\alpha}{M(\alpha)} F(x,y,t_n) + \frac{\alpha}{M(\alpha)} \int_0^{t_{n+1}} F(x,y,\tau)d\tau$$

$$y(t_{n+1}) = y(0) + \frac{1-\alpha}{M(\alpha)} H(x,y,z,t_n) + \frac{\alpha}{M(\alpha)} \int_0^{t_{n+1}} H(x,y,z,\tau)d\tau$$

$$z(t_{n+1}) = z(0) + \frac{1-\alpha}{M(\alpha)} G(y,t_n) + \frac{\alpha}{M(\alpha)} \int_0^{t_{n+1}} G(y,\tau)d\tau$$

$$x(t_n) = x(0) + \frac{1-\alpha}{M(\alpha)} F(x,y,t_{n-1}) + \frac{\alpha}{M(\alpha)} \int_0^{t_n} F(x,y,\tau)d\tau$$

$$y(t_n) = y(0) + \frac{1-\alpha}{M(\alpha)} H(x,y,z,t_{n-1}) + \frac{\alpha}{M(\alpha)} \int_0^{t_n} H(x,y,z,\tau)d\tau$$

$$z(t_n) = z(0) + \frac{1-\alpha}{M(\alpha)} G(y,t_{n-1}) + \frac{\alpha}{M(\alpha)} \int_0^{t_n} G(y,\tau)d\tau$$

Then, we obtain

$$
\begin{aligned}
x(t_{n+1}) &= x(t_n) + \frac{1-\alpha}{M(\alpha)} \left(F(x,y,t_n) - F(x,y,t_{n-1})\right) \\
&+ \frac{\alpha}{M(\alpha)} \left(\frac{3\Delta t}{2} F(x,y,t_n) - \frac{\Delta t}{2} F(x,y,t_{n-1})\right)
\end{aligned}
$$

$$
\begin{aligned}
y(t_{n+1}) &= y(t_n) + \frac{1-\alpha}{M(\alpha)} \left(H(x,y,z,t_n) - H(x,y,z,t_{n-1})\right) \\
&+ \frac{\alpha}{M(\alpha)} \left(\frac{3\Delta t}{2} H(x,y,z,t_n) - \frac{\Delta t}{2} H(x,y,z,t_{n-1})\right)
\end{aligned}
$$

$$
\begin{aligned}
z(t_{n+1}) &= z(t_n) + \frac{1-\alpha}{M(\alpha)} \left(G(y,t_n) - G(y,t_{n-1})\right) \\
&+ \frac{\alpha}{M(\alpha)} \left(\frac{3\Delta t}{2} G(y,t_n) - \frac{\Delta t}{2} G(y,t_{n-1})\right)
\end{aligned}
$$

We consider the following system with the fractal fractional derivative using power law kernel:

$$
\begin{cases}
{}^{FFP}_0 D_t^\alpha x(t) = a(y(t) - \phi(x(t))) = F(x,y,t) \\
{}^{FFP}_0 D_t^\alpha y(t) = x(t) - y(t) + z(t) = H(x,y,z,t) \\
{}^{FFP}_0 D_t^\alpha z(t) = -\lambda y(t) = G(y,t)
\end{cases}
$$

We can write the above system as

$$
\begin{cases}
{}^{RL}_0 D_t^\alpha x(t) = \beta t^{\beta-1} F(x,y,t) = K(x,y,t) \\
{}^{RL}_0 D_t^\alpha y(t) = \beta t^{\beta-1} H(x,y,z,t) = L(x,y,z,t) \\
{}^{RL}_0 D_t^\alpha z(t) = \beta t^{\beta-1} G(y,t) = M(y,t)
\end{cases}
$$

We transform the above system to the following system as

$$x(t) - x(0) = \frac{1}{\Gamma(\alpha)} \int_0^t K(x,y,\tau)(t-\tau)^{\alpha-1} d\tau$$

$$y(t) - y(0) = \frac{1}{\Gamma(\alpha)} \int_0^t L(x,y,z,\tau)(t-\tau)^{\alpha-1} d\tau$$

$$z(t) - z(0) = \frac{1}{\Gamma(\alpha)} \int_0^t M(y,\tau)(t-\tau)^{\alpha-1} d\tau$$

Then, we consider the system at $t = t_{n+1}$:

$$x(t_{n+1}) = x(0) + \frac{1}{\Gamma(\alpha)} \int_0^{t_{n+1}} K(x,y,\tau)(t_{n+1}-\tau)^{\alpha-1} d\tau$$

$$y(t_{n+1}) = y(0) + \frac{1}{\Gamma(\alpha)} \int_0^{t_{n+1}} L(x,y,z,\tau)(t_{n+1}-\tau)^{\alpha-1} d\tau$$

$$z(t_{n+1}) = z(0) + \frac{1}{\Gamma(\alpha)} \int_0^{t_{n+1}} M(y,\tau)(t_{n+1}-\tau)^{\alpha-1} d\tau$$

Then, we apply the Atangana-Toufik method and obtain

$$x^{n+1} = x^0 + \sum_{k=0}^n \left(\frac{(\Delta t)^\alpha}{\Gamma(\alpha+2)} K(x^k, y^k, t_k) \left((n+1-k)^\alpha(n-k+2+\alpha) \right. \right.$$
$$-(n-k)^\alpha(n-k+2+2\alpha))$$
$$\left. -\frac{(\Delta t)^\alpha}{\Gamma(\alpha+2)} K(x^{k-1}, y^{k-1}, t_{k-1}) \left((n+1-k)^{\alpha+1} - (n-k)^\alpha(n-k+1+\alpha)\right) \right)$$

$$y^{n+1} = y^0 + \sum_{k=0}^n \left(\frac{(\Delta t)^\alpha}{\Gamma(\alpha+2)} L(x^k, y^k, z^k, t_k) \left((n+1-k)^\alpha(n-k+2+\alpha) \right. \right.$$
$$-(n-k)^\alpha(n-k+2+2\alpha))$$
$$\left. -\frac{(\Delta t)^\alpha}{\Gamma(\alpha+2)} L(x^{k-1}, y^{k-1}, z^{k-1}, t_{k-1}) \left((n+1-k)^{\alpha+1} - (n-k)^\alpha(n-k+1+\alpha)\right) \right)$$

$$z^{n+1} = z^0 + \sum_{k=0}^n \left(\frac{(\Delta t)^\alpha}{\Gamma(\alpha+2)} M(y^k, t_k) \left((n+1-k)^\alpha(n-k+2+\alpha) \right. \right.$$
$$-(n-k)^u(n-k+2+2\alpha))$$
$$\left. -\frac{(\Delta t)^\alpha}{\Gamma(\alpha+2)} M(y^{k-1}, t_{k-1}) \left((n+1-k)^{\alpha+1} - (n-k)^\alpha(n-k+1+\alpha)\right) \right)$$

We consider the following system with the fractal fractional derivative using the generalized Mittag-Leffler function as

$$\begin{cases} {}^{FFM}_0 D_t^\alpha x(t) = a(y(t) - \phi(x(t))) = F(x,y,t) \\ {}^{FFM}_0 D_t^\alpha y(t) = x(t) - y(t) + z(t) = H(x,y,z,t) \\ {}^{FFM}_0 D_t^\alpha z(t) = -\lambda y(t) = G(y,t) \end{cases}$$

We can write the above system as

$$\begin{cases} {}^{ABR}_{0}D^{\alpha}_{t}x(t) = \beta t^{\beta-1}F(x,y,t) = K(x,y,t) \\ {}^{ABR}_{0}D^{\alpha}_{t}y(t) = \beta t^{\beta-1}H(x,y,z,t) = L(x,y,z,t) \\ {}^{ABR}_{0}D^{\alpha}_{t}z(t) = \beta t^{\beta-1}G(y,t) = M(y,t) \end{cases}$$

We transform the above system to the following system as

$$x(t) - x(0) = \frac{1-\alpha}{ABC(\alpha)}K(x,y,t) + \frac{\alpha}{ABC(\alpha)\Gamma(\alpha)}\int_0^t K(x,y,\tau)(t-\tau)^{\alpha-1}d\tau$$

$$y(t) - y(0) = \frac{1-\alpha}{ABC(\alpha)}L(x,y,z,t) + \frac{\alpha}{ABC(\alpha)\Gamma(\alpha)}\int_0^t L(x,y,z,\tau)(t-\tau)^{\alpha-1}d\tau$$

$$z(t) - z(0) = \frac{1-\alpha}{ABC(\alpha)}M(y,t) + \frac{\alpha}{ABC(\alpha)\Gamma(\alpha)}\int_0^t M(y,\tau)(t-\tau)^{\alpha-1}d\tau$$

Then, we consider the system at $t = t_{n+1}$:

$$x(t_{n+1}) = x(0) + \frac{1-\alpha}{ABC(\alpha)}K(x,y,t_n)$$

$$+ \frac{\alpha}{ABC(\alpha)\Gamma(\alpha)}\int_0^{t_{n+1}} K(x,y,\tau)(t_{n+1}-\tau)^{\alpha-1}d\tau$$

$$y(t_{n+1}) = y(0) + \frac{1-\alpha}{ABC(\alpha)}L(x,y,z,t_n)$$

$$+ \frac{\alpha}{ABC(\alpha)\Gamma(\alpha)}\int_0^{t_{n+1}} L(x,y,z,\tau)(t_{n+1}-\tau)^{\alpha-1}d\tau$$

$$z(t_{n+1}) = z(0) + \frac{1-\alpha}{ABC(\alpha)}M(y,t_n)$$

$$+ \frac{\alpha}{ABC(\alpha)\Gamma(\alpha)}\int_0^{t_{n+1}} M(y,\tau)(t_{n+1}-\tau)^{\alpha-1}d\tau$$

Then, we apply the Atangana-Toufik method and obtain

$$x^{n+1} = x^0 + \frac{1-\alpha}{ABC(\alpha)}K(x^n,y^n,t_n)$$

$$+ \frac{\alpha}{ABC(\alpha)}\sum_{k=0}^n \left(\frac{(\Delta t)^{\alpha}}{\Gamma(\alpha+2)}K(x^k,y^k,t_k)\left((n+1-k)^{\alpha}(n-k+2+\alpha)\right)\right.$$

$$-(n-k)^{\alpha}(n-k+2+2\alpha))$$

$$\left. -\frac{(\Delta t)^{\alpha}}{\Gamma(\alpha+2)}K(x^{k-1},y^{k-1},t_{k-1})\left((n+1-k)^{\alpha+1}-(n-k)^{\alpha}(n-k+1+\alpha)\right)\right)$$

$$y^{n+1} = y^0 + \frac{1-\alpha}{ABC(\alpha)} L(x^n, y^n, z^n, t_n)$$

$$+ \frac{\alpha}{ABC(\alpha)} \sum_{k=0}^{n} \left(\frac{(\Delta t)^\alpha}{\Gamma(\alpha+2)} L(x^k, y^k, z^k, t_k) \left((n+1-k)^\alpha (n-k+2+\alpha)\right) \right.$$

$$-(n-k)^\alpha (n-k+2+2\alpha))$$

$$\left. - \frac{(\Delta t)^\alpha}{\Gamma(\alpha+2)} L(x^{k-1}, y^{k-1}, z^{k-1}, t_{k-1}) \left((n+1-k)^{\alpha+1} - (n-k)^\alpha (n-k+1+\alpha)\right) \right)$$

$$z^{n+1} = z^0 + \frac{1-\alpha}{ABC(\alpha)} M(y^n, t_n)$$

$$+ \frac{\alpha}{ABC(\alpha)} \sum_{k=0}^{n} \left(\frac{(\Delta t)^\alpha}{\Gamma(\alpha+2)} M(y^k, t_k) \left((n+1-k)^\alpha (n-k+2+\alpha)\right) \right.$$

$$-(n-k)^\alpha (n-k+2+2\alpha))$$

$$\left. - \frac{(\Delta t)^\alpha}{\Gamma(\alpha+2)} M(y^{k-1}, t_{k-1}) \left((n+1-k)^{\alpha+1} - (n-k)^\alpha (n-k+1+\alpha)\right) \right)$$

We consider the following system with the fractal fractional derivative using the exponential decay kernel:

$$\begin{cases} {}^{FFE}_{0}D_t^\alpha x(t) = a\,(y(t) - \phi(x(t))) = F(x,y,t) \\ {}^{FFE}_{0}D_t^\alpha y(t) = x(t) - y(t) + z(t) = H(x,y,z,t) \\ {}^{FFE}_{0}D_t^\alpha z(t) = -\lambda y(t) = G(y,t) \end{cases}$$

We can write the above system as

$$\begin{cases} {}^{CF}_{0}D_t^\alpha x(t) = \beta t^{\beta-1} F(x,y,t) = K(x,y,t) \\ {}^{CF}_{0}D_t^\alpha y(t) = \beta t^{\beta-1} H(x,y,z,t) = L(x,y,z,t) \\ {}^{CF}_{0}D_t^\alpha z(t) = \beta t^{\beta-1} G(y,t) = M(y,t) \end{cases}$$

We transform the above system to the following system as

$$x(t) - x(0) = \frac{1-\alpha}{M(\alpha)} K(x,y,t) + \frac{\alpha}{M(\alpha)} \int_0^t K(x,y,\tau) d\tau$$

$$y(t) - y(0) = \frac{1-\alpha}{M(\alpha)} L(x,y,z,t) + \frac{\alpha}{M(\alpha)} \int_0^t L(x,y,z,\tau) d\tau$$

$$z(t) - z(0) = \frac{1-\alpha}{M(\alpha)} M(y,t) + \frac{\alpha}{M(\alpha)} \int_0^t M(y,\tau) d\tau$$

Then, we consider the system at $t = t_{n+1}$ and $t = t_n$ as

$$x(t_{n+1}) = x(0) + \frac{1-\alpha}{M(\alpha)}K(x,y,t_n) + \frac{\alpha}{M(\alpha)}\int_0^{t_{n+1}} K(x,y,\tau)d\tau$$

$$y(t_{n+1}) = y(0) + \frac{1-\alpha}{M(\alpha)}L(x,y,z,t_n) + \frac{\alpha}{M(\alpha)}\int_0^{t_{n+1}} L(x,y,z,\tau)d\tau$$

$$z(t_{n+1}) = z(0) + \frac{1-\alpha}{M(\alpha)}M(y,t_n) + \frac{\alpha}{M(\alpha)}\int_0^{t_{n+1}} M(y,\tau)d\tau$$

$$x(t_n) = x(0) + \frac{1-\alpha}{M(\alpha)}K(x,y,t_{n-1}) + \frac{\alpha}{M(\alpha)}\int_0^{t_n} K(x,y,\tau)d\tau$$

$$y(t_n) = y(0) + \frac{1-\alpha}{M(\alpha)}L(x,y,z,t_{n-1}) + \frac{\alpha}{M(\alpha)}\int_0^{t_n} L(x,y,z,\tau)d\tau$$

$$z(t_n) = z(0) + \frac{1-\alpha}{M(\alpha)}M(y,t_{n-1}) + \frac{\alpha}{M(\alpha)}\int_0^{t_n} M(y,\tau)d\tau$$

Then, we obtain

$$
\begin{aligned}
x(t_{n+1}) &= x(t_n) + \frac{1-\alpha}{M(\alpha)}\left(K(x,y,t_n) - K(x,y,t_{n-1})\right)\\
&+ \frac{\alpha}{M(\alpha)}\left(\frac{3\Delta t}{2}K(x,y,t_n) - \frac{\Delta t}{2}K(x,y,t_{n-1})\right)\\
y(t_{n+1}) &= y(t_n) + \frac{1-\alpha}{M(\alpha)}\left(L(x,y,z,t_n) - L(x,y,z,t_{n-1})\right)\\
&+ \frac{\alpha}{M(\alpha)}\left(\frac{3\Delta t}{2}L(x,y,z,t_n) - \frac{\Delta t}{2}L(x,y,z,t_{n-1})\right)\\
z(t_{n+1}) &= z(t_n) + \frac{1-\alpha}{M(\alpha)}\left(M(y,t_n) - M(y,t_{n-1})\right)\\
&+ \frac{\alpha}{M(\alpha)}\left(\frac{3\Delta t}{2}M(y,t_n) - \frac{\Delta t}{2}M(y,t_{n-1})\right)
\end{aligned}
$$

Therefore, we reach

$$x(t_{n+1}) = x(t_n) + \frac{1-\alpha}{M(\alpha)}\left(\beta t_n^{\beta-1}a\left(y(t_n) - \phi(x(t_n))\right) - \beta t_{n-1}^{\beta-1}a\left(y(t_{n-1}) - \phi(x(t_{n-1}))\right)\right)$$
$$+ \frac{\alpha}{M(\alpha)}\left(\frac{3\Delta t}{2}\beta t_n^{\beta-1}a\left(y(t_n) - \phi(x(t_n))\right) - \frac{\Delta t}{2}\beta t_{n-1}^{\beta-1}a\left(y(t_{n-1}) - \phi(x(t_{n-1}))\right)\right)$$

$$y(t_{n+1}) = y(t_n) + \frac{1-\alpha}{M(\alpha)} \left(\beta t_n^{\beta-1} \left(x(t_n) - y(t_n) + z(t_n) \right) \right.$$

$$\left. - \beta t_{n-1}^{\beta-1} \left(x(t_{n-1}) - y(t_{n-1}) + z(t_{n-1}) \right) \right)$$

$$+ \frac{\alpha}{M(\alpha)} \left(\frac{3\Delta t}{2} \beta t_n^{\beta-1} \left(x(t_n) - y(t_n) + z(t_n) \right) - \frac{\Delta t}{2} \beta t_{n-1}^{\beta-1} \left(x(t_{n-1}) - y(t_{n-1}) + z(t_{n-1}) \right) \right)$$

$$z(t_{n+1}) = z(t_n) + \frac{1-\alpha}{M(\alpha)} \left(-\beta t_n^{\beta-1} \lambda y(t_n) + \beta t_{n-1}^{\beta-1} \lambda y(t_{n-1}) \right)$$

$$+ \frac{\alpha}{M(\alpha)} \left(-\frac{3\Delta t}{2} \beta t_n^{\beta-1} \lambda y(t_n) + \frac{\Delta t}{2} \beta t_{n-1}^{\beta-1} \lambda y(t_{n-1}) \right)$$

19 Applications of the Circuit Problems

In this chapter different circuit problems with second order are considered. These classes of differential equations have been used in electrical engineering to model different types of dynamics in system circuits. Some of these equations are integral equations that are also been found very important to provide more value to circuit systems [75]. In this chapter we do not aim to obtain exact solutions for these problems, but rather provide different analyses. These models will be modified by replacing the classical derivative with fractional derivatives, and then where classical integrals are used, they will be replaced by fractional integrals. These new classes are very rich in terms of including into the mathematical model's nonlocal behaviors. An investigation underpinning the derivation of the conditions under which their exact and unique solutions exist. Due to their complexities, numerical methods will be employed to derive their solutions. In this chapter, therefore, seven problems will be considered starting with the one below.

19.1 FIRST PROBLEM

In this subsection, we take into consideration:

$$\frac{d^2 I(t)}{dt^2} = -\left(\frac{R}{L} \frac{dI(t)}{dt} + \frac{1}{LC} I(t) \right) \tag{19.1}$$

For simplicity, we define

$$A(t, I(t)) = -\left(\frac{R}{L} \frac{dI(t)}{dt} + \frac{1}{LC} I(t) \right) \tag{19.2}$$

Then, we have

$$\frac{dI'(t)}{dt} = A(t, I(t)) \tag{19.3}$$

After taking integral, we reach

$$I'(t) = I'(0) + \int_0^t A(\tau, I(\tau)) d\tau \tag{19.4}$$

Then, we obtain

$$I(t) = I(0) + \int_0^t \left(I'(0) + \int_0^\tau A(l, I(l)) dl \right) d\tau \tag{19.5}$$

$$I(t) = I(0) + I'(0)t + \int_0^t \int_0^\tau A(l, I(l)) dl d\tau \tag{19.6}$$

DOI: 10.1201/9781003359869-19

$$I(t) = I(0) + I'(0)t + \int_0^t (t - \tau)A(\tau, I(\tau))d\tau \tag{19.7}$$

Then at $t = \Delta t(n+1) = t_{n+1}$, we have

$$\begin{aligned} I(t_{n+1}) &= I(0) + I'(0)t_{n+1} + \int_0^{t_{n+1}} (t_{n+1} - \tau)A(\tau, I(\tau))d\tau \\ &= I(0) + I'(0)t_{n+1} + \sum_{j=0}^{n} \int_{t_j}^{t_{j+1}} (t_{n+1} - \tau)A(\tau, I(\tau))d\tau \end{aligned}$$

Within $[t_j, t_{j+1}]$, we can approximate

$$A(\tau, I(\tau)) \approx P_j(\tau) = A(t_j, I_j)\frac{\tau - t_{j-1}}{t_j - t_{j-1}} - A(t_{j-1}, I_{j-1})\frac{\tau - t_j}{t_j - t_{j-1}} \tag{19.8}$$

Replacing $A(\tau, I(\tau)) \approx P_j(\tau)$, we get

$$\begin{aligned} I(t_{n+1}) &= I(0) + I'(0)t_{n+1} + \sum_{j=0}^{n} \int_{t_j}^{t_{j+1}} (t_{n+1} - \tau) \\ &\quad \times \left(\frac{A(t_j, I_j)}{\Delta t}(\tau - t_{j-1}) - \frac{A(t_{j-1}, I_{j-1})}{\Delta t}(\tau - t_j) \right) d\tau \\ &= I(0) + I'(0)t_{n+1} + \sum_{j=0}^{n} \frac{A(t_j, I_j)}{\Delta t} \int_{t_j}^{t_{j+1}} (t_{n+1} - \tau)(\tau - t_{j-1})d\tau \\ &\quad - \sum_{j=0}^{n} \frac{A(t_{j-1}, I_{j-1})}{\Delta t} \int_{t_j}^{t_{j+1}} (t_{n+1} - \tau)(\tau - t_j)d\tau \\ &= I(0) + I'(0)t_{n+1} + \frac{(\Delta t)^2}{6} \sum_{j=0}^{n} (4 - 9j + 9n)A(t_j, I_j) \\ &\quad - \frac{(\Delta t)^2}{6} \sum_{j=0}^{n} (1 - 3j + 3n)A(t_{j-1}, I_{j-1}) \end{aligned}$$

19.2 SECOND PROBLEM

We consider the following problem:

$$L\frac{dI(t)}{dt} + RI(t) + \frac{1}{C} \int_0^t I(\tau)d\tau = v(t) \tag{19.9}$$

At t_{n+1}, we obtain

$$L\frac{I(t_{n+1}) - I(t_n)}{\Delta t} + RI(t_{n+1}) + \frac{1}{C} \int_0^{t_{n+1}} I(\tau)d\tau = v(t_{n+1}) \tag{19.10}$$

Then, we reach

$$L\frac{I^{n+1} - I^n}{\Delta t} + RI^{n+1} + \frac{1}{C} \sum_{j=0}^{n} \int_{t_j}^{t_{j+1}} \frac{I(t_{j+1}) + I(t_j)}{2}d\tau = v(t_{n+1}) \tag{19.11}$$

$$L\frac{I^{n+1}-I^n}{\Delta t}+RI^{n+1}+\frac{1}{C}\sum_{j=0}^{n}\frac{I(t_{j+1})+I(t_j)}{2}\Delta t=v(t_{n+1}) \tag{19.12}$$

19.3 THIRD PROBLEM

We consider the following problem:

$$L_0^C D_t^\alpha I(t)+RI(t)+\frac{1}{C_0}{}^C I_t^\alpha I(t)=v(t) \tag{19.13}$$

We can write the above equation as

$$L\frac{1}{\Gamma(1-\alpha)}\int_0^t I'(\tau)(t-\tau)^{-\alpha}d\tau+RI(t)+\frac{1}{C}\frac{1}{\Gamma(\alpha)}\int_0^t I(\tau)(t-\tau)^{\alpha-1}d\tau=v(t) \tag{19.14}$$

At t_{n+1}, we obtain

$$L\frac{1}{\Gamma(1-\alpha)}\int_0^{t_{n+1}}I'(\tau)(t_{n+1}-\tau)^{-\alpha}d\tau+RI(t_{n+1})$$
$$+\frac{1}{C}\frac{1}{\Gamma(\alpha)}\int_0^{t_{n+1}}I(\tau)(t_{n+1}-\tau)^{\alpha-1}d\tau=v(t_{n+1})$$

Then, we reach

$$L\frac{1}{\Gamma(1-\alpha)}\sum_{j=0}^{n}\int_{t_j}^{t_{j+1}}\frac{I(t_{j+1})-I(t_j)}{\Delta t}(t_{n+1}-\tau)^{-\alpha}d\tau+RI(t_{n+1})$$
$$+\frac{1}{C}\frac{1}{\Gamma(\alpha)}\sum_{j=0}^{n}\int_{t_j}^{t_{j+1}}\frac{I(t_{j+1})+I(t_j)}{2}(t_{n+1}-\tau)^{\alpha-1}d\tau=v(t_{n+1})$$

After simplification, we obtain

$$L\frac{(\Delta t)^{-\alpha}}{\Gamma(2-\alpha)}\sum_{j=0}^{n}\left(I(t_{j+1})-I(t_j)\right)\left[(n+1-j)^{1-\alpha}-(n-j)^{1-\alpha}\right]+RI(t_{n+1})$$
$$+\frac{1}{C}\frac{(\Delta t)^\alpha}{\Gamma(\alpha+1)}\sum_{j=0}^{n}\frac{I(t_{j+1})+I(t_j)}{2}\left[(n+1-j)^\alpha-(n-j)^\alpha\right]=v(t_{n+1})$$

19.4 FOURTH PROBLEM

We consider the following problem:

$$L_0^{CF}D_t^\alpha I(t)+RI^2(t)+\frac{1}{C_0}{}^{CF}I_t^\alpha I(t)=v(t) \tag{19.15}$$

We can rewrite the above equation as

$$L\frac{M(\alpha)}{1-\alpha}\int_0^t I'(\tau)\exp\left(-\frac{\alpha}{1-\alpha}(t-\tau)\right)d\tau+RI^2(t)$$
$$+\frac{1}{C}\frac{1-\alpha}{M(\alpha)}I(t)+\frac{1}{C}\frac{\alpha}{M(\alpha)}\int_0^t I(\tau)d\tau=v(t)$$

At t_{n+1}, we obtain

$$L\frac{M(\alpha)}{1-\alpha}\int_0^{t_{n+1}} I'(\tau)\exp\left(-\frac{\alpha}{1-\alpha}(t_{n+1}-\tau)\right)d\tau + RI^2(t_{n+1})$$

$$+\frac{1}{C}\frac{1-\alpha}{M(\alpha)}I(t_{n+1}) + \frac{1}{C}\frac{\alpha}{M(\alpha)}\int_0^{t_{n+1}} I(\tau)d\tau = v(t_{n+1})$$

Then, we reach

$$L\frac{M(\alpha)}{1-\alpha}\sum_{j=0}^n \int_{t_j}^{t_{j+1}} \frac{I(t_{j+1})-I(t_j)}{\Delta t}\exp\left(-\frac{\alpha}{1-\alpha}(t_{n+1}-\tau)\right)d\tau + RI^2(t_{n+1})$$

$$+\frac{1}{C}\frac{1-\alpha}{M(\alpha)}I(t_{n+1}) + \frac{1}{C}\frac{\alpha}{M(\alpha)}\sum_{j=0}^n \int_{t_j}^{t_{j+1}} \frac{I(t_{j+1})+I(t_j)}{2}d\tau = v(t_{n+1})$$

After simplification, we get

$$L\frac{M(\alpha)}{\alpha}\sum_{j=0}^n \frac{I(t_{j+1})-I(t_j)}{\Delta t}\left[\exp\left(-\frac{\alpha}{1-\alpha}(t_{n+1}-t_{j+1})\right)-\exp\left(-\frac{\alpha}{1-\alpha}(t_{n+1}-t_j)\right)\right]$$

$$+RI^2(t_{n+1}) + \frac{1}{C}\frac{1-\alpha}{M(\alpha)}I(t_{n+1}) + \frac{1}{C}\frac{\alpha}{M(\alpha)}\sum_{j0}^n \frac{I(t_{j+1})+I(t_j)}{2}(t_{j+1}-t_j) = v(t_{n+1})$$

$$L\frac{M(\alpha)}{\alpha}\sum_{j=0}^n \frac{I(t_{j+1})-I(t_j)}{\Delta t}\left[\exp\left(-\frac{\alpha}{1-\alpha}\Delta t(n-j)\right)\right.$$

$$\left.-\exp\left(-\frac{\alpha}{1-\alpha}\Delta t(n+1-j)\right)\right] + RI^2(t_{n+1}) + \frac{1}{C}\frac{1-\alpha}{M(\alpha)}I(t_{n+1})$$

$$+\frac{1}{C}\frac{\alpha}{M(\alpha)}\sum_{j=0}^n \frac{I(t_{j+1})+I(t_j)}{2}\Delta t = v(t_{n+1})$$

19.5 FIFTH PROBLEM

We consider the following problem:

$$L_0^{ABC}D_t^\alpha I(t) + RI^2(t) + \frac{1}{C_0}^{AB}I_t^\alpha I(t) = v(t) \tag{19.16}$$

We can rewrite the above equation as

$$L\frac{AB(\alpha)}{1-\alpha}\int_0^t I'(\tau)E_\alpha\left(-\frac{\alpha}{1-\alpha}(t-\tau)^\alpha\right)d\tau + RI^2(t)$$

$$+\frac{1}{C}\frac{1-\alpha}{AB(\alpha)}I(t) + \frac{1}{C}\frac{\alpha}{\Gamma(\alpha)AB(\alpha)}\int_0^t I(\tau)(t-\tau)^{\alpha-1}d\tau = v(t)$$

At t_{n+1}, we obtain

$$L\frac{AB(\alpha)}{1-\alpha}\int_0^{t_{n+1}}I'(\tau)E_\alpha\left(-\frac{\alpha}{1-\alpha}(t_{n+1}-\tau)^\alpha\right)d\tau+RI^2(t_{n+1})$$

$$+\frac{1}{C}\frac{1-\alpha}{AB(\alpha)}I(t_{n+1})+\frac{1}{C}\frac{\alpha}{\Gamma(\alpha)AB(\alpha)}\int_0^{t_{n+1}}I(\tau)(t_{n+1}-\tau)^{\alpha-1}d\tau=v(t_{n+1})$$

Then, we reach

$$L\frac{AB(\alpha)}{1-\alpha}\sum_{j=0}^n\int_{t_j}^{t_{j+1}}\frac{I(t_{j+1})-I(t_j)}{\Delta t}E_\alpha\left(-\frac{\alpha}{1-\alpha}(t_{n+1}-\tau)^\alpha\right)d\tau+RI^2(t_{n+1})$$

$$+\frac{1}{C}\frac{1-\alpha}{AB(\alpha)}I(t_{n+1})+\frac{1}{C}\frac{\alpha}{\Gamma(\alpha)AB(\alpha)}\sum_{j=0}^n\int_{t_j}^{t_{j+1}}\frac{I(t_{j+1})+I(t_j)}{2}(t_{n+1}-\tau)^{\alpha-1}d\tau$$

$$=v(t_{n+1})$$

$$L\frac{AB(\alpha)}{1-\alpha}\sum_{j=0}^n\int_{t_j}^{t_{j+1}}\frac{I(t_{j+1})-I(t_j)}{\Delta t}\sum_{k=0}^\infty\frac{\left(-\frac{\alpha}{1-\alpha}(t_{n+1}-\tau)^\alpha\right)^k}{\Gamma(\alpha k+1)}d\tau+RI^2(t_{n+1})$$

$$+\frac{1}{C}\frac{1-\alpha}{AB(\alpha)}I(t_{n+1})+\frac{1}{C}\frac{\alpha}{\Gamma(\alpha)AB(\alpha)}\sum_{j=0}^n\frac{I(t_{j+1})+I(t_j)}{2}\int_{t_j}^{t_{j+1}}(t_{n+1}-\tau)^{\alpha-1}d\tau$$

$$=v(t_{n+1})$$

$$L\frac{AB(\alpha)}{1-\alpha}\sum_{j=0}^n\int_{t_j}^{t_{j+1}}\frac{I(t_{j+1})-I(t_j)}{\Delta t}\sum_{k=0}^\infty\frac{\left(-\frac{\alpha}{1-\alpha}\right)^k}{\Gamma(\alpha k+1)}\int_{t_j}^{t_{j+1}}(t_{n+1}-\tau)^{\alpha k}d\tau+RI^2(t_{n+1})$$

$$+\frac{1}{C}\frac{1-\alpha}{AB(\alpha)}I(t_{n+1})+\frac{1}{C}\frac{(\Delta t)^\alpha}{\Gamma(\alpha)AB(\alpha)}\sum_{j=0}^n\frac{I(t_{j+1})+I(t_j)}{2}[(n+1-j)^\alpha-(n-j)^\alpha]$$

$$=v(t_{n+1})$$

$$L\frac{AB(\alpha)}{1-\alpha}\sum_{j=0}^n(I(t_{j+1})-I(t_j))\sum_{k=0}^\infty\frac{\left(-\frac{\alpha}{1-\alpha}\Delta t\right)^k}{\Gamma(\alpha k+2)}\left[(n+1-j)^{\alpha k+1}-(n-j)^{\alpha k+1}\right]$$

$$+\frac{1}{C}\frac{1-\alpha}{AB(\alpha)}I(t_{n+1})+\frac{1}{C}\frac{(\Delta t)^\alpha}{\Gamma(\alpha)AB(\alpha)}\sum_{j=0}^n\frac{I(t_{j+1})+I(t_j)}{2}[(n+1-j)^\alpha-(n-j)^\alpha]$$

$$+RI^2(t_{n+1})=v(t_{n+1})$$

Then, we obtain

$$L\frac{AB(\alpha)}{1-\alpha}\sum_{j=0}^{n}(I(t_{j+1})-I(t_j))(n+1-j)E_{\alpha,2}\left(-\frac{\alpha}{1-\alpha}(n+1-j)\Delta t\right)$$

$$-L\frac{AB(\alpha)}{1-\alpha}\sum_{j=0}^{n}(I(t_{j+1})-I(t_j))E_{\alpha,2}\left(-\frac{\alpha}{1-\alpha}(n-j)\Delta t\right)$$

$$+\frac{1}{C}\frac{1-\alpha}{AB(\alpha)}I(t_{n+1})+\frac{1}{C}\frac{(\Delta t)^{\alpha}}{\Gamma(\alpha)AB(\alpha)}\sum_{j=0}^{n}\frac{I(t_{j+1})+I(t_j)}{2}[(n+1-j)^{\alpha}-(n-j)^{\alpha}]$$

$$+RI^2(t_{n+1})=v(t_{n+1})$$

19.6   SIXTH PROBLEM

We consider the following problem

$$L_0^{ABC}D_t^{\alpha}I(t)+RI^2(t)+\frac{1}{C_0}{}^{CF}I_t^{\alpha}I(t)=v(t) \tag{19.17}$$

We can rewrite the above equation as

$$L\frac{AB(\alpha)}{1-\alpha}\int_0^t I'(\tau)E_{\alpha}\left(-\frac{\alpha}{1-\alpha}(t-\tau)^{\alpha}\right)d\tau+RI^2(t)$$

$$+\frac{1}{C}\frac{1-\alpha}{M(\alpha)}I(t)+\frac{1}{C}\frac{\alpha}{M(\alpha)}\int_0^t I(\tau)d\tau=v(t)$$

At t_{n+1}, we obtain

$$L\frac{AB(\alpha)}{1-\alpha}\int_0^{t_{n+1}} I'(\tau)E_{\alpha}\left(-\frac{\alpha}{1-\alpha}(t_{n+1}-\tau)^{\alpha}\right)d\tau+RI^2(t_{n+1})$$

$$+\frac{1}{C}\frac{1-\alpha}{M(\alpha)}I(t_{n+1})+\frac{1}{C}\frac{\alpha}{M(\alpha)}\int_0^{t_{n+1}} I(\tau)d\tau=v(t_{n+1})$$

Then, we reach

$$L\frac{AB(\alpha)}{1-\alpha}\sum_{j=0}^{n}\int_{t_j}^{t_{j+1}}\frac{I(t_{j+1})-I(t_j)}{\Delta t}E_{\alpha}\left(-\frac{\alpha}{1-\alpha}(t_{n+1}-\tau)^{\alpha}\right)d\tau$$

$$+\frac{1}{C}\frac{1-\alpha}{M(\alpha)}I(t_{n+1})+\frac{1}{C}\frac{\alpha}{M(\alpha)}\sum_{j=0}^{n}\int_{t_j}^{t_{j+1}}\frac{I(t_{j+1})+I(t_j)}{2}d\tau$$

$$+RI^2(t_{n+1})=v(t_{n+1})$$

$$L\frac{AB(\alpha)}{1-\alpha}\sum_{j=0}^{n}\int_{t_j}^{t_{j+1}}\frac{I(t_{j+1})-I(t_j)}{\Delta t}\sum_{k=0}^{\infty}\frac{\left(-\frac{\alpha}{1-\alpha}(t_{n+1}-\tau)^{\alpha}\right)^k}{\Gamma(\alpha k+1)}d\tau$$

$$+\frac{1}{C}\frac{1-\alpha}{M(\alpha)}I(t_{n+1})+\frac{1}{C}\frac{\alpha}{M(\alpha)}\sum_{j=0}^{n}\int_{t_j}^{t_{j+1}}\frac{I(t_{j+1})+I(t_j)}{2}d\tau$$

$$+RI^2(t_{n+1})=v(t_{n+1})$$

$$L\frac{AB(\alpha)}{1-\alpha}\sum_{j=0}^{n}\int_{t_j}^{t_{j+1}}\frac{I(t_{j+1})-I(t_j)}{\Delta t}\sum_{k=0}^{\infty}\frac{\left(-\frac{\alpha}{1-\alpha}\right)^k}{\Gamma(\alpha k+1)}\int_{t_j}^{t_{j+1}}(t_{n+1}-\tau)^{\alpha k}d\tau$$

$$+\frac{1}{C}\frac{1-\alpha}{M(\alpha)}I(t_{n+1})+\frac{1}{C}\frac{\alpha}{M(\alpha)}\sum_{j=0}^{n}\int_{t_j}^{t_{j+1}}\frac{I(t_{j+1})+I(t_j)}{2}d\tau$$

$$+RI^2(t_{n+1})=v(t_{n+1})$$

$$L\frac{AB(\alpha)}{1-\alpha}\sum_{j=0}^{n}(I(t_{j+1})-I(t_j))\sum_{k=0}^{\infty}\frac{\left(-\frac{\alpha}{1-\alpha}\Delta t\right)^k}{\Gamma(\alpha k+2)}\left[(n+1-j)^{\alpha k+1}-(n-j)^{\alpha k+1}\right]$$

$$+\frac{1}{C}\frac{1-\alpha}{M(\alpha)}I(t_{n+1})+\frac{1}{C}\frac{\alpha}{M(\alpha)}\sum_{j=0}^{n}\int_{t_j}^{t_{j+1}}\frac{I(t_{j+1})+I(t_j)}{2}d\tau$$

$$+RI^2(t_{n+1})=v(t_{n+1})$$

Then, we obtain

$$L\frac{AB(\alpha)}{1-\alpha}\sum_{j=0}^{n}(I(t_{j+1})-I(t_j))(n+1-j)E_{\alpha,2}\left(-\frac{\alpha}{1-\alpha}(n+1-j)\Delta t\right)$$

$$-L\frac{AB(\alpha)}{1-\alpha}\sum_{j=0}^{n}(I(t_{j+1})-I(t_j))E_{\alpha,2}\left(-\frac{\alpha}{1-\alpha}(n-j)\Delta t\right)$$

$$+\frac{1}{C}\frac{1-\alpha}{M(\alpha)}I(t_{n+1})+\frac{1}{C}\frac{\alpha}{M(\alpha)}\sum_{j=0}^{n}\frac{I(t_{j+1})+I(t_j)}{2}\Delta t$$

$$+RI^2(t_{n+1})=v(t_{n+1})$$

19.7 SEVENTH PROBLEM

We consider the following problem:

$$L_0^{ABC}D_t^{\alpha}I(t)+RI^2(t)+\frac{1}{C_0}{}_0^C I_t^{\alpha}I(t)=v(t) \tag{19.18}$$

We can rewrite the above equation as

$$L\frac{AB(\alpha)}{1-\alpha}\int_0^t I'(\tau)E_{\alpha}\left(-\frac{\alpha}{1-\alpha}(t-\tau)^{\alpha}\right)d\tau+RI^2(t)$$

$$+\frac{1}{C}\frac{1}{\Gamma(\alpha)}\int_0^t I(\tau)(t-\tau)^{\alpha-1}d\tau=v(t)$$

At t_{n+1}, we obtain

$$L\frac{AB(\alpha)}{1-\alpha}\int_0^{t_{n+1}} I'(\tau)E_{\alpha}\left(-\frac{\alpha}{1-\alpha}(t_{n+1}-\tau)^{\alpha}\right)d\tau+RI^2(t_{n+1})$$

$$+\frac{1}{C}\frac{1}{\Gamma(\alpha)}\int_0^{t_{n+1}} I(\tau)(t_{n+1}-\tau)^{\alpha-1}d\tau=v(t_{n+1})$$

Then, we reach

$$L\frac{AB(\alpha)}{1-\alpha}\sum_{j=0}^{n}\int_{t_j}^{t_{j+1}}\frac{I(t_{j+1})-I(t_j)}{\Delta t}E_\alpha\left(-\frac{\alpha}{1-\alpha}(t_{n+1}-\tau)^\alpha\right)d\tau$$
$$+\frac{1}{C}\frac{1}{\Gamma(\alpha)}\sum_{j=0}^{n}\int_{t_j}^{t_{j+1}}\frac{I(t_{j+1})+I(t_j)}{2}(t_{n+1}-\tau)^{\alpha-1}d\tau$$
$$+RI^2(t_{n+1})=v(t_{n+1})$$

$$L\frac{AB(\alpha)}{1-\alpha}\sum_{j=0}^{n}\int_{t_j}^{t_{j+1}}\frac{I(t_{j+1})-I(t_j)}{\Delta t}\sum_{k=0}^{\infty}\frac{\left(-\frac{\alpha}{1-\alpha}(t_{n+1}-\tau)^\alpha\right)^k}{\Gamma(\alpha k+1)}d\tau$$
$$+\frac{1}{C}\frac{1}{\Gamma(\alpha)}\sum_{j=0}^{n}\int_{t_j}^{t_{j+1}}\frac{I(t_{j+1})+I(t_j)}{2}(t_{n+1}-\tau)^{\alpha-1}d\tau$$
$$+RI^2(t_{n+1})=v(t_{n+1})$$

$$L\frac{AB(\alpha)}{1-\alpha}\sum_{j=0}^{n}\int_{t_j}^{t_{j+1}}\frac{I(t_{j+1})-I(t_j)}{\Delta t}\sum_{k=0}^{\infty}\frac{\left(-\frac{\alpha}{1-\alpha}\right)^k}{\Gamma(\alpha k+1)}\int_{t_j}^{t_{j+1}}(t_{n+1}-\tau)^{\alpha k}d\tau$$
$$+\frac{1}{C}\frac{(\Delta t)^\alpha}{\Gamma(\alpha+1)}\sum_{j=0}^{n}\frac{I(t_{j+1})+I(t_j)}{2}[(n+1-j)^\alpha-(n-j)^\alpha]$$
$$+RI^2(t_{n+1})=v(t_{n+1})$$

$$L\frac{AB(\alpha)}{1-\alpha}\sum_{j=0}^{n}(I(t_{j+1})-I(t_j))\sum_{k=0}^{\infty}\frac{\left(-\frac{\alpha}{1-\alpha}\Delta t\right)^k}{\Gamma(\alpha k+2)}\left[(n+1-j)^{\alpha k+1}-(n-j)^{\alpha k+1}\right]$$
$$+\frac{1}{C}\frac{(\Delta t)^\alpha}{\Gamma(\alpha+1)}\sum_{j=0}^{n}\frac{I(t_{j+1})+I(t_j)}{2}[(n+1-j)^\alpha-(n-j)^\alpha]$$
$$+RI^2(t_{n+1})=v(t_{n+1})$$

Then, we obtain

$$L\frac{AB(\alpha)}{1-\alpha}\sum_{j=0}^{n}(I(t_{j+1})-I(t_j))(n+1-j)E_{\alpha,2}\left(-\frac{\alpha}{1-\alpha}(n+1-j)\Delta t\right)$$
$$-L\frac{AB(\alpha)}{1-\alpha}\sum_{j=0}^{n}(I(t_{j+1})-I(t_j))E_{\alpha,2}\left(-\frac{\alpha}{1-\alpha}(n-j)\Delta t\right)$$
$$+\frac{1}{C}\frac{(\Delta t)^\alpha}{\Gamma(\alpha+1)}\sum_{j=0}^{n}\frac{I(t_{j+1})+I(t_j)}{2}[(n+1-j)^\alpha-(n-j)^\alpha]$$
$$+RI^2(t_{n+1})=v(t_{n+1})$$

20 Existence and Uniqueness of the Solution

While it is ideal to obtain exact solutions to differential equations as these solutions are used to depict some real-world behaviors, it is also important to note that due to the complexities of these models, existing analytical methods may have some limitations thus, they may not be able to be used for this purpose. In the last years several analytical methods were introduced aiming to provide exact solutions to several nonlinear equations. However, these methods lead to multiple solutions for one nonlinear equation, which in some instances violate the theory of uniqueness. Thus to avoid this situation, several powerful numerical schemes have been proposed in the last decades to deal with nonlinear equations with classical and fractional derivatives. However, for this process to be performed, it is usually ideal to provide at least conditions under which these equations admit unique exact solutions. In this chapter the seven differential and integral equations are considered and will be subjected to the discussion underpinning the existence and uniqueness of their solutions presented in detail. This is achieved using the Linear growth and the Lipchitz approach [76–80].

20.1 FIRST PROBLEM

In this subsection, we take into consideration the following circuit problem:

$$\frac{d^2 I(t)}{dt^2} = -\left(\frac{R}{L}\frac{dI(t)}{dt} + \frac{1}{LC}I(t)\right) \tag{20.1}$$

We integrate the above equation from 0 to t. Then, we get

$$\frac{dI(t)}{dt} = -\int_0^t \left(\frac{R}{L}\frac{dI(\tau)}{d\tau} + \frac{1}{LC}I(\tau)\right)d\tau + \frac{dI(t)}{dt}\bigg|_{t=0} \tag{20.2}$$

We define

$$F(t,I) = -\int_0^t \left(\frac{R}{L}\frac{dI(\tau)}{d\tau} + \frac{1}{LC}I(\tau)\right)d\tau + \frac{dI(t)}{dt}\bigg|_{t=0}. \tag{20.3}$$

To prove the existence and uniqueness of the solution, we verify the following:

(i) $|F(t,I)|^2 < (1+|I|^2)$ (linear growth)

(ii) $\forall I, I_1, \quad \|F(t,I) - F(t,I_1)\|_\infty^2 < k\|I - I_1\|_\infty^2$

Start with the linear growth:

$$|F(t,I)|^2 = \left| \frac{dI(t)}{dt} \right|_{t=0} - \int_0^t \left(\frac{R}{L} \frac{dI(\tau)}{d\tau} + \frac{1}{LC} I(\tau) \right) d\tau \Big|^2$$

$$\leq 2 \left| \frac{dI(t)}{dt} \right|_{t=0}^2 + 2 \left| \int_0^t \left(\frac{R}{L} \frac{dI(\tau)}{d\tau} + \frac{1}{LC} I(\tau) \right) d\tau \right|^2$$

$$\leq 2 \left| \frac{dI(t)}{dt} \right|_{t=0}^2 + 2 \left| \frac{R}{L} I(t) - \frac{R}{L} I(0) + \frac{1}{LC} \int_0^t I(\tau) d\tau \right|^2$$

$$\leq 2 \left| \frac{dI(t)}{dt} \right|_{t=0}^2 + 6 \left(\frac{R}{L} \right)^2 |I(t)|^2 + 6 \left(\frac{R}{L} \right)^2 |I(0)|^2 + 6 \left| \frac{1}{LC} \int_0^t I(\tau) d\tau \right|^2$$

$$\leq 2 \left| \frac{dI(t)}{dt} \right|_{t=0}^2 + 6 \left(\frac{R}{L} \right)^2 |I(t)|^2 + 6 \left(\frac{R}{L} \right)^2 |I(0)|^2$$

$$+ 6 \frac{1}{(LC)^2} \left| \int_0^t I(\tau) d\tau \right|^2$$

$$\leq 2 \left| \frac{dI(t)}{dt} \right|_{t=0}^2 + 6 \left(\frac{R}{L} \right)^2 |I(t)|^2 + 6 \left(\frac{R}{L} \right)^2 |I(0)|^2$$

$$+ 6 \frac{1}{(LC)^2} T \sup_{0 \leq t < T} |I(t)|^2$$

$$\leq 2 \left| \frac{dI(t)}{dt} \right|_{t=0}^2 + 6 \left(\frac{R}{L} \right)^2 |I(t)|^2 + 6 \left(\frac{R}{L} \right)^2 |I(0)|^2 + 6 \frac{1}{(LC)^2} T \| I(t) \|_\infty^2$$

$$\leq \left(2 \left| \frac{dI(t)}{dt} \right|_{t=0}^2 + 6 \left(\frac{R}{L} \right)^2 |I(0)|^2 + 6 \frac{1}{(LC)^2} T \| I(t) \|_\infty^2 \right)$$

$$\times \left(1 + \frac{6 \left(\frac{R}{L} \right)^2 |I(t)|^2}{2 \left| \frac{dI(t)}{dt} \right|_{t=0}^2 + 6 \left(\frac{R}{L} \right)^2 |I(0)|^2 + 6 \frac{1}{(LC)^2} T \| I(t) \|_\infty^2} \right)$$

If

$$\frac{6 \left(\frac{R}{L} \right)^2}{2 \left| \frac{dI(t)}{dt} \right|_{t=0}^2 + 6 \left(\frac{R}{L} \right)^2 |I(0)|^2 + 6 \frac{1}{(LC)^2} T \| I(t) \|_\infty^2} < 1$$

then, we get

$$|F(t,I)|^2 \leq k \left(1 + |I|^2 \right) \tag{20.4}$$

where

$$k = 2 \left| \frac{dI(t)}{dt} \right|_{t=0}^2 + 6 \left(\frac{R}{L} \right)^2 |I(0)|^2 + 6 \frac{1}{(LC)^2} T \| I(t) \|_\infty^2, \tag{20.5}$$

Therefore, the function satisfies the growth condition.

Now, we will verify the Lipschitz condition:

$$
\begin{aligned}
|F(t,I) - F(t,I_1)|^2 &= \left| \frac{dI(t)}{dt} \right|_{t=0} - \frac{dI_1(t)}{dt} \bigg|_{t=0} \\
&\quad - \int_0^t \left(\frac{R}{L} \left(\frac{dI(\tau)}{d\tau} - \frac{dI_1(\tau)}{d\tau} \right) + \frac{1}{LC}(I(\tau) - I_1(\tau)) \right) d\tau \bigg|^2 \\
&\leq 2 \left| \frac{dI(t)}{dt} \right|_{t=0} - \frac{dI_1(t)}{dt} \bigg|_{t=0} \bigg|^2 + 6 \left(\frac{R}{L} \right)^2 |I(t) - I_1(t)|^2 \\
&\quad + 6 \left(\frac{R}{L} \right)^2 |I(0) - I_1(0)|^2 + 6 \frac{1}{(LC)^2} \int_0^t |I(\tau) - I_1(\tau)|^2 d\tau
\end{aligned}
$$

We have that the initial condition is the same for both. Thus, we have $I(0) = I_1(0)$ and $I'(0) = I_1'(0)$. Then, we get

$$
|F(t,I) - F(t,I_1)|^2 \leq 6 \left(\frac{R}{L} \right)^2 |I(t) - I_1(t)|^2 + 6 \frac{1}{(LC)^2} \int_0^t |I(\tau) - I_1(\tau)|^2 d\tau
$$

$$
\sup_{t \in [0,T]} |F(t,I) - F(t,I_1)|^2 \leq 6 \left(\frac{R}{L} \right)^2 \sup_{t \in [0,T]} |I(t) - I_1(t)|^2
$$

$$
+ 6 \frac{1}{(LC)^2} T \sup_{t \in [0,T]} |I(t) - I_1(t)|^2
$$

$$
\|F(t,I) - F(t,I_1)\|_\infty^2 \leq \left(6 \left(\frac{R}{L} \right)^2 + 6 \frac{1}{(LC)^2} T \right) \|I(t) - I_1(t)\|_\infty^2
$$

Thus, we get

$$
\|F(t,I) - F(t,I_1)\|_\infty^2 \leq \bar{k} \|I(t) - I_1(t)\|_\infty^2
$$

where

$$
\bar{k} = 6 \left(\left(\frac{R}{L} \right)^2 + \frac{1}{(LC)^2} T \right).
$$

This shows that the function F satisfies the Lipschitz condition. Owing the fact that

$$
\frac{6 \left(\frac{R}{L} \right)^2}{2 \left| \frac{dI(t)}{dt} \right|_{t=0}^2 + 6 \left(\frac{R}{L} \right)^2 |I(0)|^2 + 6 \frac{1}{(LC)^2} T \|I(t)\|_\infty^2} < 1.
$$

then the equation admits a unique solution.

20.2   SECOND PROBLEM

In this subsection, we consider the following circuit problem:

$$L\frac{dI(t)}{dt} + RI(t) + \frac{1}{C}\int_0^t I(\tau)d\tau = v(t) \tag{20.6}$$

$$L\frac{dI(t)}{dt} = v(t) - RI(t) - \frac{1}{C}\int_0^t I(\tau)d\tau \tag{20.7}$$

$$\frac{dI(t)}{dt} = \frac{1}{L}v(t) - \frac{R}{L}I(t) - \frac{1}{LC}\int_0^t I(\tau)d\tau \tag{20.8}$$

We define

$$F(t,v,I) = \frac{1}{L}v(t) - \frac{R}{L}I(t) - \frac{1}{LC}\int_0^t I(\tau)d\tau \tag{20.9}$$

To prove the existence and uniqueness of the solution, we verify the following:

(i) $|F(t,v,I)|^2 < (1+|I|^2)$ (linear growth)

(ii) $\forall I, I_1, \quad \|F(t,v,I) - F(t,v,I_1)\|_\infty^2 < k\|I - I_1\|_\infty^2$

We start with the linear growth:

$$
\begin{aligned}
|F(t,v,I)|^2 &= \left|\frac{1}{L}v(t) - \frac{R}{L}I(t) - \frac{1}{LC}\int_0^t I(\tau)d\tau\right|^2 \\
&\leq \frac{3}{L^2}|v(t)|^2 + 3\left(\frac{R}{L}\right)^2 |I(t)|^2 + \frac{3}{(LC)^2}\left|\int_0^t I(\tau)d\tau\right|^2 \\
&\leq \frac{3}{L^2}|v(t)|^2 + 3\left(\frac{R}{L}\right)^2 |I(t)|^2 + \frac{3T}{(LC)^2}\sup_{0\leq t<T}|It)|^2 \\
&\leq \frac{3}{L^2}|v(t)|^2 + 3\left(\frac{R}{L}\right)^2 |I(t)|^2 + \frac{3T}{(LC)^2}\|I(t)\|_\infty^2 \\
&\leq \left(\frac{3}{L^2}|v(t)|^2 + \frac{3T}{(LC)^2}\|I(t)\|_\infty^2\right) \\
&\quad \times \left(1 + \frac{3\left(\frac{R}{L}\right)^2 |I(t)|^2}{\frac{3}{L^2}|v(t)|^2 + \frac{3T}{(LC)^2}\|I(t)\|_\infty^2}\right).
\end{aligned}
$$

If

$$\frac{3\left(\frac{R}{L}\right)^2}{\frac{3}{L^2}|v(t)|^2 + \frac{3T}{(LC)^2}\|I(t)\|_\infty^2} < 1$$

then, we get

$$|F(t,v,I)|^2 \leq k(1+|I|^2) \tag{20.10}$$

where

$$k = \frac{3}{L^2}|v(t)|^2 + \frac{3T}{(LC)^2}\|I(t)\|_\infty^2.$$

Therefore, the function satisfies the growth condition.

Now, we will verify the Lipschitz condition:

$$|F(t,v,I) - F(t,v,I_1)|^2 = \left| -\frac{R}{L}(I(t) - I_1(t)) - \frac{1}{LC}\int_0^t (I(\tau) - I_1(\tau))\,d\tau \right|^2$$

$$\leq 2\left(\frac{R}{L}\right)^2 |I(t) - I_1(t)|^2 + \frac{2}{(LC)^2}\int_0^t |I(\tau) - I_1(\tau)|^2\,d\tau.$$

Then, we obtain

$$\sup_{t\in[0,T]} |F(t,v,I) - F(t,v,I_1)|^2 \leq 2\left(\frac{R}{L}\right)^2 \sup_{t\in[0,T]} |I(t) - I_1(t)|^2$$

$$+ \frac{2T}{(LC)^2} \sup_{t\in[0,T]} |I(t) - I_1(t)|^2$$

$$\|F(t,v,I) - F(t,v,I_1)\|_\infty^2 \leq \left(2\left(\frac{R}{L}\right)^2 + \frac{2T}{(LC)^2}\right)\|I(t) - I_1(t)\|_\infty^2$$

Thus, we get

$$\|F(t,v,I) - F(t,v,I_1)\|_\infty^2 \leq \bar{k}\|I(t) - I_1(t)\|_\infty^2$$

where $\bar{k} = \left(2\left(\frac{R}{L}\right)^2 + \frac{2T}{(LC)^2}\right)$. This shows that the function F satisfies the Lipschitz condition. Owing the fact that

$$\frac{3\left(\frac{R}{L}\right)^2}{\frac{3}{L^2}|v(t)|^2 + \frac{3T}{(LC)^2}\|I(t)\|_\infty^2} < 1$$

then the equation admits a unique solution.

20.3 THIRD PROBLEM

We consider the following circuit problem in this subsection as

$$L_0^C D_t^\alpha I(t) + RI(t) + \frac{1}{C_0}{}^C I_t^\alpha I(t) = v(t) \tag{20.11}$$

$$_0^C D_t^\alpha I(t) = \frac{1}{L}v(t) - \frac{R}{L}I(t) - \frac{1}{LC_0}{}^C I_t^\alpha I(t) \tag{20.12}$$

We define

$$F(t,v,I) = \frac{1}{L}v(t) - \frac{R}{L}I(t) - \frac{1}{LC_0}{}^C I_t^\alpha I(t) \tag{20.13}$$

To prove the existence and uniqueness of the solution, we verify the following:

(i) $|F(t,v,I)|^2 < (1+|I|^2)$ (linear growth)

(ii) $\forall I, I_1, \quad \|F(t,v,I) - F(t,v,I_1)\|_\infty^2 < k\|I - I_1\|_\infty^2$

We start with the linear growth:

$$
\begin{aligned}
|F(t,v,I)|^2 &= \left| \frac{1}{L}v(t) - \frac{R}{L}I(t) - \frac{1}{LC}{}_0^C I_t^\alpha I(t) \right|^2 \\
&\leq \frac{3}{L^2}|v(t)|^2 + \frac{3R^2}{L^2}|I(t)|^2 + \frac{3}{(LC)^2}\left| {}_0^C I_t^\alpha I(t) \right|^2 \\
&\leq \frac{3}{L^2}|v(t)|^2 + \frac{3R^2}{L^2}|I(t)|^2 + \frac{3}{(LC)^2}\left| \frac{1}{\Gamma(\alpha)}\int_0^t (t-\tau)^{\alpha-1}I(\tau)d\tau \right|^2 \\
&\leq \frac{3}{L^2}|v(t)|^2 + \frac{3R^2}{L^2}|I(t)|^2 + \frac{3}{(LC\Gamma(\alpha))^2}\left| \int_0^t (t-\tau)^{\alpha-1}I(\tau)d\tau \right|^2 \\
&\leq \frac{3}{L^2}|v(t)|^2 + \frac{3R^2}{L^2}|I(t)|^2 + \frac{3T^{2\alpha-1}}{(LC\Gamma(\alpha))^2(2\alpha-1)}\sup_{t\in[0,T]}|I(t)|^2 \\
&\leq \frac{3}{L^2}|v(t)|^2 + \frac{3R^2}{L^2}|I(t)|^2 + \frac{3T^{2\alpha-1}}{(LC\Gamma(\alpha))^2(2\alpha-1)}\|I(t)\|_\infty^2 \\
&\leq \left(\frac{3}{L^2}|v(t)|^2 + \frac{3T^{2\alpha-1}}{(LC\Gamma(\alpha))^2(2\alpha-1)}\|I(t)\|_\infty^2 \right) \\
&\quad \times \left(1 + \frac{\frac{3R^2}{L^2}|I(t)|^2}{\frac{3}{L^2}|v(t)|^2 + \frac{3T^{2\alpha-1}}{(LC\Gamma(\alpha))^2(2\alpha-1)}\|I(t)\|_\infty^2} \right)
\end{aligned}
$$

If

$$
\frac{\frac{3R^2}{L^2}}{\frac{3}{L^2}|v(t)|^2 + \frac{3T^{2\alpha-1}}{(LC\Gamma(\alpha))^2(2\alpha-1)}\|I(t)\|_\infty^2} < 1
$$

then, we get

$$
|F(t,v,I)|^2 \leq k\left(1+|I(t)|^2\right) \tag{20.14}
$$

where

$$
k = \frac{3}{L^2}|v(t)|^2 + \frac{3T^{2\alpha-1}}{(LC\Gamma(\alpha))^2(2\alpha-1)}\|I(t)\|_\infty^2. \tag{20.15}
$$

Therefore, the function satisfies the growth condition.

Now, we will verify the Lipschitz condition:

$$|F(t,v,I) - F(t,v,I_1)|^2 = \left| -\frac{R}{L}(I(t) - I_1(t)) - \frac{1}{LC}\left({}_0^C I_t^\alpha I(t) - {}_0^C I_{1_t}^\alpha I(t)\right) \right|^2$$

$$\leq 2\left(\frac{R}{L}\right)^2 |I(t) - I_1(t)|^2 + \frac{2}{(LC)^2}\left|{}_0^C I_t^\alpha I(t) - {}_0^C I_{1_t}^\alpha I(t)\right|^2$$

$$\leq 2\left(\frac{R}{L}\right)^2 |I(t) - I_1(t)|^2$$

$$+ \frac{2}{(LC\Gamma(\alpha))^2} \left| \int_0^t |(I(\tau) - I_1(\tau))(t-\tau)^{\alpha-1}|^2 d\tau \right.$$

Then, we have

$$\sup_{t\in[0,T]} |F(t,v,I) - F(t,v,I_1)|^2 \leq 2\left(\frac{R}{L}\right)^2 \sup_{t\in[0,T]} |I(t) - I_1(t)|^2$$

$$+ \frac{2T^{2\alpha-1}}{(LC\Gamma(\alpha))^2(2\alpha - 1)} \sup_{t\in[0,T]} |I(t) - I_1(t)|^2$$

$$\|F(t,v,I) - F(t,v,I_1)\|_\infty^2 \leq 2\left(\frac{R}{L}\right)^2 \|I(t) - I_1(t)\|_\infty^2$$

$$+ \frac{2T^{2\alpha-1}}{(LC\Gamma(\alpha))^2(2\alpha - 1)} \|I(t) - I_1(t)\|_\infty^2$$

$$\leq \left(2\left(\frac{R}{L}\right)^2 + \frac{2T^{2\alpha-1}}{(LC\Gamma(\alpha))^2(2\alpha - 1)}\right) \|I(t) - I_1(t)\|_\infty^2$$

Thus, we get

$$\|F(t,v,I) - F(t,v,I_1)\|_\infty^2 \leq \bar{k} \|I(t) - I_1(t)\|_\infty^2$$

where

$$\bar{k} = \left(2\left(\frac{R}{L}\right)^2 + \frac{2T^{2\alpha-1}}{(LC\Gamma(\alpha))^2(2\alpha - 1)}\right).$$

This shows that the function F satisfies the Lipschitz condition. Owing the fact that

$$\frac{\frac{3R^2}{L^2}}{\frac{3}{L^2}|v(t)|^2 + \frac{3T^{2\alpha-1}}{(LC\Gamma(\alpha))^2(2\alpha-1)}\|I(t)\|^2} < 1$$

then the equation admits a unique solution.

20.4　FOURTH PROBLEM

We consider the following circuit problem in this subsection:

$$L_0^{CF} D_t^{\alpha} I(t) + RI^2(t) + \frac{1}{C_0}{}^{CF} I_t^{\alpha} I(t) = v(t) \tag{20.16}$$

$$_0^{CF} D_t^{\alpha} I(t) = \frac{1}{L}v(t) - \frac{R}{L}I^2(t) - \frac{1}{LC_0}{}^{CF} I_t^{\alpha} I(t) \tag{20.17}$$

We define

$$F(t,v,I) = \frac{1}{L}v(t) - \frac{R}{L}I^2(t) - \frac{1}{LC_0}{}^{CF} I_t^{\alpha} I(t) \tag{20.18}$$

To prove the existence and uniqueness of the solution, we verify the following:

(i)　$|F(t,v,I)|^2 < (1+|I|^2)$ (linear growth)

(ii)　$\forall I, I_1,\quad \|F(t,v,I) - F(t,v,I_1)\|_\infty^2 < k\|I - I_1\|_\infty^2$

We start with the linear growth:

$$
\begin{aligned}
|F(t,v,I)|^2 &= \left| \frac{1}{L}v(t) - \frac{R}{L}I^2(t) - \frac{1}{LC_0}{}^{CF} I_t^{\alpha} I(t) \right|^2 \\
&\leq 3\left(\frac{1}{L}\right)^2 |v(t)|^2 + 3\left(\frac{R}{L}\right)^2 |I^2(t)|^2 + 3\left(\frac{1}{LC}\right)^2 |_0^{CF} I_t^{\alpha} I(t)|^2 \\
&\leq 3\left(\frac{1}{L}\right)^2 |v(t)|^2 + 3\left(\frac{R}{L}\right)^2 |I^2(t)|^2 \\
&\quad + 3\left(\frac{1}{LC}\right)^2 \left| \frac{1-\alpha}{M(\alpha)}I(t) + \frac{\alpha}{M(\alpha)}\int_0^t I(\tau)d\tau \right|^2 \\
&\leq 3\left(\frac{1}{L}\right)^2 |v(t)|^2 + 3\left(\frac{R}{L}\right)^2 |I^2(t)|^2 + \frac{6(1-\alpha)^2}{(LCM(\alpha))^2}|I(t)|^2 \\
&\quad + \frac{6\alpha^2}{(LCM(\alpha))^2} \int_0^t |I(\tau)|^2 d\tau \\
&\leq 3\left(\frac{1}{L}\right)^2 |v(t)|^2 + 3\left(\frac{R}{L}\right)^2 |I^2(t)|^2 + \frac{6(1-\alpha)^2}{(LCM(\alpha))^2}|I(t)|^2 \\
&\quad + \frac{6T\alpha^2}{(LCM(\alpha))^2} \sup_{t\in[0,T]} |I(t)|^2 \\
&\leq 3\left(\frac{1}{L}\right)^2 |v(t)|^2 + 3\left(\frac{R}{L}\right)^2 \|I^2(t)\|^2 + \frac{6(1-\alpha)^2}{(LCM(\alpha))^2}|I(t)|^2 \\
&\quad + \frac{6T\alpha^2}{(LCM(\alpha))^2} \|I(t)\|_\infty^2
\end{aligned}
$$

$$\leq \left(3 \left(\frac{1}{L} \right)^2 |v(t)|^2 + 3 \left(\frac{R}{L} \right)^2 \|I^2(t)\|^2 + \frac{6T\alpha^2}{(LCM(\alpha))^2} \|I(t)\|_\infty^2 \right)$$

$$\times \left(1 + \frac{\frac{6(1-\alpha)^2}{(LCM(\alpha))^2} |I(t)|^2}{3 \left(\frac{1}{L} \right)^2 |v(t)|^2 + 3 \left(\frac{R}{L} \right)^2 \|I^2(t)\|^2 + \frac{6T\alpha^2}{(LCM(\alpha))^2} \|I(t)\|_\infty^2} \right)$$

If

$$\frac{\frac{6(1-\alpha)^2}{(LCM(\alpha))^2}}{3 \left(\frac{1}{L} \right)^2 |v(t)|^2 + 3 \left(\frac{R}{L} \right)^2 \|I^2(t)\|^2 + \frac{6T\alpha^2}{(LCM(\alpha))^2} \|I(t)\|_\infty^2} < 1$$

then, we get

$$|F(t,v,I)|^2 \leq k \left(1 + |I(t)|^2 \right) \tag{20.19}$$

where

$$k = 3 \left(\frac{1}{L} \right)^2 |v(t)|^2 + 3 \left(\frac{R}{L} \right)^2 \|I^2(t)\|^2 + \frac{6T\alpha^2}{(LCM(\alpha))^2} \|I(t)\|_\infty^2. \tag{20.20}$$

Therefore, the function satisfies the growth condition.

Now, we will verify the Lipschitz condition:

$$|F(t,v,I) - F(t,v,I_1)|^2 = \left| -\frac{R}{L} (I^2(t) - I_1^2(t)) - \frac{1}{LC} {}_0^{CF} I_t^\alpha (I(t) - I_1(t)) \right|^2$$

$$\leq 2 \left(\frac{R}{L} \right)^2 |I^2(t) - I_1^2(t)|^2 + \frac{2}{(LC)^2} \left| {}_0^{CF} I_t^\alpha (I(t) - I_1(t)) \right|^2$$

$$\leq 2 \left(\frac{R}{L} \right)^2 |I^2(t) - I_1^2(t)|^2 + \frac{2}{(LC)^2} \left| \frac{1-\alpha}{M(\alpha)} (I(t) - I_1(t)) \right.$$

$$\left. + \frac{\alpha}{M(\alpha)} \int_0^t (I(\tau) - I_1(\tau)) d\tau \right|^2$$

$$\leq 2 \left(\frac{R}{L} \right)^2 |I^2(t) - I_1^2(t)|^2 + \frac{4(1-\alpha)^2}{(LCM(\alpha))^2} |I(t) - I_1(t)|^2$$

$$+ \frac{4\alpha^2}{(LCM(\alpha))^2} \int_0^t |I(\tau) - I_1(\tau)|^2 d\tau$$

Then, we obtain

$$\sup_{t\in[0,T]} |F(t,v,I) - F(t,v,I_1)|^2 \leq 2\left(\frac{R}{L}\right)^2 \sup_{t\in[0,T]} |I^2(t) - I_1^2(t)|^2$$

$$+ \frac{4(1-\alpha)^2}{(LCM(\alpha))^2} \sup_{t\in[0,T]} |I(t) - I_1(t)|^2$$

$$+ \frac{4T\alpha^2}{(LCM(\alpha))^2} \sup_{t\in[0,T]} |I(t) - I_1(t)|^2$$

$$\|F(t,v,I) - F(t,v,I_1)\|_\infty^2 \leq 2\left(\frac{R}{L}\right)^2 \|I^2(t) - I_1^2(t)\|_\infty^2$$

$$+ \frac{4(1-\alpha)^2}{(LCM(\alpha))^2} \|I(t) - I_1(t)\|_\infty^2$$

$$+ \frac{4T\alpha^2}{(LCM(\alpha))^2} \|I(t) - I_1(t)\|_\infty^2$$

Thus, we reach

$$\|F(t,v,I) - F(t,v,I_1)\|_\infty^2 \leq \bar{k}\|I(t) - I_1(t)\|_\infty^2$$

where

$$\bar{k} = \left(2\left(\frac{R}{L}\right)^2 \|I(t) + I_1(t)\|_\infty^2 + \frac{4(1-\alpha)^2}{(LCM(\alpha))^2} + \frac{4T\alpha^2}{(LCM(\alpha))^2}\right).$$

This shows that the function F satisfies the Lipschitz condition. Owing the fact that

$$\frac{\frac{6(1-\alpha)^2}{(LCM(\alpha))^2}}{3\left(\frac{1}{L}\right)^2 |v(t)|^2 + 3\left(\frac{R}{L}\right)^2 \|I^2(t)\|^2 + \frac{6T\alpha^2}{(LCM(\alpha))^2} \|I(t)\|_\infty^2} < 1$$

then the equation admits a unique solution.

20.5 FIFTH PROBLEM

We consider the following circuit problem in this subsection:

$$L_0^{ABC} D_t^\alpha I(t) + RI^2(t) + \frac{1}{C_0} \, {}^{AB}I_t^\alpha I(t) = v(t) \tag{20.21}$$

$$ {}_0^{ABC} D_t^\alpha I(t) = \frac{1}{L} v(t) - \frac{R}{L} I^2(t) - \frac{1}{LC_0} \, {}^{AB}I_t^\alpha I(t) \tag{20.22}$$

We define

$$F(t,v,I) = \frac{1}{L} v(t) - \frac{R}{L} I^2(t) - \frac{1}{LC_0} \, {}^{AB}I_t^\alpha I(t) \tag{20.23}$$

To prove the existence and uniqueness of the solution, we verify the following:

(i) $|F(t,v,I)|^2 < (1 + |I|^2)$ (linear growth)

(ii) $\forall I, I_1, \quad \|F(t,v,I) - F(t,v,I_1)\|_\infty^2 < k\|I - I_1\|_\infty^2$

We start with the linear growth:

$$|F(t,v,I)|^2 = \left|\frac{1}{L}v(t) - \frac{R}{L}I^2(t) - \frac{1}{LC_0}{}^{AB}I_t^\alpha I(t)\right|^2$$

$$\leq \frac{3}{L^2}|v(t)|^2 + \frac{3R^2}{L^2}|I^2(t)|^2 + \frac{3}{(LC)^2}\left|_0^{AB}I_t^\alpha I(t)\right|^2$$

$$\leq \frac{3}{L^2}|v(t)|^2 + \frac{3R^2}{L^2}|I^2(t)|^2$$

$$+ \frac{3}{(LC)^2}\left|\frac{1-\alpha}{AB(\alpha)}I(t) + \frac{\alpha}{AB(\alpha)\Gamma(\alpha)}\int_0^t (t-\tau)^{\alpha-1}I(\tau)d\tau\right|^2$$

$$\leq \frac{3}{L^2}|v(t)|^2 + \frac{3R^2}{L^2}|I^2(t)|^2 + \frac{6(1-\alpha)^2}{(LCAB(\alpha))^2}|I(t)|^2$$

$$+ \frac{6\alpha^2}{(LCAB(\alpha)\Gamma(\alpha))^2}\int_0^t \left|(t-\tau)^{\alpha-1}I(\tau)\right|^2 d\tau$$

$$\leq \frac{3}{L^2}|v(t)|^2 + \frac{3R^2}{L^2}\|I^2(t)\|^2 + \frac{6(1-\alpha)^2}{(LCAB(\alpha))^2}|I(t)|^2$$

$$+ \frac{6\alpha^2 T^{2\alpha-1}}{(LCAB(\alpha)\Gamma(\alpha))^2(2\alpha-1)}\sup_{t\in[0,T]}|I(t)|^2$$

$$\leq \frac{3}{L^2}|v(t)|^2 + \frac{3R^2}{L^2}\|I^2(t)\|^2 + \frac{6(1-\alpha)^2}{(LCAB(\alpha))^2}|I(t)|^2$$

$$+ \frac{6\alpha^2 T^{2\alpha-1}}{(LCAB(\alpha)\Gamma(\alpha))^2(2\alpha-1)}\|I(t)\|_\infty^2$$

$$\leq \left(\frac{3}{L^2}|v(t)|^2 + \frac{3R^2}{L^2}\|I^2(t)\|^2 + \frac{6\alpha^2 T^{2\alpha-1}}{(LCAB(\alpha)\Gamma(\alpha))^2(2\alpha-1)}\|I(t)\|_\infty^2\right)$$

$$\times \left(1 + \frac{\frac{6(1-\alpha)^2}{(LCAB(\alpha))^2}|I(t)|^2}{\frac{3}{L^2}|v(t)|^2 + \frac{3R^2}{L^2}\|I^2(t)\|^2 + \frac{6\alpha^2 T^{2\alpha-1}}{(LCAB(\alpha)\Gamma(\alpha))^2(2\alpha-1)}\|I(t)\|_\infty^2}\right)$$

If

$$\frac{\frac{6(1-\alpha)^2}{(LCAB(\alpha))^2}}{\frac{3}{L^2}|v(t)|^2 + \frac{3R^2}{L^2}\|I^2(t)\|^2 + \frac{6\alpha^2 T^{2\alpha-1}}{(LCAB(\alpha)\Gamma(\alpha))^2(2\alpha-1)}\|I(t)\|_\infty^2} < 1$$

then, we get

$$|F(t,v,I)|^2 \leq k\left(1 + |I(t)|^2\right) \tag{20.24}$$

where

$$k = \frac{3}{L^2}|v(t)|^2 + \frac{3R^2}{L^2}\|I^2(t)\|^2 + \frac{6\alpha^2 T^{2\alpha-1}}{(LCAB(\alpha)\Gamma(\alpha))^2(2\alpha-1)}\|I(t)\|_\infty^2.$$

Therefore, the function satisfies the growth condition.

Now, we will verify the Lipschitz condition:

$$|F(t,v,I) - F(t,v,I_1)|^2 = \left| -\frac{R}{L}(I^2(t) - I_1^2(t)) - \frac{1}{LC}{}_0^{AB}I_t^\alpha (I(t) - I_1(t)) \right|^2$$

$$\leq 2\left(\frac{R}{L}\right)^2 |I^2(t) - I_1^2(t)|^2 + \frac{2}{(LC)^2} |{}_0^{AB}I_t^\alpha (I(t) - I_1(t))|^2$$

$$\leq 2\left(\frac{R}{L}\right)^2 |I^2(t) - I_1^2(t)|^2 + \frac{2}{(LC)^2} \left| \frac{1-\alpha}{AB(\alpha)}(I(t) - I_1(t)) \right.$$

$$\left. + \frac{\alpha}{AB(\alpha)\Gamma(\alpha)} \int_0^t (t-\tau)^{\alpha-1}(I(\tau) - I_1(\tau))d\tau \right|^2$$

$$\leq 2\left(\frac{R}{L}\right)^2 |I^2(t) - I_1^2(t)|^2 + \frac{4(1-\alpha)^2}{(LCAB(\alpha))^2}|I(t) - I_1(t)|^2$$

$$+ \frac{4\alpha^2}{(LCAB(\alpha)\Gamma(\alpha))^2}\int_0^t |(t-\tau)^{\alpha-1}(I(\tau) - I_1(\tau))|^2 d\tau$$

Then, we obtain

$$\sup_{t\in[0,T]} |F(t,v,I) - F(t,v,I_1)|^2 \leq 2\left(\frac{R}{L}\right)^2 \sup_{t\in[0,T]} |I^2(t) - I_1^2(t)|^2$$

$$+ \frac{4(1-\alpha)^2}{(LCAB(\alpha))^2} \sup_{t\in[0,T]} |I(t) - I_1(t)|^2$$

$$+ \frac{4\alpha^2 T^{2\alpha-1}}{(LCAB(\alpha)\Gamma(\alpha))^2(2\alpha - 1)} \sup_{t\in[0,T]} |I(t) - I_1(t)|^2$$

$$\|F(t,v,I) - F(t,v,I_1)\|_\infty^2 \leq 2\left(\frac{R}{L}\right)^2 \|I^2(t) - I_1^2(t)\|_\infty^2$$

$$+ \frac{4(1-\alpha)^2}{(LCAB(\alpha))^2} \|I(t) - I_1(t)\|_\infty^2$$

$$+ \frac{4\alpha^2 T^{2\alpha-1}}{(LCAB(\alpha)\Gamma(\alpha))^2(2\alpha - 1)} \|I(t) - I_1(t)\|_\infty^2$$

$$\leq \left(2\left(\frac{R}{L}\right)^2 \|I(t) + I_1(t)\|_\infty^2 + \frac{4(1-\alpha)^2}{(LCAB(\alpha))^2} \right.$$

$$\left. + \frac{4\alpha^2 T^{2\alpha-1}}{(LCAB(\alpha)\Gamma(\alpha))^2(2\alpha - 1)} \right) \|I(t) - I_1(t)\|_\infty^2$$

Thus, we reach

$$\|F(t,v,I) - F(t,v,I_1)\|_\infty^2 \leq \bar{k}\|I(t) - I_1(t)\|_\infty^2$$

where

$$\bar{k} = \left(2\left(\frac{R}{L}\right)^2 \|I(t)+I_1(t)\|_\infty^2 + \frac{4(1-\alpha)^2}{(LCAB(\alpha))^2} + \frac{4\alpha^2 T^{2\alpha-1}}{(LCAB(\alpha)\Gamma(\alpha))^2(2\alpha-1)}\right).$$

This shows that the function F satisfies the Lipschitz condition. Owing the fact that

$$\frac{6(1-\alpha)^2}{(LCAB(\alpha))^2}}{\frac{3}{L^2}|v(t)|^2 + \frac{3R^2}{L^2}\|I^2(t)\|^2 + \frac{6\alpha^2 T^{2\alpha-1}}{(LCAB(\alpha)\Gamma(\alpha))^2(2\alpha-1)}\|I(t)\|_\infty^2} < 1$$

then the equation admits a unique solution.

20.6   SIXTH PROBLEM

We consider the following circuit problem in this subsection:

$$L_0^{ABC}D_t^\alpha I(t) + RI^2(t) + \frac{1}{C_0}{}^{CF}I_t^\alpha I(t) = v(t) \tag{20.25}$$

$$_0^{ABC}D_t^\alpha I(t) = \frac{1}{L}v(t) - \frac{R}{L}I^2(t) - \frac{R^{CF}}{L_0}I_t^\alpha I(t) \tag{20.26}$$

We define

$$F(t,v,I) = \frac{1}{L}v(t) - \frac{R}{L}I^2(t) - \frac{1}{LC_0}{}^{CF}I_t^\alpha I(t) \tag{20.27}$$

To prove the existence and uniqueness of the solution, we verify the following:

(i) $|F(t,v,I)|^2 < (1+|I|^2)$ (linear growth)

(ii) $\forall I, I_1, \quad \|F(t,v,I) - F(t,v,I_1)\|_\infty^2 < k\|I-I_1\|_\infty^2$

We start with the linear growth:

$$|F(t,v,I)|^2 = \left|\frac{1}{L}v(t) - \frac{R}{L}I^2(t) - \frac{1}{LC_0}{}^{CF}I_t^\alpha I(t)\right|^2$$

$$\leq 3\left(\frac{1}{L}\right)^2 |v(t)|^2 + 3\left(\frac{R}{L}\right)^2 |I^2(t)|^2 + 3\left(\frac{1}{LC}\right)^2 |_0^{CF}I_t^\alpha I(t)|^2$$

$$\leq 3\left(\frac{1}{L}\right)^2 |v(t)|^2 + 3\left(\frac{R}{L}\right)^2 |I^2(t)|^2$$

$$+ 3\left(\frac{1}{LC}\right)^2 \left|\frac{1-\alpha}{M(\alpha)}I(t) + \frac{\alpha}{M(\alpha)}\int_0^t I(\tau)d\tau\right|^2$$

$$\leq 3\left(\frac{1}{L}\right)^2 |v(t)|^2 + 3\left(\frac{R}{L}\right)^2 |I^2(t)|^2 + \frac{6(1-\alpha)^2}{(LCM(\alpha))^2}|I(t)|^2$$

$$+ \frac{6\alpha^2}{(LCM(\alpha))^2}\int_0^t |I(\tau)|^2 d\tau$$

$$\leq 3\left(\frac{1}{L}\right)^2 |v(t)|^2 + 3\left(\frac{R}{L}\right)^2 |I^2(t)|^2 + \frac{6(1-\alpha)^2}{(LCM(\alpha))^2}|I(t)|^2$$

$$+\frac{6T\alpha^2}{(LCM(\alpha))^2}\sup_{t\in[0,T]}|I(t)|^2$$

$$\leq \; 3\left(\frac{1}{L}\right)^2|v(t)|^2+3\left(\frac{R}{L}\right)^2\|I^2(t)\|^2+\frac{6(1-\alpha)^2}{(LCM(\alpha))^2}|I(t)|^2$$

$$+\frac{6T\alpha^2}{(LCM(\alpha))^2}\|I(t)\|_\infty^2$$

$$\leq \; \left(3\left(\frac{1}{L}\right)^2|v(t)|^2+3\left(\frac{R}{L}\right)^2\|I^2(t)\|^2+\frac{6T\alpha^2}{(LCM(\alpha))^2}\|I(t)\|_{\bullet\bullet}^2\right)$$

$$\times\left(1+\frac{\frac{6(1-\alpha)^2}{(LCM(\alpha))^2}|I(t)|^2}{3\left(\frac{1}{L}\right)^2|v(t)|^2+3\left(\frac{R}{L}\right)^2\|I^2(t)\|^2+\frac{6T\alpha^2}{(LCM(\alpha))^2}\|I(t)\|_\infty^2}\right)$$

If

$$\frac{\frac{6(1-\alpha)^2}{(LCM(\alpha))^2}}{3\left(\frac{1}{L}\right)^2|v(t)|^2+3\left(\frac{R}{L}\right)^2\|I^2(t)\|^2+\frac{6T\alpha^2}{(LCM(\alpha))^2}\|I(t)\|_\infty^2}<1$$

then, we get

$$|F(t,v,I)|^2 \;\leq\; k\left(1+|I(t)|^2\right) \tag{20.28}$$

where

$$k=3\left(\frac{1}{L}\right)^2|v(t)|^2+3\left(\frac{R}{L}\right)^2\|I^2(t)\|^2+\frac{6T\alpha^2}{(LCM(\alpha))^2}\|I(t)\|_\infty^2. \tag{20.29}$$

Therefore, the function satisfies the growth condition.

Now, we will verify the Lipschitz condition:

$$|F(t,v,I)-F(t,v,I_1)|^2 \;=\; \left|-\frac{R}{L}(I^2(t)-I_1^2(t))-\frac{1}{LC}{}_0^{CF}I_t^\alpha(I(t)-I_1(t))\right|^2$$

$$\leq\; 2\left(\frac{R}{L}\right)^2|I^2(t)-I_1^2(t)|^2+\frac{2}{(LC)^2}|{}_0^{CF}I_t^\alpha(I(t)-I_1(t))|^2$$

$$\leq\; 2\left(\frac{R}{L}\right)^2|I^2(t)-I_1^2(t)|^2+\frac{2}{(LC)^2}\Bigg|\frac{1-\alpha}{M(\alpha)}(I(t)-I_1(t))$$

$$+\frac{\alpha}{M(\alpha)}\int_0^t(I(\tau)-I_1(\tau))\,d\tau\Bigg|^2$$

$$\leq\; 2\left(\frac{R}{L}\right)^2|I^2(t)-I_1^2(t)|^2+\frac{4(1-\alpha)^2}{(LCM(\alpha))^2}|I(t)-I_1(t)|^2$$

$$+\frac{4\alpha^2}{(LCM(\alpha))^2}\int_0^t|I(\tau)-I_1(\tau)|^2\,d\tau$$

Then, we obtain

$$
\sup_{t\in[0,T]} |F(t,v,I) - F(t,v,I_1)|^2 \leq 2\left(\frac{R}{L}\right)^2 \sup_{t\in[0,T]} |I^2(t) - I_1^2(t)|^2
$$
$$
+\frac{4(1-\alpha)^2}{(LCM(\alpha))^2}\sup_{t\in[0,T]} |I(t) - I_1(t)|^2
$$
$$
+\frac{4T\alpha^2}{(LCM(\alpha))^2}\sup_{t\in[0,T]} |I(t) - I_1(t)|^2
$$

$$
\|F(t,v,I) - F(t,v,I_1)\|_\infty^2 \leq 2\left(\frac{R}{L}\right)^2 \|I^2(t) - I_1^2(t)\|_\infty^2
$$
$$
+\frac{4(1-\alpha)^2}{(LCM(\alpha))^2}\|I(t) - I_1(t)\|_\infty^2
$$
$$
+\frac{4T\alpha^2}{(LCM(\alpha))^2}\|I(t) - I_1(t)\|_\infty^2
$$

Thus, we reach

$$
\|F(t,v,I) - F(t,v,I_1)\|_\infty^2 \leq \bar{k}\|I(t) - I_1(t)\|_\infty^2
$$

where

$$
\bar{k} = \left(2\left(\frac{R}{L}\right)^2 \|I(t) + I_1(t)\|_\infty^2 + \frac{4(1-\alpha)^2}{(LCM(\alpha))^2} + \frac{4T\alpha^2}{(LCM(\alpha))^2}\right).
$$

This shows that the function F satisfies the Lipschitz condition. Owing the fact that

$$
\frac{\frac{6(1-\alpha)^2}{(LCM(\alpha))^2}}{3\left(\frac{1}{L}\right)^2 |v(t)|^2 + 3\left(\frac{R}{L}\right)^2 \|I^2(t)\|^2 + \frac{6T\alpha^2}{(LCM(\alpha))^2}\|I(t)\|_\infty^2} < 1
$$

then the equation admits a unique solution.

20.7 SEVENTH PROBLEM

We consider the following circuit problem in this subsection:

$$
L_0^{ABC}D_t^\alpha I(t) + RI^2(t) + \frac{1}{C_0}{}_t^C I^\alpha I(t) = v(t) \tag{20.30}
$$

$$
{}_0^{ABC}D_t^\alpha I(t) = \frac{1}{L}v(t) - \frac{R}{L}I^2(t) - \frac{1}{LC_0}{}_t^C I^\alpha I(t) \tag{20.31}
$$

We define

$$
F(t,v,I) = \frac{1}{L}v(t) - \frac{R}{L}I^2(t) - \frac{1}{LC_0}{}_t^C I^\alpha I(t) \tag{20.32}
$$

To prove the existence and uniqueness of the solution, we verify the following:

(i) $|F(t,v,I)|^2 < (1+|I|^2)$ (linear growth)

(ii) $\forall I, I_1, \quad \|F(t,v,I) - F(t,v,I_1)\|_\infty^2 < k\,\|I - I_1\|_\infty^2$

We start with the linear growth:

$$
\begin{aligned}
|F(t,v,I)|^2 &= \left| \frac{1}{L}v(t) - \frac{R}{L}I^2(t) - \frac{1}{LC}\,{}_0^C I_t^\alpha I(t) \right|^2 \\[2mm]
&\leq \frac{3}{L^2}|v(t)|^2 + \frac{3R^2}{L^2}\left|I^2(t)\right|^2 + \frac{3}{(LC)^2}\left|{}_0^C I_t^\alpha I(t)\right|^2 \\[2mm]
&\leq \frac{3}{L^2}|v(t)|^2 + \frac{3R^2}{L^2}\left|I^2(t)\right|^2 + \frac{3}{(LC)^2}\left|\frac{1}{\Gamma(\alpha)}\int_0^t (t-\tau)^{\alpha-1}I(\tau)d\tau\right|^2 \\[2mm]
&\leq \frac{3}{L^2}|v(t)|^2 + \frac{3R^2}{L^2}\left|I^2(t)\right|^2 + \frac{3}{(LC\Gamma(\alpha))^2}\left|\int_0^t |(t-\tau)^{\alpha-1}I(\tau)d\tau\right|^2 \\[2mm]
&\leq \frac{3}{L^2}|v(t)|^2 + \frac{3R^2}{L^2}\left|I^2(t)\right|^2 + \frac{3T^{2\alpha-1}}{(LC\Gamma(\alpha))^2(2\alpha-1)}\sup_{t\in[0,T]}|I(t)|^2 \\[2mm]
&\leq \frac{3}{L^2}|v(t)|^2 + \frac{3R^2}{L^2}\|I(t)\|^2\,|I(t)|^2 + \frac{3T^{2\alpha-1}}{(LC\Gamma(\alpha))^2(2\alpha-1)}\|I(t)\|_\infty^2 \\[2mm]
&\leq \left(\frac{3}{L^2}|v(t)|^2 + \frac{3T^{2\alpha-1}}{(LC\Gamma(\alpha))^2(2\alpha-1)}\|I(t)\|_\infty^2\right) \\[2mm]
&\quad \times \left(1 + \frac{\frac{3R^2}{L^2}|I(t)|^2}{\frac{3}{L^2}|v(t)|^2 + \frac{3T^{2\alpha-1}}{(LC\Gamma(\alpha))^2(2\alpha-1)}\|I(t)\|_\infty^2}\right)
\end{aligned}
$$

If

$$
\frac{\frac{3R^2}{L^2}\|I(t)\|^2}{\frac{3}{L^2}|v(t)|^2 + \frac{3T^{2\alpha-1}}{(LC\Gamma(\alpha))^2(2\alpha-1)}\|I(t)\|_\infty^2} < 1
$$

then, we get

$$
|F(t,v,I)|^2 \;\leq\; k\left(1 + |I(t)|^2\right) \tag{20.33}
$$

where

$$
k = \frac{3}{L^2}|v(t)|^2 + \frac{3T^{2\alpha-1}}{(LC\Gamma(\alpha))^2(2\alpha-1)}\|I(t)\|_\infty^2. \tag{20.34}
$$

Therefore, the function satisfies the growth condition.

Now, we will verify the Lipschitz condition:

$$|F(t,v,I) - F(t,v,I_1)|^2 = \left| -\frac{R}{L}(I^2(t) - I_1^2(t)) - \frac{1}{LC}({}^C_0 I_t^\alpha I(t) - {}^C_0 I_{1_t}^\alpha I(t)) \right|^2$$

$$\leq 2\left(\frac{R}{L}\right)^2 |I^2(t) - I_1^2(t)|^2 + \frac{2}{(LC)^2}|{}^C_0 I_t^\alpha I(t) - {}^C_0 I_{1_t}^\alpha I(t)|^2$$

$$\leq 2\left(\frac{R}{L}\right)^2 |I^2(t) - I_1^2(t)|^2$$

$$+ \frac{2}{(LC\Gamma(\alpha))^2}\int_0^t |(I(\tau) - I_1(\tau))(t-\tau)^{\alpha-1}|^2 \, d\tau$$

Then, we have

$$\sup_{t\in[0,T]}|F(t,v,I) - F(t,v,I_1)|^2 \leq 2\left(\frac{R}{L}\right)^2 \sup_{t\in[0,T]}|I^2(t) - I_1^2(t)|^2$$

$$+ \frac{2T^{2\alpha-1}}{(LC\Gamma(\alpha))^2(2\alpha-1)} \sup_{t\in[0,T]}|I(t) - I_1(t)|^2$$

$$\|F(t,v,I) - F(t,v,I_1)\|_\infty^2 \leq 2\left(\frac{R}{L}\right)^2 \|I^2(t) - I_1^2(t)\|_\infty^2$$

$$+ \frac{2T^{2\alpha-1}}{(LC\Gamma(\alpha))^2(2\alpha-1)}\|I^2(t) - I_1^2(t)\|_\infty^2$$

$$\leq \left(2\left(\frac{R}{L}\right)^2 \|I(t) + I_1(t)\|_\infty^2 + \frac{2T^{2\alpha-1}}{(LC\Gamma(\alpha))^2(2\alpha-1)}\right)$$

$$\times \|I(t) - I_1(t)\|_\infty^2$$

Thus, we get

$$\|F(t,v,I) - F(t,v,I_1)\|_\infty^2 \leq \bar{k}\|I(t) - I_1(t)\|_\infty^2$$

where

$$\bar{k} = \left(2\left(\frac{R}{L}\right)^2 \|I(t) + I_1(t)\|_\infty^2 + \frac{2T^{2\alpha-1}}{(LC\Gamma(\alpha))^2(2\alpha-1)}\right).$$

This shows that the function F satisfies the Lipschitz condition. Owing the fact that

$$\frac{\frac{3R^2}{L^2}\|I(t)\|^2}{\frac{3}{L^2}|v(t)|^2 + \frac{3T^{2\alpha-1}}{(LC\Gamma(\alpha))^2(2\alpha-1)}\|I(t)\|_\infty^2} < 1$$

then the equation admits a unique solution.

21 Non-Linear Stochastic RLC Systems

While deterministic differential equations have been applied on several occasions to model some behaviors observed in nature, it has been reported that they are unable to depict stochastic processes. To solve this issue, the concept of stochastic differentiation and integration was introduced. These equations can then be viewed as deterministic with an addition of a random component. These models have been applied to several problems, for example, stock prices and thermal fluctuations. While we do not say that the inclusion of stochastic components in RLC systems will be of great value in the field, only, we are very curious to see what can happen if an RLC differential equation is connected to a stochastic type. In this chapter we will extend the come circuit model by adding a stochastic component [81–87].

$$Ri(t) + L\frac{di(t)}{dt} = V(t) \tag{21.1}$$

We added a stochastic component

$$i(t) = i(0) + \int_0^t f(\tau, i(\tau))d\tau + \int_0^t i(\tau)\omega_i dB(t) \tag{21.2}$$

where $B(t)$ is assumed to be differential such that $dB(t) = B'(t)dt$. Then, we get

$$i(t) = i(0) + \int_0^t f(\tau, i(\tau))d\tau + \int_0^t i(\tau)\omega_i \dot{B}(t)dt \tag{21.3}$$

Thus

$$i(t) = i(0) + \int_0^t f(\tau, i(\tau))d\tau + \int_0^t i(\tau)\omega dB(t) \tag{21.4}$$

where ω is the density of randomness. Here

$$f(t, i(t)) = \frac{1}{L}(V(t) - Ri(t)). \tag{21.5}$$

We first present the existence and uniqueness of the above equation by verifying

(i) Growth condition $\forall t \in D_i, \ |f(t, i(t))|^2 < k(1 + |i(t)|^2)$

(ii) The Lipschitz condition. Let i_1 and i_2 be the solutions of the equations.

$$|f(t, i_1(t)) - f(t, i_2(t))|^2 < \bar{k}|i_1 - i_2|^2$$

DOI: 10.1201/9781003359869-21

PROOF (i)   In fact

$$|f(t,i(t))|^2 = \left|\frac{1}{L}(V(t)-Ri(t))\right|^2$$

$$\leq \frac{1}{L^2}(2|V(t)|^2+2R^2|i(t)|^2)$$

$$\leq \frac{2}{L^2}\left(\sup_{t\in D_t}|V(t)|^2+R^2|i(t)|^2\right)$$

$$\leq \frac{2}{L^2}(R^2|i(t)|^2)$$

$$\leq \frac{2\|V(t)\|_\infty^2}{L^2}\left(1+\frac{R^2|i(t)|^2}{\|V(t)\|_\infty^2}\right)$$

If $R^2 < \|V(t)\|_\infty^2$, then

$$|f(t,i(t))|^2 \leq k(1+|i(t)|^2)$$

where $k=\frac{2\|V(t)\|_\infty^2}{L^2}$. Indeed

$$|i(t)w|^2 = w^2|i(t)|^2 \leq w^2\left(1+|i(t)|^2\right)$$

Here $w=\bar{k}_1$.

(ii)   Indeed

$$|f(t,i_1(t))-f(t,i_2(t))|^2 = \left|-\frac{R}{L}(i_1(t)-i_2(t))\right|^2 = \frac{R^2}{L^2}|i_1(t)-i_2(t)|^2$$

$$\leq \left(1+\frac{R^2}{L^2}\right)|i_1(t)-i_2(t)|^2 \leq k|i_1(t)-i_2(t)|^2$$

where $k=1+\frac{R^2}{L^2}$. Thus, both conditions have been verified. We can consider that the equation has a unique solution. ∎

We now present the numerical solution.

$$i(t)-i(0)=\int_0^t \frac{1}{L}(V(\tau)-Ri(\tau))\,d\tau. \tag{21.6}$$

We consider the equation at $t_{n+1}=(n+1)\Delta t$ and $t=t_n=n\Delta t$.

$$i(t_{n+1})-i(t_n)=\int_{t_n}^{t_{n+1}}\frac{1}{L}(V(\tau)-Ri(\tau))\,d\tau \tag{21.7}$$

$$+w\int_{t_n}^{t_{n+1}}i(\tau)dB(\tau).$$

(1) If $B(\tau)$ is differentiable then $dB(\tau) = B'(\tau)d\tau$. Then, we obtain

$$i(t_{n+1}) - i(t_n) = \int_{t_n}^{t_{n+1}} \frac{1}{L} (V(\tau) - Ri(\tau)) \, d\tau \qquad (21.8)$$

$$+ w \int_{t_n}^{t_{n+1}} i(\tau)B'(\tau)d\tau.$$

Using the Lagrange approximation gives

$$i(t_{n+1}) - i(t_n) = \frac{3}{2} \Delta t \frac{1}{L} (V(t_n) - Ri(t_n)) - \frac{\Delta t}{2L} (V(t_{n-1}) - Ri(t_{n-1}))$$

$$+ \frac{3}{2} \frac{\Delta t}{L} wi(t_n) \left(\frac{B(t_{n+1}) - B(t_n)}{\Delta t} \right) - \frac{\Delta t}{2L} wi(t_{n-1}) \left(\frac{B(t_n) - B(t_{n-1})}{\Delta t} \right).$$

(2) If $B(t)$ is not differentiable, then we have

$$i(t_{n+1}) - i(t_n) = \frac{3}{2} \Delta t \frac{1}{L} (V(t_n) - Ri(t_n)) - \frac{\Delta t}{2L} (V(t_{n-1}) - Ri(t_{n-1}))$$

$$= + w \sum_{i=0}^{n} i(t_j) \left(B(t_{j+1}) - B(t_j) \right) - w \sum_{i=0}^{n-1} i(t_j) \left(B(t_{j+1}) - B(t_j) \right)$$

Let us consider the following classical stochastic model.

$$V_2(t) + RC \frac{dV_2(t)}{dt} = \frac{1}{2} \left(V_1(t) - RC \frac{dV_1(t)}{dt} \right) \qquad (21.9)$$

We added a stochastic component

$$V_2(t) = V_2(0) + \int_0^t f(\tau, V(\tau))d\tau + \int_0^t V_2(\tau) \omega_V dB(t) \qquad (21.10)$$

where $B(t)$ is assumed to be differential such that $dB(t) = B'(t)dt$. Then, we get

$$V_2(t) = V_2(0) + \int_0^t f(\tau, V_2(\tau))d\tau + \int_0^t V_2(\tau) \omega_i \dot{B}(t)dt \qquad (21.11)$$

Thus

$$V_2(t) = V_2(0) + \int_0^t f(\tau, V_2(\tau))d\tau + \int_0^t V_2(\tau) \omega dB(t) \qquad (21.12)$$

where ω is the density of randomness. Here

$$f(t, V_2(t)) = \frac{1}{2RC} \left(V_1(t) - RC \frac{dV_1(t)}{dt} \right) - \frac{1}{RC} V_2(t) \qquad (21.13)$$

We first present the existence and uniqueness of the above equation by verifying

(i)   Growth condition $\forall t \in D_i$, $|f(t, V_2(t))|^2 < k(1 + |V_2(t)|^2)$

(ii) The Lipschitz condition. Let V_{2_1} and V_{2_2} be the solutions of the equations.

$$|f(t, V_{2_1}(t)) - f(t, V_{2_2}(t))|^2 < \bar{k}|V_{2_1} - V_{2_1}|^2$$

PROOF (i)   In fact

$$|f(t, V_2(t))|^2 = \left|\frac{1}{2RC}\left(V_1(t) - RC\frac{dV_1(t)}{dt}\right) - \frac{1}{RC}V_2(t)\right|^2$$

$$\leq 3\frac{1}{(2RC)^2}|V_1(t)|^2 + \frac{3}{4}\left|\frac{dV_1(t)}{dt}\right|^2 + 3\frac{1}{(RC)^2}|V_2(t)|^2$$

$$\leq 3\frac{1}{(2RC)^2}\sup_{t \in D_t}|V_1(t)|^2 + \frac{3}{4}\left|\frac{dV_1(t)}{dt}\right|^2 + 3\frac{1}{(RC)^2}|V_2(t)|^2$$

$$\leq 3\frac{1}{(2RC)^2}\|V_1(t)\|_\infty^2 + \frac{3}{4}\left\|\frac{dV_1(t)}{dt}\right\|_\infty^2 + 3\frac{1}{(RC)^2}|V_2(t)|^2$$

$$\leq \left(3\frac{1}{(2RC)^2}\|V_1(t)\|_\infty^2 + \frac{3}{4}\left\|\frac{dV_1(t)}{dt}\right\|_\infty^2\right)$$

$$\times \left(1 + \frac{3\frac{1}{(RC)^2}}{3\frac{1}{(2RC)^2}\|V_1(t)\|_\infty^2 + \frac{3}{4}\left\|\frac{dV_1(t)}{dt}\right\|_\infty^2}|V_2(t)|^2\right)$$

If $3\frac{1}{(RC)^2} < \left(3\frac{1}{(2RC)^2}\|V_1(t)\|_\infty^2 + \frac{3}{4}\left\|\frac{dV_1(t)}{dt}\right\|_\infty^2\right)$, then

$$|f(t, V_2(t))|^2 \leq k(1 + |V_2(t)|^2)$$

where $k = 3\frac{1}{(2RC)^2}\|V_1(t)\|_\infty^2 + \frac{3}{4}\left\|\frac{dV_1(t)}{dt}\right\|_\infty^2$. Indeed

$$|V_2(t)w|^2 = w^2|V_2(t)|^2 \leq w^2\left(1 + |V_2(t)|^2\right)$$

Here $w = \bar{k}_1$.
(ii)   Indeed

$$|f(t, V_{2_1}(t)) - f(t, V_{2_2}(t))|^2 = \left|-\frac{1}{RC}(V_{2_1}(t) - V_{2_2}(t))\right|^2 = \frac{1}{(RC)^2}|V_{2_1}(t) - V_{2_2}(t)|^2$$

$$\leq \left(1 + \frac{1}{(RC)^2}\right)|V_{2_1}(t) - V_{2_2}(t)|^2$$

$$\leq k|V_{2_1}(t) - V_{2_2}(t)|^2$$

where $k = 1 + \frac{1}{(RC)^2}$. Thus, both conditions have been verified. We can consider that the equation has a unique solution. We now present the numerical solution.

$$V_2(t) = V_2(0) + \int_0^t \left(\frac{1}{2RC} \left(V_1(\tau)\tau - RC\frac{dV_1(\tau)}{d\tau} \right) - \frac{1}{RC}V_2(\tau) \right) d\tau \qquad (21.14)$$

We consider the equation at $t = t_{n+1} = (n+1)\Delta t$ and $t = t_n = n\Delta t$.

$$V_2(t_{n+1}) - V_2(t_n) = \int_{t_n}^{t_{n+1}} \left(\frac{1}{2RC} \left(V_1(\tau) - RC\frac{dV_1(\tau)}{d\tau} \right) - \frac{1}{RC}V_2(\tau) \right) d\tau$$
$$+ w \int_{t_n}^{t_{n+1}} V_2(\tau)dB(\tau).$$

(1) If $B(\tau)$ is differentiable then $dB(\tau) = B'(\tau)d\tau$. Then, we obtain

$$V_2(t_{n+1}) - V_2(t_n) = \int_{t_n}^{t_{n+1}} \left(\frac{1}{2RC} \left(V_1(\tau) - RC\frac{dV_1(\tau)}{d\tau} \right) - \frac{1}{RC}V_2(\tau) \right) d\tau$$
$$+ w \int_{t_n}^{t_{n+1}} V_2(\tau)B'(\tau)d\tau.$$

Using the Lagrange approximation gives

$$V_2(t_{n+1}) - V_2(t_n) = \frac{3}{2}\Delta t \left(\frac{1}{2RC} \left(V_1(t_n) - RC\frac{V_1(t_{n+1}) - V_1(t_n)}{\Delta t} \right) - \frac{1}{RC}V_2(t_n) \right)$$
$$- \frac{\Delta t}{2RC} \left(\frac{1}{2RC} \left(V_1(t_{n-1}) - RC\frac{V_1(t_n) - V_1(t_{n-1})}{\Delta t} \right) \right.$$
$$\left. - \frac{1}{RC}V_2(t_{n-1}) \right) + \frac{3}{2}\frac{\Delta t}{2RC}wV_2(t_n) \left(\frac{B(t_{n+1}) - B(t_n)}{\Delta t} \right)$$
$$- \frac{\Delta t}{4RC}wV_2(t_{n-1}) \times \left(\frac{B(t_n) - B(t_{n-1})}{\Delta t} \right).$$

(2) If $B(t)$ is not differentiable, then we have

$$V_2(t_{n+1}) - V_2(t_n) = \frac{3}{2}\Delta t \left(\frac{1}{2RC} \left(V_1(t_n) - RC\frac{V_1(t_{n+1}) - V_1(t_n)}{\Delta t} \right) - \frac{1}{RC}V_2(t_n) \right)$$
$$- \frac{\Delta t}{4RC} \left(\frac{1}{2RC} \left(V_1(t_{n-1}) - RC\frac{V_1(t_n) - V_1(t_{n-1})}{\Delta t} \right) \right.$$
$$\left. - \frac{1}{RC}V_2(t_{n-1}) \right) + w\sum_{i=0}^n V_2(t_j) \left(B(t_{j+1}) - B(t_j) \right)$$
$$- w\sum_{i=0}^{n-1} V_2(t_j) \left(B(t_{j+1}) - B(t_j) \right)$$

∎

Let us consider the following classical stochastic model.

$$V_2(t) + R_1 C_1 \frac{dV_2(t)}{dt} = -\frac{C_1}{C_2}\left(V_1(t) - R_2 C_2 \frac{dV_1(t)}{dt}\right) \qquad (21.15)$$

We added a stochastic component

$$V_2(t) = V_2(0) + \int_0^t f(\tau, V_2(\tau))d\tau + \int_0^t V_2(\tau)\omega_v dB(t) \qquad (21.16)$$

where $B(t)$ is assumed to be differential such that $dB(t) = B'(t)dt$. Then, we get

$$V_2(t) = V_2(0) + \int_0^t f(\tau, V_2(\tau))d\tau + \int_0^t V_2(\tau)\omega_v \dot{B}(t)dt \qquad (21.17)$$

Thus

$$V_2(t) = V_2(0) + \int_0^t f(\tau, V_2(\tau))d\tau + \int_0^t V_2(\tau)\omega dB(t) \qquad (21.18)$$

where ω is the density of randomness. Here

$$f(t, V_2(t)) = -\frac{1}{R_1 C_2}\left(V_1(t) - R_2 C_2 \frac{dV_1(t)}{dt}\right) - \frac{1}{R_1 C_1}V_2(t) \qquad (21.19)$$

We first present the existence and uniqueness of the above equation by verifying

(i)   Growth condition $\forall t \in D_V$, $|f(t, V_2(t))|^2 < k(1 + |V_2(t)|^2)$

(ii) The Lipschitz condition. Let V_{2_1} and V_{2_2} be the solutions of the equations.

$$|f(t, V_{2_1}(t)) - f(t, V_{2_2}(t))|^2 < \bar{k}|V_{2_1} - V_{2_1}|^2$$

PROOF (i)   In fact

$$
\begin{aligned}
|f(t, V_2(t))|^2 &= \left| -\frac{1}{R_1 C_2}\left(V_1(t) - R_2 C_2 \frac{dV_1(t)}{dt}\right) - \frac{1}{R_1 C_1}V_2(t) \right|^2 \\
&\leq 3\frac{1}{R_1^2 C_2^2}|V_1(t)|^2 + \frac{3R_2^2}{R_1^2}\left|\frac{dV_1(t)}{dt}\right|^2 + 3\frac{1}{(R_1 C_1)^2}|V_2(t)|^2 \\
&\quad + 3\frac{1}{(R_1 C_1)^2}|V_2(t)|^2 \\
&\leq 3\frac{1}{(R_1 C_2)^2}\sup_{t\in D_t}|V_1(t)|^2 + \frac{3R_2^2}{R_1^2}\left|\frac{dV_1(t)}{dt}\right|^2 \\
&\quad + 3\frac{1}{(R_1 C_1)^2}|V_2(t)|^2 \\
&\leq 3\frac{1}{(R_1 C_2)^2}\|V_1(t)\|_\infty^2 + \frac{3R_2^2}{R_1^2}\left\|\frac{dV_1(t)}{dt}\right\|_\infty^2 \\
&\quad + 3\frac{1}{(R_1 C_1)^2}|V_2(t)|^2
\end{aligned}
$$

$$\leq \left(3\frac{1}{(R_1C_2)^2}\|V_1(t)\|_\infty^2 + \frac{3R_2^2}{R_1^2}\left\|\frac{dV_1(t)}{dt}\right\|_\infty^2\right)$$

$$\times\left(1 + \frac{3\frac{1}{(R_1C_1)^2}}{3\frac{1}{(R_1C_2)^2}\|V_1(t)\|_\infty^2 + \frac{3R_2^2}{R_1^2}\left\|\frac{dV_1(t)}{dt}\right\|_\infty^2}|V_2(t)|^2\right)$$

If $\dfrac{3\frac{1}{(R_1C_1)^2}}{3\frac{1}{(R_1C_2)^2}\|V_1(t)\|_\infty^2 + \frac{3R_2^2}{R_1^2}\left\|\frac{dV_1(t)}{dt}\right\|_\infty^2} < 1$ then

$$|f(t,V_2(t))|^2 \leq k(1+|V_2(t)|^2)$$

where $k = 3\frac{1}{(R_1C_1)^2}\|V_1(t)\|_\infty^2 + \frac{3R_2^2}{R_1^2}\left\|\frac{dV_1(t)}{dt}\right\|_\infty^2$. Indeed

$$|V_2(t)w|^2 = w^2|V_2(t)|^2 \leq w^2\left(1+|V_2(t)|^2\right)$$

Here $w = \bar{k}_1$.
(ii)   Indeed

$$|f(t,V_{2_1}(t)) - f(t,V_{2_2}(t))|^2 = \left|-\frac{1}{R_1C_1}(V_{2_1}(t) - V_{2_2}(t))\right|^2$$

$$= \frac{1}{(R_1C_1)^2}|V_{2_1}(t) - V_{2_2}(t)|^2$$

$$\leq \left(1 + \frac{1}{(R_1C_1)^2}\right)|V_{2_1}(t) - V_{2_2}(t)|^2$$

$$\leq k|V_{2_1}(t) - V_{2_2}(t)|^2$$

where $k = 1 + \frac{1}{(R_1C_1)^2}$. Thus, both conditions have been verified. We can consider that the equation has a unique solution. We now present the numerical solution.

$$V_2(t) = V_2(0) - \int_0^t \left(\frac{1}{R_1C_2}\left(V_1(\tau) - R_2C_2\frac{dV_1(\tau)}{d\tau}\right) + \frac{1}{R_1C_1}V_2(\tau)\right)d\tau \quad (21.20)$$

We consider the equation at $t = t_{n+1} = (n+1)\Delta t$ and $t = t_n = n\Delta t$.

$$V_2(t_{n+1}) - V_2(t_n) = -\int_{t_n}^{t_{n+1}}\left(\frac{1}{R_1C_2}\left(V_1(\tau) - R_2C_2\frac{dV_1(\tau)}{d\tau}\right) + \frac{1}{R_1C_1}V_2(\tau)\right)d\tau$$

$$+ w\int_{t_n}^{t_{n+1}}V_2(\tau)dB(\tau).$$

(1) If $B(\tau)$ is differentiable then $dB(\tau) = B'(\tau)d\tau$. Then, we obtain

$$V_2(t_{n+1}) - V_2(t_n) = -\int_{t_n}^{t_{n+1}} \left(\frac{1}{R_1C_2}\left(V_1(\tau) - R_2C_2\frac{dV_1(\tau)}{d\tau} \right) + \frac{1}{R_1C_1}V_2(\tau) \right) d\tau$$
$$+ w\int_{t_n}^{t_{n+1}} V_2(\tau)B'(\tau)d\tau.$$

Using the Lagrange approximation gives

$$V_2(t_{n+1}) - V_2(t_n) = -\frac{3}{2}\Delta t \left(\frac{1}{R_1C_2}\left(V_1(t_n) - R_2C_2\frac{V_1(t_{n+1}) - V_1(t_n)}{\Delta t} \right) - \frac{1}{R_1C_1}V_2(t_n) \right)$$
$$+ \frac{\Delta t}{2}\left(\frac{1}{R_1C_2}\left(V_1(t_{n-1}) - R_2C_2\frac{V_1(t_n) - V_1(t_{n-1})}{\Delta t} \right) - \frac{1}{R_1C_1}V_2(t_n) \right)$$
$$+ \frac{3}{2}\frac{\Delta t}{R_1C_1}wV_2(t_n)\left(\frac{B(t_{n+1}) - B(t_n)}{\Delta t} \right) - \frac{\Delta t}{2R_1C_1}wV_2(t_{n-1})$$
$$\times \left(\frac{B(t_n) - B(t_{n-1})}{\Delta t} \right).$$

(2) If $B(t)$ is not differentiable, then we have

$$V_2(t_{n+1}) - V_2(t_n) = -\frac{3}{2}\Delta t \left(\frac{1}{R_1C_2}\left(V_1(t_n) - R_2C_2\frac{V_1(t_{n+1}) - V_1(t_n)}{\Delta t} \right) - \frac{1}{R_1C_1}V_2(t_n) \right)$$
$$+ \frac{\Delta t}{2}\left(\frac{1}{R_1C_2}\left(V_1(t_{n-1}) - R_2C_2\frac{V_1(t_n) - V_1(t_{n-1})}{\Delta t} \right) - \frac{1}{RC}V_2(t_{n-1}) \right)$$
$$+ w\sum_{i=0}^{n} V_2(t_j)\left(B(t_{j+1}) - B(t_j) \right)$$
$$- w\sum_{i=0}^{n-1} V_2(t_j)\left(B(t_{j+1}) - B(t_j) \right)$$

■

Let us consider the following classical stochastic model.

$$V_2(t) + (R_1 + R_2)C\frac{dV_2(t)}{dt} = V_1(t) + R_2C\frac{dV_1(t)}{dt} \qquad (21.21)$$

We added a stochastic component

$$V_2(t) = V_2(0) + \int_0^t f(\tau, V_2(\tau))d\tau + \int_0^t V_2(\tau)\omega_V dB(t) \qquad (21.22)$$

where $B(t)$ is assumed to be differential such that $dB(t) = B'(t)dt$. Then, we get

$$V_2(t) = V_2(0) + \int_0^t f(\tau, V_2(\tau))d\tau + \int_0^t V_2(\tau)\omega_V \dot{B}(t)dt \qquad (21.23)$$

Thus

$$V_2(t) = V_2(0) + \int_0^t f(\tau, V_2(\tau))d\tau + \int_0^t V_2(\tau)\omega dB(t) \qquad (21.24)$$

where ω is the density of randomness. Here

$$f(t, V_2(t)) = \frac{1}{(R_1+R_2)C}V_1(t) + \frac{R_2}{R_1+R_2}\frac{dV_1(t)}{dt} - \frac{1}{(R_1+R_2)C}V_2(t) \qquad (21.25)$$

We first present the existence and uniqueness of the above equation by verifying

(i)   Growth condition $\forall t \in D_V$, $|f(t, V_2(t))|^2 < k(1 + |V_2(t)|^2)$

(ii) The Lipschitz condition. Let V_{2_1} and V_{2_2} be the solutions of the equations.

$$|f(t, V_{2_1}(t)) - f(t, V_{2_2}(t))|^2 < \bar{k}|V_{2_1} - V_{2_1}|^2$$

PROOF (i)   In fact

$$
\begin{aligned}
|f(t, V_2(t))|^2 \;=\; & \left| \frac{1}{(R_1+R_2)C}V_1(t) + \frac{R_2}{R_1+R_2}\frac{dV_1(t)}{dt} - \frac{1}{(R_1+R_2)C}V_2(t) \right|^2 \\[2mm]
\leq\; & 3\frac{1}{((R_1+R_2)C)^2}|V_1(t)|^2 + \frac{3R_2^2}{(R_1+R_2)^2}\left|\frac{dV_1(t)}{dt}\right|^2 \\[1mm]
& + 3\frac{1}{((R_1+R_2)C)^2}|V_2(t)|^2 \\[2mm]
\leq\; & 3\frac{1}{((R_1+R_2)C)^2}\sup_{t\in D_t}|V_1(t)|^2 + \frac{3R_2^2}{(R_1+R_2)^2}\sup_{t\in D_t}\left|\frac{dV_1(t)}{dt}\right|^2 \\[1mm]
& + 3\frac{1}{((R_1+R_2)C)^2}|V_2(t)|^2 \\[2mm]
<\; & 3\frac{1}{((R_1+R_2)C)^2}\|V_1(t)\|_\infty^2 + \frac{3R_2^2}{(R_1+R_2)^2}\left\|\frac{dV_1(t)}{dt}\right\|_\infty^2 \\[1mm]
& + 3\frac{1}{((R_1+R_2)C)^2}|V_2(t)|^2 \\[2mm]
\leq\; & \left(3\frac{1}{((R_1+R_2)C)^2}\|V_1(t)\|_\infty^2 + \frac{3R_2^2}{(R_1+R_2)^2}\left\|\frac{dV_1(t)}{dt}\right\|_\infty^2 \right) \\[2mm]
& \times \left(1 + \frac{3\frac{1}{((R_1+R_2)C)^2}}{3\frac{1}{((R_1+R_2)C)^2}\|V_1(t)\|_\infty^2 + \frac{3R_2^2}{(R_1R_2)^2}\left\|\frac{dV_1(t)}{dt}\right\|_\infty^2}|V_2(t)|^2 \right)
\end{aligned}
$$

If $\dfrac{3\frac{1}{((R_1+R_2)C)^2}}{3\frac{1}{((R_1+R_2)C)^2}\|V_1(t)\|_\infty^2+\frac{3R_2^2}{(R_1R_2)^2}\left\|\frac{dV_1(t)}{dt}\right\|_\infty^2}<1$ then

$$|f(t,V_2(t))|^2 \;\leq\; k(1+|V_2(t)|^2)$$

where $k=3\dfrac{1}{((R_1+R_2)C)^2}\,\|V_1(t)\|_\infty^2+\dfrac{3R_2^2}{(R_1+R_2)^2}\left\|\dfrac{dV_1(t)}{dt}\right\|_\infty^2$. Indeed

$$|V_2(t)w|^2 = w^2\,|V_2(t)|^2 \leq w^2\left(1+|V_2(t)|^2\right)$$

Here $w=\bar{k}_1$.

(ii)   Indeed

$$\begin{aligned}
|f(t,V_{2_1}(t))-f(t,V_{2_2}(t))|^2 &= \left|-\frac{1}{(R_1+R_2)C}(V_{2_1}(t)-V_{2_2}(t))\right|^2\\
&= \frac{1}{((R_1+R_2)C)^2}|V_{2_1}(t)-V_{2_2}(t)|^2\\
&\leq \left(1+\frac{1}{((R_1+R_2)C)^2}\right)|V_{2_1}(t)-V_{2_2}(t)|^2\\
&\leq k|V_{2_1}(t)-V_{2_2}(t)|^2
\end{aligned}$$

where $k=1+\dfrac{1}{((R_1+R_2)C)^2}$. Thus, both conditions have been verified. We can consider that the equation has a unique solution. We now present the numerical solution.

$$V_2(t)=V_2(0)+\int_0^t\left(\frac{1}{(R_1+R_2)C}V_1(\tau)+\frac{R_2}{R_1+R_2}\frac{dV_1(\tau)}{dt}-\frac{1}{(R_1+R_2)C}V_2(\tau)\right)d\tau \tag{21.26}$$

We consider the equation at $t=t_{n+1}=(n+1)\Delta t$ and $t=t_n=n\Delta t$.

$$\begin{aligned}
&V_2(t_{n+1})-V_2(t_n)\\
&=\int_{t_n}^{t_{n+1}}\left(\frac{1}{(R_1+R_2)C}V_1(\tau)+\frac{R_2}{R_1+R_2}\frac{dV_1(\tau)}{dt}-\frac{1}{(R_1+R_2)C}V_2(\tau)\right)d\tau\\
&\quad+w\int_{t_n}^{t_{n+1}}V_2(\tau)dB(\tau).
\end{aligned}$$

(1)   If $B(\tau)$ is differentiable then $dB(\tau)=B'(\tau)d\tau$. Then, we obtain

$$\begin{aligned}
&V_2(t_{n+1})-V_2(t_n)\\
&=\int_{t_n}^{t_{n+1}}\left(\frac{1}{(R_1+R_2)C}V_1(\tau)+\frac{R_2}{R_1+R_2}\frac{dV_1(\tau)}{dt}-\frac{1}{(R_1+R_2)C}V_2(\tau)\right)d\tau\\
&\quad+w\int_{t_n}^{t_{n+1}}V_2(\tau)B'(\tau)d\tau.
\end{aligned}$$

Using the Lagrange approximation gives

$$
V_2(t_{n+1}) - V_2(t_n) = \frac{3}{2}\Delta t \left(\frac{1}{(R_1+R_2)C}V_1(t_n) + \left(\frac{R_2}{R_1+R2} \right) \frac{V_1(t_{n+1}) - V_1(t_n)}{\Delta t} \right.
$$

$$
\left. -\frac{1}{(R_1+R_2)C}V_2(t_n) \right)
$$

$$
-\frac{\Delta t}{2} \left(\frac{1}{(R_1+R_2)C}V_1(t_{n-1}) + \left(\frac{R_2}{R_1+R2} \right) \frac{V_1(t_n) - V_1(t_{n-1})}{\Delta t} \right.
$$

$$
\left. -\frac{1}{(R_1+R_2)C}V_2(t_{n-1}) \right)
$$

$$
+\frac{3}{2}\frac{\Delta t}{(R_1+R_2)C}wV_2(t_n)\left(\frac{B(t_{n+1}) - B(t_n)}{\Delta t} \right)
$$

$$
-\frac{\Delta t}{2(R_1+R_2)C}wV_2(t_{n-1})\left(\frac{B(t_n) - B(t_{n-1})}{\Delta t} \right).
$$

(2) If $B(t)$ is not differentiable, then we have

$$
V_2(t_{n+1}) - V_2(t_n) = \frac{3}{2}\Delta t \left(\frac{1}{(R_1+R_2)C}V_1(t_n) + \left(\frac{R_2}{R_1+R2} \right) \frac{V_1(t_{n+1}) - V_1(t_n)}{\Delta t} \right.
$$

$$
\left. -\frac{1}{(R_1+R_2)C}V_2(t_n) \right)
$$

$$
-\frac{\Delta t}{2} \left(\frac{1}{(R_1+R_2)C}V_1(t_{n-1}) + \left(\frac{R_2}{R_1+R2} \right) \frac{V_1(t_n) - V_1(t_{n-1})}{\Delta t} \right.
$$

$$
\left. -\frac{1}{(R_1+R_2)C}V_2(t_{n-1}) \right)
$$

$$
+w\sum_{i=0}^{n} V_2(t_j)\left(B(t_{j+1}) - B(t_j) \right)
$$

$$
-w\sum_{i=0}^{n-1} V_2(t_j)\left(B(t_{j+1}) - B(t_j) \right)
$$

∎

Let us consider the following classical stochastic model.

$$
\frac{dV_2(t)}{dt} + \left(\frac{R_1+R_2}{R_1R_2C} \right)V_2(t) = \frac{dV_1(t)}{dt} + \frac{1}{R_1C}V_1(t) \tag{21.27}
$$

We added a stochastic component

$$
V_2(t) = V_2(0) + \int_0^t f(\tau, V_2(\tau))d\tau + \int_0^t V_2(\tau)\omega_V dB(t) \tag{21.28}
$$

where $B(t)$ is assumed to be differential such that $dB(t) = B'(t)dt$. Then, we get

$$
V_2(t) = V_2(0) + \int_0^t f(\tau, V_2(\tau))d\tau + \int_0^t V_2(\tau)\omega_V \dot{B}(t)dt \tag{21.29}
$$

Thus

$$V_2(t) = V_2(0) + \int_0^t f(\tau, V_2(\tau)) d\tau + \int_0^t V_2(\tau) \omega dB(t) \qquad (21.30)$$

where ω is the density of randomness. Here

$$f(t, V_2(t)) = \frac{dV_1(t)}{dt} + \frac{1}{R_1 C} V_1(t) - \left(\frac{R_1 + R_2}{R_1 R_2 C}\right) V_2(t) \qquad (21.31)$$

We first present the existence and uniqueness of the above equation by verifying

(i) Growth condition $\forall t \subset D_V$, $|f(t, V_2(t))|^2 < k(1 + |V_2(t)|^2)$

(ii) The Lipschitz condition. Let V_{2_1} and V_{2_2} be the solutions of the equations.

$$|f(t, V_{2_1}(t)) - f(t, V_{2_2}(t))|^2 < \bar{k} |V_{2_1} - V_{2_1}|^2$$

PROOF (i)   In fact

$$
\begin{aligned}
|f(t, V_2(t))|^2 &= \left| \frac{dV_1(t)}{dt} + \frac{1}{R_1 C} V_1(t) - \left(\frac{R_1 + R_2}{R_1 R_2 C}\right) V_2(t) \right|^2 \\[2mm]
&\leq 3 \left| \frac{dV_1(t)}{dt} \right|^2 + \frac{3}{(R_1 C)^2} |V_1(t)|^2 + 3 \frac{(R_1 + R_2)^2}{(R_1 R_2 C)^2} |V_2(t)|^2 \\[2mm]
&\leq 3 \sup_{t \in D_t} \left| \frac{dV_1(t)}{dt} \right|^2 + \frac{3}{(R_1 C)^2} \sup_{t \in D_t} |V_1(t)|^2 + 3 \frac{(R_1 + R_2)^2}{(R_1 R_2 C)^2} |V_2(t)|^2 \\[2mm]
&\leq 3 \left\| \frac{dV_1(t)}{dt} \right\|_\infty^2 + \frac{3}{(R_1 C)^2} \|V_1(t)\|_\infty^2 + 3 \frac{(R_1 + R_2)^2}{(R_1 R_2 C)^2} |V_2(t)|^2 \\[2mm]
&\leq \left(3 \left\| \frac{dV_1(t)}{dt} \right\|_\infty^2 + \frac{3}{(R_1 C)^2} \|V_1(t)\|_\infty^2 \right) \\[2mm]
&\quad \times \left(1 + \frac{3 \frac{(R_1+R_2)^2}{(R_1 R_2 C)^2}}{3 \left\| \frac{dV_1(t)}{dt} \right\|_\infty^2 + \frac{3}{(R_1 C)^2} \|V_1(t)\|_\infty^2} \left\| \frac{dV_1(t)}{dt} \right\|_\infty^2 |V_2(t)|^2 \right)
\end{aligned}
$$

If $3\frac{(R_1+R_2)^2}{(R_1 R_2 C)^2} < \left(3 \left\| \frac{dV_1(t)}{dt} \right\|_\infty^2 + \frac{3}{(R_1 C)^2} \|V_1(t)\|_\infty^2 \right)$, then

$$|f(t, V_2(t))|^2 \leq k(1 + |V_2(t)|^2)$$

where $k = 3 \left\| \frac{dV_1(t)}{dt} \right\|_\infty^2 + \frac{3}{(R_1 C)^2} \|V_1(t)\|_\infty^2$. Indeed

$$|V_2(t)w|^2 = w^2 |V_2(t)|^2 \leq w^2 \left(1 + |V_2(t)|^2 \right)$$

Here $w = \bar{k}_1$.

(ii) Indeed

$$
\begin{aligned}
|f(t, V_{2_1}(t)) - f(t, V_{2_2}(t))|^2 &= \left| -\frac{(R_1 + R_2)}{(R_1 R_2 C)} (V_{2_1}(t) - V_{2_2}(t)) \right|^2 \\
&= \frac{(R_1 + R_2)^2}{(R_1 R_2 C)^2} |V_{2_1}(t) - V_{2_2}(t)|^2 \\
&\leq \left(1 + \frac{(R_1 + R_2)^2}{(R_1 R_2 C)^2} \right) |V_{2_1}(t) - V_{2_2}(t)|^2 \\
&\leq k |V_{2_1}(t) - V_{2_2}(t)|^2
\end{aligned}
$$

where $k = 1 + \frac{(R_1 + R_2)^2}{(R_1 R_2 C)^2}$. Thus, both conditions have been verified. We can consider that the equation has a unique solution. We now present the numerical solution.

$$
V_2(t) = V_2(0) + \int_0^t \left(\frac{dV_1(\tau)}{d\tau} + \frac{1}{R_1 C} V_1(\tau) - \left(\frac{R_1 + R_2}{R_1 R_2 C} \right) V_2(\tau) \right) d\tau \qquad (21.32)
$$

We consider the equation at $t = t_{n+1} = (n+1)\Delta t$ and $t = t_n = n\Delta t$.

$$
\begin{aligned}
V_2(t_{n+1}) - V_2(t_n) &= \int_{t_n}^{t_{n+1}} \left(\frac{dV_1(\tau)}{d\tau} + \frac{1}{R_1 C} V_1(\tau) - \left(\frac{R_1 + R_2}{R_1 R_2 C} \right) V_2(\tau) \right) d\tau \\
&+ w \int_{t_n}^{t_{n+1}} V_2(\tau) dB(\tau).
\end{aligned}
$$

(1) If $B(\tau)$ is differentiable then $dB(\tau) = B'(\tau) d\tau$. Then, we obtain

$$
\begin{aligned}
V_2(t_{n+1}) - V_2(t_n) &= \int_{t_n}^{t_{n+1}} \left(\frac{dV_1(\tau)}{d\tau} + \frac{1}{R_1 C} V_1(\tau) - \left(\frac{R_1 + R_2}{R_1 R_2 C} \right) V_2(\tau) \right) d\tau \\
&+ w \int_{t_n}^{t_{n+1}} V_2(\tau) B'(\tau) d\tau.
\end{aligned}
$$

Using the Lagrange approximation gives

$$
\begin{aligned}
V_2(t_{n+1}) - V_2(t_n) &= \frac{3}{2} \Delta t \left(\frac{1}{R_1 C} V_1(t_n) + \frac{V_1(t_{n+1}) - V_1(t_n)}{\Delta t} \right. \\
&\qquad \left. - \frac{R_1 + R_2}{R_1 R_2 C} V_2(t_n) \right) \\
&\quad - \frac{\Delta t}{2} \left(\frac{1}{R_1 C} V_1(t_{n-1}) + \frac{V_1(t_n) - V_1(t_{n-1})}{\Delta t} \right. \\
&\qquad \left. - \frac{R_1 + R_2}{R_1 R_2 C} V_2(t_{n-1}) \right) \\
&\quad + \frac{3\Delta t}{2} w V_2(t_n) \left(\frac{B(t_{n+1}) - B(t_n)}{\Delta t} \right) \\
&\quad - \frac{\Delta t}{2} w V_2(t_{n-1}) \left(\frac{B(t_n) - B(t_{n-1})}{\Delta t} \right).
\end{aligned}
$$

(2) If $B(t)$ is not differentiable, then we have

$$
\begin{aligned}
V_2(t_{n+1}) - V_2(t_n) \;=\; & \frac{3}{2}\Delta t \left(\frac{1}{R_1 C} V_1(t_n) + \frac{V_1(t_{n+1}) - V_1(t_n)}{\Delta t} \right. \\
& \left. - \frac{R_1 + R_2}{R_1 R_2 C} V_2(t_n) \right) \\
& - \frac{\Delta t}{2} \left(\frac{1}{R_1 C} V_1(t_{n-1}) + \frac{V_1(t_n) - V_1(t_{n-1})}{\Delta t} \right. \\
& \left. - \frac{R_1 + R_2}{R_1 R_2 C} V_2(t_{n-1}) \right) \\
& + w \sum_{i=0}^{n} V_2(t_j) \left(B(t_{j+1}) - B(t_j) \right) \\
& - w \sum_{i=0}^{n-1} V_2(t_j) \left(B(t_{j+1}) - B(t_j) \right)
\end{aligned}
$$

∎

Let us consider the following classical stochastic model.

$$
\frac{dV_2(t)}{dt} + 4 \times 10^6 V_2(t) = -4 \times 10^7 V_1(t) \tag{21.33}
$$

We added a stochastic component

$$
V_2(t) = V_2(0) + \int_0^t f(\tau, V_2(\tau)) d\tau + \int_0^t V_2(\tau) \omega_v dB(t) \tag{21.34}
$$

where $B(t)$ is assumed to be differential such that $dB(t) = B'(t)dt$. Then, we get

$$
V_2(t) = V_2(0) + \int_0^t f(\tau, V_2(\tau)) d\tau + \int_0^t V_2(\tau) \omega_v \dot{B}(t) dt \tag{21.35}
$$

Thus

$$
V_2(t) = V_2(0) + \int_0^t f(\tau, V_2(\tau)) d\tau + \int_0^t V_2(\tau) \omega dB(t) \tag{21.36}
$$

where ω is the density of randomness. Here

$$
f(t, V_2(t)) = -4 \times 10^7 V_1(t) - 4 \times 10^6 V_2(t) \tag{21.37}
$$

We first present the existence and uniqueness of the above equation by verifying

(i) Growth condition $\forall t \in D_V$, $|f(t, V_2(t))|^2 < k(1 + |V_2(t)|^2)$

(ii) The Lipschitz condition. Let V_{2_1} and V_{2_2} be the solutions of the equations.

$$
|f(t, V_{2_1}(t)) - f(t, V_{2_2}(t))|^2 < \bar{k} |V_{2_1} - V_{2_1}|^2
$$

PROOF (i)   In fact

$$
\begin{aligned}
|f(t,V_2(t))|^2 &= \left| -4 \times 10^7 V_1(t) - 4 \times 10^6 V_2(t) \right|^2 \\
&\leq 2(4 \times 10^7)^2 |V_1(t)|^2 + 2(4 \times 10^6)^2 |V_2(t)|^2 \\
&\leq 2(4 \times 10^7)^2 \sup_{t \in D_t} |V_1(t)|^2 + 2(4 \times 10^6)^2 |V_2(t)|^2 \\
&\leq 2(4 \times 10^7)^2 \|V_1(t)\|_\infty^2 + 2(4 \times 10^6)^2 |V_2(t)|^2 \\
&\leq \left(2(410^7)^2 \|V_1(t)\|_\infty^2 \right) \\
&\quad \times \left(1 + \frac{2(4 \times 10^6)^2 |V_2(t)|^2}{2(410^7)^2 \|V_1(t)\|_\infty^2} \right)
\end{aligned}
$$

If $2(410^6)^2 < 2(410^7)^2 \|V_1(t)\|_\infty^2$, then

$$
|f(t,V_2(t))|^2 \leq k(1 + |V_2(t)|^2)
$$

where $k = 2(410^7)^2 \|V_1(t)\|_\infty^2$. Indeed

$$
|V_2(t)w|^2 = w^2 |V_2(t)|^2 \leq w^2 \left(1 + |V_2(t)|^2 \right)
$$

Here $w = \bar{k}_1$.

(ii)   Indeed

$$
\begin{aligned}
|f(t,V_{2_1}(t)) - f(t,V_{2_2}(t))|^2 &= \left| -4 \times 10^6 \left(V_{2_1}(t) - V_{2_2}(t) \right) \right|^2 \\
&= (4 \times 10^6)^2 |V_{2_1}(t) - V_{2_2}(t)|^2 \\
&\leq \left(1 + (4 \times 10^6)^2 \right) |V_{2_1}(t) - V_{2_2}(t)|^2 \\
&\leq k |V_{2_1}(t) - V_{2_2}(t)|^2
\end{aligned}
$$

where $k = 1 + (4 \times 10^6)^2$. Thus, both conditions have been verified. We can consider that the equation has a unique solution. We now present the numerical solution.

$$
V_2(t) = V_2(0) - \int_0^t \left(4 \times 10^7 V_1(\tau) + 4 \times 10^6 V_2(\tau) \right) d\tau \tag{21.38}
$$

We consider the equation at $t = t_{n+1} - (n+1)\Delta t$ and $t = t_n = n\Delta t$.

$$
\begin{aligned}
V_2(t_{n+1}) - V_2(t_n) &= -\int_{t_n}^{t_{n+1}} \left(4 \times 10^7 V_1(\tau) + 4 \times 10^6 V_2(\tau) \right) d\tau \\
&\quad + w \int_{t_n}^{t_{n+1}} V_2(\tau) dB(\tau).
\end{aligned}
$$

(1) If $B(\tau)$ is differentiable then $dB(\tau) = B'(\tau)d\tau$. Then, we obtain

$$V_2(t_{n+1}) - V_2(t_n) = \int_{t_n}^{t_{n+1}} \left(4 \times 10^7 V_1(\tau) + 4 \times 10^6 V_2(\tau)\right) d\tau$$

$$+ w \int_{t_n}^{t_{n+1}} V_2(\tau) B'(\tau) d\tau.$$

Using the Lagrange approximation gives

$$V_2(t_{n+1}) - V_2(t_n) = \frac{3}{2} \Delta t \left(4 \times 10^7 V_1(t_n) + 4 \times 10^6 V_2(t_n)\right)$$

$$- \frac{\Delta t}{2} \left(4 \times 10^7 V_1(t_{n-1}) + 4 \times 10^6 V_2(t_{n-1})\right)$$

$$+ \frac{3\Delta t}{2} w V_2(t_n) \left(\frac{B(t_{n+1}) - B(t_n)}{\Delta t}\right)$$

$$- \frac{\Delta t}{2} w V_2(t_{n-1}) \left(\frac{B(t_n) - B(t_{n-1})}{\Delta t}\right).$$

(2) If $B(t)$ is not differentiable, then we have

$$V_2(t_{n+1}) - V_2(t_n) = \frac{3}{2} \Delta t \left(4 \times 10^7 V_1(t_n) + 4 \times 10^6 V_2(t_n)\right)$$

$$- \frac{\Delta t}{2} \left(4 \times 10^7 V_1(t_{n-1}) + 4 \times 10^6 V_2(t_{n-1})\right)$$

$$+ w \sum_{i=0}^{n} V_2(t_j) \left(B(t_{j+1}) - B(t_j)\right)$$

$$- w \sum_{i=0}^{n-1} V_2(t_j) \left(B(t_{j+1}) - B(t_j)\right)$$

∎

Let us consider the following classical stochastic model.

$$R_1 R_2 (C_1 + C_2) \frac{dV_2(t)}{dt} + (R_1 + R_2) V_2(t) = R_1 R_2 C_1 \frac{dV_1(t)}{dt} + R_2 V_1(t) \qquad (21.39)$$

We added a stochastic component

$$V_2(t) = V_2(0) + \int_0^t f(\tau, V_2(\tau)) d\tau + \int_0^t V_2(\tau) \omega_V dB(t) \qquad (21.40)$$

where $B(t)$ is assumed to be differential such that $dB(t) = B'(t)dt$. Then, we get

$$V_2(t) = V_2(0) + \int_0^t f(\tau, V_2(\tau)) d\tau + \int_0^t V_2(\tau) \omega_V \dot{B}(t) dt \qquad (21.41)$$

Thus

$$V_2(t) = V_2(0) + \int_0^t f(\tau, V_2(\tau)) d\tau + \int_0^t V_2(\tau) \omega dB(t) \qquad (21.42)$$

where ω is the density of randomness. Here

$$f(t, V_2(t)) = \frac{C_1}{C_1 + C_2} \frac{dV_1(t)}{dt} + \frac{1}{R_1(C_1 + C_2)} V_1(t) - \frac{R_1 + R_2}{R_1 R_2(C_1 + C_2)} V_2(t) \quad (21.43)$$

We first present the existence and uniqueness of the above equation by verifying

(i)   Growth condition $\forall t \in D_V$, $|f(t, V_2(t))|^2 < k(1 + |V_2(t)|^2)$

(ii) The Lipschitz condition. Let V_{2_1} and V_{2_2} be the solutions of the equations.

$$|f(t, V_{2_1}(t)) - f(t, V_{2_2}(t))|^2 < \bar{k}|V_{2_1} - V_{2_1}|^2$$

PROOF (i)   In fact

$$|f(t, V_2(t))|^2$$

$$= \left| \frac{C_1}{C_1 + C_2} \frac{dV_1(t)}{dt} + \frac{1}{R_1(C_1 + C_2)} V_1(t) - \left(\frac{R_1 + R_2}{R_1 R_2(C_1 + C_2)} \right) V_2(t) \right|^2$$

$$\leq 3 \left(\frac{C_1}{C_1 + C_2} \right)^2 \left| \frac{dV_1(t)}{dt} \right|^2 + 3 \left(\frac{1}{R_1(C_1 + C_2)} \right)^2 |V_1(t)|^2 + 3 \frac{(R_1 + R_2)^2}{(R_1 R_2 C)^2} |V_2(t)|^2$$

$$\leq \left(\frac{C_1}{C_1 + C_2} \right)^2 \sup_{t \in D_t} \left| \frac{dV_1(t)}{dt} \right|^2 7 + 3 \left(\frac{1}{R_1(C_1 + C_2)} \right)^2 \sup_{t \in D_t} |V_1(t)|^2$$

$$+ 3 \frac{(R_1 + R_2)^2}{(R_1 R_2 C)^2} |V_2(t)|^2$$

$$\leq 3 \left(\frac{C_1}{C_1 + C_2} \right)^2 \left\| \frac{dV_1(t)}{dt} \right\|_\infty^2 + 3 \left(\frac{1}{R_1(C_1 + C_2)} \right)^2 \|V_1(t)\|_\infty^2 + 3 \frac{(R_1 + R_2)^2}{(R_1 R_2 C)^2} |V_2(t)|^2$$

$$\leq \left(3 \left(\frac{C_1}{C_1 + C_2} \right)^2 \left\| \frac{dV_1(t)}{dt} \right\|_\infty^2 + 3 \left(\frac{1}{R_1(C_1 + C_2)} \right)^2 \|V_1(t)\|_\infty^2 \right)$$

$$\times \left(1 + \frac{3 \frac{(R_1 + R_2)^2}{(R_1 R_2 C)^2}}{3 \left(\frac{C_1}{C_1 + C_2} \right)^2 \left\| \frac{dV_1(t)}{dt} \right\|_\infty^2 + 3 \left(\frac{1}{R_1(C_1 + C_2)} \right)^2 \|V_1(t)\|_\infty^2} \|V_1(t)\|_\infty^2 |V_2(t)|^2 \right)$$

If $3 \frac{(R_1 + R_2)^2}{(R_1 R_2 C)^2} < 3 \left(\frac{C_1}{C_1 + C_2} \right)^2 \left\| \frac{dV_1(t)}{dt} \right\|_\infty^2 + 3 \left(\frac{1}{R_1(C_1 + C_2)} \right)^2 \|V_1(t)\|_\infty^2$, then

$$|f(t, V_2(t))|^2 \leq k(1 + |V_2(t)|^2)$$

where $k = 3 \left(\frac{C_1}{C_1 + C_2} \right)^2 \left\| \frac{dV_1(t)}{dt} \right\|_\infty^2 + 3 \left(\frac{1}{R_1(C_1 + C_2)} \right)^2 \|V_1(t)\|_\infty^2$. Indeed

$$|V_2(t)w|^2 = w^2 |V_2(t)|^2 \leq w^2 \left(1 + |V_2(t)|^2 \right)$$

Here $w = \bar{k}_1$.

(ii)   Indeed

$$
\begin{aligned}
|f(t, V_{2_1}(t)) - f(t, V_{2_2}(t))|^2 &= \left| -\frac{(R_1 + R_2)}{(R_1 R_2 (C_1 + C_2))} (V_{2_1}(t) - V_{2_2}(t)) \right|^2 \\
&= \frac{(R_1 + R_2)^2}{(R_1 R_2 (C_1 + C_2))^2} |V_{2_1}(t) - V_{2_2}(t)|^2 \\
&\leq \left(1 + \frac{(R_1 + R_2)^2}{(R_1 R_2 (C_1 + C_2))^2} \right) |V_{2_1}(t) - V_{2_2}(t)|^2 \\
&\leq k |V_{2_1}(t) - V_{2_2}(t)|^2
\end{aligned}
$$

where $k = 1 + \frac{(R_1 + R_2)^2}{(R_1 R_2 (C_1 + C_2))^2}$. Thus, both conditions have been verified. We can consider that the equation has a unique solution. We now present the numerical solution.

$$
\begin{aligned}
V_2(t) = V_2(0) \\
+ \int_0^t \left(\frac{C_1}{C_1 + C_2} \frac{dV_1(\tau)}{dt} + \frac{1}{R_1(C_1 + C_2)} V_1(\tau) - \frac{R_1 + R_2}{R_1 R_2(C_1 + C_2)} V_2(\tau) \right) d\tau
\end{aligned}
$$

$$(21.44)$$

We consider the equation at $t = t_{n+1} = (n+1)\Delta t$ and $t = t_n = n\Delta t$.

$$
\begin{aligned}
V_2(t_{n+1}) - V_2(t_n) \\
= \int_{t_n}^{t_{n+1}} \left(\frac{C_1}{C_1 + C_2} \frac{dV_1(\tau)}{dt} + \frac{1}{R_1(C_1 + C_2)} V_1(\tau) - \frac{R_1 + R_2}{R_1 R_2(C_1 + C_2)} V_2(\tau) \right) d\tau \\
+ w \int_{t_n}^{t_{n+1}} V_2(\tau) dB(\tau).
\end{aligned}
$$

(1)   If $B(\tau)$ is differentiable then $dB(\tau) = B'(\tau) d\tau$. Then, we obtain

$$
\begin{aligned}
V_2(t_{n+1}) - V_2(t_n) \\
= \int_{t_n}^{t_{n+1}} \left(\frac{C_1}{C_1 + C_2} \frac{dV_1(\tau)}{dt} + \frac{1}{R_1(C_1 + C_2)} V_1(\tau) - \frac{R_1 + R_2}{R_1 R_2(C_1 + C_2)} V_2(\tau) \right) d\tau \\
+ w \int_{t_n}^{t_{n+1}} V_2(\tau) B'(\tau) d\tau.
\end{aligned}
$$

Using the Lagrange approximation gives

$$
\begin{aligned}
V_2(t_{n+1}) - V_2(t_n) = {} & \frac{3}{2}\Delta t \left(\frac{1}{R_1(C_1 + C_2)} V_1(t_n) + \frac{C_1}{C_1 + C_2} \frac{V_1(t_{n+1}) - V_1(t_n)}{\Delta t} \right. \\
& \left. - \frac{R_1 + R_2}{R_1 R_2(C_1 + C_2)} V_2(t_n) \right) \\
& - \frac{\Delta t}{2} \left(\frac{1}{R_1(C_1 + C_2)} V_1(t_{n-1}) + \frac{C_1}{C_1 + C_2} \frac{V_1(t_n) - V_1(t_{n-1})}{\Delta t} \right. \\
& \left. - \frac{R_1 + R_2}{R_1 R_2(C_1 + C_2)} V_2(t_{n-1}) \right) \\
& + \frac{3\Delta t}{2(R_1 R_2(C_1 + C_2))} w V_2(t_n) \left(\frac{B(t_{n+1}) - B(t_n)}{\Delta t} \right) \\
& - \frac{\Delta t}{2(R_1 R_2(C_1 + C_2))} w V_2(t_{n-1}) \left(\frac{B(t_n) - B(t_{n-1})}{\Delta t} \right).
\end{aligned}
$$

(2) If $B(t)$ is not differentiable, then we have

$$
\begin{aligned}
V_2(t_{n+1}) - V_2(t_n) \;=\; & \frac{3}{2}\Delta t \left(\frac{1}{R_1(C_1+C_2)}V_1(t_n) + \frac{C_1}{C_1+C_2}\frac{V_1(t_{n+1})-V_1(t_n)}{\Delta t} \right. \\
& \left. - \frac{R_1+R_2}{R_1R_2(C_1+C_2)}V_2(t_n) \right) \\
& - \frac{\Delta t}{2} \left(\frac{1}{R_1(C_1+C_2)}V_1(t_{n-1}) + \frac{C_1}{C_1+C_2}\frac{V_1(t_n)-V_1(t_{n-1})}{\Delta t} \right. \\
& \left. - \frac{R_1+R_2}{R_1R_2(C_1+C_2)}V_2(t_{n-1}) \right) \\
& + w\sum_{i=0}^{n} V_2(t_j)\left(B(t_{j+1}) - B(t_j) \right) \\
& - w\sum_{i=0}^{n-1} V_2(t_j)\left(B(t_{j+1}) - B(t_j) \right)
\end{aligned}
$$

∎

We consider the following problem:

$$
\frac{1}{RC}\frac{di(t)}{dt} + \frac{d^2i(t)}{dt^2} = \frac{V(t)}{RCL} + \frac{1}{L}\frac{dV(t)}{dt} \tag{21.45}
$$

In this case, we assume that the function $V(t)$ is knomn. Then, for simplicity we assume

$$
F(t,V(t)) = \frac{V(t)}{RCL} + \frac{1}{L}\frac{dV(t)}{dt}. \tag{21.46}
$$

Then, the equation becomes

$$
\frac{1}{RC}\frac{di(t)}{dt} = F(t,V(t)) - \frac{d^2i(t)}{dt^2} \tag{21.47}
$$

$$
\frac{di(t)}{dt} - RCF(t,V(t)) - RC\frac{d^2i(t)}{dt^2} = H(t,i(t)) \tag{21.48}
$$

We can now convert the above equation into integral equation and the randomness to obtain

$$
i(t) - i(0) = \int_0^t H(\tau,i(\tau))d\tau + \sigma \int_0^t i(\tau)dB(\tau) \tag{21.49}
$$

Again we present the existence and uniqueness of the above problem by verifying:

(i)   Growth condition $\forall t \in D_V$, $|f(t,i(t))|^2 < k(1+|i(t)|^2)$
(ii) The Lipschitz condition. Let i_1 and i_2 be the solutions of the equations.

$$
|f(t,i_1(t)) - f(t,i_2(t))|^2 < \bar{k}|i_1 - i_2|^2
$$

$$|H(t,i(t))|^2 = \left|RC\left[F(t,V(t)) - \frac{d^2i(t)}{dt^2}\right]\right|^2$$

$$= (RC)^2\left|F(t,V(t)) - \frac{d^2i(t)}{dt^2}\right|^2$$

$$\leq (RC)^2\left[2|F(t,V(t))|^2 + 2\left|\frac{d^2i(t)}{dt^2}\right|^2\right]$$

$$\leq 2(RC)^2\left[\sup_{t\in D_F}|F(t,V(t))|^2 + \sup_{t\in D_{i_{tt}}}\left|\frac{d^2i(t)}{dt^2}\right|^2\right]$$

$$\leq 2(RC)^2\left[\|F(t,V(t))\|_\infty^2 + \left\|\frac{d^2i(t)}{dt^2}\right\|_\infty^2\right](1+|i(t)|^2)$$

$$\leq k\left(1+|i(t)|^2\right)$$

$$|H(t,i_1(t)) - H(t,i_2(t))|^2 = (RC)^2\left|\frac{d^2}{dt^2}(i_1(t) - i_2(t))\right|^2$$

We asuming that $\frac{d^2}{dt^2}(i_1(t) - i_2(t))$ is bounded. Then, we obtain

$$|H(t,i_1(t)) - H(t,i_2(t))|^2 = (RC)^2\left|\frac{d^2}{dt^2}(i_1(t) - i_2(t))\right|^2$$

$$\leq (1+(RC)^2)\sup_{t\in D_{\frac{d^2}{dt^2}i_1 - i_2}}\left|\frac{d^2}{dt^2}(i_1(t) - i_2(t))\right|^2$$

$$\times |i_1(t) - i_2(t)|^2$$

$$\leq k|i_1(t) - i_2(t)|^2$$

Under the conditions the above were derived, we can conclude that the equation has unique solution. We now present the numerical solution at $t = t_{n+1}$ and $t = t_n$. Then, we have

$$i(t_{n+1}) - i(t_n) = \frac{3}{2}\Delta t H(t_n, i(t_n)) - \frac{1}{2}\Delta t H(t_{n-1}, i(t_{n-1}))$$

$$+ \frac{3}{2}\Delta t\,\sigma i(t_n)\frac{B(t_{n+1}) - B(t_n)}{\Delta t}$$

$$- \frac{1}{2}\Delta t\,\sigma i(t_{n-1})\frac{B(t_n) - B(t_{n-1})}{\Delta t}$$

If $B(t)$ is not differentiable, then

$$
\begin{aligned}
i(t_{n+1}) - i(t_n) &= \frac{3}{2}\Delta t H(t_n, i(t_n)) - \frac{1}{2}\Delta t H(t_{n-1}, i(t_{n-1})) \\
&\quad + \sigma RC \sum_{i=0}^{n} i(t_j)\left(B(t_{j+1}) - B(t_j)\right) \\
&\quad - RC\sigma \sum_{i=0}^{n-1} i(t_j)\left(B(t_{j+1}) - B(t_j)\right)
\end{aligned}
$$

Here

$$
H(t_n, i(t_n)) = RCF(t_n, V(t_n)) - RC\left(\frac{i(t_{n+1}) - 2i(t_n) + i(t_{n-1})}{(\Delta t)^2}\right)
$$

$$
F(t_n, V(t_n)) = \frac{V(t_n)}{RCL} + \frac{1}{L}\frac{V(t_{n+1}) - V(t_n)}{\Delta t}
$$

We consider the following problem:

$$
\frac{R_1}{CL}i(t) + \frac{1}{C}\frac{di(t)}{dt} + \frac{V(t)}{CL} = \frac{d^2V(t)}{dt^2} + \left(\frac{R_1}{L} + \frac{1}{R_2C}\right)\frac{dV(t)}{dt} + \frac{R_1 + R_2}{R_2CL}V(t) \quad (21.50)
$$

In this case we assume that the function $i(t)$ is known. Then, for simplicity we assume

$$
F(t, i(t)) = \frac{R_1}{CL}i(t) + \frac{1}{C}\frac{di(t)}{dt} \quad (21.51)
$$

Then, the equation becomes

$$
\left(\frac{R_1}{L} + \frac{1}{R_2C}\right)\frac{dV(t)}{dt} = F(t, i(t)) + \frac{V(t)}{CL} - \frac{R_1 + R_2}{R_2CL}V(t) - \frac{d^2V(t)}{dt^2} \quad (21.52)
$$

$$
\frac{dV(t)}{dt} = \frac{1}{\left(\frac{R_1}{L} + \frac{1}{R_2C}\right)}\left(F(t, i(t)) + \frac{V(t)}{CL} - \frac{R_1 + R_2}{R_2CL}V(t) - \frac{d^2V(t)}{dt^2}\right) \quad (21.53)
$$

$$
\frac{dV(t)}{dt} = \frac{LR_2C}{R_1R_2C \mid L}\left(F(t, i(t)) + \left(\frac{1}{CL} - \frac{R_1 + R_2}{R_2CL}\right)V(t) - \frac{d^2V(t)}{dt^2}\right) \quad (21.54)
$$

We can now convert the above equation into integral equation and the randomness to obtain

$$
V(t) - V(0) = \int_0^t H(\tau, V(\tau))d\tau + \sigma \int_0^t V(\tau)dB(\tau) \quad (21.55)
$$

Again we present the existence and uniqueness of the above problem by verifying:

(i)   Growth condition $\forall t \in D_V$, $|f(t, V(t))|^2 < k(1 + |V(t)|^2)$

(ii)   The Lipschitz condition. Let V_1 and V_2 be the solutions of the equations.

$$|f(t, V_1(t)) - f(t, V_2(t))|^2 < \bar{k}|V_1 - V_2|^2$$

$$
\begin{aligned}
|H(t, V(t))|^2 &= \left| \frac{LR_2C}{R_1R_2C + L} \left(F(t, i(t)) + \left(\frac{1}{CL} - \frac{R_1 + R_2}{R_2CL} \right) V(t) - \frac{d^2V(t)}{dt^2} \right) \right|^2 \\
&= \left(\frac{LR_2C}{R_1R_2C + L} \right)^2 \left| F(t, i(t)) + \left(\frac{1}{CL} - \frac{R_1 + R_2}{R_2CL} \right) V(t) - \frac{d^2V(t)}{dt^2} \right|^2 \\
&\leq \left(\frac{LR_2C}{R_1R_2C + L} \right)^2 \left[3|F(t, i(t))|^2 + 3 \left(\frac{1}{CL} - \frac{R_1 + R_2}{R_2CL} \right)^2 |V(t)|^2 \right. \\
&\quad \left. + 3 \left| \frac{d^2V(t)}{dt} \right|^2 \right] \\
&\leq \left(\frac{LR_2C}{R_1R_2C + L} \right)^2 \left[3 \sup_{t \in D_F} |F(t, i(t))|^2 + 3 \left(\frac{1}{CL} - \frac{R_1 + R_2}{R_2CL} \right)^2 |V(t)|^2 \right. \\
&\quad \left. + 3 \sup_{t \in D_V} \left| \frac{d^2V(t)}{dt} \right|^2 \right] \\
&\leq \left(\frac{LR_2C}{R_1R_2C + L} \right)^2 \left[3\|F(t, i(t))\|_\infty^2 + 3 \left(\frac{1}{CL} - \frac{R_1 + R_2}{R_2CL} \right)^2 |V(t)|^2 \right. \\
&\quad \left. + 3 \left\| \frac{d^2V(t)}{dt} \right\|_\infty^2 \right] \\
&\leq \left(\frac{LR_2C}{R_1R_2C + L} \right)^2 3 \left(\|F(t, i(t))\|_\infty^2 + \left\| \frac{d^2V(t)}{dt} \right\|_\infty^2 \right) \\
&\quad \times \left[1 + \frac{3 \left(\frac{1}{CL} - \frac{R_1 + R_2}{R_2CL} \right)^2 |V(t)|^2}{\|F(t, i(t))\|_\infty^2 + \left\| \frac{d^2V(t)}{dt} \right\|_\infty^2} \right]
\end{aligned}
$$

If

$$3 \left(\frac{1}{CL} - \frac{R_1 + R_2}{R_2CL} \right)^2 < \|F(t, i(t))\|_\infty^2 + \left\| \frac{d^2V(t)}{dt} \right\|_\infty^2$$

Then, we reach

$$|H(t, V(t))|^2 \leq k \left[1 + |V(t)|^2 \right]$$

where

$$k = \left(\frac{LR_2C}{R_1R_2C+L}\right)^2 3 \left(\|F(t,i(t))\|_\infty^2 + \left\|\frac{d^2V(t)}{dt}\right\|_\infty^2\right)$$

$$|H(t,v_1(t)) - H(t,v_2(t))|^2 = \left|\left(\frac{1}{CL} - \frac{R_1+R_2}{R_2CL}\right)\frac{LR_2C}{R_1R_2C+L}(V_1(t) - V_2(t))\right.$$
$$\left. - \frac{LR_2C}{R_1R_2C+L}\frac{d^2}{dt^2}(V_1(t) - V_2(t))\right|^2$$

We asuming that $\frac{d^2}{dt^2}(V_1(t) - V_2(t))$ is bounded. Then, we obtain

$$|H(t,v_1(t)) - H(t,v_2(t))|^2 = \left|\left(\frac{1}{CL} - \frac{R_1+R_2}{R_2CL}\right)\frac{LR_2C}{R_1R_2C+L}(V_1(t) - V_2(t))\right.$$
$$\left. - \frac{LR_2C}{R_1R_2C+L}\frac{d^2}{dt^2}(V_1(t) - V_2(t))\right|^2$$

$$\leq 2\left[\left(\frac{1}{CL} - \frac{R_1+R_2}{R_2CL}\right)\frac{LR_2C}{R_1R_2C+L}\right]^2 |(V_1(t) - V_2(t))|^2$$
$$+ 2\left(\frac{LR_2C}{R_1R_2C+L}\right)^2 \left|\frac{d^2}{dt^2}(V_1(t) - V_2(t))\right|^2$$

$$\leq 2\left[\left(\frac{1}{CL} - \frac{R_1+R_2}{R_2CL}\right)\frac{LR_2C}{R_1R_2C+L}\right]^2 |(V_1(t) - V_2(t))|^2$$
$$+ 2\left(\frac{LR_2C}{R_1R_2C+L}\right)^2 \sup_{t\in D_{V_{tt}}}\left|\frac{d^2}{dt^2}(V_1(t) - V_2(t))\right|^2$$

$$\leq 2\left[\left(\frac{1}{CL} - \frac{R_1+R_2}{R_2CL}\right)\frac{LR_2C}{R_1R_2C+L}\right]^2 |V_1(t) - V_2(t)|^2$$
$$+ 2\left(\frac{LR_2C}{R_1R_2C+L}\right)^2 \left\|\frac{d^2}{dt^2}(V_1(t) - V_2(t))\right\|_\infty^2$$
$$\leq k|V_1(t) - V_2(t)|^2$$

Under the conditions the above were derived, we can conclude that the equation has unique solution. We now present the numerical solution at $t = t_{n+1}$ and $t = t_n$. Then, we have

$$V(t_{n+1}) - V(t_n) = \frac{3}{2}\Delta t H(t_n, V(t_n)) - \frac{1}{2}\Delta t H(t_{n-1}, V(t_{n-1}))$$
$$+ \frac{3}{2}\Delta t \frac{LR_2C}{R_1R_2C+L}\sigma V(t_n)\frac{B(t_{n+1}) - B(t_n)}{\Delta t}$$
$$- \frac{1}{2}\Delta t \frac{LR_2C}{R_1R_2C+L}\sigma V(t_{n-1})\frac{B(t_n) - B(t_{n-1})}{\Delta t}$$

If $B(t)$ is not differentiable, then

$$
V(t_{n+1}) - V(t_n) = \frac{3}{2}\Delta t H(t_n, V(t_n)) - \frac{1}{2}\Delta t H(t_{n-1}, V(t_{n-1}))
$$
$$
+ \sigma \frac{LR_2 C}{R_1 R_2 C + L} \sum_{i=0}^{n} V(t_j)\left(B(t_{j+1}) - B(t_j)\right)
$$
$$
- \sigma \frac{LR_2 C}{R_1 R_2 C + L} \sum_{i=0}^{n-1} V(t_j)\left(B(t_{j+1}) - B(t_j)\right)
$$

Here

$$
H(t_n, V(t_n)) = \frac{LR_2 C}{R_1 R_2 C + L}\left(F(t_n, i(t_n)) + \left(\frac{1}{CL} - \frac{R_1 + R_2}{R_2 CL}\right)V(t_n)\right.
$$
$$
\left. - \frac{V(t_{n+1}) - 2V(t_n) + V(t_{n-1})}{(\Delta t)^2}\right)
$$

$$
F(t_n, i(t_n)) = \frac{R_1}{CL} i_n(t) + \frac{1}{C}\frac{i(t_{n+1}) - i(t_n)}{\Delta t}
$$

We consider the following problem:

$$
\frac{1}{CL}i(t) + \frac{1}{L}\frac{dV(t)}{dt} + \frac{V(t)}{R_1 CL} = \frac{d^2 i(t)}{dt^2} + \left(\frac{R_2}{L} + \frac{1}{R_1 C}\right)\frac{di(t)}{dt} + \frac{R_1 + R_2}{R_1 CL}i(t) \quad (21.56)
$$

In this case we assume that the function $V(t)$ is known. Then, for simplicity we assume

$$
F(t, V(t)) = \frac{1}{L}\frac{dV(t)}{dt} + \frac{V(t)}{R_1 CL} \quad (21.57)
$$

Then, the equation becomes

$$
\left(\frac{R_2}{L} + \frac{1}{R_1 C}\right)\frac{di(t)}{dt} = F(t, V(t)) + \frac{1}{CL}i(t) - \frac{d^2 i(t)}{dt^2} - \frac{R_1 + R_2}{R_1 CL}i(t) \quad (21.58)
$$

$$
\frac{di(t)}{dt} = \frac{1}{\left(\frac{R_2}{L} + \frac{1}{R_1 C}\right)}\left(F(t, V(t)) + \frac{1}{CL}i(t) - \frac{d^2 i(t)}{dt^2} - \frac{R_1 + R_2}{R_1 CL}i(t)\right) \quad (21.59)
$$

$$
\frac{di(t)}{dt} = \frac{R_1 LC}{R_1 R_2 C + L}\left(F(t, V(t)) + \frac{1}{CL}i(t) - \frac{d^2 i(t)}{dt^2} - \frac{R_1 + R_2}{R_1 CL}i(t)\right) \quad (21.60)
$$

We can now convert the above equation into integral equation and the randomness to obtain

$$
i(t) - i(0) = \int_0^t H(\tau, i(\tau))d\tau + \sigma \int_0^t i(\tau)dB(\tau) \quad (21.61)
$$

Again we present the existence and uniqueness of the above problem by verifying:

(i) Growth condition $\forall t \in D_i$, $|f(t,i(t))|^2 < k(1+|i(t)|^2)$

(ii) The Lipschitz condition. Let i_1 and i_2 be the solutions of the equations.

$$|f(t,i_1(t)) - f(t,i_2(t))|^2 < \bar{k}|i_1 - i_2|^2$$

$$
\begin{aligned}
|H(t,i(t))|^2 &= \left| \frac{R_1 LC}{R_1 R_2 C + L} \left(F(t,V(t)) + \frac{1}{CL}i(t) - \frac{d^2 i(t)}{dt^2} - \frac{R_1 + R_2}{R_1 CL}i(t) \right) \right|^2 \\
&= \left(\frac{R_1 LC}{R_1 R_2 C + L} \right)^2 \left| F(t,V(t)) + \left(\frac{1}{CL} - \frac{R_1 + R_2}{R_1 CL} \right) i(t) - \frac{d^2 i(t)}{dt^2} \right|^2 \\
&\leq 3\left(\frac{R_1 LC}{R_1 R_2 C + L} \right)^2 \left[|F(t,V(t))|^2 \right. \\
&\quad + \left. \left(\frac{1}{CL} - \frac{R_1 + R_2}{R_1 CL} \right)^2 |i(t)|^2 + \left| \frac{d^2 i(t)}{dt^2} \right|^2 \right] \\
&\leq 3\left(\frac{R_1 LC}{R_1 R_2 C + L} \right)^2 \left[\sup_{t \in D_F} |F(t,V(t))|^2 \right. \\
&\quad + \left. \left(\frac{1}{CL} - \frac{R_1 + R_2}{R_1 CL} \right)^2 |i(t)|^2 + \sup_{t \in D_F} \left| \frac{d^2 i(t)}{dt^2} \right|^2 \right] \\
&\leq 3\left(\frac{R_1 LC}{R_1 R_2 C + L} \right)^2 \left[\sup_{t \in D_F} |F(t,V(t))|^2 \right. \\
&\quad + \left. \left(\frac{1}{CL} - \frac{R_1 + R_2}{R_1 CL} \right)^2 |i(t)|^2 + \sup_{t \in D_F} \left| \frac{d^2 i(t)}{dt^2} \right|^2 \right] \\
&\leq 3\left(\frac{R_1 LC}{R_1 R_2 C + L} \right)^2 \left[\|F(t,V(t))\|_\infty^2 \right. \\
&\quad + \left. \left(\frac{1}{CL} - \frac{R_1 + R_2}{R_1 CL} \right)^2 |i(t)|^2 + \left\| \frac{d^2 i(t)}{dt^2} \right\|_\infty^2 \right]
\end{aligned}
$$

$$
\begin{aligned}
\leq\ & 3\left(\frac{R_1 LC}{R_1 R_2 C + L} \right)^2 \left(\|F(t,V(t))\|_\infty^2 + \left\| \frac{d^2 i(t)}{dt^2} \right\|_\infty^2 \right) \\
& \times \left(1 + \frac{\left(\frac{1}{CL} - \frac{R_1 + R_2}{R_1 CL} \right)^2 |i(t)|^2}{\|F(t,V(t))\|_\infty^2 + \left\| \frac{d^2 i(t)}{dt^2} \right\|_\infty^2} \right)
\end{aligned}
$$

If

$$\left(\frac{1}{CL}-\frac{R_1+R_2}{R_1CL}\right)^2 < \|F(t,V(t))\|_{\infty}^2 + \left\|\frac{d^2i(t)}{dt^2}\right\|_{\infty}^2$$

Then, we reach

$$|H(t,i(t))|^2 \le k\left[1+|i(t)|^2\right]$$

where

$$k = 3\left(\frac{R_1LC}{R_1R_2C+L}\right)^2\left(\|F(t,V(t))\|_{\infty}^2 + \left\|\frac{d^2i(t)}{dt^2}\right\|_{\infty}^2\right)$$

$$
\begin{aligned}
|H(t,i_1(t))-H(t,i_2(t))|^2 &= \left|\left(\frac{R_1LC}{R_1R_2C+L}\right)\left(\frac{1}{CL}-\frac{R_1+R_2}{R_1CL}\right)(i_1(t)-i_2(t))\right.\\
&\quad \left.-\left(\frac{R_1LC}{R_1R_2C+L}\right)\frac{d^2}{dt^2}(i_1(t)-i_2(t))\right|^2
\end{aligned}
$$

We assuming that $\frac{d^2}{dt^2}(i_1(t)-i_2(t))$ is bounded. Then, we obtain

$$
\begin{aligned}
|H(t,i_1(t))-H(t,i_2(t))|^2 &= \left|\left(\frac{R_1LC}{R_1R_2C+L}\right)\left(\frac{1}{CL}-\frac{R_1+R_2}{R_1CL}\right)(i_1(t)-i_2(t))\right.\\
&\quad \left.-\left(\frac{R_1LC}{R_1R_2C+L}\right)\frac{d^2}{dt^2}(i_1(t)-i_2(t))\right|^2\\
&\le 2\left(\frac{R_1LC}{R_1R_2C+L}\right)\left(\frac{1}{CL}-\frac{R_1+R_2}{R_1CL}\right)^2|i_1(t)-i_2(t)|^2\\
&\quad +2\left(\frac{R_1LC}{R_1R_2C+L}\right)^2\left|\frac{d^2}{dt^2}(i_1(t)-i_2(t))\right|^2\\
&\le 2\left(\frac{R_1LC}{R_1R_2C+L}\right)\left(\frac{1}{CL}-\frac{R_1+R_2}{R_1CL}\right)^2|i_1(t)-i_2(t)|^2\\
&\quad +2\left(\frac{R_1LC}{R_1R_2C+L}\right)^2\sup_{t\in D_{i_{tt}}}\left|\frac{d^2}{dt^2}(i_1(t)-i_2(t))\right|^2\\
&\le 2\left(\frac{R_1LC}{R_1R_2C+L}\right)\left(\frac{1}{CL}-\frac{R_1+R_2}{R_1CL}\right)^2|i_1(t)-i_2(t)|^2\\
&\quad +2\left(\frac{R_1LC}{R_1R_2C+L}\right)^2\left\|\frac{d^2}{dt^2}(i_1(t)-i_2(t))\right\|_{\infty}^2\\
&\le k|i_1(t)-i_2(t)|^2
\end{aligned}
$$

Under the conditions the above were derived, we can conclude that the equation has unique solution. We now present the numerical solution at $t = t_{n+1}$ and $t = t_n$. Then, we have

$$
\begin{aligned}
i(t_{n+1}) - i(t_n) &= \frac{3}{2}\Delta t H(t_n, i(t_n)) - \frac{1}{2}\Delta t H(t_{n-1}, i(t_{n-1})) \\
&+ \frac{3}{2}\Delta t \frac{R_1 L C}{R_1 R_2 C + L} \sigma i(t_n) \frac{B(t_{n+1}) - B(t_n)}{\Delta t} \\
&- \frac{1}{2}\Delta t \frac{R_1 L C}{R_1 R_2 C + L} \sigma i(t_{n-1}) \frac{B(t_n) - B(t_{n-1})}{\Delta t}
\end{aligned}
$$

If $B(t)$ is not differentiable, then

$$
\begin{aligned}
i(t_{n+1}) - i(t_n) &= \frac{3}{2}\Delta t H(t_n, i(t_n)) - \frac{1}{2}\Delta t H(t_{n-1}, i(t_{n-1})) \\
&+ \sigma \frac{R_1 L C}{R_1 R_2 C + L} \sum_{i=0}^{n} i(t_j) \left(B(t_{j+1}) - B(t_j) \right) \\
&- \sigma \frac{R_1 L C}{R_1 R_2 C + L} \sum_{i=0}^{n-1} i(t_j) \left(B(t_{j+1}) - B(t_j) \right)
\end{aligned}
$$

Here

$$
\begin{aligned}
H(t_n, i(t_n)) &= \frac{1}{\left(\frac{R_2}{L} + \frac{1}{R_1 C} \right)} \left(F(t_n, V(t_n)) + \frac{1}{CL} i(t_n) - \frac{R_1 + R_2}{R_1 CL} i(t_n) \right. \\
&\left. - \frac{i(t_{n+1}) - 2i(t_n) + i(t_{n-1})}{(\Delta t)^2} \right)
\end{aligned}
$$

$$
F(t_n, V(t_n)) = \frac{1}{L} \frac{V(t_{n+1}) - V(t_n)}{\Delta t} + \frac{V(t_n)}{R_1 CL}
$$

We now consider the following equation

$$
L\frac{dI(t)}{dt} + RI(t) + \frac{1}{C}\int_0^t I(\tau)d\tau = V(t) \tag{21.62}
$$

Here, the function $V(t)$ is well known. We assume that the solution $I(t)$ is bounded in its domain, $V(t)$ is also bounded. We firstly convert the above equation as

$$
\frac{dI(t)}{dt} = \frac{V(t)}{L} - \frac{R}{L}I(t) - \frac{1}{CL}\int_0^t I(\tau)d\tau \tag{21.63}
$$

We now convert the above equation into integral equation as

$$
I(t) - I(0) = \frac{1}{L}\int_0^t \left(V(\tau) - RI(\tau) - \frac{1}{C}\int_0^\tau I(l)dl \right)d\tau
$$

We add the stochastic component to have

$$
\begin{aligned}
I(t) - I(0) \;=\; & \frac{1}{L}\int_0^t \left(v(\tau) - RI(\tau) - \frac{1}{C}\int_0^\tau I(l)dl \right) d\tau \\
& + \sigma \int_0^t I(\tau)dB(\tau)
\end{aligned}
$$

Here

$$
F(t,I(t)) \;=\; \frac{1}{L}\left(v(t) - RI(t) - \frac{1}{C}\int_0^t I(\tau)d\tau \right)
$$

$$
G(t,I(t)) \;=\; \sigma I(t)
$$

Of course for all $t \in D_I$, we have

$$
|G(t,I(t))|^2 \le k\left(1 + |I(t)|^2\right)
$$

So

$$
|G(t,I(t)) - G(t,J(t))|^2 \le \underline{k}\,|I(t) - J(t)|^2
$$

We have to show that $F(t,I(t))$ also verifies these conditons.

$$
\begin{aligned}
|F(t,I(t))|^2 \;=\; & \left| \frac{1}{L}\left(v(t) - RI(t) - \frac{1}{C}\int_0^t I(\tau)d\tau \right) \right|^2 \\[2mm]
\le\; & \frac{3}{L^2}|V(t)|^2 + \frac{3R^2}{L^2}|I(t)|^2 + \frac{3}{(LC)^2}\left| \int_0^t I(\tau)d\tau \right|^2 \\[2mm]
\le\; & \frac{3}{L^2}|V(t)|^2 + \frac{3R^2}{L^2}|I(t)|^2 + \frac{3T}{(LC)^2}\int_0^t |I(\tau)|^2\,d\tau \\[2mm]
\le\; & \frac{3}{L^2}\sup_{t\in D_V}|V(t)|^2 + \frac{3R^2}{L^2}|I(t)|^2 + \frac{3T}{(LC)^2}\int_0^t \sup_{0\le\xi\tau}|I(\xi)|^2\,d\tau \\[2mm]
\le\; & \frac{3}{L^2}\|V(t)\|_\infty^2 + \frac{3R^2}{L^2}|I(t)|^2 + \frac{3T^2}{(LC)^2}\|I(\xi)\|_\infty^2
\end{aligned}
$$

$$
\begin{aligned}
\le\; & \frac{3}{L^2}\|V(t)\|_\infty^2 + \frac{3R^2}{L^2}|I(t)|^2 + \frac{3T^2}{(LC)^2}M \\[2mm]
\le\; & \left(\frac{3}{L^2}\|V(t)\|_\infty^2 + \frac{3T^2}{(LC)^2}M \right) \\[2mm]
& \times \left(1 + \frac{\frac{3R^2}{L^2}|I(t)|^2}{\frac{3}{L^2}\|V(t)\|_\infty^2 + \frac{3T^2}{(LC)^2}M} \right)
\end{aligned}
$$

Here, $\|I\|_\infty^2 < M$ since $I(t)$ is assumed to be bounded. Then

$$
|F(t,I(t))|^2 \;\le\; k\left(1 + |I(t)|^2\right)
$$

under the condition that

$$\frac{\frac{3R^2}{L^2}}{\frac{3}{L^2}\|V(t)\|_\infty^2 + \frac{3T^2}{(LC)^2}M} < 1.$$

and

$$k = \frac{3}{L^2}\|V(t)\|_\infty^2 + \frac{3T^2}{(LC)^2}M.$$

Also

$$
\begin{aligned}
|F(t,I_1(t)) - F(t,I_2(t))|^2 &= \left| \frac{R}{L}(I_1(t) - I_2(t)) - \frac{1}{LC}\int_0^t (I_1(\tau) - I_2(\tau))d\tau \right|^2 \\
&\leq 2\frac{R^2}{L^2}|I_1(t) - I_2(t)|^2 + 2\frac{1}{(LC)^2}\left|\int_0^t (I_1(\tau) - I_2(\tau))d\tau\right|^2 \\
&\leq 2\frac{R^2}{L^2}|I_1(t) - I_2(t)|^2 + 2\frac{T}{(LC)^2}\int_0^t |I_1(\tau) - I_2(\tau)|^2 d\tau \\
&\leq 2\frac{R^2}{L^2}\sup_{t\in D_{I_1,I_2}}|I_1(t) - I_2(t)|^2 + 2\frac{T^2}{(LC)^2}\sup_{t\in D_{I_1,I_2}}|I_1(t) - I_2(t)|^2 \\
&\leq \left(2\frac{R^2}{L^2} + 2\frac{T^2}{(LC)^2}\right)\sup_{t\in D_{I_1,I_2}}|I_1(t) - I_2(t)|^2 \\
&\leq K\|I_1(t) - I_2(t)\|_\infty^2
\end{aligned}
$$

where

$$K = 2\frac{R^2}{L^2} + 2\frac{T^2}{(LC)^2}.$$

Therefore, under the conditions described above, we have unique solution. We can now present the numerical solution of the stochastic equation. Thus at $t = t_{n+1} = (n+1)\Delta t$, we get

$$
\begin{aligned}
I(t_{n+1}) - I(0) &= \frac{1}{L}\int_0^{t_{n+1}} F(\tau,I(\tau))d\tau + \int_0^{t_{n+1}} G(\tau,I(\tau))d\tau \\
&= \frac{1}{L}\sum_{j=0}^n \int_{t_j}^{t_{j+1}} F(\tau,I(\tau))d\tau + \sum_{j=0}^n \int_{t_j}^{t_{j+1}} G(\tau,I(\tau))d\tau
\end{aligned}
$$

Using the Lagrange within $[t_j, t_{j+1}]$, we can approximate the functions $F(\tau,I(\tau))$ and $G(\tau,I(\tau))$ to obtain

$$
\begin{aligned}
I(t_{n+1}) - I(0) &= \frac{1}{L}\sum_{j=1}^n \left(\frac{3}{2}\Delta t F(t_j,I(t_j)) - \frac{1}{2}\Delta t F(t_{j-1},I(t_{j-1}))\right) \\
&+ \sum_{j=1}^n \left(\frac{3}{2}\Delta t G(t_j,I(t_j)) - \frac{1}{2}\Delta t G(t_{j-1},I(t_{j-1}))\right)
\end{aligned}
$$

where

$$
\begin{aligned}
F(t_j, I(t_j)) &= \frac{1}{L}\left(v(t_j) - RI(t_j) - \frac{1}{C}\int_0^{t_j} I(\tau)d\tau \right) \\
&= \frac{1}{L}\left(v(t_j) - RI(t_j) - \frac{1}{C}\sum_{k=0}^{j-1}\int_{t_k}^{t_{k+1}} I(\tau)d\tau \right) \\
&= \frac{1}{L}\left(v(t_j) - RI(t_j) - \frac{1}{C}\sum_{k=0}^{j-1}\Delta t I(t_k) \right)
\end{aligned}
$$

We consider the following differential equation

$$
\frac{di}{dt} = -\frac{R}{L}i(t) + \frac{1}{L}V(t) = f(t, i(t)) \tag{21.64}
$$

We convert the above differential equation to the fractional stochastic differential equation as: In power-law case we have

$$
i(t) - i(0) = \frac{1}{\Gamma(\alpha)}\int_0^t f(\tau, i(\tau))(t-\tau)^{\alpha-1}d\tau + \frac{G}{\Gamma(\alpha)}\int_0^t i(\tau)(t-\tau)^{\alpha-1}dB(\tau) \tag{21.65}
$$

We define a mapping

$$
\begin{aligned}
|\Gamma(i(t))| &= \left| \frac{1}{\Gamma(\alpha)}\int_0^t f(\tau, i(\tau))(t-\tau)^{\alpha-1}d\tau + \frac{G}{\Gamma(\alpha)}\int_0^t i(\tau)(t-\tau)^{\alpha-1}dB(\tau) \right| \\
&< \frac{1}{\Gamma(\alpha)}\left[\|f(\cdot,i)\|_\infty \frac{T^\alpha}{\alpha} + GL\|i\|_\infty \frac{T^\alpha}{\alpha} \right] \\
&< \frac{T^\alpha}{\Gamma(\alpha+1)}(M_1 + M_2) \\
&< \frac{T^\alpha}{\Gamma(\alpha+1)}M
\end{aligned}
$$

We have

$$
\begin{aligned}
|\Gamma(i) - \Gamma(i_1)| &= \frac{1}{\Gamma(\alpha)}\left| \int_0^t (f(\tau, i(\tau)) - f(\tau, i_1(\tau)))(t-\tau)^{\alpha-1}d\tau \right. \\
&\quad \left. + \frac{G}{\Gamma(\alpha)}\int_0^t (i(\tau) - i_1(\tau))(t-\tau)^{\alpha-1}dB(\tau) \right. \\
&< \frac{1}{\Gamma(\alpha)}\left[|(f(\tau, i(\tau)) - f(\tau, i_1(\tau)))(t-\tau)^{\alpha-1}d\tau| \right. \\
&\quad \left. G|(i(\tau) - i_1(\tau))(t-\tau)^{\alpha-1}dB(\tau)| \right] \\
&< \frac{1}{\Gamma(\alpha)}\left[\frac{R}{L}\sup_{t\in[0,T]}|i(t) - i_1(t)|\frac{T^\alpha}{\alpha} + GL\frac{T^\alpha}{\alpha}\sup_{t\in[0,T]}|i(t) - i_1(t)| \right] \\
&< \frac{T^\alpha}{\Gamma(\alpha+1)}\left(\frac{R}{L} + GL \right)\|i - i_1\|_\infty < L_1\|i - i_1\|_\infty
\end{aligned}
$$

We assume that $B(t)$ is Lipschitzian on $[0,T]$ with constant L, then

$$
\begin{aligned}
|\Gamma(i(t))| \;=\;& \frac{1}{\Gamma(\alpha)}\left| \int_0^t f(\tau,i(\tau))(t-\tau)^{\alpha-1}d\tau + \frac{G}{\Gamma(\alpha)}\int_0^t i(\tau)(t-\tau)^{\alpha-1}dB(\tau) \right| \\
\leq\;& \frac{1}{\Gamma(\alpha)}\left[\left| \int_0^t f(\tau,i(\tau))(t-\tau)^{\alpha-1}d\tau \right| \right. \\
& \left. G\left| \int_0^t i(\tau)(t-\tau)^{\alpha-1}dB(\tau) \right| \right] \\
<\;& \frac{1}{\Gamma(\alpha)}\left[\int_0^t |f(\tau,i(\tau))|\,(t-\tau)^{\alpha-1}d\tau \right. \\
& \left. G\int_0^t |i(\tau)|\,(t-\tau)^{\alpha-1}dB(\tau) \right] \\
<\;& \frac{1}{\Gamma(\alpha)}\left[\int_0^t \sup_{l\in[0,\tau]} |f(l,i(l))|\,(t-\tau)^{\alpha-1}d\tau \right. \\
& \left. G\int_0^t \sup_{l\in[0,\tau]} |i(l)|\,(t-\tau)^{\alpha-1}dB(\tau) \right]
\end{aligned}
$$

This shows that our mapping has a unique solution. We now present the numerical solution of the equation

$$
\begin{aligned}
i(t_{n+1}) \;=\;& i(0) + \frac{1}{\Gamma(\alpha)}\int_0^{t_{n+1}} f(\tau,i(\tau))(t_{n+1}-\tau)^{\alpha-1}d\tau \\
& + \frac{G}{\Gamma(\alpha)}\int_0^{t_{n+1}} i(\tau)(t_{n+1}-\tau)^{\alpha-1}dB(\tau) \\
\;=\;& i(0) + \frac{1}{\Gamma(\alpha)}\sum_{j=0}^{n}\int_{t_j}^{t_{j+1}} f(\tau,i(\tau))(t_{n+1}-\tau)^{\alpha-1}d\tau \\
& + \frac{G}{\Gamma(\alpha)}\sum_{j=0}^{n}\int_{t_j}^{t_{j+1}} i(\tau)(t_{n+1}-\tau)^{\alpha-1}dB(\tau) \\
\;=\;& i(0) + \frac{(\Delta t)^{\alpha}}{\Gamma(\alpha+1)}\sum_{j=2}^{n} f(t_{j-2},i^{j-2})\left[(n-j+1)^{\alpha}-(n-j)^{\alpha}\right] \\
& + \frac{(\Delta t)^{\alpha}}{\Gamma(\alpha+2)}\sum_{j=2}^{n}\left[f(t_{j-1},i^{j-1}) - f(t_{j-2},i^{j-2}) \right] \\
& \times\left[(n-j+1)^{\alpha}(n-j+3+2\alpha)-(n-j^{\alpha})(n-j+3+3\alpha)\right] \\
& + \frac{(\Delta t)^{\alpha}}{2\Gamma(\alpha+3)}\left[f(t_j,i^j) - 2f(t_{j-1},i^{j-1}) + f(t_{j-2},i^{j-2}) \right] \\
& \times\left[(n-j+1)^{\alpha}\left(2(n-j)^2+(3\alpha+10)(n-j)+2\alpha^2+9\alpha+12\right) \right. \\
& \left. -(n-j)^{\alpha}\left(2(n-j)^2+(5\alpha+10)(n-j)+6\alpha^2+18\alpha+12\right)\right] \\
& + \frac{G(\Delta t)^{\alpha}}{\Gamma(\alpha+1)}\sum_{j=0}^{n} i(t_j)\left[(n-j+1)^{\alpha}-(n-j)^{\alpha}\right]\left(B(t_{j+1})-B(t_j)\right)
\end{aligned}
$$

Then, we obtain

$$
\begin{aligned}
i(t_{n+1}) \;=\; & i(0) + \frac{(\Delta t)^\alpha}{\Gamma(\alpha+1)} \sum_{j=2}^{n} \left(-\frac{R}{L}i(t_{j-2}) + \frac{1}{L}V(t_{j-2}) \right) \\
& \times \left[(n-j+1)^\alpha - (n-j)^\alpha \right] \\
& + \frac{(\Delta t)^\alpha}{\Gamma(\alpha+2)} \sum_{j=2}^{n} \left[\left(-\frac{R}{L}i(t_{j-1}) + \frac{1}{L}V(t_{j-1}) \right) \right. \\
& \left. - \left(-\frac{R}{L}i(t_{j-2}) + \frac{1}{L}V(t_{j-2}) \right) \right] \\
& \times \left[(n-j+1)^\alpha(n-j+3+2\alpha) - (n-j^\alpha)(n-j+3+3\alpha) \right] \\
& + \frac{(\Delta t)^\alpha}{2\Gamma(\alpha+3)} \left[\left(-\frac{R}{L}i(t_j) + \frac{1}{L}V(t_j) \right) \right. \\
& \left. -2 \left(-\frac{R}{L}i(t_{j-1}) + \frac{1}{L}V(t_{j-1}) \right) + \left(-\frac{R}{L}i(t_{j-2}) + \frac{1}{L}V(t_{j-2}) \right) \right] \\
& \times \left[(n-j+1)^\alpha \left(2(n-j)^2 + (3\alpha+10)(n-j) + 2\alpha^2 + 9\alpha + 12 \right) \right. \\
& \left. -(n-j)^\alpha \left(2(n-j)^2 + (5\alpha+10)(n-j) + 6\alpha^2 + 18\alpha + 12 \right) \right] \\
& + \frac{G(\Delta t)^\alpha}{\Gamma(\alpha+1)} \sum_{j=0}^{n} i(t_j) \left[(n-j+1)^\alpha - (n-j)^\alpha \right] \left(B(t_{j+1}) - B(t_j) \right)
\end{aligned}
$$

We consider the following differential equation

$$
\frac{dV_2}{dt} = -\frac{1}{RC}V_2(t) + \frac{1}{2RC}V_1(t) - \frac{dV_1(t)}{dt} = f(t, V_2(t)) \tag{21.66}
$$

We convert the above differential equation to the fractional stochastic differential equation as: In power-law case we have

$$
V_2(t) - V_2(0) = \frac{1}{\Gamma(\alpha)} \int_0^t f(\tau, V_2(\tau))(t-\tau)^{\alpha-1} d\tau + \frac{G}{\Gamma(\alpha)} \int_0^t V_2(\tau)(t-\tau)^{\alpha-1} dB(\tau) \tag{21.67}
$$

We define a mapping

$$
\begin{aligned}
|\Gamma(V_2(t))| &= \left| \frac{1}{\Gamma(\alpha)} \int_0^t f(\tau, V_2(\tau))(t-\tau)^{\alpha-1} d\tau + \frac{G}{\Gamma(\alpha)} \int_0^t V_2(\tau)(t-\tau)^{\alpha-1} dB(\tau) \right| \\
&< \frac{1}{\Gamma(\alpha)} \left[\|f(\cdot, V_2)\|_\infty \frac{T^\alpha}{\alpha} + GL\|V_2\|_\infty \frac{T^\alpha}{\alpha} \right] \\
&< \frac{T^\alpha}{\Gamma(\alpha+1)} (M_1 + M_2) \\
&< \frac{T^\alpha}{\Gamma(\alpha+1)} M
\end{aligned}
$$

We have

$$|\Gamma(V_2) - \Gamma(V_{2_1})| = \frac{1}{\Gamma(\alpha)} \left| \int_0^t (f(\tau, V_2(\tau)) - f(\tau, V_{2_1}(\tau)))(t-\tau)^{\alpha-1} d\tau \right|$$
$$+ \frac{G}{\Gamma(\alpha)} \int_0^t (V_2(\tau) - V_{2_1}(\tau))(t-\tau)^{\alpha-1} dB(\tau)$$
$$< \frac{1}{\Gamma(\alpha)} \left[|(f(\tau, V_2(\tau)) - f(\tau, V_{2_1}(\tau)))(t-\tau)^{\alpha-1} d\tau| \right.$$
$$\left. G|(V_2(\tau) - V_{2_1}(\tau))(t-\tau)^{\alpha-1} dB(\tau)| \right]$$
$$< \frac{1}{\Gamma(\alpha)} \left[\frac{1}{RC} \sup_{t \in [0,T]} |V_2(t) - V_{2_1}(t)| \frac{T^\alpha}{\alpha} \right.$$
$$\left. + GL \frac{T^\alpha}{\alpha} \sup_{t \in [0,T]} |V_2(t) - V_{2_1}(t)| \right]$$
$$< \frac{T^\alpha}{\Gamma(\alpha+1)} \left(\frac{1}{RC} + GL \right) \|V_2 - V_{2_1}\|_\infty < L_1 \|V_2 - V_{2_1}\|_\infty$$

We assume that $B(t)$ is Lipschitzian on $[0, T]$ with constant L, then

$$|\Gamma(V_2(t))| = \frac{1}{\Gamma(\alpha)} \left| \int_0^t f(\tau, V_2(\tau))(t-\tau)^{\alpha-1} d\tau + \frac{G}{\Gamma(\alpha)} \int_0^t V_2(\tau)(t-\tau)^{\alpha-1} dB(\tau) \right|$$
$$\leq \frac{1}{\Gamma(\alpha)} \left[\left| \int_0^t f(\tau, V_2(\tau))(t-\tau)^{\alpha-1} d\tau \right| \right.$$
$$\left. G \left| \int_0^t V_2(\tau)(t-\tau)^{\alpha-1} dB(\tau) \right| \right]$$
$$< \frac{1}{\Gamma(\alpha)} \left[\int_0^t |f(\tau, V_2(\tau))|(t-\tau)^{\alpha-1} d\tau \right.$$
$$\left. G \int_0^t |V_2(\tau)|(t-\tau)^{\alpha-1} dB(\tau) \right]$$
$$< \frac{1}{\Gamma(\alpha)} \left[\int_0^t \sup_{l \in [0,\tau]} |f(l, V_2(l))|(t-\tau)^{\alpha-1} d\tau \right.$$
$$\left. G \int_0^t \sup_{l \in [0,\tau]} |V_2(l)|(t-\tau)^{\alpha-1} dB(\tau) \right]$$

This shows that our mapping has a unique solution. We now present the numerical solution of the equation

$$V_2(t_{n+1}) = V_2(0) + \frac{1}{\Gamma(\alpha)} \int_0^{t_{n+1}} f(\tau, V_2(\tau))(t_{n+1} - \tau)^{\alpha-1} d\tau$$
$$+ \frac{G}{\Gamma(\alpha)} \int_0^{t_{n+1}} V_2(\tau)(t_{n+1} - \tau)^{\alpha-1} dB(\tau)$$
$$= V_2(0) + \frac{1}{\Gamma(\alpha)} \sum_{j=0}^n \int_{t_j}^{t_{j+1}} f(\tau, V_2(\tau))(t_{n+1} - \tau)^{\alpha-1} d\tau$$

$$+\frac{G}{\Gamma(\alpha)}\sum_{j=0}^{n}\int_{t_j}^{t_{j+1}}V_2(\tau)(t_{n+1}-\tau)^{\alpha-1}dB(\tau)$$

$$=V_2(0)+\frac{(\Delta t)^\alpha}{\Gamma(\alpha+1)}\sum_{j=2}^{n}f(t_{j-2},V_2^{j-2})\left[(n-j+1)^\alpha-(n-j)^\alpha\right]$$

$$+\frac{(\Delta t)^\alpha}{\Gamma(\alpha+2)}\sum_{j=2}^{n}\left[f(t_{j-1},V_2^{j-1})-f(t_{j-2},V_2^{j-2})\right]$$

$$\times\left[(n-j+1)^\alpha(n-j+3+2\alpha)-(n-j^\alpha)(n-j+3+3\alpha)\right]$$

$$+\frac{(\Delta t)^\alpha}{2\Gamma(\alpha+3)}\left[f(t_j,V_2^j)-2f(t_{j-1},V_2^{j-1})+f(t_{j-2},V_2^{j-2})\right]$$

$$\times\left[(n-j+1)^\alpha\left(2(n-j)^2+(3\alpha+10)(n-j)+2\alpha^2+9\alpha+12\right)\right.$$

$$\left.-(n-j)^\alpha\left(2(n-j)^2+(5\alpha+10)(n-j)+6\alpha^2+18\alpha+12\right)\right]$$

$$+\frac{G(\Delta t)^\alpha}{\Gamma(\alpha+1)}\sum_{j=0}^{n}i(t_j)\left[(n-j+1)^\alpha-(n-j)^\alpha\right]\left(B(t_{j+1})-B(t_j)\right)$$

Then, we obtain

$$V_2(t_{n+1})=V_2(0)+\frac{(\Delta t)^\alpha}{\Gamma(\alpha+1)}\sum_{j=2}^{n}\left(-\frac{1}{RC}V_2(t_{j-2})+\frac{1}{2RC}V_1(t_{j-2})-\frac{dV_1(t_{j-2})}{dt}\right)$$

$$\times\left[(n-j+1)^\alpha-(n-j)^\alpha\right]$$

$$+\frac{(\Delta t)^\alpha}{\Gamma(\alpha+2)}\sum_{j=2}^{n}\left[\left(-\frac{1}{RC}V_2(t_{j-1})+\frac{1}{2RC}V_1(t_{j-1})-\frac{dV_1(t_{j-1})}{dt}\right)\right.$$

$$\left.-\left(-\frac{1}{RC}V_2(t_{j-2})+\frac{1}{2RC}V_1(t_{j-2})-\frac{dV_1(t_{j-2})}{dt}\right)\right]$$

$$\times\left[(n-j+1)^\alpha(n-j+3+2\alpha)-(n-j^\alpha)(n-j+3+3\alpha)\right]$$

$$+\frac{(\Delta t)^\alpha}{2\Gamma(\alpha+3)}\left[\left(-\frac{1}{RC}V_2(t_j)+\frac{1}{2RC}V_1(t_j)-\frac{dV_1(t_j)}{dt}\right)\right.$$

$$-2\left(-\frac{1}{RC}V_2(t_{j-1})+\frac{1}{2RC}V_1(t_{j-1})-\frac{dV_1(t_{j-1})}{dt}\right)$$

$$\left.+\left(-\frac{1}{RC}V_2(t_{j-2})+\frac{1}{2RC}V_1(t_{j-2})-\frac{dV_1(t_{j-2})}{dt}\right)\right]$$

$$\times\left[(n-j+1)^\alpha\left(2(n-j)^2+(3\alpha+10)(n-j)+2\alpha^2+9\alpha+12\right)\right.$$

$$\left.-(n-j)^\alpha\left(2(n-j)^2+(5\alpha+10)(n-j)+6\alpha^2+18\alpha+12\right)\right]$$

$$+\frac{G(\Delta t)^\alpha}{\Gamma(\alpha+1)}\sum_{j=0}^{n}i(t_j)\left[(n-j+1)^\alpha-(n-j)^\alpha\right]\left(B(t_{j+1})-B(t_j)\right)$$

We consider the following differential equation

$$\frac{dV_2}{dt}=-\frac{1}{R_1C_1}V_2(t)-\frac{1}{R_1C_2}V_1(t)+\frac{R_2}{R_1}\frac{dV_1(t)}{dt}=f(t,V_2(t)) \tag{21.68}$$

We convert the above differential equation to the fractional stochastic differential equation as: In power-law case we have

$$V_2(t) - V_2(0) = \frac{1}{\Gamma(\alpha)} \int_0^t f(\tau, V_2(\tau))(t-\tau)^{\alpha-1} d\tau + \frac{G}{\Gamma(\alpha)} \int_0^t V_2(\tau)(t-\tau)^{\alpha-1} dB(\tau)$$

(21.69)

We define a mapping

$$|\Gamma(V_2(t))| = \left| \frac{1}{\Gamma(\alpha)} \int_0^t f(\tau, V_2(\tau))(t-\tau)^{\alpha-1} d\tau + \frac{G}{\Gamma(\alpha)} \int_0^t V_2(\tau)(t-\tau)^{\alpha-1} dB(\tau) \right|$$

$$< \frac{1}{\Gamma(\alpha)} \left[\|f(\cdot, V_2)\|_\infty \frac{T^\alpha}{\alpha} + GL\|V_2\|_\infty \frac{T^\alpha}{\alpha} \right]$$

$$< \frac{T^\alpha}{\Gamma(\alpha+1)} (M_1 + M_2)$$

$$< \frac{T^\alpha}{\Gamma(\alpha+1)} M$$

We have

$$|\Gamma(V_2) - \Gamma(V_{2_1})| = \frac{1}{\Gamma(\alpha)} \left| \int_0^t (f(\tau, V_2(\tau)) - f(\tau, V_{2_1}(\tau)))(t-\tau)^{\alpha-1} d\tau \right.$$

$$+ \frac{G}{\Gamma(\alpha)} \int_0^t (V_2(\tau) - V_{2_1}(\tau))(t-\tau)^{\alpha-1} dB(\tau)$$

$$< \frac{1}{\Gamma(\alpha)} \left[|(f(\tau, V_2(\tau)) - f(\tau, V_{2_1}(\tau)))(t-\tau)^{\alpha-1} d\tau| \right.$$

$$\left. G|(V_2(\tau) - V_{2_1}(\tau))(t-\tau)^{\alpha-1} dB(\tau)| \right]$$

$$< \frac{1}{\Gamma(\alpha)} \left[\frac{1}{R_1 C_1} \sup_{t \in [0,T]} |V_2(t) - V_{2_1}(t)| \frac{T^\alpha}{\alpha} \right.$$

$$\left. + GL \frac{T^\alpha}{\alpha} \sup_{t \in [0,T]} |V_2(t) - V_{2_1}(t)| \right]$$

$$< \frac{T^\alpha}{\Gamma(\alpha+1)} \left(\frac{1}{R_1 C_1} + GL \right) \|V_2 - V_{2_1}\|_\infty < L_1 \|V_2 - V_{2_1}\|_\infty$$

We assume that $B(t)$ is Lipschitzian on $[0, T]$ with constant L, then

$$|\Gamma(V_2(t))| = \frac{1}{\Gamma(\alpha)} \left| \int_0^t f(\tau, V_2(\tau))(t-\tau)^{\alpha-1} d\tau + \frac{G}{\Gamma(\alpha)} \int_0^t V_2(\tau)(t-\tau)^{\alpha-1} dB(\tau) \right|$$

$$\leq \frac{1}{\Gamma(\alpha)} \left[\left| \int_0^t f(\tau, V_2(\tau))(t-\tau)^{\alpha-1} d\tau \right| \right.$$

$$\left. G \left| \int_0^t V_2(\tau)(t-\tau)^{\alpha-1} dB(\tau) \right| \right]$$

$$< \frac{1}{\Gamma(\alpha)} \left[\int_0^t |f(\tau, V_2(\tau))| (t-\tau)^{\alpha-1} d\tau \right.$$

$$G \int_0^t |V_2(\tau)| (t - \tau)^{\alpha-1} dB(\tau) \Bigg]$$

$$< \frac{1}{\Gamma(\alpha)} \Bigg[\int_0^t \sup_{l \in [0,\tau]} |f(l, V_2(l))| (t - \tau)^{\alpha-1} d\tau$$

$$G \int_0^t \sup_{l \in [0,\tau]} |V_2(l)| (t - \tau)^{\alpha-1} dB(\tau) \Bigg]$$

This shows that our mapping has a unique solution. We now present the numerical solution of the equation

$$
\begin{aligned}
V_2(t_{n+1}) &= V_2(0) + \frac{1}{\Gamma(\alpha)} \int_0^{t_{n+1}} f(\tau, V_2(\tau))(t_{n+1} - \tau)^{\alpha-1} d\tau \\
&\quad + \frac{G}{\Gamma(\alpha)} \int_0^{t_{n+1}} V_2(\tau)(t_{n+1} - \tau)^{\alpha-1} dB(\tau) \\
&= V_2(0) + \frac{1}{\Gamma(\alpha)} \sum_{j=0}^{n} \int_{t_j}^{t_{j+1}} f(\tau, V_2(\tau))(t_{n+1} - \tau)^{\alpha-1} d\tau \\
&\quad + \frac{G}{\Gamma(\alpha)} \sum_{j=0}^{n} \int_{t_j}^{t_{j+1}} V_2(\tau)(t_{n+1} - \tau)^{\alpha-1} dB(\tau) \\
&= V_2(0) + \frac{(\Delta t)^\alpha}{\Gamma(\alpha+1)} \sum_{j=2}^{n} f(t_{j-2}, V_2^{j-2}) [(n-j+1)^\alpha - (n-j)^\alpha] \\
&\quad + \frac{(\Delta t)^\alpha}{\Gamma(\alpha+2)} \sum_{j=2}^{n} \left[f(t_{j-1}, V_2^{j-1}) - f(t_{j-2}, V_2^{j-2}) \right] \\
&\quad \times [(n-j+1)^\alpha (n-j+3+2\alpha) - (n-j^\alpha)(n-j+3+3\alpha)] \\
&\quad + \frac{(\Delta t)^\alpha}{2\Gamma(\alpha+3)} \left[f(t_j, V_2^j) - 2f(t_{j-1}, V_2^{j-1}) + f(t_{j-2}, V_2^{j-2}) \right] \\
&\quad \times [(n-j+1)^\alpha (2(n-j)^2 + (3\alpha+10)(n-j) + 2\alpha^2 + 9\alpha + 12) \\
&\quad - (n-j)^\alpha (2(n-j)^2 + (5\alpha+10)(n-j) + 6\alpha^2 + 18\alpha + 12)] \\
&\quad + \frac{G(\Delta t)^\alpha}{\Gamma(\alpha+1)} \sum_{j=0}^{n} i(t_j) [(n-j+1)^\alpha - (n-j)^\alpha] (B(t_{j+1}) - B(t_j))
\end{aligned}
$$

Then, we obtain

$$
\begin{aligned}
V_2(t_{n+1}) &= V_2(0) + \frac{(\Delta t)^\alpha}{\Gamma(\alpha+1)} \sum_{j=2}^{n} \left(-\frac{1}{R_1 C_1} V_2(t_{j-2}) - \frac{1}{R_1 C_2} V_1(t_{j-2}) + \frac{R_2}{R_1} \frac{dV_1(t_{j-2})}{dt} \right) \\
&\quad \times [(n-j+1)^\alpha - (n-j)^\alpha] \\
&\quad + \frac{(\Delta t)^\alpha}{\Gamma(\alpha+2)} \sum_{j=2}^{n} \left[\left(-\frac{1}{R_1 C_1} V_2(t_{j-1}) - \frac{1}{R_1 C_2} V_1(t_{j-1}) + \frac{R_2}{R_1} \frac{dV_1(t_{j-1})}{dt} \right) \right.
\end{aligned}
$$

$$-\left(-\frac{1}{R_1C_1}V_2(t_{j-2})-\frac{1}{R_1C_2}V_1(t_{j-2})+\frac{R_2}{R_1}\frac{dV_1(t_{j-2})}{dt}\right)\Bigg]$$

$$\times\left[(n-j+1)^\alpha(n-j+3+2\alpha)-(n-j^\alpha)(n-j+3+3\alpha)\right]$$

$$+\frac{(\Delta t)^\alpha}{2\Gamma(\alpha+3)}\left[\left(-\frac{1}{R_1C_1}V_2(t_j)-\frac{1}{R_1C_2}V_1(t_j)+\frac{R_2}{R_1}\frac{dV_1(t_j)}{dt}\right)\right.$$

$$-2\left(-\frac{1}{R_1C_1}V_2(t_{j-1})-\frac{1}{R_1C_2}V_1(t_{j-1})+\frac{R_2}{R_1}\frac{dV_1(t_{j-1})}{dt}\right)$$

$$+\left(-\frac{1}{R_1C_1}V_2(t_{j-2})-\frac{1}{R_1C_2}V_1(t_{j-2})+\frac{R_2}{R_1}\frac{dV_1(t_{j-2})}{dt}\right)\Bigg]$$

$$\times\left[(n-j+1)^\alpha\left(2(n-j)^2+(3\alpha+10)(n-j)+2\alpha^2+9\alpha+12\right)\right.$$

$$-(n-j)^\alpha\left(2(n-j)^2+(5\alpha+10)(n-j)+6\alpha^2+18\alpha+12\right)\Big]$$

$$+\frac{G(\Delta t)^\alpha}{\Gamma(\alpha+1)}\sum_{j=0}^{n}i(t_j)\left[(n-j+1)^\alpha-(n-j)^\alpha\right]\left(B(t_{j+1})-B(t_j)\right)$$

We consider the following differential equation

$$\frac{dV_2}{dt}=-\frac{1}{(R_1+R_2)C}V_2(t)+\frac{1}{(R_1+R_2)C}V_1(t)+\frac{R_2}{R_1+R_2}\frac{dV_1(t)}{dt}=f(t,V_2(t))$$

(21.70)

We convert the above differential equation to the fractional stochastic differential equation as: In power-law case we have

$$V_2(t)-V_2(0)=\frac{1}{\Gamma(\alpha)}\int_0^t f(\tau,V_2(\tau))(t-\tau)^{\alpha-1}d\tau+\frac{G}{\Gamma(\alpha)}\int_0^t V_2(\tau)(t-\tau)^{\alpha-1}dB(\tau)$$

(21.71)

We define a mapping

$$\begin{aligned}|\Gamma(V_2(t))| &= \left|\frac{1}{\Gamma(\alpha)}\int_0^t f(\tau,V_2(\tau))(t-\tau)^{\alpha-1}d\tau\right.\\ &\qquad\left.+\frac{G}{\Gamma(\alpha)}\int_0^t V_2(\tau)(t-\tau)^{\alpha-1}dB(\tau)\right|\\ &< \frac{1}{\Gamma(\alpha)}\left[\|f(.,V_2)\|_\infty\frac{T^\alpha}{\alpha}+GL\|V_2\|_\infty\frac{T^\alpha}{\alpha}\right]\\ &< \frac{T^\alpha}{\Gamma(\alpha+1)}(M_1+M_2)\\ &< \frac{T^\alpha}{\Gamma(\alpha+1)}M\end{aligned}$$

We have

$$|\Gamma(V_2) - \Gamma(V_{2_1})| = \frac{1}{\Gamma(\alpha)} \left| \int_0^t (f(\tau, V_2(\tau)) - f(\tau, V_{2_1}(\tau))) (t - \tau)^{\alpha - 1} d\tau \right.$$

$$+ \frac{G}{\Gamma(\alpha)} \int_0^t (V_2(\tau) - V_{2_1}(\tau)) (t - \tau)^{\alpha - 1} dB(\tau)$$

$$< \frac{1}{\Gamma(\alpha)} \left[|(f(\tau, V_2(\tau)) - f(\tau, V_{2_1}(\tau))) (t - \tau)^{\alpha - 1} d\tau| \right.$$

$$G |(V_2(\tau) - V_{2_1}(\iota)) (t - \tau)^{\alpha - 1} dB(\tau)| \right]$$

$$< \frac{1}{\Gamma(\alpha)} \left[\frac{1}{(R_1 + R_2)C} \sup_{t \in [0,T]} |V_2(t) - V_{2_1}(t)| \frac{T^\alpha}{\alpha} \right.$$

$$+ GL \frac{T^\alpha}{\alpha} \sup_{t \in [0,T]} |V_2(t) - V_{2_1}(t)| \right]$$

$$< \frac{T^\alpha}{\Gamma(\alpha + 1)} \left(\frac{1}{(R_1 + R_2)C} + GL \right) \|V_2 - V_{2_1}\|_\infty < L_1 \|V_2 - V_{2_1}\|_\infty$$

We assume that $B(t)$ is Lipschitzian on $[0, T]$ with constant L, then

$$|\Gamma(V_2(t))| = \frac{1}{\Gamma(\alpha)} \left| \int_0^t f(\tau, V_2(\tau))(t - \tau)^{\alpha - 1} d\tau + \frac{G}{\Gamma(\alpha)} \int_0^t V_2(\tau)(t - \tau)^{\alpha - 1} dB(\tau) \right|$$

$$\leq \frac{1}{\Gamma(\alpha)} \left[\left| \int_0^t f(\tau, V_2(\tau))(t - \tau)^{\alpha - 1} d\tau \right| \right.$$

$$+ G \left| \int_0^t V_2(\tau)(t - \tau)^{\alpha - 1} dB(\tau) \right| \right]$$

$$< \frac{1}{\Gamma(\alpha)} \left[\int_0^t |f(\tau, V_2(\tau))| (t - \tau)^{\alpha - 1} d\tau \right.$$

$$+ G \int_0^t |V_2(\tau)| (t - \tau)^{\alpha - 1} dB(\tau) \right]$$

$$< \frac{1}{\Gamma(\alpha)} \left[\int_0^t \sup_{l \in [0,\tau]} |f(l, V_2(l))| (t - \tau)^{\alpha - 1} d\tau \right.$$

$$+ G \int_0^t \sup_{l \in [0,\tau]} |V_2(l)| (t - \tau)^{\alpha - 1} dB(\tau) \right]$$

This shows that our mapping has a unique solution. We now present the numerical solution of the equation

$$V_2(t_{n+1}) = V_2(0) + \frac{1}{\Gamma(\alpha)} \int_0^{t_{n+1}} f(\tau, V_2(\tau))(t_{n+1} - \tau)^{\alpha - 1} d\tau$$

$$+ \frac{G}{\Gamma(\alpha)} \int_0^{t_{n+1}} V_2(\tau)(t_{n+1} - \tau)^{\alpha - 1} dB(\tau)$$

$$= V_2(0) + \frac{1}{\Gamma(\alpha)} \sum_{j=0}^{n} \int_{t_j}^{t_{j+1}} f(\tau, V_2(\tau))(t_{n+1} - \tau)^{\alpha-1} d\tau$$

$$+ \frac{G}{\Gamma(\alpha)} \sum_{j=0}^{n} \int_{t_j}^{t_{j+1}} V_2(\tau)(t_{n+1} - \tau)^{\alpha-1} dB(\tau)$$

$$= V_2(0) + \frac{(\Delta t)^\alpha}{\Gamma(\alpha+1)} \sum_{j=2}^{n} f(t_{j-2}, V_2^{j-2}) [(n-j+1)^\alpha - (n-j)^\alpha]$$

$$+ \frac{(\Delta t)^\alpha}{\Gamma(\alpha+2)} \sum_{j=2}^{n} \left[f(t_{j-1}, V_2^{j-1}) - f(t_{j-2}, V_2^{j-2}) \right]$$

$$\times [(n-j+1)^\alpha(n-j+3+2\alpha) - (n-j^\alpha)(n-j+3+3\alpha)]$$

$$+ \frac{(\Delta t)^\alpha}{2\Gamma(\alpha+3)} \left[f(t_j, V_2^j) - 2f(t_{j-1}, V_2^{j-1}) + f(t_{j-2}, V_2^{j-2}) \right]$$

$$\times [(n-j+1)^\alpha (2(n-j)^2 + (3\alpha+10)(n-j) + 2\alpha^2 + 9\alpha + 12)$$

$$- (n-j)^\alpha (2(n-j)^2 + (5\alpha+10)(n-j) + 6\alpha^2 + 18\alpha + 12)]$$

$$+ \frac{G(\Delta t)^\alpha}{\Gamma(\alpha+1)} \sum_{j=0}^{n} i(t_j) [(n-j+1)^\alpha - (n-j)^\alpha] (B(t_{j+1}) - B(t_j))$$

Then, we obtain

$$V_2(t_{n+1}) = V_2(0) + \frac{(\Delta t)^\alpha}{\Gamma(\alpha+1)} \sum_{j=2}^{n} \left(-\frac{1}{(R_1+R_2)C} V_2(t_{j-2}) + \frac{1}{(R_1+R_2)C} V_1(t_{j-2}) \right.$$

$$+ \frac{R_2}{R_1+R_2} \frac{dV_1(t_{j-2})}{dt} \bigg) \times [(n-j+1)^\alpha - (n-j)^\alpha]$$

$$+ \frac{(\Delta t)^\alpha}{\Gamma(\alpha+2)} \sum_{j=2}^{n} \left[\left(-\frac{1}{(R_1+R_2)C} V_2(t_{j-1}) + \frac{1}{(R_1+R_2)C} V_1(t_{j-1}) \right. \right.$$

$$+ \frac{R_2}{R_1+R_2} \frac{dV_1(t_{j-1})}{dt} \bigg)$$

$$- \left(-\frac{1}{(R_1+R_2)C} V_2(t_{j-2}) + \frac{1}{(R_1+R_2)C} V_1(t_{j-2}) + \frac{R_2}{R_1+R_2} \frac{dV_1(t_{j-2})}{dt} \right) \bigg]$$

$$\times [(n-j+1)^\alpha(n-j+3+2\alpha) - (n-j^\alpha)(n-j+3+3\alpha)]$$

$$+ \frac{(\Delta t)^\alpha}{2\Gamma(\alpha+3)} \left[\left(-\frac{1}{(R_1+R_2)C} V_2(t_j) + \frac{1}{(R_1+R_2)C} V_1(t_j) + \frac{R_2}{R_1+R_2} \frac{dV_1(t_j)}{dt} \right) \right.$$

$$-2 \left(-\frac{1}{(R_1+R_2)C} V_2(t_{j-1}) + \frac{1}{(R_1+R_2)C} V_1(t_{j-1}) + \frac{R_2}{R_1+R_2} \frac{dV_1(t_{j-1})}{dt} \right)$$

$$+ \left(-\frac{1}{(R_1+R_2)C}V_2(t_{j-2}) + \frac{1}{(R_1+R_2)C}V_1(t_{j-2}) + \frac{R_2}{R_1+R_2}\frac{dV_1(t_{j-2})}{dt} \right) \Big]$$

$$\times \left[(n-j+1)^\alpha \left(2(n-j)^2 + (3\alpha+10)(n-j) + 2\alpha^2 + 9\alpha + 12 \right) \right.$$
$$\left. -(n-j)^\alpha \left(2(n-j)^2 + (5\alpha+10)(n-j) + 6\alpha^2 + 18\alpha + 12 \right) \right]$$

$$+ \frac{G(\Delta t)^\alpha}{\Gamma(\alpha+1)} \sum_{j=0}^{n} i(t_j) \left[(n-j+1)^\alpha - (n-j)^\alpha \right] (B(t_{j+1}) - B(t_j))$$

We consider the following differential equation

$$\frac{dV_2}{dt} = -\frac{R_1+R_2}{R_1R_2C}V_2(t) + \frac{1}{R_1C_1}V_1(t) + \frac{dV_1(t)}{dt} = f(t, V_2(t)) \tag{21.72}$$

We convert the above differential equation to the fractional stochastic differential equation as: In power-law case we have

$$V_2(t) - V_2(0) = \frac{1}{\Gamma(\alpha)} \int_0^t f(\tau, V_2(\tau))(t-\tau)^{\alpha-1} d\tau + \frac{G}{\Gamma(\alpha)} \int_0^t V_2(\tau)(t-\tau)^{\alpha-1} dB(\tau) \tag{21.73}$$

We define a mapping

$$|\Gamma(V_2(t))| = \left| \frac{1}{\Gamma(\alpha)} \int_0^t f(\tau, V_2(\tau))(t-\tau)^{\alpha-1} d\tau + \frac{G}{\Gamma(\alpha)} \int_0^t V_2(\tau)(t-\tau)^{\alpha-1} dB(\tau) \right|$$

$$< \frac{1}{\Gamma(\alpha)} \left[\|f(\cdot, V_2)\|_\infty \frac{T^\alpha}{\alpha} + GL\|V_2\|_\infty \frac{T^\alpha}{\alpha} \right]$$

$$< \frac{T^\alpha}{\Gamma(\alpha+1)} (M_1 + M_2)$$

$$< \frac{T^\alpha}{\Gamma(\alpha+1)} M$$

We have

$$|\Gamma(V_2) - \Gamma(V_{2_1})| = \frac{1}{\Gamma(\alpha)} \left| \int_0^t (f(\tau, V_2(\tau)) - f(\tau, V_{2_1}(\tau)))(t-\tau)^{\alpha-1} d\tau \right.$$

$$+ \frac{G}{\Gamma(\alpha)} \int_0^t (V_2(\tau) - V_{2_1}(\tau))(t-\tau)^{\alpha-1} dB(\tau)$$

$$< \frac{1}{\Gamma(\alpha)} \left[|(f(\tau, V_2(\tau)) - f(\tau, V_{2_1}(\tau)))(t-\tau)^{\alpha-1} d\tau| \right.$$

$$\left. + G|(V_2(\tau) - V_{2_1}(\tau))(t-\tau)^{\alpha-1} dB(\tau)| \right]$$

$$< \frac{1}{\Gamma(\alpha)} \left[\frac{R_1+R_2}{R_1R_2C} \sup_{t\in[0,T]} |V_2(t) - V_{2_1}(t)| \frac{T^\alpha}{\alpha} \right.$$

$$\left. + GL\frac{T^\alpha}{\alpha} \sup_{t\in[0,T]} |V_2(t) - V_{2_1}(t)| \right]$$

$$< \frac{T^\alpha}{\Gamma(\alpha+1)} \left(\frac{R_1+R_2}{R_1R_2C} + GL \right) \|V_2 - V_{2_1}\|_\infty < L_1 \|V_2 - V_{2_1}\|_\infty$$

We assume that $B(t)$ is Lipschitzian on $[0,T]$ with constant L, then

$$|\Gamma(V_2(t))| = \frac{1}{\Gamma(\alpha)}\left|\int_0^t f(\tau,V_2(\tau))(t-\tau)^{\alpha-1}d\tau + \frac{G}{\Gamma(\alpha)}\int_0^t V_2(\tau)(t-\tau)^{\alpha-1}dB(\tau)\right|$$

$$\leq \frac{1}{\Gamma(\alpha)}\left[\left|\int_0^t f(\tau,V_2(\tau))(t-\tau)^{\alpha-1}d\tau\right|\right.$$

$$\left.+G\left|\int_0^t V_2(\tau)(t-\tau)^{\alpha-1}dB(\tau)\right|\right]$$

$$< \frac{1}{\Gamma(\alpha)}\left[\int_0^t |f(\tau,V_2(\tau))|(t-\tau)^{\alpha-1}d\tau\right.$$

$$\left.+G\int_0^t |V_2(\tau)|(t-\tau)^{\alpha-1}dB(\tau)\right]$$

$$< \frac{1}{\Gamma(\alpha)}\left[\int_0^t \sup_{l\in[0,\tau]} |f(l,V_2(l))|(t-\tau)^{\alpha-1}d\tau\right.$$

$$\left.+G\int_0^t \sup_{l\in[0,\tau]} |V_2(l)|(t-\tau)^{\alpha-1}dB(\tau)\right]$$

This shows that our mapping has a unique solution. We now present the numerical solution of the equation

$$
\begin{aligned}
V_2(t_{n+1}) &= V_2(0) + \frac{1}{\Gamma(\alpha)}\int_0^{t_{n+1}} f(\tau,V_2(\tau))(t_{n+1}-\tau)^{\alpha-1}d\tau \\
&\quad + \frac{G}{\Gamma(\alpha)}\int_0^{t_{n+1}} V_2(\tau)(t_{n+1}-\tau)^{\alpha-1}dB(\tau) \\
&= V_2(0) + \frac{1}{\Gamma(\alpha)}\sum_{j=0}^n \int_{t_j}^{t_{j+1}} f(\tau,V_2(\tau))(t_{n+1}-\tau)^{\alpha-1}d\tau \\
&\quad + \frac{G}{\Gamma(\alpha)}\sum_{j=0}^n \int_{t_j}^{t_{j+1}} V_2(\tau)(t_{n+1}-\tau)^{\alpha-1}dB(\tau) \\
&= V_2(0) + \frac{(\Delta t)^\alpha}{\Gamma(\alpha+1)}\sum_{j=2}^n f(t_{j-2},V_2^{j-2})\left[(n-j+1)^\alpha - (n-j)^\alpha\right] \\
&\quad + \frac{(\Delta t)^\alpha}{\Gamma(\alpha+2)}\sum_{j=2}^n \left[f(t_{j-1},V_2^{j-1}) - f(t_{j-2},V_2^{j-2})\right] \\
&\quad \times \left[(n-j+1)^\alpha(n-j+3+2\alpha) - (n-j^\alpha)(n-j+3+3\alpha)\right] \\
&\quad + \frac{(\Delta t)^\alpha}{2\Gamma(\alpha+3)}\left[f(t_j,V_2^j) - 2f(t_{j-1},V_2^{j-1}) + f(t_{j-2},V_2^{j-2})\right] \\
&\quad \times \left[(n-j+1)^\alpha\left(2(n-j)^2 + (3\alpha+10)(n-j) + 2\alpha^2 + 9\alpha + 12\right)\right. \\
&\quad \left. -(n-j)^\alpha\left(2(n-j)^2 + (5\alpha+10)(n-j) + 6\alpha^2 + 18\alpha + 12\right)\right] \\
&\quad + \frac{G(\Delta t)^\alpha}{\Gamma(\alpha+1)}\sum_{j=0}^n i(t_j)\left[(n-j+1)^\alpha - (n-j)^\alpha\right](B(t_{j+1}) - B(t_j))
\end{aligned}
$$

Then, we obtain

$$
\begin{aligned}
V_2(t_{n+1} =\,&V_2(0) + \frac{(\Delta t)^\alpha}{\Gamma(\alpha+1)} \sum_{j=2}^{n} \left(-\frac{R_1+R_2}{R_1 R_2 C} V_2(t_{j-2}) + \frac{1}{R_1 C_1} V_1(t_{j-2}) + \frac{dV_1(t_{j-2})}{dt} \right) \\
&\times [(n-j+1)^\alpha - (n-j)^\alpha] \\
&+ \frac{(\Delta t)^\alpha}{\Gamma(\alpha+2)} \sum_{j=2}^{n} \left[\left(-\frac{R_1+R_2}{R_1 R_2 C} V_2(t_{j-1}) + \frac{1}{R_1 C_1} V_1(t_{j-1}) + \frac{dV_1(t_{j-1})}{dt} \right) \right. \\
&\left. - \left(-\frac{R_1+R_2}{R_1 R_2 C} V_2(t_{j-2}) + \frac{1}{R_1 C_1} V_1(t_{j-2}) + \frac{dV_1(t_{j-2})}{dt} \right) \right] \\
&\times [(n-j+1)^\alpha (n-j+3+2\alpha) - (n-j^\alpha)(n-j+3+3\alpha)] \\
&+ \frac{(\Delta t)^\alpha}{2\Gamma(\alpha+3)} \left[\left(-\frac{R_1+R_2}{R_1 R_2 C} V_2(t_j) + \frac{1}{R_1 C_1} V_1(t_j) + \frac{dV_1(t_j)}{dt} \right) \right. \\
&-2 \left(-\frac{R_1+R_2}{R_1 R_2 C} V_2(t_{j-1}) + \frac{1}{R_1 C_1} V_1(t_{j-1}) + \frac{dV_1(t_{j-1})}{dt} \right) \\
&\left. + \left(-\frac{R_1+R_2}{R_1 R_2 C} V_2(t_{j-2}) + \frac{1}{R_1 C_1} V_1(t_{j-2}) + \frac{dV_1(t_{j-2})}{dt} \right) \right] \\
&\times \big[(n-j+1)^\alpha \left(2(n-j)^2 + (3\alpha+10)(n-j) + 2\alpha^2 + 9\alpha + 12 \right) \\
&- (n-j)^\alpha \left(2(n-j)^2 + (5\alpha+10)(n-j) + 6\alpha^2 + 18\alpha + 12 \right) \big] \\
&+ \frac{G(\Delta t)^\alpha}{\Gamma(\alpha+1)} \sum_{j=0}^{n} i(t_j) \left[(n-j+1)^\alpha - (n-j)^\alpha \right] \left(B(t_{j+1}) - B(t_j) \right)
\end{aligned}
$$

We consider the following differential equation

$$
\frac{dV_2}{dt} = -4 \times 10^6 V_2(t) - 4 \times 10^7 V_1(t) = f(t, V_2(t)) \tag{21.74}
$$

We convert the above differential equation to the fractional stochastic differential equation as: In power-law case we have

$$
V_2(t) - V_2(0) = \frac{1}{\Gamma(\alpha)} \int_0^t f(\tau, V_2(\tau))(t-\tau)^{\alpha-1} d\tau + \frac{G}{\Gamma(\alpha)} \int_0^t V_2(\tau)(t-\tau)^{\alpha-1} dB(\tau) \tag{21.75}
$$

We define a mapping

$$
\begin{aligned}
|\Gamma(V_2(t))| &= \left| \frac{1}{\Gamma(\alpha)} \int_0^t f(\tau, V_2(\tau))(t-\tau)^{\alpha-1} d\tau + \frac{G}{\Gamma(\alpha)} \int_0^t V_2(\tau)(t-\tau)^{\alpha-1} dB(\tau) \right| \\
&< \frac{1}{\Gamma(\alpha)} \left[\|f(\cdot, V_2)\|_\infty \frac{T^\alpha}{\alpha} + GL\|V_2\|_\infty \frac{T^\alpha}{\alpha} \right] \\
&< \frac{T^\alpha}{\Gamma(\alpha+1)} (M_1 + M_2) \\
&< \frac{T^\alpha}{\Gamma(\alpha+1)} M
\end{aligned}
$$

We have

$$
\begin{aligned}
|\Gamma(V_2) - \Gamma(V_{2_1})| &= \frac{1}{\Gamma(\alpha)} \left| \int_0^t (f(\tau, V_2(\tau)) - f(\tau, V_{2_1}(\tau)))(t-\tau)^{\alpha-1} d\tau \right| \\
&\quad + \frac{G}{\Gamma(\alpha)} \int_0^t (V_2(\tau) - V_{2_1}(\tau))(t-\tau)^{\alpha-1} dB(\tau) \\
&< \frac{1}{\Gamma(\alpha)} \left[|(f(\tau, V_2(\tau)) - f(\tau, V_{2_1}(\tau)))(t-\tau)^{\alpha-1} d\tau| \right. \\
&\qquad\qquad \left. G|(V_2(\tau) - V_{2_1}(\tau))(t-\tau)^{\alpha-1} dB(\tau)| \right] \\
&< \frac{1}{\Gamma(\alpha)} \left[4 \times 10^6 \sup_{t \in [0,T]} |V_2(t) - V_{2_1}(t)| \frac{T^\alpha}{\alpha} \right.
\end{aligned}
$$

$$
\left. + GL\frac{T^\alpha}{\alpha} \sup_{t \in [0,T]} |V_2(t) - V_{2_1}(t)| \right]
$$

$$
< \frac{T^\alpha}{\Gamma(\alpha+1)} \left(4 \times 10^6 + GL \right) \|V_2 - V_{2_1}\|_\infty < L_1 \|V_2 - V_{2_1}\|_\infty
$$

We assume that $B(t)$ is Lipschitzian on $[0,T]$ with constant L, then

$$
\begin{aligned}
|\Gamma(V_2(t))| &= \frac{1}{\Gamma(\alpha)} \left| \int_0^t f(\tau, V_2(\tau))(t-\tau)^{\alpha-1} d\tau + \frac{G}{\Gamma(\alpha)} \int_0^t V_2(\tau)(t-\tau)^{\alpha-1} dB(\tau) \right| \\
&\leq \frac{1}{\Gamma(\alpha)} \left[\left| \int_0^t f(\tau, V_2(\tau))(t-\tau)^{\alpha-1} d\tau \right| \right. \\
&\qquad\qquad \left. + G \left| \int_0^t V_2(\tau)(t-\tau)^{\alpha-1} dB(\tau) \right| \right] \\
&< \frac{1}{\Gamma(\alpha)} \left[\int_0^t |f(\tau, V_2(\tau))|(t-\tau)^{\alpha-1} d\tau \right. \\
&\qquad\qquad \left. + G \int_0^t |V_2(\tau)|(t-\tau)^{\alpha-1} dB(\tau) \right] \\
&< \frac{1}{\Gamma(\alpha)} \left[\int_0^t \sup_{l \in [0,\tau]} |f(l, V_2(l))|(t-\tau)^{\alpha-1} d\tau \right. \\
&\qquad\qquad \left. + G \int_0^t \sup_{l \in [0,\tau]} |V_2(l)|(t-\tau)^{\alpha-1} dB(\tau) \right]
\end{aligned}
$$

This shows that our mapping has a unique solution. We now present the numerical solution of the equation

$$
\begin{aligned}
V_2(t_{n+1}) &= V_2(0) + \frac{1}{\Gamma(\alpha)} \int_0^{t_{n+1}} f(\tau, V_2(\tau))(t_{n+1}-\tau)^{\alpha-1} d\tau \\
&\quad + \frac{G}{\Gamma(\alpha)} \int_0^{t_{n+1}} V_2(\tau)(t_{n+1}-\tau)^{\alpha-1} dB(\tau)
\end{aligned}
$$

$$= V_2(0) + \frac{1}{\Gamma(\alpha)} \sum_{j=0}^{n} \int_{t_j}^{t_{j+1}} f(\tau, V_2(\tau))(t_{n+1} - \tau)^{\alpha-1} d\tau$$

$$+ \frac{G}{\Gamma(\alpha)} \sum_{j=0}^{n} \int_{t_j}^{t_{j+1}} V_2(\tau)(t_{n+1} - \tau)^{\alpha-1} dB(\tau)$$

$$= V_2(0) + \frac{(\Delta t)^\alpha}{\Gamma(\alpha+1)} \sum_{j=2}^{n} f(t_{j-2}, V_2^{j-2}) \left[(n-j+1)^\alpha - (n-j)^\alpha \right]$$

$$+ \frac{(\Delta t)^\alpha}{\Gamma(\alpha+2)} \sum_{j=2}^{n} \left[f(t_{j-1}, V_2^{j-1}) - f(t_{j-2}, V_2^{j-2}) \right]$$

$$\times \left[(n-j+1)^\alpha (n-j+3+2\alpha) - (n-j^\alpha)(n-j+3+3\alpha) \right]$$

$$+ \frac{(\Delta t)^\alpha}{2\Gamma(\alpha+3)} \left[f(t_j, V_2^j) - 2f(t_{j-1}, V_2^{j-1}) + f(t_{j-2}, V_2^{j-2}) \right]$$

$$\times \left[(n-j+1)^\alpha \left(2(n-j)^2 + (3\alpha+10)(n-j) + 2\alpha^2 + 9\alpha + 12 \right) \right.$$

$$\left. -(n-j)^\alpha \left(2(n-j)^2 + (5\alpha+10)(n-j) + 6\alpha^2 + 18\alpha + 12 \right) \right]$$

$$+ \frac{G(\Delta t)^\alpha}{\Gamma(\alpha+1)} \sum_{j=0}^{n} i(t_j) \left[(n-j+1)^\alpha - (n-j)^\alpha \right] (B(t_{j+1}) - B(t_j))$$

Then, we obtain

$$V_2(t_{n+1}) = V_2(0) + \frac{(\Delta t)^\alpha}{\Gamma(\alpha+1)} \sum_{j=2}^{n} \left(-4 \times 10^6 V_2(t_{j-2}) - 4 \times 10^7 V_1(t_{j-2}) \right)$$

$$\times \left[(n-j+1)^\alpha - (n-j)^\alpha \right]$$

$$+ \frac{(\Delta t)^\alpha}{\Gamma(\alpha+2)} \sum_{j=2}^{n} \left[\left(-4 \times 10^6 V_2(t_{j-1}) - 4 \times 10^7 V_1(t_{j-1}) \right) \right.$$

$$\left. - \left(-4 \times 10^6 V_2(t_{j-2}) - 4 \times 10^7 V_1(t_{j-2}) \right) \right]$$

$$\times \left[(n-j+1)^\alpha (n-j+3+2\alpha) - (n-j^\alpha)(n-j+3+3\alpha) \right]$$

$$+ \frac{(\Delta t)^\alpha}{2\Gamma(\alpha+3)} \left[\left(-4 \times 10^6 V_2(t_j) - 4 \times 10^7 V_1(t_j) \right) \right.$$

$$-2 \left(-4 \times 10^6 V_2(t_{j-1}) - 4 \times 10^7 V_1(t_{j-1}) \right)$$

$$\left. + \left(-4 \times 10^6 V_2(t_{j-2}) - 4 \times 10^7 V_1(t_{j-2}) \right) \right]$$

$$\times \left[(n-j+1)^\alpha \left(2(n-j)^2 + (3\alpha+10)(n-j) + 2\alpha^2 + 9\alpha + 12 \right) \right.$$

$$\left. -(n-j)^\alpha \left(2(n-j)^2 + (5\alpha+10)(n-j) + 6\alpha^2 + 18\alpha + 12 \right) \right]$$

$$+ \frac{G(\Delta t)^\alpha}{\Gamma(\alpha+1)} \sum_{j=0}^{n} i(t_j) \left[(n-j+1)^\alpha - (n-j)^\alpha \right] (B(t_{j+1}) - B(t_j))$$

We consider the following differential equation

$$\frac{dV_2}{dt} = -\frac{R_1+R_2}{R_1R_2(C_1+C_2)}V_2(t) + \frac{C_1}{C_1+C_2}\frac{dV_1(t)}{dt} + \frac{1}{R_1(C_1+C_2)}V_1(t) = f(t,V_2(t))$$

(21.76)

We convert the above differential equation to the fractional stochastic differential equation as: In power-law case we have

$$V_2(t) - V_2(0) = \frac{1}{\Gamma(\alpha)}\int_0^t f(\tau,V_2(\tau))(t-\tau)^{\alpha-1}d\tau + \frac{G}{\Gamma(\alpha)}\int_0^t V_2(\tau)(t-\tau)^{\alpha-1}dB(\tau)$$

(21.77)

We define a mapping

$$|\Gamma(V_2(t))| = \left|\frac{1}{\Gamma(\alpha)}\int_0^t f(\tau,V_2(\tau))(t-\tau)^{\alpha-1}d\tau + \frac{G}{\Gamma(\alpha)}\int_0^t V_2(\tau)(t-\tau)^{\alpha-1}dB(\tau)\right|$$

$$< \frac{1}{\Gamma(\alpha)}\left[\|f(.,V_2)\|_\infty\frac{T^\alpha}{\alpha} + GL\|V_2\|_\infty\frac{T^\alpha}{\alpha}\right]$$

$$< \frac{T^\alpha}{\Gamma(\alpha+1)}(M_1+M_2)$$

$$< \frac{T^\alpha}{\Gamma(\alpha+1)}M$$

We have

$$|\Gamma(V_2) - \Gamma(V_{2_1})| = \frac{1}{\Gamma(\alpha)}\left|\int_0^t (f(\tau,V_2(\tau)) - f(\tau,V_{2_1}(\tau)))(t-\tau)^{\alpha-1}d\tau\right.$$

$$+ \frac{G}{\Gamma(\alpha)}\int_0^t (V_2(\tau) - V_{2_1}(\tau))(t-\tau)^{\alpha-1}dB(\tau)$$

$$< \frac{1}{\Gamma(\alpha)}\left[|(f(\tau,V_2(\tau)) - f(\tau,V_{2_1}(\tau)))(t-\tau)^{\alpha-1}d\tau|\right.$$

$$\left. + G|(V_2(\tau) - V_{2_1}(\tau))(t-\tau)^{\alpha-1}dB(\tau)|\right]$$

$$< \frac{1}{\Gamma(\alpha)}\left[\frac{R_1+R_2}{R_1R_2(C_1+C_2)}\sup_{t\in[0,T]}|V_2(t) - V_{2_1}(t)|\frac{T^\alpha}{\alpha}\right.$$

$$\left. + GL\frac{T^\alpha}{\alpha}\sup_{t\in[0,T]}|V_2(t) - V_{2_1}(t)|\right]$$

$$< \frac{T^\alpha}{\Gamma(\alpha+1)}\left(\frac{R_1+R_2}{R_1R_2(C_1+C_2)} + GL\right)\|V_2 - V_{2_1}\|_\infty < L_1\|V_2 - V_{2_1}\|_\infty$$

We assume that $B(t)$ is Lipschitzian on $[0,T]$ with constant L, then

$$|\Gamma(V_2(t))| = \frac{1}{\Gamma(\alpha)}\left|\int_0^t f(\tau,V_2(\tau))(t-\tau)^{\alpha-1}d\tau + \frac{G}{\Gamma(\alpha)}\int_0^t V_2(\tau)(t-\tau)^{\alpha-1}dB(\tau)\right|$$

$$\leq \frac{1}{\Gamma(\alpha)}\left[\left|\int_0^t f(\tau,V_2(\tau))(t-\tau)^{\alpha-1}d\tau\right|\right.$$

$$+ G \left| \int_0^t V_2(\tau)(t-\tau)^{\alpha-1} dB(\tau) \right| \Bigg]$$

$$< \frac{1}{\Gamma(\alpha)} \left[\int_0^t |f(\tau, V_2(\tau))| (t-\tau)^{\alpha-1} d\tau \right.$$

$$+ G \int_0^t |V_2(\tau)| (t-\tau)^{\alpha-1} dB(\tau) \Bigg] < \qquad \frac{1}{\Gamma(\alpha)} \left[\int_0^t \sup_{l \in [0,\tau]} |f(l, V_2(l))| (t-\tau)^{\alpha-1} d\tau \right.$$

$$+ G \int_0^t \sup_{l \in [0,\tau]} |V_2(l)| (t-\tau)^{\alpha-1} dB(\tau) \Bigg]$$

This shows that our mapping has a unique solution. We now present the numerical solution of the equation

$$
\begin{aligned}
V_2(t_{n+1}) \;=\; & V_2(0) + \frac{1}{\Gamma(\alpha)} \int_0^{t_{n+1}} f(\tau, V_2(\tau))(t_{n+1}-\tau)^{\alpha-1} d\tau \\
& + \frac{G}{\Gamma(\alpha)} \int_0^{t_{n+1}} V_2(\tau)(t_{n+1}-\tau)^{\alpha-1} dB(\tau) \\
=\; & V_2(0) + \frac{1}{\Gamma(\alpha)} \sum_{j=0}^{n} \int_{t_j}^{t_{j+1}} f(\tau, V_2(\tau))(t_{n+1}-\tau)^{\alpha-1} d\tau \\
& + \frac{G}{\Gamma(\alpha)} \sum_{j=0}^{n} \int_{t_j}^{t_{j+1}} V_2(\tau)(t_{n+1}-\tau)^{\alpha-1} dB(\tau) \\
=\; & V_2(0) + \frac{(\Delta t)^\alpha}{\Gamma(\alpha+1)} \sum_{j=2}^{n} f(t_{j-2}, V_2^{j-2}) [(n-j+1)^\alpha - (n-j)^\alpha] \\
& + \frac{(\Delta t)^\alpha}{\Gamma(\alpha+2)} \sum_{j=2}^{n} \left[f(t_{j-1}, V_2^{j-1}) - f(t_{j-2}, V_2^{j-2}) \right] \\
& \times [(n-j+1)^\alpha (n-j+3+2\alpha) - (n-j^\alpha)(n-j+3+3\alpha)] \\
& + \frac{(\Delta t)^\alpha}{2\Gamma(\alpha+3)} \left[f(t_j, V_2^j) - 2f(t_{j-1}, V_2^{j-1}) + f(t_{j-2}, V_2^{j-2}) \right] \\
& \times \big[(n-j+1)^\alpha \left(2(n-j)^2 + (3\alpha+10)(n-j) + 2\alpha^2 + 9\alpha + 12 \right) \\
& - (n-j)^\alpha \left(2(n-j)^2 + (5\alpha+10)(n-j) + 6\alpha^2 + 18\alpha + 12 \right) \big] \\
& + \frac{G(\Delta t)^\alpha}{\Gamma(\alpha+1)} \sum_{j=0}^{n} i(t_j) [(n-j+1)^\alpha - (n-j)^\alpha] \left(B(t_{j+1}) - B(t_j) \right)
\end{aligned}
$$

Then, we obtain

$$
\begin{aligned}
V_2(t_{n+1}) =\; & V_2(0) + \frac{(\Delta t)^\alpha}{\Gamma(\alpha+1)} \sum_{j=2}^{n} \left(-\frac{R_1+R_2}{R_1 R_2 (C_1+C_2)} V_2(t_{j-2}) + \frac{C_1}{C_1+C_2} \frac{dV_1(t_{j-2})}{dt} \right. \\
& + \frac{1}{R_1(C_1+C_2)} V_1(t_{j-2}) \Bigg) [(n-j+1)^\alpha - (n-j)^\alpha]
\end{aligned}
$$

$$+\frac{(\Delta t)^{\alpha}}{\Gamma(\alpha+2)}\sum_{j=2}^{n}\left[\left(-\frac{R_1+R_2}{R_1R_2(C_1+C_2)}V_2(t_{j-1})+\frac{C_1}{C_1+C_2}\frac{dV_1(t_{j-1})}{dt}\right.\right.$$

$$\left.+\frac{1}{R_1(C_1+C_2)}V_1(t_{j-1})\right)$$

$$-\left(-\frac{R_1+R_2}{R_1R_2(C_1+C_2)}V_2(t_{j-2})+\frac{C_1}{C_1+C_2}\frac{dV_1(t_{j-2})}{dt}\right.$$

$$\left.\left.+\frac{1}{R_1(C_1+C_2)}V_1(t_{j-2})\right)\right]$$

$$\times\left[(n-j+1)^{\alpha}(n-j+3+2\alpha)-(n-j^{\alpha})(n-j+3+3\alpha)\right]$$

$$+\frac{(\Delta t)^{\alpha}}{2\Gamma(\alpha+3)}\left[\left(-\frac{R_1+R_2}{R_1R_2(C_1+C_2)}V_2(t_j)+\frac{C_1}{C_1+C_2}\frac{dV_1(t_j)}{dt}\right.\right.$$

$$\left.+\frac{1}{R_1(C_1+C_2)}V_1(t_j)\right)$$

$$-2\left(-\frac{R_1+R_2}{R_1R_2(C_1+C_2)}V_2(t_{j-1})+\frac{C_1}{C_1+C_2}\frac{dV_1(t_{j-1})}{dt}\right.$$

$$\left.+\frac{1}{R_1(C_1+C_2)}V_1(t_{j-1})\right)$$

$$+\left(-\frac{R_1+R_2}{R_1R_2(C_1+C_2)}V_2(t_{j-2})+\frac{C_1}{C_1+C_2}\frac{dV_1(t_{j-2})}{dt}\right.$$

$$\left.\left.+\frac{1}{R_1(C_1+C_2)}V_1(t_{j-2})\right)\right]$$

$$\times\left[(n-j+1)^{\alpha}\left(2(n-j)^2+(3\alpha+10)(n-j)+2\alpha^2+9\alpha+12\right)\right.$$

$$\left.-(n-j)^{\alpha}\left(2(n-j)^2+(5\alpha+10)(n-j)+6\alpha^2+18\alpha+12\right)\right]$$

$$+\frac{G(\Delta t)^{\alpha}}{\Gamma(\alpha+1)}\sum_{j=0}^{n}i(t_j)\left[(n-j+1)^{\alpha}-(n-j)^{\alpha}\right]\left(B(t_{j+1})-B(t_j)\right)$$

We consider the following differential equation

$$\frac{dV}{dt}=-\frac{1}{C}i(t)-\frac{1}{R_1C}V(t)+A\frac{d^2i(t)}{dt^2}+\left(R_2+\frac{L}{R_1C}\right)\frac{di(t)}{dt}+\frac{R_1+R_2}{R_1C}i(t)=f(t,V(t))$$

(21.78)

We convert the above differential equation to the fractional stochastic differential equation as: In power-law case we have

$$V(t)-V(0)=\frac{1}{\Gamma(\alpha)}\int_0^t f(\tau,V(\tau))(t-\tau)^{\alpha-1}d\tau+\frac{G}{\Gamma(\alpha)}\int_0^t V(\tau)(t-\tau)^{\alpha-1}dB(\tau)$$

(21.79)

We define a mapping

$$
\begin{aligned}
|\Gamma(V(t))| &= \left| \frac{1}{\Gamma(\alpha)} \int_0^t f(\tau,V(\tau))(t-\tau)^{\alpha-1}d\tau + \frac{G}{\Gamma(\alpha)} \int_0^t V(\tau)(t-\tau)^{\alpha-1}dB(\tau) \right| \\
&< \frac{1}{\Gamma(\alpha)} \left[\|f(.,V)\|_\infty \frac{T^\alpha}{\alpha} + GL\|V\|_\infty \frac{T^\alpha}{\alpha} \right] \\
&< \frac{T^\alpha}{\Gamma(\alpha+1)} (M_1 + M_2) \\
&< \frac{T^\alpha}{\Gamma(\alpha+1)} M
\end{aligned}
$$

We have

$$
\begin{aligned}
|\Gamma(V) - \Gamma(V_1)| &= \frac{1}{\Gamma(\alpha)} \left| \int_0^t (f(\tau,V(\tau)) - f(\tau,V_1(\tau)))(t-\tau)^{\alpha-1}d\tau \right. \\
&\quad + \frac{G}{\Gamma(\alpha)} \int_0^t (V(\tau) - V_1(\tau))(t-\tau)^{\alpha-1}dB(\tau) \\
&< \frac{1}{\Gamma(\alpha)} \left[|(f(\tau,V(\tau)) - f(\tau,V_1(\tau)))(t-\tau)^{\alpha-1}d\tau| \right. \\
&\quad + G|(V(\tau) - V_1(\tau))(t-\tau)^{\alpha-1}dB(\tau)| \Big] \\
&< \frac{1}{\Gamma(\alpha)} \left[\frac{1}{R_1 C} \sup_{t\in[0,T]} |V(t) - V_1(t)| \frac{T^\alpha}{\alpha} \right. \\
&\quad + GL\frac{T^\alpha}{\alpha} \sup_{t\in[0,T]} |V(t) - V_1(t)| \Big] \\
&< \frac{T^\alpha}{\Gamma(\alpha+1)} \left(\frac{1}{R_1 C} + GL \right) \|V - V_1\|_\infty < L_1 \|V - V_1\|_\infty
\end{aligned}
$$

We assume that $B(t)$ is Lipschitzian on $[0,T]$ with constant L, then

$$
\begin{aligned}
|\Gamma(V(t))| &= \frac{1}{\Gamma(\alpha)} \left| \int_0^t f(\tau,V(\tau))(t-\tau)^{\alpha-1}d\tau + \frac{G}{\Gamma(\alpha)} \int_0^t V(\tau)(t-\tau)^{\alpha-1}dB(\tau) \right| \\
&\leq \frac{1}{\Gamma(\alpha)} \left[\left| \int_0^t f(\tau,V(\tau))(t-\tau)^{\alpha-1}d\tau \right| \right. \\
&\quad + G \left| \int_0^t V(\tau)(t-\tau)^{\alpha-1}dB(\tau) \right| \Big] \\
&< \frac{1}{\Gamma(\alpha)} \left[\int_0^t |f(\tau,V(\tau))|(t-\tau)^{\alpha-1}d\tau \right. \\
&\quad + G \int_0^t |V(\tau)|(t-\tau)^{\alpha-1}dB(\tau) \Big]
\end{aligned}
$$

$$< \frac{1}{\Gamma(\alpha)} \left[\int_0^t \sup_{l \in [0,\tau]} |f(l,V(l))| (t-\tau)^{\alpha-1} d\tau \right.$$

$$\left. + G \int_0^t \sup_{l \in [0,\tau]} |V(l)| (t-\tau)^{\alpha-1} dB(\tau) \right]$$

This shows that our mapping has a unique solution. We now present the numerical solution of the equation

$$
\begin{aligned}
V(t_{n+1}) &= V(0) + \frac{1}{\Gamma(\alpha)} \int_0^{t_{n+1}} f(\tau,V(\tau))(t_{n+1}-\tau)^{\alpha-1} d\tau \\
&\quad + \frac{G}{\Gamma(\alpha)} \int_0^{t_{n+1}} V(\tau)(t_{n+1}-\tau)^{\alpha-1} dB(\tau) \\
&= V(0) + \frac{1}{\Gamma(\alpha)} \sum_{j=0}^{n} \int_{t_j}^{t_{j+1}} f(\tau,V(\tau))(t_{n+1}-\tau)^{\alpha-1} d\tau \\
&\quad + \frac{G}{\Gamma(\alpha)} \sum_{j=0}^{n} \int_{t_j}^{t_{j+1}} V(\tau)(t_{n+1}-\tau)^{\alpha-1} dB(\tau) \\
&= V(0) + \frac{(\Delta t)^\alpha}{\Gamma(\alpha+1)} \sum_{j=2}^{n} f(t_{j-2},V^{j-2}) \left[(n-j+1)^\alpha - (n-j)^\alpha \right] \\
&\quad + \frac{(\Delta t)^\alpha}{\Gamma(\alpha+2)} \sum_{j=2}^{n} \left[f(t_{j-1},V^{j-1}) - f(t_{j-2},V^{j-2}) \right] \\
&\quad \times \left[(n-j+1)^\alpha (n-j+3+2\alpha) - (n-j^\alpha)(n-j+3+3\alpha) \right] \\
&\quad + \frac{(\Delta t)^\alpha}{2\Gamma(\alpha+3)} \left[f(t_j,V_2^j) - 2f(t_{j-1},V_2^{j-1}) + f(t_{j-2},V_2^{j-2}) \right] \\
&\quad \times \left[(n-j+1)^\alpha \left(2(n-j)^2 + (3\alpha+10)(n-j) + 2\alpha^2 + 9\alpha + 12 \right) \right. \\
&\quad \left. - (n-j)^\alpha \left(2(n-j)^2 + (5\alpha+10)(n-j) + 6\alpha^2 + 18\alpha + 12 \right) \right] \\
&\quad + \frac{G(\Delta t)^\alpha}{\Gamma(\alpha+1)} \sum_{j=0}^{n} i(t_j) \left[(n-j+1)^\alpha - (n-j)^\alpha \right] \left(B(t_{j+1}) - B(t_j) \right)
\end{aligned}
$$

Then, we obtain

$$
\begin{aligned}
V(t_{n+1}) &= V(0) + \frac{(\Delta t)^\alpha}{\Gamma(\alpha+1)} \sum_{j=2}^{n} \left(-\frac{1}{C} i(t_{j-2}) - \frac{1}{R_1 C} V(t_{j-2}) + A \frac{d^2 i(t_{j-2})}{dt^2} \right. \\
&\quad \left. + \left(R_2 + \frac{L}{R_1 C} \right) \frac{di(t_{j-2})}{dt} + \frac{R_1+R_2}{R_1 C} i(t_{j-2}) \right) \left[(n-j+1)^\alpha - (n-j)^\alpha \right] \\
&\quad + \frac{(\Delta t)^\alpha}{\Gamma(\alpha+2)} \sum_{j=2}^{n} \left[\left(-\frac{1}{C} i(t_{j-1}) - \frac{1}{R_1 C} V(t_{j-1}) + A \frac{d^2 i(t_{j-1})}{dt^2} \right. \right.
\end{aligned}
$$

$$+\left(R_2+\frac{L}{R_1C}\right)\frac{di(t_{j-1})}{dt}+\frac{R_1+R_2}{R_1C}i(t_{j-1})\Bigg)$$

$$-\left(-\frac{1}{C}i(t_{j-2})-\frac{1}{R_1C}V(t_{j-2})+A\frac{d^2i(t_{j-2})}{dt^2}\right.$$

$$\left.+\left(R_2+\frac{L}{R_1C}\right)\frac{di(t_{j-2})}{dt}+\frac{R_1+R_2}{R_1C}i(t_{j-2})\right)\Bigg]$$

$$\times\left[(n-j+1)^\alpha(n\quad j\mid 3+2\alpha)-(n-j^\alpha)(n-j+3+3\alpha)\right]$$

$$+\frac{(\Delta t)^\alpha}{2\Gamma(\alpha+3)}\Bigg[\left(-\frac{1}{C}i(t_j)-\frac{1}{R_1C}V(t_j)+A\frac{d^2i(t_j)}{dt^2}\right.$$

$$\left.+\left(R_2+\frac{L}{R_1C}\right)\frac{di(t_j)}{dt}+\frac{R_1+R_2}{R_1C}i(t_j)\right)$$

$$-2\left(-\frac{1}{C}i(t_{j-1})-\frac{1}{R_1C}V(t_{j-1})+A\frac{d^2i(t_{j-1})}{dt^2}\right.$$

$$\left.+\left(R_2+\frac{L}{R_1C}\right)\frac{di(t_{j-1})}{dt}+\frac{R_1+R_2}{R_1C}i(t_{j-1})\right)$$

$$+\left(-\frac{1}{C}i(t_{j-2})-\frac{1}{R_1C}V(t_{j-2})+A\frac{d^2i(t_{j-2})}{dt^2}\right.$$

$$\left.+\left(R_2+\frac{L}{R_1C}\right)\frac{di(t_{j-2})}{dt}+\frac{R_1+R_2}{R_1C}i(t_{j-2})\right)\Bigg]$$

$$\times\left[(n-j+1)^\alpha\left(2(n-j)^2+(3\alpha+10)(n-j)+2\alpha^2+9\alpha+12\right)\right.$$

$$\left.-(n-j)^\alpha\left(2(n-j)^2+(5\alpha+10)(n-j)+6\alpha^2+18\alpha+12\right)\right]$$

$$+\frac{G(\Delta t)^\alpha}{\Gamma(\alpha+1)}\sum_{j=0}^{n}i(t_j)\left[(n-j+1)^\alpha-(n-j)^\alpha\right]\left(B(t_{j+1})-B(t_j)\right)$$

We consider the following differential equation

$$\frac{di}{dt}=-RC\frac{d^2i(t)}{dt^2}+\frac{1}{L}V(t)+\frac{RC}{L}\frac{dV(t)}{dt}=f(t,i(t)) \qquad (21.80)$$

We convert the above differential equation to the fractional stochastic differential equation as: In power-law case we have

$$i(t)-i(0)=\frac{1}{\Gamma(\alpha)}\int_0^t f(\tau,i(\tau))(t-\tau)^{\alpha-1}d\tau+\frac{G}{\Gamma(\alpha)}\int_0^t i(\tau)(t-\tau)^{\alpha-1}dB(\tau)$$

$$(21.81)$$

We define a mapping

$$
\begin{aligned}
|\Gamma(i(t))| &= \left| \frac{1}{\Gamma(\alpha)} \int_0^t f(\tau, i(\tau))(t-\tau)^{\alpha-1} d\tau + \frac{G}{\Gamma(\alpha)} \int_0^t i(\tau)(t-\tau)^{\alpha-1} dB(\tau) \right| \\
&< \frac{1}{\Gamma(\alpha)} \left[\|f(.,i)\|_\infty \frac{T^\alpha}{\alpha} + GL \|i\|_\infty \frac{T^\alpha}{\alpha} \right] \\
&< \frac{T^\alpha}{\Gamma(\alpha+1)} (M_1 + M_2) \\
&< \frac{T^\alpha}{\Gamma(\alpha+1)} M
\end{aligned}
$$

We have

$$
\begin{aligned}
|\Gamma(i) - \Gamma(i_1)| &= \frac{1}{\Gamma(\alpha)} \left| \int_0^t (f(\tau, i(\tau)) - f(\tau, i_1(\tau)))(t-\tau)^{\alpha-1} d\tau \right. \\
&\quad \left. + \frac{G}{\Gamma(\alpha)} \int_0^t (i(\tau) - i_1(\tau))(t-\tau)^{\alpha-1} dB(\tau) \right| \\
&< \frac{1}{\Gamma(\alpha)} \left[\left| (f(\tau, i(\tau)) - f(\tau, i_1(\tau)))(t-\tau)^{\alpha-1} d\tau \right| \right. \\
&\quad \left. + G \left| (i(\tau) - i_1(\tau))(t-\tau)^{\alpha-1} dB(\tau) \right| \right] \\
&< \frac{1}{\Gamma(\alpha)} \left[RC \sup_{t \in [0,T]} |i^2(t) - i_1^2(t)| \frac{T^\alpha}{\alpha} \right. \\
&\quad \left. + GL \frac{T^\alpha}{\alpha} \sup_{t \in [0,T]} |i(t) - i_1(t)| \right] \\
&< \frac{T^\alpha}{\Gamma(\alpha+1)} (RC \|i+i_1\|_\infty + GL) \|i - i_1\|_\infty < L_1 \|i - i_1\|_\infty
\end{aligned}
$$

We assume that $B(t)$ is Lipschitzian on $[0,T]$ with constant L, then

$$
\begin{aligned}
|\Gamma(i(t))| &= \frac{1}{\Gamma(\alpha)} \left| \int_0^t f(\tau, i(\tau))(t-\tau)^{\alpha-1} d\tau + \frac{G}{\Gamma(\alpha)} \int_0^t i(\tau)(t-\tau)^{\alpha-1} dB(\tau) \right| \\
&\leq \frac{1}{\Gamma(\alpha)} \left[\left| \int_0^t f(\tau, i(\tau))(t-\tau)^{\alpha-1} d\tau \right| \right. \\
&\quad \left. G \left| \int_0^t i(\tau)(t-\tau)^{\alpha-1} dB(\tau) \right| \right] \\
&< \frac{1}{\Gamma(\alpha)} \left[\int_0^t |f(\tau, i(\tau))| (t-\tau)^{\alpha-1} d\tau \right. \\
&\quad \left. G \int_0^t |i(\tau)| (t-\tau)^{\alpha-1} dB(\tau) \right]
\end{aligned}
$$

$$< \frac{1}{\Gamma(\alpha)} \left[\int_0^t \sup_{l \in [0,\tau]} |f(l, i(l))| (t-\tau)^{\alpha-1} d\tau \right.$$

$$\left. + G \int_0^t \sup_{l \in [0,\tau]} |i(l)| (t-\tau)^{\alpha-1} dB(\tau) \right]$$

This shows that our mapping has a unique solution. We now present the numerical solution of the equation

$$
\begin{aligned}
i(t_{n+1}) &= i(0) + \frac{1}{\Gamma(\alpha)} \int_0^{t_{n+1}} f(\tau, i(\tau))(t_{n+1} - \tau)^{\alpha-1} d\tau \\
&\quad + \frac{G}{\Gamma(\alpha)} \int_0^{t_{n+1}} i(\tau)(t_{n+1} - \tau)^{\alpha-1} dB(\tau) \\
&= i(0) + \frac{1}{\Gamma(\alpha)} \sum_{j=0}^{n} \int_{t_j}^{t_{j+1}} f(\tau, i(\tau))(t_{n+1} - \tau)^{\alpha-1} d\tau \\
&\quad + \frac{G}{\Gamma(\alpha)} \sum_{j=0}^{n} \int_{t_j}^{t_{j+1}} i(\tau)(t_{n+1} - \tau)^{\alpha-1} dB(\tau) \\
&= i(0) + \frac{(\Delta t)^{\alpha}}{\Gamma(\alpha+1)} \sum_{j=2}^{n} f(t_{j-2}, i^{j-2}) [(n-j+1)^{\alpha} - (n-j)^{\alpha}] \\
&\quad + \frac{(\Delta t)^{\alpha}}{\Gamma(\alpha+2)} \sum_{j=2}^{n} [f(t_{j-1}, i^{j-1}) - f(t_{j-2}, i^{j-2})] \\
&\quad \times [(n-j+1)^{\alpha}(n-j+3+2\alpha) - (n-j^{\alpha})(n-j+3+3\alpha)] \\
&\quad + \frac{(\Delta t)^{\alpha}}{2\Gamma(\alpha+3)} \left[f(t_j, i_2^j) - 2f(t_{j-1}, i_2^{j-1}) + f(t_{j-2}, i_2^{j-2}) \right] \\
&\quad \times \left[(n-j+1)^{\alpha} \left(2(n-j)^2 + (3\alpha+10)(n-j) + 2\alpha^2 + 9\alpha + 12 \right) \right. \\
&\quad \left. - (n-j)^{\alpha} \left(2(n-j)^2 + (5\alpha+10)(n-j) + 6\alpha^2 + 18\alpha + 12 \right) \right] \\
&\quad + \frac{G(\Delta t)^{\alpha}}{\Gamma(\alpha+1)} \sum_{j=0}^{n} i(t_j) [(n-j+1)^{\alpha} - (n-j)^{\alpha}] \left(B(t_{j+1}) - B(t_j) \right)
\end{aligned}
$$

Then, we obtain

$$
\begin{aligned}
i(t_{n+1}) &= i(0) + \frac{(\Delta t)^{\alpha}}{\Gamma(\alpha+1)} \sum_{j=2}^{n} \left(-RC \frac{d^2 i(t_{j-2})}{dt^2} + \frac{1}{L} V(t_{j-2}) + \frac{RC}{L} \frac{dV(t_{j-2})}{dt} \right) \\
&\quad \times [(n-j+1)^{\alpha} - (n-j)^{\alpha}] \\
&\quad + \frac{(\Delta t)^{\alpha}}{\Gamma(\alpha+2)} \sum_{j=2}^{n} \left[\left(-RC \frac{d^2 i(t_{j-1})}{dt^2} + \frac{1}{L} V(t_{j-1}) + \frac{RC}{L} \frac{dV(t_{j-1})}{dt} \right) \right. \\
&\quad \left. - \left(-RC \frac{d^2 i(t_{j-2})}{dt^2} + \frac{1}{L} V(t_{j-2}) + \frac{RC}{L} \frac{dV(t_{j-2})}{dt} \right) \right]
\end{aligned}
$$

$$\times [(n-j+1)^\alpha(n-j+3+2\alpha)-(n-j^\alpha)(n-j+3+3\alpha)]$$
$$+\frac{(\Delta t)^\alpha}{2\Gamma(\alpha+3)}\left[\left(-RC\frac{d^2i(t_j)}{dt^2}+\frac{1}{L}V(t_j)+\frac{RC}{L}\frac{dV(t_j)}{dt}\right)\right.$$
$$-2\left(-RC\frac{d^2i(t_{j-1})}{dt^2}+\frac{1}{L}V(t_{j-1})+\frac{RC}{L}\frac{dV(t_{j-1})}{dt}\right)$$
$$+\left.\left(-RC\frac{d^2i(t_{j-2})}{dt^2}+\frac{1}{L}V(t_{j-2})+\frac{RC}{L}\frac{dV(t_{j-2})}{dt}\right)\right]$$
$$\times [(n-j+1)^\alpha\left(2(n-j)^2+(3\alpha+10)(n-j)+2\alpha^2+9\alpha+12\right)$$
$$-(n-j)^\alpha\left(2(n-j)^2+(5\alpha+10)(n-j)+6\alpha^2+18\alpha+12\right)]$$
$$+\frac{G(\Delta t)^\alpha}{\Gamma(\alpha+1)}\sum_{j=0}^{n}i(t_j)\left[(n-j+1)^\alpha-(n-j)^\alpha\right]\left(B(t_{j+1})-B(t_j)\right)$$

We consider the following differential equation

$$\frac{di}{dt}=-\frac{R_1}{A}i(t)+\frac{R_1}{AR_2}V(t)+C\frac{d^2V(t)}{dt^2}+\left(\frac{CR_1}{A}+\frac{1}{R_2}\right)\frac{dV(t)}{dt}=f(t,i(t)) \quad (21.82)$$

We convert the above differential equation to the fractional stochastic differential equation as: In power-law case we have

$$i(t)-i(0)=\frac{1}{\Gamma(\alpha)}\int_0^t f(\tau,i(\tau))(t-\tau)^{\alpha-1}d\tau+\frac{G}{\Gamma(\alpha)}\int_0^t i(\tau)(t-\tau)^{\alpha-1}dB(\tau)$$
$$(21.83)$$

We define a mapping

$$|\Gamma(i(t))| = \left|\frac{1}{\Gamma(\alpha)}\int_0^t f(\tau,i(\tau))(t-\tau)^{\alpha-1}d\tau+\frac{G}{\Gamma(\alpha)}\int_0^t i(\tau)(t-\tau)^{\alpha-1}dB(\tau)\right|$$
$$< \frac{1}{\Gamma(\alpha)}\left[\|f(.,i)\|_\infty\frac{T^\alpha}{\alpha}+GL\|i\|_\infty\frac{T^\alpha}{\alpha}\right]$$
$$< \frac{T^\alpha}{\Gamma(\alpha+1)}(M_1+M_2)$$
$$< \frac{T^\alpha}{\Gamma(\alpha+1)}M$$

We have

$$|\Gamma(i)-\Gamma(i_1)| = \frac{1}{\Gamma(\alpha)}\left|\int_0^t (f(\tau,i(\tau))-f(\tau,i_1(\tau)))(t-\tau)^{\alpha-1}d\tau\right.$$
$$+\frac{G}{\Gamma(\alpha)}\int_0^t (i(\tau)-i_1(\tau))(t-\tau)^{\alpha-1}dB(\tau)$$

$$< \frac{1}{\Gamma(\alpha)} \left[|(f(\tau, i(\tau)) - f(\tau, i_1(\tau))) (t - \tau)^{\alpha-1} d\tau| \right.$$

$$\left. + G |(i(\tau) - i_1(\tau)) (t - \tau)^{\alpha-1} dB(\tau)| \right]$$

$$< \frac{1}{\Gamma(\alpha)} \left[\frac{R_1}{A} \sup_{t \in [0,T]} |i(t) - i_1(t)| \frac{T^\alpha}{\alpha} + GL \frac{T^\alpha}{\alpha} \sup_{t \in [0,T]} |i(t) - i_1(t)| \right]$$

$$< \frac{T^\alpha}{\Gamma(\alpha+1)} \left(\frac{R_1}{A} + GL \right) \|i - i_1\|_\infty < L_1 \|i - i_1\|_\infty$$

We assume that $B(t)$ is Lipschitzian on $[0, T]$ with constant L, then

$$|\Gamma(i(t))| = \frac{1}{\Gamma(\alpha)} \left| \int_0^t f(\tau, i(\tau)) (t - \tau)^{\alpha-1} d\tau + \frac{G}{\Gamma(\alpha)} \int_0^t i(\tau) (t - \tau)^{\alpha-1} dB(\tau) \right|$$

$$\leq \frac{1}{\Gamma(\alpha)} \left[\left| \int_0^t f(\tau, i(\tau)) (t - \tau)^{\alpha-1} d\tau \right| \right.$$

$$\left. G \left| \int_0^t i(\tau) (t - \tau)^{\alpha-1} dB(\tau) \right| \right]$$

$$< \frac{1}{\Gamma(\alpha)} \left[\int_0^t |f(\tau, i(\tau))| (t - \tau)^{\alpha-1} d\tau \right.$$

$$\left. G \int_0^t |i(\tau)| (t - \tau)^{\alpha-1} dB(\tau) \right]$$

$$< \frac{1}{\Gamma(\alpha)} \left[\int_0^t \sup_{l \in [0,\tau]} |f(l, i(l))| (t - \tau)^{\alpha-1} d\tau \right.$$

$$\left. + G \int_0^t \sup_{l \in [0,\tau]} |i(l)| (t - \tau)^{\alpha-1} dB(\tau) \right]$$

This shows that our mapping has a unique solution. We now present the numerical solution of the equation

$$i(t_{n+1}) = i(0) + \frac{1}{\Gamma(\alpha)} \int_0^{t_{n+1}} f(\tau, i(\tau)) (t_{n+1} - \tau)^{\alpha-1} d\tau$$

$$+ \frac{G}{\Gamma(\alpha)} \int_0^{t_{n+1}} i(\tau) (t_{n+1} - \tau)^{\alpha-1} dB(\tau)$$

$$= i(0) + \frac{1}{\Gamma(\alpha)} \sum_{j=0}^n \int_{t_j}^{t_{j+1}} f(\tau, i(\tau)) (t_{n+1} - \tau)^{\alpha-1} d\tau$$

$$+ \frac{G}{\Gamma(\alpha)} \sum_{j=0}^n \int_{t_j}^{t_{j+1}} i(\tau) (t_{n+1} - \tau)^{\alpha-1} dB(\tau)$$

$$= i(0) + \frac{(\Delta t)^\alpha}{\Gamma(\alpha+1)} \sum_{j=2}^n f(t_{j-2}, i^{j-2}) [(n - j + 1)^\alpha - (n - j)^\alpha]$$

$$
\begin{aligned}
&+\frac{(\Delta t)^{\alpha}}{\Gamma(\alpha+2)} \sum_{j=2}^{n}\left[f(t_{j-1},i^{j-1})-f(t_{j-2},i^{j-2})\right] \\
&\times\left[(n-j+1)^{\alpha}(n-j+3+2\alpha)-(n-j^{\alpha})(n-j+3+3\alpha)\right] \\
&+\frac{(\Delta t)^{\alpha}}{2\Gamma(\alpha+3)}\left[f(t_{j},i^{j})-2f(t_{j-1},i^{j-1})+f(t_{j-2},i^{j-2})\right] \\
&\times\left[(n-j+1)^{\alpha}\left(2(n-j)^{2}+(3\alpha+10)(n-j)+2\alpha^{2}+9\alpha+12\right)\right. \\
&\left.-(n-j)^{\alpha}\left(2(n-j)^{2}+(5\alpha+10)(n-j)+6\alpha^{2}+18\alpha+12\right)\right] \\
&+\frac{G(\Delta t)^{\alpha}}{\Gamma(\alpha+1)} \sum_{j=0}^{n} i(t_{j})\left[(n-j+1)^{\alpha}-(n-j)^{\alpha}\right]\left(B(t_{j+1})-B(t_{j})\right)
\end{aligned}
$$

Then, we obtain

$$
\begin{aligned}
i(t_{n+1}) =\ & i(0)+\frac{(\Delta t)^{\alpha}}{\Gamma(\alpha+1)} \sum_{j=2}^{n}\left(-\frac{R_{1}}{A}i(t_{j-2})+\frac{R_{1}}{AR_{2}}V(t_{j-2})+C\frac{d^{2}V(t_{j-2})}{dt^{2}}\right. \\
&+\left(\frac{CR_{1}}{A}+\frac{1}{R_{2}}\right)\frac{dV(t_{j-2})}{dt}\bigg)\left[(n-j+1)^{\alpha}-(n-j)^{\alpha}\right] \\
&+\frac{(\Delta t)^{\alpha}}{\Gamma(\alpha+2)} \sum_{j=2}^{n}\left[\left(-\frac{R_{1}}{A}i(t_{j-1})+\frac{R_{1}}{AR_{2}}V(t_{j-1})+C\frac{d^{2}V(t_{j-1})}{dt^{2}}\right.\right. \\
&+\left(\frac{CR_{1}}{A}+\frac{1}{R_{2}}\right)\frac{dV(t_{j-2})}{dt}\bigg) \\
&-\left(-\frac{R_{1}}{A}i(t_{j-1})+\frac{R_{1}}{AR_{2}}V(t_{j-2})+C\frac{d^{2}V(t_{j-2})}{dt^{2}}\right. \\
&+\left(\frac{CR_{1}}{A}+\frac{1}{R_{2}}\right)\frac{dV(t_{j-2})}{dt}\bigg)\bigg] \\
&\times\left[(n-j+1)^{\alpha}(n-j+3+2\alpha)-(n-j^{\alpha})(n-j+3+3\alpha)\right] \\
&+\frac{(\Delta t)^{\alpha}}{2\Gamma(\alpha+3)}\left[\left(-\frac{R_{1}}{A}i(t_{j})+\frac{R_{1}}{AR_{2}}V(t_{j})+C\frac{d^{2}V(t_{j})}{dt^{2}}\right.\right. \\
&+\left(\frac{CR_{1}}{A}+\frac{1}{R_{2}}\right)\frac{dV(t_{j})}{dt}\bigg) \\
&-2\left(-\frac{R_{1}}{A}i(t_{j-1})+\frac{R_{1}}{AR_{2}}V(t_{j-1})+C\frac{d^{2}V(t_{j-1})}{dt^{2}}\right. \\
&+\left(\frac{CR_{1}}{A}+\frac{1}{R_{2}}\right)\frac{dV(t_{j-1})}{dt}\bigg) \\
&+\left(-\frac{R_{1}}{A}i(t_{j-2})+\frac{R_{1}}{AR_{2}}V(t_{j-2})+C\frac{d^{2}V(t_{j-2})}{dt^{2}}\right. \\
&+\left(\frac{CR_{1}}{A}+\frac{1}{R_{2}}\right)\frac{dV(t_{j-2})}{dt}\bigg)\bigg]
\end{aligned}
$$

$$\times \left[(n-j+1)^{\alpha} \left(2(n-j)^2 + (3\alpha+10)(n-j) + 2\alpha^2 + 9\alpha + 12 \right) \right.$$
$$- (n-j)^{\alpha} \left(2(n-j)^2 + (5\alpha+10)(n-j) + 6\alpha^2 + 18\alpha + 12 \right) \right]$$
$$+ \frac{G(\Delta t)^{\alpha}}{\Gamma(\alpha+1)} \sum_{j=0}^{n} i(t_j) \left[(n-j+1)^{\alpha} - (n-j)^{\alpha} \right] \left(B(t_{j+1}) - B(t_j) \right)$$

We consider the following differential equation

$$\frac{di}{dt} = -\frac{R}{A} i(t) - \frac{1}{AC} \int_0^t i(y) dy + \frac{1}{A} V(t) = f(t, i(t)) \qquad (21.84)$$

We convert the above differential equation to the fractional stochastic differential equation as: In power-law case we have

$$i(t) - i(0) = \frac{1}{\Gamma(\alpha)} \int_0^t f(\tau, i(\tau))(t-\tau)^{\alpha-1} d\tau + \frac{G}{\Gamma(\alpha)} \int_0^t i(\tau)(t-\tau)^{\alpha-1} dB(\tau)$$
$$(21.85)$$

We define a mapping

$$|\Gamma(i(t))| = \left| \frac{1}{\Gamma(\alpha)} \int_0^t f(\tau,i(\tau))(t-\tau)^{\alpha-1} d\tau + \frac{G}{\Gamma(\alpha)} \int_0^t i(\tau)(t-\tau)^{\alpha-1} dB(\tau) \right|$$
$$< \frac{1}{\Gamma(\alpha)} \left[\|f(.,i)\|_{\infty} \frac{T^{\alpha}}{\alpha} + GL\|i\|_{\infty} \frac{T^{\alpha}}{\alpha} \right]$$
$$< \frac{T^{\alpha}}{\Gamma(\alpha+1)} (M_1 + M_2)$$
$$< \frac{T^{\alpha}}{\Gamma(\alpha+1)} M$$

We have

$$|\Gamma(i) - \Gamma(i_1)| = \frac{1}{\Gamma(\alpha)} \left| \int_0^t (f(\tau,i(\tau)) - f(\tau,i_1(\tau)))(t-\tau)^{\alpha-1} d\tau \right.$$
$$+ \frac{G}{\Gamma(\alpha)} \int_0^t (i(\tau) - i_1(\tau))(t-\tau)^{\alpha-1} dB(\tau)$$
$$< \frac{1}{\Gamma(\alpha)} \left[|(f(\tau,i(\tau)) - f(\tau,i_1(\tau)))(t-\tau)^{\alpha-1} d\tau| \right.$$
$$+ G|(i(\tau) - i_1(\tau))(t-\tau)^{\alpha-1} dB(\tau)| \right]$$
$$< \frac{1}{\Gamma(\alpha)} \left[\left(\frac{R}{A} + \frac{1}{AC} \right) \sup_{t \in [0,T]} |i(t) - i_1(t)| \frac{T^{\alpha}}{\alpha} \right.$$

$$|\Gamma(i) - \Gamma(i_1)| = \frac{1}{\Gamma(\alpha)} \left| \int_0^t \left(f(\tau, i(\tau)) - f(\tau, i_1(\tau)) \right) (t - \tau)^{\alpha - 1} d\tau \right.$$

$$+ \frac{G}{\Gamma(\alpha)} \int_0^t (i(\tau) - i_1(\tau))(t - \tau)^{\alpha - 1} dB(\tau)$$

$$< \frac{1}{\Gamma(\alpha)} \left[\left| (f(\tau, i(\tau)) - f(\tau, i_1(\tau))) (t - \tau)^{\alpha - 1} d\tau \right| \right.$$

$$\left. + G \left| (i(\tau) - i_1(\tau))(t - \tau)^{\alpha - 1} dB(\tau) \right| \right]$$

$$< \frac{1}{\Gamma(\alpha)} \left[\left(\frac{R}{A} + \frac{1}{AC} \right) \sup_{t \in [0,T]} |i(t) - i_1(t)| \frac{T^\alpha}{\alpha} \right.$$

$$\left. + GL \frac{T^\alpha}{\alpha} \sup_{t \in [0,T]} |i(t) - i_1(t)| \right]$$

$$< \frac{T^\alpha}{\Gamma(\alpha + 1)} \left(\left(\frac{R}{A} + \frac{1}{AC} \right) + GL \right) \|i - i_1\|_\infty < L_1 \|i - i_1\|_\infty$$

We assume that $B(t)$ is Lipschitzian on $[0,T]$ with constant L, then

$$|\Gamma(i(t))| = \frac{1}{\Gamma(\alpha)} \left| \int_0^t f(\tau, i(\tau))(t - \tau)^{\alpha - 1} d\tau + \frac{G}{\Gamma(\alpha)} \int_0^t i(\tau)(t - \tau)^{\alpha - 1} dB(\tau) \right|$$

$$\leq \frac{1}{\Gamma(\alpha)} \left[\left| \int_0^t f(\tau, i(\tau))(t - \tau)^{\alpha - 1} d\tau \right| \right.$$

$$\left. G \left| \int_0^t i(\tau)(t - \tau)^{\alpha - 1} dB(\tau) \right| \right]$$

$$< \frac{1}{\Gamma(\alpha)} \left[\int_0^t |f(\tau, i(\tau))| (t - \tau)^{\alpha - 1} d\tau \right.$$

$$\left. G \int_0^t |i(\tau)| (t - \tau)^{\alpha - 1} dB(\tau) \right]$$

$$< \frac{1}{\Gamma(\alpha)} \left[\int_0^t \sup_{l \in [0,\tau]} |f(l, i(l))| (t - \tau)^{\alpha - 1} d\tau \right.$$

$$\left. + G \int_0^t \sup_{l \in [0,\tau]} |i(l)| (t - \tau)^{\alpha - 1} dB(\tau) \right]$$

This shows that our mapping has a unique solution. We now present the numerical solution of the equation

$$i(t_{n+1}) = i(0) + \frac{1}{\Gamma(\alpha)} \int_0^{t_{n+1}} f(\tau, i(\tau))(t_{n+1} - \tau)^{\alpha - 1} d\tau$$

$$+ \frac{G}{\Gamma(\alpha)} \int_0^{t_{n+1}} i(\tau)(t_{n+1} - \tau)^{\alpha - 1} dB(\tau)$$

$$
\begin{aligned}
= \ & i(0) + \frac{1}{\Gamma(\alpha)} \sum_{j=0}^{n} \int_{t_j}^{t_{j+1}} f(\tau, i(\tau))(t_{n+1} - \tau)^{\alpha-1} d\tau \\
& + \frac{G}{\Gamma(\alpha)} \sum_{j=0}^{n} \int_{t_j}^{t_{j+1}} i(\tau)(t_{n+1} - \tau)^{\alpha-1} dB(\tau) \\
= \ & i(0) + \frac{(\Delta t)^{\alpha}}{\Gamma(\alpha+1)} \sum_{j=2}^{n} f(t_{j-2}, i^{j-2}) \left[(n-j+1)^{\alpha} - (n-j)^{\alpha} \right] \\
& + \frac{(\Delta t)^{\alpha}}{\Gamma(\alpha+2)} \sum_{j=2}^{n} \left[f(t_{j-1}, i^{j-1}) - f(t_{j-2}, i^{j-2}) \right] \\
& \times \left[(n-j+1)^{\alpha}(n-j+3+2\alpha) - (n-j^{\alpha})(n-j+3+3\alpha) \right] \\
& + \frac{(\Delta t)^{\alpha}}{2\Gamma(\alpha+3)} \left[f(t_j, i^j) - 2f(t_{j-1}, i^{j-1}) + f(t_{j-2}, i^{j-2}) \right] \\
& \times \left[(n-j+1)^{\alpha} \left(2(n-j)^2 + (3\alpha+10)(n-j) + 2\alpha^2 + 9\alpha + 12 \right) \right. \\
& \left. - (n-j)^{\alpha} \left(2(n-j)^2 + (5\alpha+10)(n-j) + 6\alpha^2 + 18\alpha + 12 \right) \right] \\
& + \frac{G(\Delta t)^{\alpha}}{\Gamma(\alpha+1)} \sum_{j=0}^{n} i(t_j) \left[(n-j+1)^{\alpha} - (n-j)^{\alpha} \right] \left(B(t_{j+1}) - B(t_j) \right)
\end{aligned}
$$

Then, we obtain

$$
\begin{aligned}
i(t_{n+1}) = \ & i(0) + \frac{(\Delta t)^{\alpha}}{\Gamma(\alpha+1)} \sum_{j=2}^{n} \left(-\frac{R}{A} i(t_{j-2}) - \frac{1}{AC} \int_0^{t_{j-2}} i(y) dy + \frac{1}{A} V(t_{j-2}) \right) \\
& \times \left[(n-j+1)^{\alpha} - (n-j)^{\alpha} \right] \\
& + \frac{(\Delta t)^{\alpha}}{\Gamma(\alpha+2)} \sum_{j=2}^{n} \left[\left(-\frac{R}{A} i(t_{j-1}) - \frac{1}{AC} \int_0^{t_{j-1}} i(y) dy + \frac{1}{A} V(t_{j-1}) \right) \right. \\
& \left. - \left(-\frac{R}{A} i(t_{j-2}) - \frac{1}{AC} \int_0^{t_{j-2}} i(y) dy + \frac{1}{A} V(t_{j-2}) \right) \right] \\
& \times \left[(n-j+1)^{\alpha}(n-j+3+2\alpha) - (n-j^{\alpha})(n-j+3+3\alpha) \right] \\
& + \frac{(\Delta t)^{\alpha}}{2\Gamma(\alpha+3)} \left[\left(-\frac{R}{A} i(t_j) - \frac{1}{AC} \int_0^{t_j} i(y) dy + \frac{1}{A} V(t_j) \right) \right. \\
& - 2 \left(-\frac{R}{A} i(t_{j-1}) - \frac{1}{AC} \int_0^{t_{j-1}} i(y) dy + \frac{1}{A} V(t_{j-1}) \right) \\
& \left. + \left(-\frac{R}{A} i(t_{j-2}) - \frac{1}{AC} \int_0^{t_{j-2}} i(y) dy + \frac{1}{A} V(t_{j-2}) \right) \right] \\
& \times \left[(n-j+1)^{\alpha} \left(2(n-j)^2 + (3\alpha+10)(n-j) + 2\alpha^2 + 9\alpha + 12 \right) \right. \\
& \left. - (n-j)^{\alpha} \left(2(n-j)^2 + (5\alpha+10)(n-j) + 6\alpha^2 + 18\alpha + 12 \right) \right] \\
& + \frac{G(\Delta t)^{\alpha}}{\Gamma(\alpha+1)} \sum_{j=0}^{n} i(t_j) \left[(n-j+1)^{\alpha} - (n-j)^{\alpha} \right] \left(B(t_{j+1}) - B(t_j) \right)
\end{aligned}
$$

We consider the following differential equation

$$\frac{di}{dt} = -\frac{R}{L}i(t) + \frac{1}{L}V(t) = f(t, i(t)) \tag{21.86}$$

We convert the above differential equation to the fractional stochastic differential equation as: In exponential-decay case we have

$$i(t) - i(0) = \frac{1-\alpha}{M(\alpha)}f(t, i(t)) + \frac{\alpha}{M(\alpha)}\int_0^t f(\tau, i(\tau))d\tau + \frac{G\alpha}{M(\alpha)}\int_0^t i(\tau)dB(\tau) \tag{21.87}$$

We define a mapping

$$
\begin{aligned}
|\Gamma(i(t))| &= \left| \frac{1-\alpha}{M(\alpha)}f(t, i(t)) + \frac{\alpha}{M(\alpha)}\int_0^t f(\tau, i(\tau))d\tau \right. \\
&\quad \left. + \frac{G\alpha}{M(\alpha)}\int_0^t i(\tau)dB(\tau) \right| \\
&< \frac{1-\alpha}{M(\alpha)}\sup_{t\in[0,T]}|f(t, i(t))| + \frac{\alpha}{M(\alpha)}\int_0^t \sup_{\tau\in[0,t]}|f(\tau, i(\tau))|d\tau \\
&\quad + \frac{G\alpha}{M(\alpha)}\int_0^t \sup_{\tau\in[0,t]}|i(\tau)|dB(\tau) \\
&< \frac{1-\alpha}{M(\alpha)}\|f(.,i)\|_\infty + T\frac{\alpha}{M(\alpha)}\|f(.,i)\|_\infty + GL\|i\|_\infty \\
&< M_1 + M_2 + M_3 \\
&< M
\end{aligned}
$$

We have

$$
\begin{aligned}
|\Gamma(i) - \Gamma(i_1)| &= \left| \frac{1-\alpha}{M(\alpha)}(f(t, i(t)) - f(t, i_1(t))) \right. \\
&\quad + \frac{\alpha}{M(\alpha)}\int_0^t (f(\tau, i(\tau)) - f(\tau, i_1(\tau)))d\tau \\
&\quad \left. + \frac{G\alpha}{M(\alpha)}\int_0^t (i(\tau) - i_1(\tau))dB(\tau) \right| \\
&< \frac{1-\alpha}{M(\alpha)}|f(t, i(t)) - f(t, i_1(t))| \\
&\quad + \frac{\alpha}{M(\alpha)}\int_0^t |f(\tau, i(\tau)) - f(\tau, i_1(\tau))|d\tau \\
&\quad + \frac{G\alpha}{M(\alpha)}\int_0^t |i(\tau) - i_1(\tau)|dB(\tau) \\
&< \frac{1-\alpha}{M(\alpha)}\sup_{t\in[0,T]}|f(t, i(t)) - f(t, i_1(t))|
\end{aligned}
$$

$$+\frac{\alpha}{M(\alpha)}\int_0^t \sup_{\tau\in[0,t]}|f(\tau,i(\tau))-f(\tau,i_1(\tau))|\,d\tau$$

$$+\frac{G\alpha}{M(\alpha)}\int_0^t \sup_{\tau\in[0,t]}|i(\tau)-i_1(\tau)|\,dB(\tau)$$

$$< \frac{1-\alpha}{M(\alpha)}\frac{R}{L}\|i(t)-i_1(t)\|_\infty$$

$$+\frac{\alpha}{M(\alpha)}\frac{TR}{L}\|i(t))\quad i_1(t)\|_\infty + GL\|i(t)-i_1(t)\|_\infty$$

$$< \left(\frac{1-\alpha}{M(\alpha)}\frac{R}{L}+\frac{\alpha}{M(\alpha)}\frac{TR}{L}+GL\right)\|i(t)-i_1(t)\|_\infty$$

$$< L_1\|i(t)-i_1(t)\|_\infty$$

We assume that $B(t)$ is Lipschitzian on $[0,T]$ with constant L, then

$$|\Gamma(i(t))| = \left|\frac{1-\alpha}{M(\alpha)}f(t,i(t))+\frac{\alpha}{M(\alpha)}\int_0^t f(\tau,i(\tau))d\tau\right.$$
$$\left.+\frac{G\alpha}{M(\alpha)}\int_0^t i(\tau)dB(\tau)\right|$$

$$< \frac{1-\alpha}{M(\alpha)}|f(t,i(t))|$$

$$+\frac{\alpha}{M(\alpha)}\int_0^t |f(\tau,i(\tau))|\,d\tau$$

$$+\frac{G\alpha}{M(\alpha)}\int_0^t |i(\tau)|\,dB(\tau)$$

$$< \frac{1-\alpha}{M(\alpha)}\sup_{l\in[0,t]}|f(l,i(l))|$$

$$+\frac{\alpha}{M(\alpha)}\int_0^t \sup_{l\in[0,\tau]}|f(l,i(l))|\,d\tau$$

$$+\frac{G\alpha}{M(\alpha)}\int_0^t \sup_{l\in[0,\tau]}|i(l)|\,dB(\tau)$$

This shows that our mapping has a unique solution. We now present the numerical solution of the equation

$$i(t_{n+1}) = i(0)+\frac{1-\alpha}{M(\alpha)}f(t_n,i(t_n))+\frac{\alpha}{M(\alpha)}\int_0^{t_{n+1}} f(\tau,i(\tau))d\tau$$
$$+\frac{G\alpha}{M(\alpha)}\int_0^{t_{n+1}} i(\tau)dB(\tau)$$

and

$$
\begin{aligned}
i(t_n) &= i(0) + \frac{1-\alpha}{M(\alpha)} f(t_{n-1}, i(t_{n-1})) + \frac{\alpha}{M(\alpha)} \int_0^{t_n} f(\tau, i(\tau)) d\tau \\
&\quad + \frac{G\alpha}{M(\alpha)} \int_0^{t_n} i(\tau) dB(\tau)
\end{aligned}
$$

Thus, we reach

$$
\begin{aligned}
i(t_{n+1}) &= i(n) + \frac{1-\alpha}{M(\alpha)} \left(f(t_n, i(t_n)) - f(t_{n-1}, i(t_{n-1})) \right) \\
&\quad + \frac{\alpha}{M(\alpha)} \int_{t_n}^{t_{n+1}} f(\tau, i(\tau)) d\tau + \frac{G\alpha}{M(\alpha)} \int_{t_n}^{t_{n+1}} i(\tau) dB(\tau) \\
&= i(n) + \frac{1-\alpha}{M(\alpha)} \left(f(t_n, i(t_n)) - f(t_{n-1}, i(t_{n-1})) \right) \\
&\quad + \frac{\alpha}{M(\alpha)} \sum_{j=2}^{n} \int_{t_j}^{t_{j+1}} f(\tau, i(\tau)) d\tau + \frac{G\alpha}{M(\alpha)} \sum_{j=2}^{n} \int_{t_j}^{t_{j+1}} i(\tau) dB(\tau) \\
&= i(n) + \frac{1-\alpha}{M(\alpha)} \left(f(t_n, i(t_n)) - f(t_{n-1}, i(t_{n-1})) \right) \\
&\quad + \frac{\alpha}{M(\alpha)} \sum_{j=2}^{n} \left[-(\Delta t) \frac{4}{3} f(t_{j-1}, i^{j-1}) + (\Delta t) \frac{5}{12} f(t_{j-2}, i^{j-2}) \right. \\
&\quad \left. + (\Delta t) \frac{23}{12} f(t_j, i^j) \right] \\
&\quad + \frac{G\alpha}{M(\alpha)} \sum_{j=2}^{n} i(t_j) \left(B(t_{j+1}) - B(t_j) \right)
\end{aligned}
$$

Then, we obtain

$$
\begin{aligned}
i(t_{n+1}) &= i(n) + \frac{1-\alpha}{M(\alpha)} \left(\left(-\frac{R}{L} i(t_n) + \frac{1}{L} V(t_n) \right) \right. \\
&\quad \left. - \left(-\frac{R}{L} i(t_{n-1}) + \frac{1}{L} V(t_{n-1}) \right) \right) \\
&\quad + \frac{\alpha}{M(\alpha)} \sum_{j=2}^{n} \left[-(\Delta t) \frac{4}{3} \left(-\frac{R}{L} i(t_{j-1}) + \frac{1}{L} V(t_{j-1}) \right) \right. \\
&\quad + (\Delta t) \frac{5}{12} \left(-\frac{R}{L} i(t_{j-2}) + \frac{1}{L} V(t_{j-2}) \right) \\
&\quad \left. + (\Delta t) \frac{23}{12} \left(-\frac{R}{L} i(t_j) + \frac{1}{L} V(t_j) \right) \right] \\
&\quad + \frac{G\alpha}{M(\alpha)} \sum_{j=2}^{n} i(t_j) \left(B(t_{j+1}) - B(t_j) \right)
\end{aligned}
$$

We consider the following differential equation

$$\frac{dV_2}{dt} = -\frac{1}{RC}V_2(t) + \frac{1}{2RC}V_1(t) - \frac{dV_1(t)}{dt} = f(t, V_2(t)) \qquad (21.88)$$

We convert the above differential equation to the fractional stochastic differential equation as: In exponential-decay case we have

$$V_2(t) - V_2(0) = \frac{1-\alpha}{M(\alpha)}f(t, V_2(t)) + \frac{\alpha}{M(\alpha)}\int_0^t f(\tau, V_2(\tau))d\tau + \frac{G\alpha}{M(\alpha)}\int_0^t V_2(\tau)dB(\tau)$$

$$(21.89)$$

We define a mapping

$$
\begin{aligned}
|\Gamma(V_2(t))| &= \left| \frac{1-\alpha}{M(\alpha)}f(t, V_2(t)) + \frac{\alpha}{M(\alpha)}\int_0^t f(\tau, V_2(\tau))d\tau \right. \\
&\quad \left. + \frac{G\alpha}{M(\alpha)}\int_0^t V_2(\tau)dB(\tau) \right| \\
&< \frac{1-\alpha}{M(\alpha)}\sup_{t\in[0,T]}|f(t, V_2(t))| + \frac{\alpha}{M(\alpha)}\int_0^t \sup_{\tau\in[0,t]}|f(\tau, V_2(\tau))|d\tau \\
&\quad + \frac{G\alpha}{M(\alpha)}\int_0^t \sup_{\tau\in[0,t]}|V_2(\tau)|dB(\tau) \\
&< \frac{1-\alpha}{M(\alpha)}\|f(.,i)\|_\infty + T\frac{\alpha}{M(\alpha)}\|f(.,V_2)\|_\infty + GL\|V_2\|_\infty \\
&< M_1 + M_2 + M_3 \\
&< M
\end{aligned}
$$

We have

$$
\begin{aligned}
|\Gamma(V_2) - \Gamma(V_{2_1})| &= \left| \frac{1-\alpha}{M(\alpha)}(f(t, V_2(t)) - f(t, V_{2_1}(t))) \right. \\
&\quad + \frac{\alpha}{M(\alpha)}\int_0^t (f(\tau, V_2(\tau)) - f(\tau, V_{2_1}(\tau)))d\tau \\
&\quad \left. + \frac{G\alpha}{M(\alpha)}\int_0^t (V_2(\tau) - V_{2_1}(\tau))dB(\tau) \right| \\
&< \frac{1-\alpha}{M(\alpha)}|f(t, V_2(t)) - f(t, V_{2_1}(t))| \\
&\quad + \frac{\alpha}{M(\alpha)}\int_0^t |f(\tau, V_2(\tau)) - f(\tau, V_{2_1}(\tau))|d\tau \\
&\quad + \frac{G\alpha}{M(\alpha)}\int_0^t |V_2(\tau) - V_{2_1}(\tau)|dB(\tau) \\
&< \frac{1-\alpha}{M(\alpha)}\sup_{t\in[0,T]}|f(t, V_2(t)) - f(t, V_{2_1}(t))|
\end{aligned}
$$

$$+\frac{\alpha}{M(\alpha)}\int_0^t \sup_{\tau\in[0,t]}|f(\tau,V_2(\tau))-f(\tau,V_{2_1}(\tau))|\,d\tau$$

$$+\frac{G\alpha}{M(\alpha)}\int_0^t \sup_{\tau\in[0,t]}|V_2(\tau)-V_{2_1}(\tau)|\,dB(\tau)$$

$$< \frac{1-\alpha}{M(\alpha)}\frac{1}{RC}\|V_2(t)-V_{2_1}(t)\|_\infty$$

$$+\frac{\alpha}{M(\alpha)}\frac{T}{RC}\|V_2(t))-V_{2_1}(t)\|_\infty + GL\|V_2(t)-V_{2_1}(t)\|_\infty$$

$$< \left(\frac{1-\alpha}{M(\alpha)}\frac{1}{RC}+\frac{\alpha}{M(\alpha)}\frac{T}{RC}+GL\right)\|V_2(t)-V_{2_1}(t)\|_\infty$$

$$< L_1\|V_2(t)-V_{2_1}(t)\|_\infty$$

We assume that $B(t)$ is Lipschitzian on $[0,T]$ with constant L, then

$$
\begin{aligned}
|\Gamma(V_2(t))| &= \left|\frac{1-\alpha}{M(\alpha)}f(t,V_2(t))+\frac{\alpha}{M(\alpha)}\int_0^t f(\tau,V_2(\tau))d\tau\right. \\
&\qquad \left. +\frac{G\alpha}{M(\alpha)}\int_0^t V_2(\tau)dB(\tau)\right| \\[2mm]
&< \frac{1-\alpha}{M(\alpha)}|f(t,V_2(t))| \\[2mm]
&\qquad +\frac{\alpha}{M(\alpha)}\int_0^t |f(\tau,V_2(\tau))|\,d\tau \\[2mm]
&\qquad +\frac{G\alpha}{M(\alpha)}\int_0^t |V_2(\tau)|\,dB(\tau) \\[2mm]
&< \frac{1-\alpha}{M(\alpha)}\sup_{l\in[0,t]}|f(l,V_2(l))| \\[2mm]
&\qquad +\frac{\alpha}{M(\alpha)}\int_0^t \sup_{l\in[0,\tau]}|f(l,V_2(l))|\,d\tau \\[2mm]
&\qquad +\frac{G\alpha}{M(\alpha)}\int_0^t \sup_{l\in[0,\tau]}|V_2(l)|\,dB(\tau)
\end{aligned}
$$

This shows that our mapping has a unique solution. We now present the numerical solution of the equation

$$
\begin{aligned}
V_2(t_{n+1}) &= V_2(0)+\frac{1-\alpha}{M(\alpha)}f(t_n,V_2(t_n))+\frac{\alpha}{M(\alpha)}\int_0^{t_{n+1}} f(\tau,i(\tau))d\tau \\[2mm]
&\qquad +\frac{G\alpha}{M(\alpha)}\int_0^{t_{n+1}} V_2(\tau)dB(\tau)
\end{aligned}
$$

and

$$
\begin{aligned}
V_2(t_n) &= V_2(0) + \frac{1-\alpha}{M(\alpha)} f(t_{n-1}, V_2(t_{n-1})) + \frac{\alpha}{M(\alpha)} \int_0^{t_n} f(\tau, V_2(\tau))d\tau \\
&\quad + \frac{G\alpha}{M(\alpha)} \int_0^{t_n} V_2(\tau)dB(\tau)
\end{aligned}
$$

Thus, we reach

$$
\begin{aligned}
V_2(t_{n+1}) &= V_2(n) + \frac{1-\alpha}{M(\alpha)} \left(f(t_n, V_2(t_n)) - f(t_{n-1}, V_2(t_{n-1})) \right) \\
&\quad + \frac{\alpha}{M(\alpha)} \int_{t_n}^{t_{n+1}} f(\tau, V_2(\tau))d\tau + \frac{G\alpha}{M(\alpha)} \int_{t_n}^{t_{n+1}} V_2(\tau)dB(\tau) \\
&= i(n) + \frac{1-\alpha}{M(\alpha)} \left(f(t_n, V_2(t_n)) - f(t_{n-1}, V_2(t_{n-1})) \right) \\
&\quad + \frac{\alpha}{M(\alpha)} \sum_{j=2}^{n} \int_{t_j}^{t_{j+1}} f(\tau, V_2(\tau))d\tau + \frac{G\alpha}{M(\alpha)} \sum_{j=2}^{n} \int_{t_j}^{t_{j+1}} i(\tau)dB(\tau) \\
&= i(n) + \frac{1-\alpha}{M(\alpha)} \left(f(t_n, V_2(t_n)) - f(t_{n-1}, V_2(t_{n-1})) \right) \\
&\quad + \frac{\alpha}{M(\alpha)} \sum_{j=2}^{n} \left[-(\Delta t)\frac{4}{3} f(t_{j-1}, i^{j-1}) + (\Delta t)\frac{5}{12} f(t_{j-2}, i^{j-2}) \right. \\
&\quad \left. + (\Delta t)\frac{23}{12} f(t_j, i^j) \right] \\
&\quad + \frac{G\alpha}{M(\alpha)} \sum_{j=2}^{n} i(t_j) \left(B(t_{j+1}) - B(t_j) \right)
\end{aligned}
$$

Then, we obtain

$$
\begin{aligned}
V_2(t_{n+1}) &= V_2(n) + \frac{1-\alpha}{M(\alpha)} \left(\left(-\frac{1}{RC} V_2(t_n) + \frac{1}{2RC} V_1(t_n) - \frac{dV_1(t_n)}{dt} \right) \right. \\
&\quad \left. - \left(-\frac{1}{RC} V_2(t_{n-1}) + \frac{1}{2RC} V_1(t_{n-1}) - \frac{dV_1(t_{n-1})}{dt} \right) \right) \\
&\quad + \frac{\alpha}{M(\alpha)} \sum_{j=2}^{n} \left[-(\Delta t)\frac{4}{3} \left(-\frac{1}{RC} V_2(t_{j-1}) + \frac{1}{2RC} V_1(t_{j-1}) - \frac{dV_1(t_{j-1})}{dt} \right) \right. \\
&\quad + (\Delta t)\frac{5}{12} \left(-\frac{1}{RC} V_2(t_{j-2}) + \frac{1}{2RC} V_1(t_{j-2}) - \frac{dV_1(t_{j-2})}{dt} \right) \\
&\quad \left. + (\Delta t)\frac{23}{12} \left(-\frac{1}{RC} V_2(t_j) + \frac{1}{2RC} V_1(t_j) - \frac{dV_1(t_j)}{dt} \right) \right] \\
&\quad + \frac{G\alpha}{M(\alpha)} \sum_{j=2}^{n} V_2(t_j) \left(B(t_{j+1}) - B(t_j) \right)
\end{aligned}
$$

We consider the following differential equation

$$\frac{dV_2}{dt} = -\frac{1}{R_1 C_1} V_2(t) - \frac{1}{R_1 C_2} V_1(t) + \frac{R_2}{R_1} \frac{dV_1(t)}{dt} = f(t, V_2(t)) \qquad (21.90)$$

We convert the above differential equation to the fractional stochastic differential equation as: In exponential-decay case we have

$$V_2(t) - V_2(0) = \frac{1-\alpha}{M(\alpha)} f(t, V_2(t)) + \frac{\alpha}{M(\alpha)} \int_0^t f(\tau, V_2(\tau)) d\tau + \frac{G\alpha}{M(\alpha)} \int_0^t V_2(\tau) dB(\tau)$$

$$(21.91)$$

We define a mapping

$$
\begin{aligned}
|\Gamma(V_2(t))| \;=\; & \left| \frac{1-\alpha}{M(\alpha)} f(t, V_2(t)) + \frac{\alpha}{M(\alpha)} \int_0^t f(\tau, V_2(\tau)) d\tau \right. \\
& \left. + \frac{G\alpha}{M(\alpha)} \int_0^t V_2(\tau) dB(\tau) \right| \\
<\; & \frac{1-\alpha}{M(\alpha)} \sup_{t\in[0,T]} |f(t, V_2(t))| + \frac{\alpha}{M(\alpha)} \int_0^t \sup_{\tau\in[0,t]} |f(\tau, V_2(\tau))| d\tau \\
& + \frac{G\alpha}{M(\alpha)} \int_0^t \sup_{\tau\in[0,t]} |V_2(\tau)| dB(\tau) \\
<\; & \frac{1-\alpha}{M(\alpha)} \|f(.,i)\|_\infty + T \frac{\alpha}{M(\alpha)} \|f(.,V_2)\|_\infty + GL \|V_2\|_\infty \\
<\; & M_1 + M_2 + M_3 \\
<\; & M
\end{aligned}
$$

We have

$$
\begin{aligned}
|\Gamma(V_2) - \Gamma(V_{2_1})| \;=\; & \left| \frac{1-\alpha}{M(\alpha)} \left(f(t, V_2(t)) - f(t, V_{2_1}(t)) \right) \right. \\
& + \frac{\alpha}{M(\alpha)} \int_0^t \left(f(\tau, V_2(\tau)) - f(\tau, V_{2_1}(\tau)) \right) d\tau \\
& \left. + \frac{G\alpha}{M(\alpha)} \int_0^t \left(V_2(\tau) - V_{2_1}(\tau) \right) dB(\tau) \right| \\
<\; & \frac{1-\alpha}{M(\alpha)} |f(t, V_2(t)) - f(t, V_{2_1}(t))| \\
& + \frac{\alpha}{M(\alpha)} \int_0^t |f(\tau, V_2(\tau)) - f(\tau, V_{2_1}(\tau))| d\tau \\
& + \frac{G\alpha}{M(\alpha)} \int_0^t |V_2(\tau) - V_{2_1}(\tau)| dB(\tau) \\
<\; & \frac{1-\alpha}{M(\alpha)} \sup_{t\in[0,T]} |f(t, V_2(t)) - f(t, V_{2_1}(t))|
\end{aligned}
$$

$$+\frac{\alpha}{M(\alpha)}\int_0^t \sup_{\tau\in[0,t]}|f(\tau,V_2(\tau))-f(\tau,V_{2_1}(\tau))|d\tau$$

$$+\frac{G\alpha}{M(\alpha)}\int_0^t \sup_{\tau\in[0,t]}|V_2(\tau)-V_{2_1}(\tau)|dB(\tau)$$

$$<\frac{1-\alpha}{M(\alpha)}\frac{1}{R_1C_1}\|V_2(t)-V_{2_1}(t)\|_\infty$$

$$+\frac{\alpha}{M(\alpha)}\frac{T}{R_1C_1}\|V_2(t))-V_{2_1}(t)\|_\infty+GL\|V_2(t)-V_{2_1}(t)\|_\infty$$

$$<\left(\frac{1-\alpha}{M(\alpha)}\frac{1}{R_1C_1}+\frac{\alpha}{M(\alpha)}\frac{T}{R_1C_1}+GL\right)\|V_2(t)-V_{2_1}(t)\|_\infty$$

$$<L_1\|V_2(t)-V_{2_1}(t)\|_\infty$$

We assume that $B(t)$ is Lipschitzian on $[0,T]$ with constant L, then

$$|\Gamma(V_2(t))| = \left|\frac{1-\alpha}{M(\alpha)}f(t,V_2(t))+\frac{\alpha}{M(\alpha)}\int_0^t f(\tau,V_2(\tau))d\tau\right.$$

$$\left.+\frac{G\alpha}{M(\alpha)}\int_0^t V_2(\tau)dB(\tau)\right|$$

$$<\frac{1-\alpha}{M(\alpha)}|f(t,V_2(t))|$$

$$+\frac{\alpha}{M(\alpha)}\int_0^t |f(\tau,V_2(\tau))|d\tau$$

$$+\frac{G\alpha}{M(\alpha)}\int_0^t |V_2(\tau)|dB(\tau)$$

$$<\frac{1-\alpha}{M(\alpha)}\sup_{l\in[0,t]}|f(l,V_2(l))|$$

$$+\frac{\alpha}{M(\alpha)}\int_0^t \sup_{l\in[0,\tau]}|f(l,V_2(l))|d\tau$$

$$+\frac{G\alpha}{M(\alpha)}\int_0^t \sup_{l\in[0,\tau]}|V_2(l)|dB(\tau)$$

This shows that our mapping has a unique solution. We now present the numerical solution of the equation

$$V_2(t_{n+1}) = i(0)+\frac{1-\alpha}{M(\alpha)}f(t_n,V_2(t_n))+\frac{\alpha}{M(\alpha)}\int_0^{t_{n+1}} f(\tau,i(\tau))d\tau$$

$$+\frac{G\alpha}{M(\alpha)}\int_0^{t_{n+1}} V_2(\tau)dB(\tau)$$

and

$$
\begin{aligned}
V_2(t_n) &= V_2(0) + \frac{1-\alpha}{M(\alpha)} f(t_{n-1}, V_2(t_{n-1})) + \frac{\alpha}{M(\alpha)} \int_0^{t_n} f(\tau, V_2(\tau)) d\tau \\
&\quad + \frac{G\alpha}{M(\alpha)} \int_0^{t_n} V_2(\tau) dB(\tau)
\end{aligned}
$$

Thus, we reach

$$
\begin{aligned}
V_2(t_{n+1}) &= V_2(n) + \frac{1-\alpha}{M(\alpha)} \left(f(t_n, V_2(t_n)) - f(t_{n-1}, V_2(t_{n-1})) \right) \\
&\quad + \frac{\alpha}{M(\alpha)} \int_{t_n}^{t_{n+1}} f(\tau, V_2(\tau)) d\tau + \frac{G\alpha}{M(\alpha)} \int_{t_n}^{t_{n+1}} V_2(\tau) dB(\tau) \\
&= i(n) + \frac{1-\alpha}{M(\alpha)} \left(f(t_n, V_2(t_n)) - f(t_{n-1}, V_2(t_{n-1})) \right) \\
&\quad + \frac{\alpha}{M(\alpha)} \sum_{j=2}^{n} \int_{t_j}^{t_{j+1}} f(\tau, V_2(\tau)) d\tau + \frac{G\alpha}{M(\alpha)} \sum_{j=2}^{n} \int_{t_j}^{t_{j+1}} i(\tau) dB(\tau) \\
&= i(n) + \frac{1-\alpha}{M(\alpha)} \left(f(t_n, V_2(t_n)) - f(t_{n-1}, V_2(t_{n-1})) \right) \\
&\quad + \frac{\alpha}{M(\alpha)} \sum_{j=2}^{n} \left[-(\Delta t)\frac{4}{3} f(t_{j-1}, V_2^{j-1}) + (\Delta t)\frac{5}{12} f(t_{j-2}, V_2^{j-2}) \right. \\
&\quad \left. + (\Delta t)\frac{23}{12} f(t_j, V_2^{j}) \right] \\
&\quad + \frac{G\alpha}{M(\alpha)} \sum_{j=2}^{n} V_2(t_j) \left(B(t_{j+1}) - B(t_j) \right)
\end{aligned}
$$

Then, we obtain

$$
\begin{aligned}
V_2(t_{n+1}) &= V_2(n) + \frac{1-\alpha}{M(\alpha)} \left(\left(-\frac{1}{R_1 C_1} V_2(t_n) - \frac{1}{R_1 C_2} V_1(t_n) + \frac{R_2}{R_1} \frac{dV_1(t_n)}{dt} \right) \right. \\
&\quad \left. - \left(-\frac{1}{R_1 C_1} V_2(t_{n-1}) - \frac{1}{R_1 C_2} V_1(t_{n-1}) + \frac{R_2}{R_1} \frac{dV_1(t_{n-1})}{dt} \right) \right) \\
&\quad + \frac{\alpha}{M(\alpha)} \sum_{j=2}^{n} \left[-(\Delta t)\frac{4}{3} \left(-\frac{1}{R_1 C_1} V_2(t_{j-1}) - \frac{1}{R_1 C_2} V_1(t_{j-1}) + \frac{R_2}{R_1} \frac{dV_1(t_{j-1})}{dt} \right) \right. \\
&\quad + (\Delta t)\frac{5}{12} \left(-\frac{1}{R_1 C_1} V_2(t_{j-2}) - \frac{1}{R_1 C_2} V_1(t_{j-2}) + \frac{R_2}{R_1} \frac{dV_1(t_{j-2})}{dt} \right) \\
&\quad \left. + (\Delta t)\frac{23}{12} \left(-\frac{1}{R_1 C_1} V_2(t_j) - \frac{1}{R_1 C_2} V_1(t_j) + \frac{R_2}{R_1} \frac{dV_1(t_j)}{dt} \right) \right] \\
&\quad + \frac{G\alpha}{M(\alpha)} \sum_{j=2}^{n} i(t_j) \left(B(t_{j+1}) - B(t_j) \right)
\end{aligned}
$$

We consider the following differential equation

$$\frac{dV_2}{dt} = -\frac{1}{(R_1+R_2)C}V_2(t) + \frac{1}{(R_1+R_2)C}V_1(t) + \frac{R_2}{R_1+R_2}\frac{dV_1(t)}{dt} = f(t,V_2(t))$$

(21.92)

We convert the above differential equation to the fractional stochastic differential equation as: In exponential-decay case we have

$$V_2(t) - V_2(0) = \frac{1}{M(\alpha)}\frac{\alpha}{} f(t,V_2(t)) + \frac{\alpha}{M(\alpha)}\int_0^t f(\tau,V_2(\tau))d\tau + \frac{G\alpha}{M(\alpha)}\int_0^t V_2(\tau)dB(\tau)$$

(21.93)

We define a mapping

$$
\begin{aligned}
|\Gamma(V_2(t))| &= \left| \frac{1-\alpha}{M(\alpha)} f(t,V_2(t)) + \frac{\alpha}{M(\alpha)}\int_0^t f(\tau,V_2(\tau))d\tau \right. \\
&\quad \left. + \frac{G\alpha}{M(\alpha)}\int_0^t V_2(\tau)dB(\tau) \right| \\
&< \frac{1-\alpha}{M(\alpha)}\sup_{t\in[0,T]}|f(t,V_2(t))| + \frac{\alpha}{M(\alpha)}\int_0^t \sup_{\tau\in[0,t]}|f(\tau,V_2(\tau))|d\tau \\
&\quad + \frac{G\alpha}{M(\alpha)}\int_0^t \sup_{\tau\in[0,t]}|V_2(\tau)|dB(\tau) \\
&< \frac{1-\alpha}{M(\alpha)}\|f(\cdot,i)\|_\infty + T\frac{\alpha}{M(\alpha)}\|f(\cdot,V_2)\|_\infty + GL\|V_2\|_\infty \\
&< M_1 + M_2 + M_3 \\
&< M
\end{aligned}
$$

We have

$$
\begin{aligned}
&|\Gamma(V_2) - \Gamma(V_{2_1})| \\
&= \left| \frac{1-\alpha}{M(\alpha)}(f(t,V_2(t)) - f(t,V_{2_1}(t))) \right. \\
&\quad + \frac{\alpha}{M(\alpha)}\int_0^t (f(\tau,V_2(\tau)) - f(\tau,V_{2_1}(\tau)))d\tau \\
&\quad \left. + \frac{G\alpha}{M(\alpha)}\int_0^t (V_2(\tau) - V_{2_1}(\tau))dB(\tau) \right| \\
&< \frac{1-\alpha}{M(\alpha)}|f(t,V_2(t)) - f(t,V_{2_1}(t))| \\
&\quad + \frac{\alpha}{M(\alpha)}\int_0^t |f(\tau,V_2(\tau)) - f(\tau,V_{2_1}(\tau))|d\tau \\
&\quad + \frac{G\alpha}{M(\alpha)}\int_0^t |V_2(\tau) - V_{2_1}(\tau)|dB(\tau)
\end{aligned}
$$

$$< \frac{1-\alpha}{M(\alpha)} \sup_{t\in[0,T]} |f(t,V_2(t)) - f(t,V_{2_1}(t))|$$

$$+ \frac{\alpha}{M(\alpha)} \int_0^t \sup_{\tau\in[0,t]} |f(\tau,V_2(\tau)) - f(\tau,V_{2_1}(\tau))| d\tau$$

$$+ \frac{G\alpha}{M(\alpha)} \int_0^t \sup_{\tau\in[0,t]} |V_2(\tau) - V_{2_1}(\tau)| dB(\tau)$$

$$< \frac{1-\alpha}{M(\alpha)} \frac{1}{(R_1+R_2)C} \|V_2(t) - V_{2_1}(t)\|_\infty$$

$$+ \frac{\alpha}{M(\alpha)} \frac{T}{(R_1+R_2)C} \|V_2(t)) - V_{2_1}(t)\|_\infty + GL\|V_2(t) - V_{2_1}(t)\|_\infty$$

$$< \left(\frac{1-\alpha}{M(\alpha)} \frac{1}{(R_1+R_2)C} + \frac{\alpha}{M(\alpha)} \frac{T}{(R_1+R_2)C} + GL \right) \|V_2(t) - V_{2_1}(t)\|_\infty$$

$$< L_1 \|V_2(t) - V_{2_1}(t)\|_\infty$$

We assume that $B(t)$ is Lipschitzian on $[0,T]$ with constant L, then

$$
\begin{aligned}
|\Gamma(V_2(t))| &= \left| \frac{1-\alpha}{M(\alpha)} f(t,V_2(t)) + \frac{\alpha}{M(\alpha)} \int_0^t f(\tau,V_2(\tau)) d\tau \right. \\
&\quad \left. + \frac{G\alpha}{M(\alpha)} \int_0^t V_2(\tau) dB(\tau) \right| \\
&< \frac{1-\alpha}{M(\alpha)} |f(t,V_2(t))| \\
&\quad + \frac{\alpha}{M(\alpha)} \int_0^t |f(\tau,V_2(\tau))| d\tau \\
&\quad + \frac{G\alpha}{M(\alpha)} \int_0^t |V_2(\tau)| dB(\tau) \\
&< \frac{1-\alpha}{M(\alpha)} \sup_{l\in[0,t]} |f(l,V_2(l))| \\
&\quad + \frac{\alpha}{M(\alpha)} \int_0^t \sup_{l\in[0,\tau]} |f(l,V_2(l))| d\tau \\
&\quad + \frac{G\alpha}{M(\alpha)} \int_0^t \sup_{l\in[0,\tau]} |V_2(l)| dB(\tau)
\end{aligned}
$$

This shows that our mapping has a unique solution. We now present the numerical solution of the equation

$$
\begin{aligned}
V_2(t_{n+1}) &= V_2(0) + \frac{1-\alpha}{M(\alpha)} f(t_n,V_2(t_n)) + \frac{\alpha}{M(\alpha)} \int_0^{t_{n+1}} f(\tau,i(\tau)) d\tau \\
&\quad + \frac{G\alpha}{M(\alpha)} \int_0^{t_{n+1}} V_2(\tau) dB(\tau)
\end{aligned}
$$

and

$$
\begin{aligned}
V_2(t_n) \;=\;& V_2(0) + \frac{1-\alpha}{M(\alpha)} f(t_{n-1}, V_2(t_{n-1})) + \frac{\alpha}{M(\alpha)} \int_0^{t_n} f(\tau, V_2(\tau)) d\tau \\
& + \frac{G\alpha}{M(\alpha)} \int_0^{t_n} V_2(\tau) dB(\tau)
\end{aligned}
$$

Thus, we reach

$$
\begin{aligned}
V_2(t_{n+1}) \;=\;& V_2(n) + \frac{1-\alpha}{M(\alpha)} \left(f(t_n, V_2(t_n)) - f(t_{n-1}, V_2(t_{n-1})) \right) \\
& + \frac{\alpha}{M(\alpha)} \int_{t_n}^{t_{n+1}} f(\tau, V_2(\tau)) d\tau + \frac{G\alpha}{M(\alpha)} \int_{t_n}^{t_{n+1}} V_2(\tau) dB(\tau) \\
\;=\;& i(n) + \frac{1-\alpha}{M(\alpha)} \left(f(t_n, V_2(t_n)) - f(t_{n-1}, V_2(t_{n-1})) \right) \\
& + \frac{\alpha}{M(\alpha)} \sum_{j=2}^{n} \int_{t_j}^{t_{j+1}} f(\tau, V_2(\tau)) d\tau + \frac{G\alpha}{M(\alpha)} \sum_{j=2}^{n} \int_{t_j}^{t_{j+1}} i(\tau) dB(\tau) \\
\;=\;& i(n) + \frac{1-\alpha}{M(\alpha)} \left(f(t_n, V_2(t_n)) - f(t_{n-1}, V_2(t_{n-1})) \right) \\
& + \frac{\alpha}{M(\alpha)} \sum_{j=2}^{n} \left[-(\Delta t)\frac{4}{3} f(t_{j-1}, V_2^{j-1}) + (\Delta t)\frac{5}{12} f(t_{j-2}, V_2^{j-2}) \right. \\
& \left. + (\Delta t)\frac{23}{12} f(t_j, V_2^j) \right] \\
& + \frac{G\alpha}{M(\alpha)} \sum_{j=2}^{n} V_2(t_j) \left(B(t_{j+1}) - B(t_j) \right)
\end{aligned}
$$

Then, we obtain

$V_2(t_{n+1})$

$$
\begin{aligned}
=\;& V_2(n) + \frac{1-\alpha}{M(\alpha)} \left(\left(-\frac{1}{(R_1+R_2)C} V_2(t_n) + \frac{1}{(R_1+R_2)C} V_1(t_n) + \frac{R_2}{R_1+R_2} \frac{dV_1(t_n)}{dt} \right) \right. \\
& \left. - \left(-\frac{1}{(R_1+R_2)C} V_2(t_{n-1}) + \frac{1}{(R_1+R_2)C} V_1(t_{n-1}) + \frac{R_2}{R_1+R_2} \frac{dV_1(t_{n-1})}{dt} \right) \right) \\
& + \frac{\alpha}{M(\alpha)} \sum_{j=2}^{n} \left[-(\Delta t)\frac{4}{3} \left(-\frac{1}{(R_1+R_2)C} V_2(t_{j-1}) + \frac{1}{(R_1+R_2)C} V_1(t_{j-1}) \right. \right. \\
& \left. + \frac{R_2}{R_1+R_2} \frac{dV_1(t_{j-1})}{dt} \right) + (\Delta t)\frac{5}{12} \left(-\frac{1}{(R_1+R_2)C} V_2(t_{j-2}) + \frac{1}{(R_1+R_2)C} V_1(t_{j-2}) \right.
\end{aligned}
$$

$$+\frac{R_2}{R_1+R_2}\frac{dV_1(t_{j-2})}{dt}\Big) + (\Delta t)\frac{23}{12}\left(-\frac{1}{(R_1+R_2)C}V_2(t_j) + \frac{1}{(R_1+R_2)C}V_1(t_j)\right.$$

$$+\frac{R_2}{R_1+R_2}\frac{dV_1(t_j)}{dt}\Big) + \frac{G\alpha}{M(\alpha)}\sum_{j=2}^{n} i(t_j)\left(B(t_{j+1}-B(t_j))\right)\Big]$$

We consider the following differential equation

$$\frac{dV_2}{dt} = -\frac{R_1+R_2}{R_1R_2C}V_2(t) + \frac{1}{R_1C_1}V_1(t) + \frac{dV_1(t)}{dt} = f(t, V_2(t)) \qquad (21.94)$$

We convert the above differential equation to the fractional stochastic differential equation as: In exponential-decay case we have

$$V_2(t) - V_2(0) = \frac{1-\alpha}{M(\alpha)}f(t, V_2(t)) + \frac{\alpha}{M(\alpha)}\int_0^t f(\tau, V_2(\tau))d\tau + \frac{G\alpha}{M(\alpha)}\int_0^t V_2(\tau)dB(\tau)$$

$$(21.95)$$

We define a mapping

$$\begin{aligned}
|\Gamma(V_2(t))| &= \left|\frac{1-\alpha}{M(\alpha)}f(t, V_2(t)) + \frac{\alpha}{M(\alpha)}\int_0^t f(\tau, V_2(\tau))d\tau\right. \\
&\quad \left. + \frac{G\alpha}{M(\alpha)}\int_0^t V_2(\tau)dB(\tau)\right| \\
&< \frac{1-\alpha}{M(\alpha)}\sup_{t\in[0,T]}|f(t, V_2(t))| + \frac{\alpha}{M(\alpha)}\int_0^t \sup_{\tau\in[0,t]}|f(\tau, V_2(\tau))|d\tau \\
&\quad + \frac{G\alpha}{M(\alpha)}\int_0^t \sup_{\tau\in[0,t]}|V_2(\tau)|dB(\tau) \\
&< \frac{1-\alpha}{M(\alpha)}\|f(.,i)\|_\infty + T\frac{\alpha}{M(\alpha)}\|f(.,V_2)\|_\infty + GL\|V_2\|_\infty \\
&< M_1 + M_2 + M_3 \\
&< M
\end{aligned}$$

We have

$$\begin{aligned}
|\Gamma(V_2) - \Gamma(V_{2_1})| &= \left|\frac{1-\alpha}{M(\alpha)}(f(t, V_2(t)) - f(t, V_{2_1}(t)))\right. \\
&\quad + \frac{\alpha}{M(\alpha)}\int_0^t (f(\tau, V_2(\tau)) - f(\tau, V_{2_1}(\tau)))d\tau \\
&\quad \left. + \frac{G\alpha}{M(\alpha)}\int_0^t (V_2(\tau) - V_{2_1}(\tau))dB(\tau)\right| \\
&< \frac{1-\alpha}{M(\alpha)}|f(t, V_2(t)) - f(t, V_{2_1}(t))| \\
&\quad + \frac{\alpha}{M(\alpha)}\int_0^t |f(\tau, V_2(\tau)) - f(\tau, V_{2_1}(\tau))|d\tau
\end{aligned}$$

$$+\frac{G\alpha}{M(\alpha)}\int_0^t |V_2(\tau) - V_{2_1}(\tau)|\, dB(\tau)$$

$$<\frac{1-\alpha}{M(\alpha)}\sup_{t\in[0,T]} |f(t,V_2(t)) - f(t,V_{2_1}(t))|$$

$$+\frac{\alpha}{M(\alpha)}\int_0^t \sup_{\tau\in[0,t]} |f(\tau,V_2(\tau)) - f(\tau,V_{2_1}(\tau))|\, d\tau$$

$$\left| \frac{G\alpha}{M(\alpha)}\int_0^t \sup_{\tau\in[0,t]} |V_2(\tau) - V_{2_1}(\tau)|\, dB(\tau) \right.$$

$$<\frac{1-\alpha}{M(\alpha)}\frac{R_1+R_2}{R_1 R_2 C}\|V_2(t) - V_{2_1}(t)\|_\infty$$

$$+\frac{\alpha}{M(\alpha)}\frac{T(R_1+R_2)}{R_1 R_2 C}\|V_2(t)) - V_{2_1}(t)\|_\infty + GL\|V_2(t) - V_{2_1}(t)\|_\infty$$

$$<\left(\frac{1-\alpha}{M(\alpha)}\frac{R_1+R_2}{R_1 R_2 C} + \frac{\alpha}{M(\alpha)}\frac{T(R_1+R_2)}{R_1 R_2 C} + GL\right)\|V_2(t) - V_{2_1}(t)\|_\infty$$

$$<L_1\|V_2(t) - V_{2_1}(t)\|_\infty$$

We assume that $B(t)$ is Lipschitzian on $[0,T]$ with constant L, then

$$|\Gamma(V_2(t))| = \left| \frac{1-\alpha}{M(\alpha)}f(t,V_2(t)) + \frac{\alpha}{M(\alpha)}\int_0^t f(\tau,V_2(\tau))d\tau \right.$$

$$\left. +\frac{G\alpha}{M(\alpha)}\int_0^t V_2(\tau)dB(\tau) \right|$$

$$< \frac{1-\alpha}{M(\alpha)}|f(t,V_2(t))|$$

$$+\frac{\alpha}{M(\alpha)}\int_0^t |f(\tau,V_2(\tau))|\, d\tau$$

$$+\frac{G\alpha}{M(\alpha)}\int_0^t |V_2(\tau)|\, dB(\tau)$$

$$< \frac{1-\alpha}{M(\alpha)}\sup_{l\in[0,t]}|f(l,V_2(l))|$$

$$+\frac{\alpha}{M(\alpha)}\int_0^t \sup_{l\in[0,\tau]}|f(l,V_2(l))|\, d\tau$$

$$+\frac{G\alpha}{M(\alpha)}\int_0^t \sup_{l\in[0,\tau]}|V_2(l)|\, dB(\tau)$$

This shows that our mapping has a unique solution. We now present the numerical solution of the equation

$$
\begin{aligned}
V_2(t_{n+1}) &= V_2(0) + \frac{1-\alpha}{M(\alpha)} f(t_n, V_2(t_n)) + \frac{\alpha}{M(\alpha)} \int_0^{t_{n+1}} f(\tau, i(\tau)) d\tau \\
&+ \frac{G\alpha}{M(\alpha)} \int_0^{t_{n+1}} V_2(\tau) dB(\tau)
\end{aligned}
$$

and

$$
\begin{aligned}
V_2(t_n) &= V_2(0) + \frac{1-\alpha}{M(\alpha)} f(t_{n-1}, V_2(t_{n-1})) + \frac{\alpha}{M(\alpha)} \int_0^{t_n} f(\tau, V_2(\tau)) d\tau \\
&+ \frac{G\alpha}{M(\alpha)} \int_0^{t_n} i(\tau) dB(\tau)
\end{aligned}
$$

Thus, we reach

$$
\begin{aligned}
V_2(t_{n+1}) &= V_2(n) + \frac{1-\alpha}{M(\alpha)} \left(f(t_n, V_2(t_n)) - f(t_{n-1}, V_2(t_{n-1})) \right) \\
&+ \frac{\alpha}{M(\alpha)} \int_{t_n}^{t_{n+1}} f(\tau, V_2(\tau)) d\tau + \frac{G\alpha}{M(\alpha)} \int_{t_n}^{t_{n+1}} V_2(\tau) dB(\tau) \\
&= i(n) + \frac{1-\alpha}{M(\alpha)} \left(f(t_n, V_2(t_n)) - f(t_{n-1}, V_2(t_{n-1})) \right) \\
&+ \frac{\alpha}{M(\alpha)} \sum_{j=2}^n \int_{t_j}^{t_{j+1}} f(\tau, V_2(\tau)) d\tau + \frac{G\alpha}{M(\alpha)} \sum_{j=2}^n \int_{t_j}^{t_{j+1}} i(\tau) dB(\tau) \\
&= i(n) + \frac{1-\alpha}{M(\alpha)} \left(f(t_n, V_2(t_n)) - f(t_{n-1}, V_2(t_{n-1})) \right) \\
&+ \frac{\alpha}{M(\alpha)} \sum_{j=2}^n \left[-(\Delta t) \frac{4}{3} f(t_{j-1}, V_2^{j-1}) + (\Delta t) \frac{5}{12} f(t_{j-2}, V_2^{j-2}) \right. \\
&\left. + (\Delta t) \frac{23}{12} f(t_j, V_2^j) \right] \\
&+ \frac{G\alpha}{M(\alpha)} \sum_{j-2}^n V_2(t_j) \left(B(t_{j+1}) - B(t_j) \right)
\end{aligned}
$$

Then, we obtain

$$
\begin{aligned}
V_2(t_{n+1}) &= V_2(n) + \frac{1-\alpha}{M(\alpha)} \left(\left(-\frac{R_1+R_2}{R_1 R_2 C} V_2(t_n) + \frac{1}{R_1 C_1} V_1(t_n) + \frac{dV_1(t_n)}{dt} \right) \right. \\
&- \left. \left(-\frac{R_1+R_2}{R_1 R_2 C} V_2(t_{n-1}) + \frac{1}{R_1 C_1} V_1(t_{n-1}) + \frac{dV_1(t_{n-1})}{dt} \right) \right) \\
&+ \frac{\alpha}{M(\alpha)} \sum_{j=2}^n \left[-(\Delta t) \frac{4}{3} \left(-\frac{R_1+R_2}{R_1 R_2 C} V_2(t_{j-1}) + \frac{1}{R_1 C_1} V_1(t_{j-1}) + \frac{dV_1(t_{j-1})}{dt} \right) \right.
\end{aligned}
$$

$$+(\Delta t)\frac{5}{12}\left(-\frac{R_1+R_2}{R_1R_2C}V_2(t_{j-2})+\frac{1}{R_1C_1}V_1(t_{j-2})+\frac{dV_1(t_{j-2})}{dt}\right)$$

$$+(\Delta t)\frac{23}{12}\left(-\frac{R_1+R_2}{R_1R_2C}V_2(t_j)+\frac{1}{R_1C_1}V_1(t_j)+\frac{dV_1(t_j)}{dt}\right)\Bigg]$$

$$+\frac{G\alpha}{M(\alpha)}\sum_{j=2}^{n}i(t_j)\left(B(t_{j+1})-B(t_j)\right)$$

We consider the following differential equation

$$\frac{dV_2}{dt}=-4\times10^6V_2(t)-4\times10^7V_1(t)=f(t,V_2(t)) \qquad (21.96)$$

We convert the above differential equation to the fractional stochastic differential equation as: In exponential-decay case we have

$$V_2(t)-V_2(0)=\frac{1-\alpha}{M(\alpha)}f(t,V_2(t))+\frac{\alpha}{M(\alpha)}\int_0^t f(\tau,V_2(\tau))d\tau+\frac{G\alpha}{M(\alpha)}\int_0^t V_2(\tau)dB(\tau)$$

$$(21.97)$$

We define a mapping

$$\begin{aligned}
|\Gamma(V_2(t))| &= \left|\frac{1-\alpha}{M(\alpha)}f(t,V_2(t))+\frac{\alpha}{M(\alpha)}\int_0^t f(\tau,V_2(\tau))d\tau\right.\\
&\quad \left.+\frac{G\alpha}{M(\alpha)}\int_0^t V_2(\tau)dB(\tau)\right|\\
&< \frac{1-\alpha}{M(\alpha)}\sup_{t\in[0,T]}|f(t,V_2(t))|+\frac{\alpha}{M(\alpha)}\int_0^t\sup_{\tau\in[0,t]}|f(\tau,V_2(\tau))|d\tau\\
&\quad +\frac{G\alpha}{M(\alpha)}\int_0^t\sup_{\tau\in[0,t]}|V_2(\tau)|dB(\tau)\\
&< \frac{1-\alpha}{M(\alpha)}\|f(\cdot,i)\|_\infty+T\frac{\alpha}{M(\alpha)}\|f(\cdot,V_2)\|_\infty+GL\|V_2\|_\infty\\
&< M_1+M_2+M_3\\
&< M
\end{aligned}$$

We have

$$\begin{aligned}
|\Gamma(V_2)-\Gamma(V_{2_1})| &= \left|\frac{1-\alpha}{M(\alpha)}\left(f(t,V_2(t))-f(t,V_{2_1}(t))\right)\right.\\
&\quad +\frac{\alpha}{M(\alpha)}\int_0^t\left(f(\tau,V_2(\tau))-f(\tau,V_{2_1}(\tau))\right)d\tau\\
&\quad \left.+\frac{G\alpha}{M(\alpha)}\int_0^t\left(V_2(\tau)-V_{2_1}(\tau)\right)dB(\tau)\right|\\
&< \frac{1-\alpha}{M(\alpha)}|f(t,V_2(t))-f(t,V_{2_1}(t))|
\end{aligned}$$

$$+ \frac{\alpha}{M(\alpha)} \int_0^t |f(\tau, V_2(\tau)) - f(\tau, V_{2_1}(\tau))| d\tau$$

$$+ \frac{G\alpha}{M(\alpha)} \int_0^t |V_2(\tau) - V_{2_1}(\tau)| dB(\tau)$$

$$< \frac{1-\alpha}{M(\alpha)} \sup_{t \in [0,T]} |f(t, V_2(t)) - f(t, V_{2_1}(t))|$$

$$+ \frac{\alpha}{M(\alpha)} \int_0^t \sup_{\tau \in [0,t]} |f(\tau, V_2(\tau)) - f(\tau, V_{2_1}(\tau))| d\tau$$

$$+ \frac{G\alpha}{M(\alpha)} \int_0^t \sup_{\tau \in [0,t]} |V_2(\tau) - V_{2_1}(\tau)| dB(\tau)$$

$$< \frac{1-\alpha}{M(\alpha)} 4 \times 10^6 \|V_2(t) - V_{2_1}(t)\|_\infty$$

$$+ \frac{\alpha}{M(\alpha)} T 4 \times 10^6 \|V_2(t)) - V_{2_1}(t)\|_\infty + GL \|V_2(t) - V_{2_1}(t)\|_\infty$$

$$< \left(\frac{1-\alpha}{M(\alpha)} 4 \times 10^6 + \frac{\alpha}{M(\alpha)} T 4 \times 10^6 + GL \right) \|V_2(t) - V_{2_1}(t)\|_\infty$$

$$< L_1 \|V_2(t) - V_{2_1}(t)\|_\infty$$

We assume that $B(t)$ is Lipschitzian on $[0,T]$ with constant L, then

$$
\begin{aligned}
|\Gamma(V_2(t))| &= \left| \frac{1-\alpha}{M(\alpha)} f(t, V_2(t)) + \frac{\alpha}{M(\alpha)} \int_0^t f(\tau, V_2(\tau)) d\tau \right. \\
&\quad \left. + \frac{G\alpha}{M(\alpha)} \int_0^t V_2(\tau) dB(\tau) \right| \\
&< \frac{1-\alpha}{M(\alpha)} |f(t, V_2(t))| \\
&\quad + \frac{\alpha}{M(\alpha)} \int_0^t |f(\tau, V_2(\tau))| d\tau \\
&\quad + \frac{G\alpha}{M(\alpha)} \int_0^t |V_2(\tau)| dB(\tau) \\
&< \frac{1-\alpha}{M(\alpha)} \sup_{l \in [0,t]} |f(l, V_2(l))| \\
&\quad + \frac{\alpha}{M(\alpha)} \int_0^t \sup_{l \in [0,\tau]} |f(l, V_2(l))| d\tau \\
&\quad + \frac{G\alpha}{M(\alpha)} \int_0^t \sup_{l \in [0,\tau]} |V_2(l)| dB(\tau)
\end{aligned}
$$

This shows that our mapping has a unique solution. We now present the numerical solution of the equation

$$
\begin{aligned}
V_2(t_{n+1}) &= V_2(0) + \frac{1-\alpha}{M(\alpha)} f(t_n, V_2(t_n)) + \frac{\alpha}{M(\alpha)} \int_0^{t_{n+1}} f(\tau, i(\tau))d\tau \\
&\quad + \frac{G\alpha}{M(\alpha)} \int_0^{t_{n+1}} V_2(\tau)dB(\tau)
\end{aligned}
$$

and

$$
\begin{aligned}
V_2(t_n) &= V_2(0) + \frac{1-\alpha}{M(\alpha)} f(t_{n-1}, V_2(t_{n-1})) + \frac{\alpha}{M(\alpha)} \int_0^{t_n} f(\tau, V_2(\tau))d\tau \\
&\quad + \frac{G\alpha}{M(\alpha)} \int_0^{t_n} V_2(\tau)dB(\tau)
\end{aligned}
$$

Thus, we reach

$$
\begin{aligned}
V_2(t_{n+1}) &= V_2(n) + \frac{1-\alpha}{M(\alpha)} \left(f(t_n, V_2(t_n)) - f(t_{n-1}, V_2(t_{n-1})) \right) \\
&\quad + \frac{\alpha}{M(\alpha)} \int_{t_n}^{t_{n+1}} f(\tau, V_2(\tau))d\tau + \frac{G\alpha}{M(\alpha)} \int_{t_n}^{t_{n+1}} V_2(\tau)dB(\tau) \\
&= i(n) + \frac{1-\alpha}{M(\alpha)} \left(f(t_n, V_2(t_n)) - f(t_{n-1}, V_2(t_{n-1})) \right) \\
&\quad + \frac{\alpha}{M(\alpha)} \sum_{j=2}^{n} \int_{t_j}^{t_{j+1}} f(\tau, V_2(\tau))d\tau + \frac{G\alpha}{M(\alpha)} \sum_{j=2}^{n} \int_{t_j}^{t_{j+1}} i(\tau)dB(\tau) \\
&= i(n) + \frac{1-\alpha}{M(\alpha)} \left(f(t_n, V_2(t_n)) - f(t_{n-1}, V_2(t_{n-1})) \right) \\
&\quad + \frac{\alpha}{M(\alpha)} \sum_{j=2}^{n} \left[-(\Delta t)\frac{4}{3} f(t_{j-1}, V_2^{j-1}) + (\Delta t)\frac{5}{12} f(t_{j-2}, V_2^{j-2}) \right. \\
&\quad \left. + (\Delta t)\frac{23}{12} f(t_j, V_2^{j}) \right] \\
&\quad + \frac{G\alpha}{M(\alpha)} \sum_{j=2}^{n} V_2(t_j) \left(B(t_{j+1}) - B(t_j) \right)
\end{aligned}
$$

Then, we obtain

$$
\begin{aligned}
V_2(t_{n+1}) &= V_2(n) + \frac{1-\alpha}{M(\alpha)} \left(\left(-4 \times 10^6 V_2(t_n) - 4 \times 10^7 V_1(t_n) \right) \right. \\
&\quad \left. - \left(-4 \times 10^6 V_2(t_{n-1}) - 4 \times 10^7 V_1(t_{n-1}) \right) \right) \\
&\quad + \frac{\alpha}{M(\alpha)} \sum_{j=2}^{n} \left[-(\Delta t)\frac{4}{3} \left(-4 \times 10^6 V_2(t_{j-1}) - 4 \times 10^7 V_1(t_{j-1}) \right) \right. \\
&\quad + (\Delta t)\frac{5}{12} \left(-4 \times 10^6 V_2(t_{j-2}) - 4 \times 10^7 V_1(t_{j-2}) \right)
\end{aligned}
$$

$$+(\Delta t)\frac{23}{12}\left(-4\times 10^6 V_2(t_j)-4\times 10^7 V_1(t_j)\right)\Bigg]$$

$$+\frac{G\alpha}{M(\alpha)}\sum_{j=2}^{n}i(t_j)\left(B(t_{j+1}-B(t_j))\right)$$

We consider the following differential equation

$$\frac{dV_2}{dt}=-\frac{R_1+R_2}{R_1 R_2(C_1+C_2)}V_2(t)+\frac{C_1}{C_1+C_2}\frac{dV_1(t)}{dt}+\frac{1}{R_1(C_1+C_2)}V_1(t)=f(t,V_2(t))$$

$$(21.98)$$

We convert the above differential equation to the fractional stochastic differential equation as: In exponential-decay case we have

$$V_2(t)-V_2(0)=\frac{1-\alpha}{M(\alpha)}f(t,V_2(t))+\frac{\alpha}{M(\alpha)}\int_0^t f(\tau,V_2(\tau))d\tau+\frac{G\alpha}{M(\alpha)}\int_0^t V_2(\tau)dB(\tau)$$

$$(21.99)$$

We define a mapping

$$
\begin{aligned}
|\Gamma(V_2(t))| &= \left|\frac{1-\alpha}{M(\alpha)}f(t,V_2(t))+\frac{\alpha}{M(\alpha)}\int_0^t f(\tau,V_2(\tau))d\tau\right. \\
&\qquad \left.+\frac{G\alpha}{M(\alpha)}\int_0^t V_2(\tau)dB(\tau)\right| \\
&< \frac{1-\alpha}{M(\alpha)}\sup_{t\in[0,T]}|f(t,V_2(t))|+\frac{\alpha}{M(\alpha)}\int_0^t \sup_{\tau\in[0,t]}|f(\tau,V_2(\tau))|d\tau \\
&\qquad +\frac{G\alpha}{M(\alpha)}\int_0^t \sup_{\tau\in[0,t]}|V_2(\tau)|dB(\tau) \\
&< \frac{1-\alpha}{M(\alpha)}\|f(.,i)\|_\infty+T\frac{\alpha}{M(\alpha)}\|f(.,V_2)\|_\infty+GL\|V_2\|_\infty \\
&< M_1+M_2+M_3 \\
&< M
\end{aligned}
$$

We have

$$
\begin{aligned}
|\Gamma(V_2)-\Gamma(V_{2_1})| &= \left|\frac{1-\alpha}{M(\alpha)}\left(f(t,V_2(t))-f(t,V_{2_1}(t))\right)\right. \\
&\qquad +\frac{\alpha}{M(\alpha)}\int_0^t \left(f(\tau,V_2(\tau))-f(\tau,V_{2_1}(\tau))\right)d\tau \\
&\qquad \left.+\frac{G\alpha}{M(\alpha)}\int_0^t \left(V_2(\tau)-V_{2_1}(\tau)\right)dB(\tau)\right| \\
&< \frac{1-\alpha}{M(\alpha)}|f(t,V_2(t))-f(t,V_{2_1}(t))|
\end{aligned}
$$

$$+\frac{\alpha}{M(\alpha)}\int_0^t |f(\tau,V_2(\tau))-f(\tau,V_{2_1}(\tau))|\,d\tau$$

$$+\frac{G\alpha}{M(\alpha)}\int_0^t |V_2(\tau)-V_{2_1}(\tau)|\,dB(\tau)$$

$$<\frac{1-\alpha}{M(\alpha)}\sup_{t\in[0,T]}|f(t,V_2(t))-f(t,V_{2_1}(t))|$$

$$+\frac{\alpha}{M(\alpha)}\int_0^t \sup_{\tau\in[0,t]}|f(\tau,V_2(\tau))-f(\tau,V_{2_1}(\tau))|\,d\tau$$

$$+\frac{G\alpha}{M(\alpha)}\int_0^t \sup_{\tau\in[0,t]}|V_2(\tau)-V_{2_1}(\tau)|\,dB(\tau)$$

$$<\frac{1-\alpha}{M(\alpha)}\frac{R_1+R_2}{R_1R_2(C_1+C_2)}\|V_2(t)-V_{2_1}(t)\|_\infty$$

$$+\frac{\alpha}{M(\alpha)}T\frac{R_1+R_2}{R_1R_2(C_1+C_2)}\|V_2(t))-V_{2_1}(t)\|_\infty$$

$$+GL\|V_2(t)-V_{2_1}(t)\|_\infty$$

$$<\left(\frac{1-\alpha}{M(\alpha)}\frac{R_1+R_2}{R_1R_2(C_1+C_2)}+\frac{\alpha}{M(\alpha)}T\frac{R_1+R_2}{R_1R_2(C_1+C_2)}+GL\right)$$

$$\times\|V_2(t)-V_{2_1}(t)\|_\infty$$

$$<L_1\|V_2(t)-V_{2_1}(t)\|_\infty$$

We assume that $B(t)$ is Lipschitzian on $[0,T]$ with constant L, then

$$\begin{aligned}|\Gamma(V_2(t))| &= \left|\frac{1-\alpha}{M(\alpha)}f(t,V_2(t))+\frac{\alpha}{M(\alpha)}\int_0^t f(\tau,V_2(\tau))d\tau\right.\\ &\left.+\frac{G\alpha}{M(\alpha)}\int_0^t V_2(\tau)dB(\tau)\right|\\ &<\frac{1-\alpha}{M(\alpha)}|f(t,V_2(t))|\\ &+\frac{\alpha}{M(\alpha)}\int_0^t |f(\tau,V_2(\tau))|\,d\tau\\ &+\frac{G\alpha}{M(\alpha)}\int_0^t |V_2(\tau)|\,dB(\tau)\\ &<\frac{1-\alpha}{M(\alpha)}\sup_{l\in[0,t]}|f(l,V_2(l))|\\ &+\frac{\alpha}{M(\alpha)}\int_0^t \sup_{l\in[0,\tau]}|f(l,V_2(l))|\,d\tau\\ &+\frac{G\alpha}{M(\alpha)}\int_0^t \sup_{l\in[0,\tau]}|V_2(l)|\,dB(\tau)\end{aligned}$$

This shows that our mapping has a unique solution. We now present the numerical solution of the equation

$$
\begin{aligned}
V_2(t_{n+1}) &= V_2(0) + \frac{1-\alpha}{M(\alpha)} f(t_n, V_2(t_n)) + \frac{\alpha}{M(\alpha)} \int_0^{t_{n+1}} f(\tau, i(\tau)) d\tau \\
&\quad + \frac{G\alpha}{M(\alpha)} \int_0^{t_{n+1}} V_2(\tau) dB(\tau)
\end{aligned}
$$

and

$$
\begin{aligned}
V_2(t_n) &= V_2(0) + \frac{1-\alpha}{M(\alpha)} f(t_{n-1}, V_2(t_{n-1})) + \frac{\alpha}{M(\alpha)} \int_0^{t_n} f(\tau, V_2(\tau)) d\tau \\
&\quad + \frac{G\alpha}{M(\alpha)} \int_0^{t_n} V_2(\tau) dB(\tau)
\end{aligned}
$$

Thus, we reach

$$
\begin{aligned}
V_2(t_{n+1}) &= V_2(n) + \frac{1-\alpha}{M(\alpha)} \left(f(t_n, V_2(t_n)) - f(t_{n-1}, V_2(t_{n-1})) \right) \\
&\quad + \frac{\alpha}{M(\alpha)} \int_{t_n}^{t_{n+1}} f(\tau, V_2(\tau)) d\tau + \frac{G\alpha}{M(\alpha)} \int_{t_n}^{t_{n+1}} V_2(\tau) dB(\tau) \\
&= i(n) + \frac{1-\alpha}{M(\alpha)} \left(f(t_n, V_2(t_n)) - f(t_{n-1}, V_2(t_{n-1})) \right) \\
&\quad + \frac{\alpha}{M(\alpha)} \sum_{j=2}^{n} \int_{t_j}^{t_{j+1}} f(\tau, V_2(\tau)) d\tau + \frac{G\alpha}{M(\alpha)} \sum_{j=2}^{n} \int_{t_j}^{t_{j+1}} i(\tau) dB(\tau) \\
&= i(n) + \frac{1-\alpha}{M(\alpha)} \left(f(t_n, V_2(t_n)) - f(t_{n-1}, V_2(t_{n-1})) \right) \\
&\quad + \frac{\alpha}{M(\alpha)} \sum_{j=2}^{n} \left[-(\Delta t) \frac{4}{3} f(t_{j-1}, V_2^{j-1}) + (\Delta t) \frac{5}{12} f(t_{j-2}, V_2^{j-2}) \right. \\
&\quad \left. + (\Delta t) \frac{23}{12} f(t_j, V_2^j) \right] \\
&\quad + \frac{G\alpha}{M(\alpha)} \sum_{j=2}^{n} V_2(t_j) \left(B(t_{j+1}) - B(t_j) \right)
\end{aligned}
$$

Then, we obtain

$$
\begin{aligned}
V_2(t_{n+1}) &= V_2(n) + \frac{1-\alpha}{M(\alpha)} \left(\left(-\frac{R_1 + R_2}{R_1 R_2 (C_1 + C_2)} V_2(t_n) + \frac{C_1}{C_1 + C_2} \frac{dV_1(t_n)}{dt} \right. \right. \\
&\quad \left. + \frac{1}{R_1(C_1 + C_2)} V_1(t_n) \right) \\
&\quad \left. - \left(-\frac{R_1 + R_2}{R_1 R_2 (C_1 + C_2)} V_2(t_{n-1}) + \frac{C_1}{C_1 + C_2} \frac{dV_1(t_{n-1})}{dt} \right. \right.
\end{aligned}
$$

$$+\frac{1}{R_1(C_1+C_2)}V_1(t_{n-1})\Bigg)\Bigg)$$

$$+\frac{\alpha}{M(\alpha)}\sum_{j=2}^{n}\Bigg[-(\Delta t)\frac{4}{3}\Bigg(-\frac{R_1+R_2}{R_1R_2(C_1+C_2)}V_2(t_{j-1})+\frac{C_1}{C_1+C_2}\frac{dV_1(t_{j-1})}{dt}$$

$$+\frac{1}{R_1(C_1+C_2)}V_1(t_{j-1})\Bigg)$$

$$+(\Delta t)\frac{5}{12}\Bigg(-\frac{R_1+R_2}{R_1R_2(C_1+C_2)}V_2(t_{j-2})+\frac{C_1}{C_1+C_2}\frac{dV_1(t_{j-2})}{dt}$$

$$+\frac{1}{R_1(C_1+C_2)}V_1(t_{j-2})\Bigg)$$

$$+(\Delta t)\frac{23}{12}\Bigg(-\frac{R_1+R_2}{R_1R_2(C_1+C_2)}V_2(t_j)+\frac{C_1}{C_1+C_2}\frac{dV_1(t_j)}{dt}$$

$$+\frac{1}{R_1(C_1+C_2)}V_1(t_j)\Bigg)\Bigg]$$

$$+\frac{G\alpha}{M(\alpha)}\sum_{j=2}^{n}i(t_j)\left(B(t_{j+1}-B(t_j))\right)$$

We consider the following differential equation

$$\frac{dV}{dt}=-\frac{1}{C}i(t)-\frac{1}{R_1C}V(t)+A\frac{d^2i(t)}{dt^2}+\left(R_2+\frac{L}{R_1C}\right)\frac{di(t)}{dt}+\frac{R_1+R_2}{R_1C}i(t)=f(t,V(t))$$

$$(21.100)$$

We convert the above differential equation to the fractional stochastic differential equation as: In exponential-decay case we have

$$V(t)-V(0)=\frac{1-\alpha}{M(\alpha)}f(t,V(t))+\frac{\alpha}{M(\alpha)}\int_0^t f(\tau,V(\tau))d\tau+\frac{G\alpha}{M(\alpha)}\int_0^t V(\tau)dB(\tau)$$

$$(21.101)$$

We define a mapping

$$
\begin{aligned}
|\Gamma(V(t))| \;=\; & \left|\frac{1-\alpha}{M(\alpha)}f(t,V(t))+\frac{\alpha}{M(\alpha)}\int_0^t f(\tau,V(\tau))d\tau\right.\\
& \left.+\frac{G\alpha}{M(\alpha)}\int_0^t V(\tau)dB(\tau)\right|\\
\;<\; & \frac{1-\alpha}{M(\alpha)}\sup_{t\in[0,T]}|f(t,V(t))|+\frac{\alpha}{M(\alpha)}\int_0^t \sup_{\tau\in[0,t]}|f(\tau,V(\tau))|d\tau\\
& +\frac{G\alpha}{M(\alpha)}\int_0^t \sup_{\tau\in[0,t]}|V(\tau)|dB(\tau)
\end{aligned}
$$

$$< \frac{1-\alpha}{M(\alpha)}\|f(.,i)\|_\infty + T\frac{\alpha}{M(\alpha)}\|f(.,V)\|_\infty + GL\|V\|_\infty$$
$$< M_1 + M_2 + M_3$$
$$< M$$

We have

$$
\begin{aligned}
|\Gamma(V) - \Gamma(V_1)| &= \left| \frac{1-\alpha}{M(\alpha)}(f(t,V(t)) - f(t,V_1(t))) \right. \\
&\quad + \frac{\alpha}{M(\alpha)}\int_0^t (f(\tau,V(\tau)) - f(\tau,V_1(\tau)))\,d\tau \\
&\quad + \left. \frac{G\alpha}{M(\alpha)}\int_0^t (V(\tau) - V_1(\tau))\,dB(\tau) \right| \\
&< \frac{1-\alpha}{M(\alpha)}|f(t,V(t)) - f(t,V_1(t))| \\
&\quad + \frac{\alpha}{M(\alpha)}\int_0^t |f(\tau,V(\tau)) - f(\tau,V_1(\tau))|\,d\tau \\
&\quad + \frac{G\alpha}{M(\alpha)}\int_0^t |V(\tau) - V_1(\tau)|\,dB(\tau) \\
&< \frac{1-\alpha}{M(\alpha)}\sup_{t\in[0,T]} |f(t,V(t)) - f(t,V_1(t))| \\
&\quad + \frac{\alpha}{M(\alpha)}\int_0^t \sup_{\tau\in[0,t]} |f(\tau,V(\tau)) - f(\tau,V_1(\tau))|\,d\tau \\
&\quad + \frac{G\alpha}{M(\alpha)}\int_0^t \sup_{\tau\in[0,t]} |V(\tau) - V_1(\tau)|\,dB(\tau) \\
&< \frac{1-\alpha}{M(\alpha)}\frac{1}{C}i(t) - \frac{1}{R_1C}\|V(t) - V_1(t)\|_\infty \\
&\quad + \frac{\alpha}{M(\alpha)}T\frac{1}{C}i(t) - \frac{1}{R_1C}\|V(t)) - V_1(t)\|_\infty \\
&\quad + GL\|V(t) - V_1(t)\|_\infty \\
&< \left(\frac{1-\alpha}{M(\alpha)}\frac{1}{C}i(t) - \frac{1}{R_1C} + \frac{\alpha}{M(\alpha)}T\frac{1}{C}i(t) - \frac{1}{R_1C} + GL \right) \\
&\quad \times \|V(t) - V_1(t)\|_\infty \\
&< L_1\|V(t) - V_1(t)\|_\infty
\end{aligned}
$$

We assume that $B(t)$ is Lipschitzian on $[0,T]$ with constant L, then

$$
\begin{aligned}
|\Gamma(V(t))| &= \left| \frac{1-\alpha}{M(\alpha)}f(t,V(t)) + \frac{\alpha}{M(\alpha)}\int_0^t f(\tau,V(\tau))d\tau \right. \\
&\quad + \left. \frac{G\alpha}{M(\alpha)}\int_0^t V(\tau)dB(\tau) \right|
\end{aligned}
$$

$$< \frac{1-\alpha}{M(\alpha)} |f(t,V(t))|$$

$$+\frac{\alpha}{M(\alpha)} \int_0^t |f(\tau,V(\tau))| d\tau$$

$$+\frac{G\alpha}{M(\alpha)} \int_0^t |V(\tau)| dB(\tau)$$

$$< \frac{1-\alpha}{M(\alpha)} \sup_{l\in[0,t]} |f(l,V(l))|$$

$$+\frac{\alpha}{M(\alpha)} \int_0^t \sup_{l\in[0,\tau]} |f(l,V(l))| d\tau$$

$$+\frac{G\alpha}{M(\alpha)} \int_0^t \sup_{l\in[0,\tau]} |V(l)| dB(\tau)$$

This shows that our mapping has a unique solution. We now present the numerical solution of the equation

$$V(t_{n+1}) = V(0) + \frac{1-\alpha}{M(\alpha)} f(t_n,V(t_n)) + \frac{\alpha}{M(\alpha)} \int_0^{t_{n+1}} f(\tau,i(\tau)) d\tau$$

$$+\frac{G\alpha}{M(\alpha)} \int_0^{t_{n+1}} V(\tau) dB(\tau)$$

and

$$V_2(t_n) = V_2(0) + \frac{1-\alpha}{M(\alpha)} f(t_{n-1},V(t_{n-1})) + \frac{\alpha}{M(\alpha)} \int_0^{t_n} f(\tau,V_2(\tau)) d\tau$$

$$+\frac{G\alpha}{M(\alpha)} \int_0^{t_n} V(\tau) dB(\tau)$$

Thus, we reach

$$V(t_{n+1}) = V(n) + \frac{1-\alpha}{M(\alpha)} (f(t_n,V(t_n)) - f(t_{n-1},V(t_{n-1})))$$

$$+\frac{\alpha}{M(\alpha)} \int_{t_n}^{t_{n+1}} f(\tau,V(\tau)) d\tau + \frac{G\alpha}{M(\alpha)} \int_{t_n}^{t_{n+1}} V(\tau) dB(\tau)$$

$$= i(n) + \frac{1-\alpha}{M(\alpha)} (f(t_n,V(t_n)) - f(t_{n-1},V(t_{n-1})))$$

$$+\frac{\alpha}{M(\alpha)} \sum_{j=2}^n \int_{t_j}^{t_{j+1}} f(\tau,V(\tau)) d\tau + \frac{G\alpha}{M(\alpha)} \sum_{j=2}^n \int_{t_j}^{t_{j+1}} i(\tau) dB(\tau)$$

$$= i(n) + \frac{1-\alpha}{M(\alpha)} (f(t_n,V(t_n)) - f(t_{n-1},V(t_{n-1})))$$

$$+\frac{\alpha}{M(\alpha)} \sum_{j=2}^n \left[-(\Delta t)\frac{4}{3} f(t_{j-1},V^{j-1}) + (\Delta t)\frac{5}{12} f(t_{j-2},V^{j-2}) \right]$$

$$+(\Delta t)\frac{23}{12}f(t_j,V^j)\bigg]$$

$$+\frac{G\alpha}{M(\alpha)}\sum_{j=2}^{n}V(t_j)\left(B(t_{j+1}-B(t_j))\right)$$

Then, we obtain

$$
\begin{aligned}
V(t_{n+1}) \;=\; & V(n)+\frac{1-\alpha}{M(\alpha)}\left(\left(-\frac{1}{C}i(t)-\frac{1}{R_1C}V(t_n)+A\frac{d^2i(t_n)}{dt^2}\right.\right.\\
& +\left(R_2+\frac{L}{R_1C}\right)\frac{di(t_n)}{dt}+\frac{R_1+R_2}{R_1C}i(t_n)\bigg)\\
& -\left(-\frac{1}{C}i(t)-\frac{1}{R_1C}V(t_{n-1})+A\frac{d^2i(t_{n-1})}{dt^2}\right.\\
& +\left.\left(R_2+\frac{L}{R_1C}\right)\frac{di(t_{n-1})}{dt}+\frac{R_1+R_2}{R_1C}i(t_{n-1})\bigg)\right)\\
& +\frac{\alpha}{M(\alpha)}\sum_{j=2}^{n}\left[-(\Delta t)\frac{4}{3}\left(-\frac{1}{C}i(t)-\frac{1}{R_1C}V(t_{j-1})+A\frac{d^2i(t_{j-1})}{dt^2}\right.\right.\\
& +\left(R_2+\frac{L}{R_1C}\right)\frac{di(t_{j-1})}{dt}+\frac{R_1+R_2}{R_1C}i(t_{j-1})\bigg)\\
& +(\Delta t)\frac{5}{12}\left(-\frac{1}{C}i(t)-\frac{1}{R_1C}V(t_{j-2})+A\frac{d^2i(t_{j-2})}{dt^2}\right.\\
& +\left(R_2+\frac{L}{R_1C}\right)\frac{di(t_{j-2})}{dt}+\frac{R_1+R_2}{R_1C}i(t_{j-2})\bigg)\\
& +(\Delta t)\frac{23}{12}\left(-\frac{1}{C}i(t)-\frac{1}{R_1C}V(t_j)+A\frac{d^2i(t_j)}{dt^2}\right.\\
& +\left.\left(R_2+\frac{L}{R_1C}\right)\frac{di(t_j)}{dt}+\frac{R_1+R_2}{R_1C}i(t_j)\bigg)\right]\\
& +\frac{G\alpha}{M(\alpha)}\sum_{j=2}^{n}i(t_j)\left(B(t_{j+1}-B(t_j))\right)
\end{aligned}
$$

We consider the following differential equation

$$\frac{dV}{dt}=-\frac{1}{C}i(t)-\frac{1}{R_1C}V(t)+A\frac{d^2i(t)}{dt^2}+\left(R_2+\frac{L}{R_1C}\right)\frac{di(t)}{dt}+\frac{R_1+R_2}{R_1C}i(t)=f(t,V(t))$$

$$(21.102)$$

We convert the above differential equation to the fractional stochastic differential equation as: In exponential-decay case we have

$$V(t) - V(0) = \frac{1-\alpha}{M(\alpha)} f(t, V(t)) + \frac{\alpha}{M(\alpha)} \int_0^t f(\tau, V(\tau)) d\tau + \frac{G\alpha}{M(\alpha)} \int_0^t V(\tau) dB(\tau)$$

$$(21.103)$$

We define a mapping

$$
\begin{aligned}
|\Gamma(V(t))| &= \left| \frac{1-\alpha}{M(\alpha)} f(t, V(t)) + \frac{\alpha}{M(\alpha)} \int_0^t f(\tau, V(\tau)) d\tau \right. \\
&\quad \left. + \frac{G\alpha}{M(\alpha)} \int_0^t V(\tau) dB(\tau) \right| \\
&< \frac{1-\alpha}{M(\alpha)} \sup_{t \in [0,T]} |f(t, V(t))| + \frac{\alpha}{M(\alpha)} \int_0^t \sup_{\tau \in [0,t]} |f(\tau, V(\tau))| d\tau \\
&\quad + \frac{G\alpha}{M(\alpha)} \int_0^t \sup_{\tau \in [0,t]} |V(\tau)| dB(\tau) \\
&< \frac{1-\alpha}{M(\alpha)} \|f(.,i)\|_\infty + T \frac{\alpha}{M(\alpha)} \|f(.,V)\|_\infty + GL \|V\|_\infty \\
&< M_1 + M_2 + M_3 \\
&< M
\end{aligned}
$$

We have

$$
\begin{aligned}
|\Gamma(V) - \Gamma(V_1)| &= \left| \frac{1-\alpha}{M(\alpha)} (f(t, V(t)) - f(t, V_1(t))) \right. \\
&\quad + \frac{\alpha}{M(\alpha)} \int_0^t (f(\tau, V(\tau)) - f(\tau, V_1(\tau))) d\tau \\
&\quad \left. + \frac{G\alpha}{M(\alpha)} \int_0^t (V(\tau) - V_1(\tau)) dB(\tau) \right| \\
&< \frac{1-\alpha}{M(\alpha)} |f(t, V(t)) - f(t, V_1(t))| \\
&\quad + \frac{\alpha}{M(\alpha)} \int_0^t |f(\tau, V(\tau)) - f(\tau, V_1(\tau))| d\tau \\
&\quad + \frac{G\alpha}{M(\alpha)} \int_0^t |V(\tau) - V_1(\tau)| dB(\tau) \\
&< \frac{1-\alpha}{M(\alpha)} \sup_{t \in [0,T]} |f(t, V(t)) - f(t, V_1(t))| \\
&\quad + \frac{\alpha}{M(\alpha)} \int_0^t \sup_{\tau \in [0,t]} |f(\tau, V(\tau)) - f(\tau, V_1(\tau))| d\tau \\
&\quad + \frac{G\alpha}{M(\alpha)} \int_0^t \sup_{\tau \in [0,t]} |V(\tau) - V_1(\tau)| dB(\tau) \\
&< \frac{1-\alpha}{M(\alpha)} \left| \frac{1}{C} i(t) - \frac{1}{R_1 C} \right| \|V(t) - V_1(t)\|_\infty
\end{aligned}
$$

$$+\frac{\alpha}{M(\alpha)}T\frac{1}{C}i(t)-\frac{1}{R_1 C}\|V(t))-V_1(t)\|_\infty$$

$$+GL\|V(t)-V_1(t)\|_\infty$$

$$<\quad \left(\frac{1-\alpha}{M(\alpha)}\frac{1}{C}i(t)-\frac{1}{R_1 C}+\frac{\alpha}{M(\alpha)}T\frac{1}{C}i(t)-\frac{1}{R_1 C}+GL\right)$$

$$\times\|V(t)-V_1(t)\|_\infty$$

$$<\quad L_1\|V(t)-V_1(t)\|_\infty$$

We assume that $B(t)$ is Lipschitzian on $[0,T]$ with constant L, then

$$
\begin{aligned}
|\Gamma(V(t))| \quad &= \quad \left|\frac{1-\alpha}{M(\alpha)}f(t,V(t))+\frac{\alpha}{M(\alpha)}\int_0^t f(\tau,V(\tau))d\tau\right. \\
&\qquad \left. +\frac{G\alpha}{M(\alpha)}\int_0^t V(\tau)dB(\tau)\right| \\
&< \quad \frac{1-\alpha}{M(\alpha)}|f(t,V(t))| \\
&\qquad +\frac{\alpha}{M(\alpha)}\int_0^t |f(\tau,V(\tau))|d\tau \\
&\qquad +\frac{G\alpha}{M(\alpha)}\int_0^t |V(\tau)|dB(\tau) \\
&< \quad \frac{1-\alpha}{M(\alpha)}\sup_{l\in[0,t]}|f(l,V(l))| \\
&\qquad +\frac{\alpha}{M(\alpha)}\int_0^t \sup_{l\in[0,\tau]}|f(l,V(l))|d\tau \\
&\qquad +\frac{G\alpha}{M(\alpha)}\int_0^t \sup_{l\in[0,\tau]}|V(l)|dB(\tau)
\end{aligned}
$$

This shows that our mapping has a unique solution. We now present the numerical solution of the equation

$$
\begin{aligned}
V(t_{n+1}) \quad &= \quad V(0)+\frac{1-\alpha}{M(\alpha)}f(t_n,V(t_n))+\frac{\alpha}{M(\alpha)}\int_0^{t_{n+1}} f(\tau,i(\tau))d\tau \\
&\qquad +\frac{G\alpha}{M(\alpha)}\int_0^{t_{n+1}} V(\tau)dB(\tau)
\end{aligned}
$$

and

$$
\begin{aligned}
V_2(t_n) \quad &= \quad V_2(0)+\frac{1-\alpha}{M(\alpha)}f(t_{n-1},V(t_{n-1}))+\frac{\alpha}{M(\alpha)}\int_0^{t_n} f(\tau,V_2(\tau))d\tau \\
&\qquad +\frac{G\alpha}{M(\alpha)}\int_0^{t_n} V(\tau)dB(\tau)
\end{aligned}
$$

Thus, we reach

$$
\begin{aligned}
V(t_{n+1}) &= V(n) + \frac{1-\alpha}{M(\alpha)}\left(f(t_n, V(t_n)) - f(t_{n-1}, V(t_{n-1}))\right) \\
&\quad + \frac{\alpha}{M(\alpha)}\int_{t_n}^{t_{n+1}} f(\tau, V(\tau))d\tau + \frac{G\alpha}{M(\alpha)}\int_{t_n}^{t_{n+1}} V(\tau)dB(\tau) \\
&= i(n) + \frac{1-\alpha}{M(\alpha)}\left(f(t_n, V(t_n)) - f(t_{n-1}, V(t_{n-1}))\right) \\
&\quad + \frac{\alpha}{M(\alpha)}\sum_{j=2}^{n}\int_{t_j}^{t_{j+1}} f(\tau, V(\tau))d\tau + \frac{G\alpha}{M(\alpha)}\sum_{j=2}^{n}\int_{t_j}^{t_{j+1}} i(\tau)dB(\tau) \\
&= i(n) + \frac{1-\alpha}{M(\alpha)}\left(f(t_n, V(t_n)) - f(t_{n-1}, V(t_{n-1}))\right) \\
&\quad + \frac{\alpha}{M(\alpha)}\sum_{j=2}^{n}\left[-(\Delta t)\frac{4}{3}f(t_{j-1}, V^{j-1}) + (\Delta t)\frac{5}{12}f(t_{j-2}, V^{j-2})\right. \\
&\quad \left. + (\Delta t)\frac{23}{12}f(t_j, V^j)\right] \\
&\quad + \frac{G\alpha}{M(\alpha)}\sum_{j=2}^{n} V(t_j)\left(B(t_{j+1}) - B(t_j)\right)
\end{aligned}
$$

Then, we obtain

$$
\begin{aligned}
V(t_{n+1}) &= V(n) + \frac{1-\alpha}{M(\alpha)}\left(\left(-\frac{1}{C}i(t) - \frac{1}{R_1 C}V(t_n) + A\frac{d^2 i(t_n)}{dt^2}\right.\right. \\
&\quad + \left(R_2 + \frac{L}{R_1 C}\right)\frac{di(t_n)}{dt} + \frac{R_1 + R_2}{R_1 C}i(t_n)\Big) \\
&\quad - \left(-\frac{1}{C}i(t) - \frac{1}{R_1 C}V(t_{n-1}) + A\frac{d^2 i(t_{n-1})}{dt^2}\right. \\
&\quad \left.\left. + \left(R_2 + \frac{L}{R_1 C}\right)\frac{di(t_{n-1})}{dt} + \frac{R_1 + R_2}{R_1 C}i(t_{n-1})\right)\right) \\
&\quad + \frac{\alpha}{M(\alpha)}\sum_{j=2}^{n}\left[-(\Delta t)\frac{4}{3}\left(-\frac{1}{C}i(t) - \frac{1}{R_1 C}V(t_{j-1}) + A\frac{d^2 i(t_{j-1})}{dt^2}\right.\right. \\
&\quad \left. + \left(R_2 + \frac{L}{R_1 C}\right)\frac{di(t_{j-1})}{dt} + \frac{R_1 + R_2}{R_1 C}i(t_{j-1})\right) \\
&\quad + (\Delta t)\frac{5}{12}\left(-\frac{1}{C}i(t) - \frac{1}{R_1 C}V(t_{j-2}) + A\frac{d^2 i(t_{j-2})}{dt^2}\right. \\
&\quad \left. + \left(R_2 + \frac{L}{R_1 C}\right)\frac{di(t_{j-2})}{dt} + \frac{R_1 + R_2}{R_1 C}i(t_{j-2})\right) \\
&\quad + (\Delta t)\frac{23}{12}\left(-\frac{1}{C}i(t) - \frac{1}{R_1 C}V(t_j) + A\frac{d^2 i(t_j)}{dt^2}\right.
\end{aligned}
$$

$$+\left(R_2+\frac{L}{R_1C}\right)\frac{di(t_j)}{dt}+\frac{R_1+R_2}{R_1C}i(t_j)\right)\bigg]$$

$$+\frac{G\alpha}{M(\alpha)}\sum_{j=2}^{n}i(t_j)\left(B(t_{j+1}-B(t_j))\right)$$

We consider the following differential equation

$$\frac{di}{dt}=-RC\frac{d^2i(t)}{dt^2}+\frac{1}{L}V(t)+\frac{RC}{L}\frac{dV(t)}{dt}=f(t,i(t)) \tag{21.104}$$

We convert the above differential equation to the fractional stochastic differential equation as: In exponential-decay case we have

$$i(t)-i(0)=\frac{1-\alpha}{M(\alpha)}f(t,i(t))+\frac{\alpha}{M(\alpha)}\int_0^t f(\tau,i(\tau))d\tau+\frac{G\alpha}{M(\alpha)}\int_0^t i(\tau)dB(\tau)$$
$$\tag{21.105}$$

We define a mapping

$$
\begin{aligned}
|\Gamma(i(t))| &= \left|\frac{1-\alpha}{M(\alpha)}f(t,i(t))+\frac{\alpha}{M(\alpha)}\int_0^t f(\tau,i(\tau))d\tau\right. \\
&\quad \left.+\frac{G\alpha}{M(\alpha)}\int_0^t i(\tau)dB(\tau)\right| \\
&< \frac{1-\alpha}{M(\alpha)}\sup_{t\in[0,T]}|f(t,i(t))|+\frac{\alpha}{M(\alpha)}\int_0^t \sup_{\tau\in[0,t]}|f(\tau,i(\tau))|d\tau \\
&\quad +\frac{G\alpha}{M(\alpha)}\int_0^t \sup_{\tau\in[0,t]}|i(\tau)|dB(\tau) \\
&< \frac{1-\alpha}{M(\alpha)}\|f(.,i)\|_\infty+T\frac{\alpha}{M(\alpha)}\|f(.,i)\|_\infty+GL\|i\|_\infty \\
&< M_1+M_2+M_3 \\
&< M
\end{aligned}
$$

We have

$$
\begin{aligned}
|\Gamma(i)-\Gamma(i_1)| &= \left|\frac{1-\alpha}{M(\alpha)}(f(t,i(t))-f(t,i_1(t)))\right. \\
&\quad +\frac{\alpha}{M(\alpha)}\int_0^t (f(\tau,i(\tau))-f(\tau,i_1(\tau)))d\tau \\
&\quad \left.+\frac{G\alpha}{M(\alpha)}\int_0^t (i(\tau)-i_1(\tau))dB(\tau)\right| \\
&< \frac{1-\alpha}{M(\alpha)}|f(t,i(t))-f(t,i_1(t))| \\
&\quad +\frac{\alpha}{M(\alpha)}\int_0^t |f(\tau,i(\tau))-f(\tau,i_1(\tau))|d\tau
\end{aligned}
$$

$$+\frac{G\alpha}{M(\alpha)}\int_0^t |i(\tau)-i_1(\tau)|\,dB(\tau)$$

$$< \frac{1-\alpha}{M(\alpha)}\sup_{t\in[0,T]} |f(t,i(t))-f(t,i_1(t))|$$

$$+\frac{\alpha}{M(\alpha)}\int_0^t \sup_{\tau\in[0,t]} |f(\tau,i(\tau))-f(\tau,i_1(\tau))|\,d\tau$$

$$+\frac{G\alpha}{M(\alpha)}\int_0^t \sup_{\tau\in[0,t]} |i(\tau)-i_1(\tau)|\,dB(\tau)$$

$$< \frac{1-\alpha}{M(\alpha)}RC\|i(t)-i_1(t)\|_\infty$$

$$+\frac{\alpha}{M(\alpha)}TRC\|i(t))-i_1(t)\|_\infty+GL\|i(t)-i_1(t)\|_\infty$$

$$< \left(\frac{1-\alpha}{M(\alpha)}RC+\frac{\alpha}{M(\alpha)}TRC+GL\right)\|i(t)-i_1(t)\|_\infty$$

$$< L_1\|i(t)-i_1(t)\|_\infty$$

We assume that $B(t)$ is Lipschitzian on $[0,T]$ with constant L, then

$$|\Gamma(i(t))| = \left|\frac{1-\alpha}{M(\alpha)}f(t,i(t))+\frac{\alpha}{M(\alpha)}\int_0^t f(\tau,i(\tau))d\tau\right.$$

$$\left.+\frac{G\alpha}{M(\alpha)}\int_0^t i(\tau)dB(\tau)\right|$$

$$< \frac{1-\alpha}{M(\alpha)}|f(t,i(t))|$$

$$+\frac{\alpha}{M(\alpha)}\int_0^t |f(\tau,i(\tau))|\,d\tau$$

$$+\frac{G\alpha}{M(\alpha)}\int_0^t |i(\tau)|\,dB(\tau)$$

$$< \frac{1-\alpha}{M(\alpha)}\sup_{l\in[0,t]} |f(l,i(l))|$$

$$+\frac{\alpha}{M(\alpha)}\int_0^t \sup_{l\in[0,\tau]} |f(l,i(l))|\,d\tau$$

$$+\frac{G\alpha}{M(\alpha)}\int_0^t \sup_{l\in[0,\tau]} |i(l)|\,dB(\tau)$$

This shows that our mapping has a unique solution. We now present the numerical solution of the equation

$$i(t_{n+1}) = i(0) + \frac{1-\alpha}{M(\alpha)} f(t_n, i(t_n)) + \frac{\alpha}{M(\alpha)} \int_0^{t_{n+1}} f(\tau, i(\tau)) d\tau$$
$$+ \frac{G\alpha}{M(\alpha)} \int_0^{t_{n+1}} i(\tau) dB(\tau)$$

and

$$i(t_n) = i(0) + \frac{1-\alpha}{M(\alpha)} f(t_{n-1}, i(t_{n-1})) + \frac{\alpha}{M(\alpha)} \int_0^{t_n} f(\tau, i(\tau)) d\tau$$
$$+ \frac{G\alpha}{M(\alpha)} \int_0^{t_n} i(\tau) dB(\tau)$$

Thus, we reach

$$i(t_{n+1}) = i(n) + \frac{1-\alpha}{M(\alpha)} (f(t_n, i(t_n)) - f(t_{n-1}, i(t_{n-1})))$$
$$+ \frac{\alpha}{M(\alpha)} \int_{t_n}^{t_{n+1}} f(\tau, i(\tau)) d\tau + \frac{G\alpha}{M(\alpha)} \int_{t_n}^{t_{n+1}} i(\tau) dB(\tau)$$
$$= i(n) + \frac{1-\alpha}{M(\alpha)} (f(t_n, i(t_n)) - f(t_{n-1}, i(t_{n-1})))$$
$$+ \frac{\alpha}{M(\alpha)} \sum_{j=2}^{n} \int_{t_j}^{t_{j+1}} f(\tau, i(\tau)) d\tau + \frac{G\alpha}{M(\alpha)} \sum_{j=2}^{n} \int_{t_j}^{t_{j+1}} i(\tau) dB(\tau)$$
$$= i(n) + \frac{1-\alpha}{M(\alpha)} (f(t_n, i(t_n)) - f(t_{n-1}, i(t_{n-1})))$$
$$+ \frac{\alpha}{M(\alpha)} \sum_{j=2}^{n} \left[-(\Delta t) \frac{4}{3} f(t_{j-1}, i^{j-1}) + (\Delta t) \frac{5}{12} f(t_{j-2}, i^{j-2}) \right.$$
$$\left. + (\Delta t) \frac{23}{12} f(t_j, i^j) \right]$$
$$+ \frac{G\alpha}{M(\alpha)} \sum_{j=2}^{n} i(t_j) \left(B(t_{j+1}) - B(t_j) \right)$$

Then, we obtain

$$i(t_{n+1}) = i(n) + \frac{1-\alpha}{M(\alpha)} \left(\left(-RC \frac{d^2 i(t_n)}{dt^2} + \frac{1}{L} V(t_n) + \frac{RC}{L} \frac{dV(t_n)}{dt} \right) \right.$$
$$\left. - \left(-RC \frac{d^2 i(t_{n-1})}{dt^2} + \frac{1}{L} V(t_{n-1}) + \frac{RC}{L} \frac{dV(t_{n-1})}{dt} \right) \right)$$
$$+ \frac{\alpha}{M(\alpha)} \sum_{j=2}^{n} \left[-(\Delta t) \frac{4}{3} \left(-RC \frac{d^2 i(t_{j-1})}{dt^2} + \frac{1}{L} V(t_{j-1}) + \frac{RC}{L} \frac{dV(t_{j-1})}{dt} \right) \right.$$

$$+(\Delta t)\frac{5}{12}\left(-RC\frac{d^2 i(t_{j-2})}{dt^2}+\frac{1}{L}V(t_{j-2})+\frac{RC}{L}\frac{dV(t_{j-2})}{dt}\right)$$

$$+(\Delta t)\frac{23}{12}\left(-RC\frac{d^2 i(t_j)}{dt^2}+\frac{1}{L}V(t_j)+\frac{RC}{L}\frac{dV(t_j)}{dt}\right)\Bigg]$$

$$+\frac{G\alpha}{M(\alpha)}\sum_{j=2}^{n}i(t_j)\left(B(t_{j+1}-B(t_j))\right)$$

We consider the following differential equation

$$\frac{di}{dt}=-\frac{R_1}{A}i(t)+\frac{R_1}{AR_2}V(t)+C\frac{d^2 V(t)}{dt^2}+\left(\frac{CR_1}{A}+\frac{1}{R_2}\right)\frac{dV(t)}{dt}=f(t,i(t)) \quad (21.106)$$

We convert the above differential equation to the fractional stochastic differential equation as: In exponential-decay case we have

$$i(t)-i(0)=\frac{1-\alpha}{M(\alpha)}f(t,i(t))+\frac{\alpha}{M(\alpha)}\int_0^t f(\tau,i(\tau))d\tau+\frac{G\alpha}{M(\alpha)}\int_0^t i(\tau)dB(\tau)$$

$$(21.107)$$

We define a mapping

$$\begin{aligned}|\Gamma(i(t))| &= \left|\frac{1-\alpha}{M(\alpha)}f(t,i(t))+\frac{\alpha}{M(\alpha)}\int_0^t f(\tau,i(\tau))d\tau\right.\\ &\quad \left.+\frac{G\alpha}{M(\alpha)}\int_0^t i(\tau)dB(\tau)\right|\\ &< \frac{1-\alpha}{M(\alpha)}\sup_{t\in[0,T]}|f(t,i(t))|+\frac{\alpha}{M(\alpha)}\int_0^t \sup_{\tau\in[0,t]}|f(\tau,i(\tau))|d\tau\\ &\quad +\frac{G\alpha}{M(\alpha)}\int_0^t \sup_{\tau\in[0,t]}|i(\tau)|dB(\tau)\\ &< \frac{1-\alpha}{M(\alpha)}\|f(.,i)\|_\infty+T\frac{\alpha}{M(\alpha)}\|f(.,i)\|_\infty+GL\|i\|_\infty\\ &< M_1+M_2+M_3\\ &< M\end{aligned}$$

We have

$$\begin{aligned}|\Gamma(i)-\Gamma(i_1)| &= \left|\frac{1-\alpha}{M(\alpha)}\left(f(t,i(t))-f(t,i_1(t))\right)\right.\\ &\quad +\frac{\alpha}{M(\alpha)}\int_0^t \left(f(\tau,i(\tau))-f(\tau,i_1(\tau))\right)d\tau\\ &\quad \left.+\frac{G\alpha}{M(\alpha)}\int_0^t \left(i(\tau)-i_1(\tau)\right)dB(\tau)\right|\\ &< \frac{1-\alpha}{M(\alpha)}|f(t,i(t))-f(t,i_1(t))|\end{aligned}$$

$$+\frac{\alpha}{M(\alpha)}\int_0^t |f(\tau,i(\tau))-f(\tau,i_1(\tau))|\,d\tau$$

$$+\frac{G\alpha}{M(\alpha)}\int_0^t |i(\tau)-i_1(\tau)|\,dB(\tau)$$

$$<\quad \frac{1-\alpha}{M(\alpha)}\sup_{t\in[0,T]}|f(t,i(t))-f(t,i_1(t))|$$

$$+\frac{\alpha}{M(\alpha)}\int_0^t \sup_{\tau\in[0,t]}|f(\tau,i(\tau))-f(\tau,i_1(\tau))|\,d\tau$$

$$+\frac{G\alpha}{M(\alpha)}\int_0^t \sup_{\tau\in[0,t]}|i(\tau)-i_1(\tau)|\,dB(\tau)$$

$$<\quad \frac{1-\alpha}{M(\alpha)}\frac{R_1}{A}\|i(t)-i_1(t)\|_\infty$$

$$+\frac{\alpha}{M(\alpha)}T\frac{R_1}{A}\|i(t))-i_1(t)\|_\infty+GL\|i(t)-i_1(t)\|_\infty$$

$$<\quad \left(\frac{1-\alpha}{M(\alpha)}\frac{R_1}{A}+\frac{\alpha}{M(\alpha)}T\frac{R_1}{A}+GL\right)\|i(t)-i_1(t)\|_\infty$$

$$<\quad L_1\|i(t)-i_1(t)\|_\infty$$

We assume that $B(t)$ is Lipschitzian on $[0,T]$ with constant L, then

$$|\Gamma(i(t))| \;=\; \left|\frac{1-\alpha}{M(\alpha)}f(t,i(t))+\frac{\alpha}{M(\alpha)}\int_0^t f(\tau,i(\tau))d\tau\right.$$

$$\left.+\frac{G\alpha}{M(\alpha)}\int_0^t i(\tau)dB(\tau)\right|$$

$$<\quad \frac{1-\alpha}{M(\alpha)}|f(t,i(t))|$$

$$+\frac{\alpha}{M(\alpha)}\int_0^t |f(\tau,i(\tau))|\,d\tau$$

$$+\frac{G\alpha}{M(\alpha)}\int_0^t |i(\tau)|\,dB(\tau)$$

$$<\quad \frac{1-\alpha}{M(\alpha)}\sup_{l\in[0,t]}|f(l,i(l))|$$

$$+\frac{\alpha}{M(\alpha)}\int_0^t \sup_{l\in[0,\tau]}|f(l,i(l))|\,d\tau$$

$$+\frac{G\alpha}{M(\alpha)}\int_0^t \sup_{l\in[0,\tau]}|i(l)|\,dB(\tau)$$

This shows that our mapping has a unique solution. We now present the numerical solution of the equation

$$
\begin{aligned}
i(t_{n+1}) &= i(0) + \frac{1-\alpha}{M(\alpha)} f(t_n, i(t_n)) + \frac{\alpha}{M(\alpha)} \int_0^{t_{n+1}} f(\tau, i(\tau)) d\tau \\
&\quad + \frac{G\alpha}{M(\alpha)} \int_0^{t_{n+1}} i(\tau) dB(\tau)
\end{aligned}
$$

and

$$
\begin{aligned}
i(t_n) &= i(0) + \frac{1-\alpha}{M(\alpha)} f(t_{n-1}, i(t_{n-1})) + \frac{\alpha}{M(\alpha)} \int_0^{t_n} f(\tau, i(\tau)) d\tau \\
&\quad + \frac{G\alpha}{M(\alpha)} \int_0^{t_n} i(\tau) dB(\tau)
\end{aligned}
$$

Thus, we reach

$$
\begin{aligned}
i(t_{n+1}) &= i(n) + \frac{1-\alpha}{M(\alpha)} \left(f(t_n, i(t_n)) - f(t_{n-1}, i(t_{n-1})) \right) \\
&\quad + \frac{\alpha}{M(\alpha)} \int_{t_n}^{t_{n+1}} f(\tau, i(\tau)) d\tau + \frac{G\alpha}{M(\alpha)} \int_{t_n}^{t_{n+1}} i(\tau) dB(\tau) \\
&= i(n) + \frac{1-\alpha}{M(\alpha)} \left(f(t_n, i(t_n)) - f(t_{n-1}, i(t_{n-1})) \right) \\
&\quad + \frac{\alpha}{M(\alpha)} \sum_{j=2}^n \int_{t_j}^{t_{j+1}} f(\tau, i(\tau)) d\tau + \frac{G\alpha}{M(\alpha)} \sum_{j=2}^n \int_{t_j}^{t_{j+1}} i(\tau) dB(\tau) \\
&= i(n) + \frac{1-\alpha}{M(\alpha)} \left(f(t_n, i(t_n)) - f(t_{n-1}, i(t_{n-1})) \right) \\
&\quad + \frac{\alpha}{M(\alpha)} \sum_{j=2}^n \left[-(\Delta t) \frac{4}{3} f(t_{j-1}, i^{j-1}) + (\Delta t) \frac{5}{12} f(t_{j-2}, i^{j-2}) \right. \\
&\quad \left. + (\Delta t) \frac{23}{12} f(t_j, i^j) \right] \\
&\quad + \frac{G\alpha}{M(\alpha)} \sum_{j=2}^n i(t_j) \left(B(t_{j+1} - B(t_j)) \right)
\end{aligned}
$$

Then, we obtain

$$
\begin{aligned}
i(t_{n+1}) &= i(n) + \frac{1-\alpha}{M(\alpha)} \left(\left(-\frac{R_1}{A} i(t_n) + \frac{R_1}{AR_2} V(t_n) + C \frac{d^2 V(t_n)}{dt^2} \right. \right. \\
&\quad \left. \left. + \left(\frac{CR_1}{A} + \frac{1}{R_2} \right) \frac{dV(t_n)}{dt} \right) \right. \\
&\quad \left. - \left(-\frac{R_1}{A} i(t_{n-1}) + \frac{R_1}{AR_2} V(t_{n-1}) + C \frac{d^2 V(t_{n-1})}{dt^2} \right. \right.
\end{aligned}
$$

$$+\left(\frac{CR_1}{A}+\frac{1}{R_2}\right)\frac{dV(t_{n-1})}{dt}\Bigg)\Bigg)$$

$$+\frac{\alpha}{M(\alpha)}\sum_{j=2}^{n}\left[-(\Delta t)\frac{4}{3}\left(-\frac{R_1}{A}i(t_{j-1})+\frac{R_1}{AR_2}V(t_{j-1})+C\frac{d^2V(t_{j-1})}{dt^2}\right.\right.$$

$$+\left(\frac{CR_1}{A}+\frac{1}{R_2}\right)\frac{dV(t_{j-1})}{dt}\right)$$

$$+(\Delta t)\frac{5}{12}\left(-\frac{R_1}{A}i(t_{j-2})+\frac{R_1}{AR_2}V(t_{j-2})+C\frac{d^2V(t_{j-2})}{dt^2}\right.$$

$$+\left(\frac{CR_1}{A}+\frac{1}{R_2}\right)\frac{dV(t_{j-2})}{dt}\right)$$

$$+(\Delta t)\frac{23}{12}\left(-\frac{R_1}{A}i(t_j)+\frac{R_1}{AR_2}V(t_j)+C\frac{d^2V(t_j)}{dt^2}\right.$$

$$+\left(\frac{CR_1}{A}+\frac{1}{R_2}\right)\frac{dV(t_j)}{dt}\right)\Bigg]$$

$$+\frac{G\alpha}{M(\alpha)}\sum_{j=2}^{n}i(t_j)\left(B(t_{j+1})-B(t_j)\right)$$

We consider the following differential equation

$$\frac{di}{dt}=-\frac{R}{A}i(t)-\frac{1}{AC}\int_0^t i(y)dy+\frac{1}{A}V(t)=f(t,i(t)) \qquad (21.108)$$

We convert the above differential equation to the fractional stochastic differential equation as: In exponential-decay case we have

$$i(t)-i(0)=\frac{1-\alpha}{M(\alpha)}f(t,i(t))+\frac{\alpha}{M(\alpha)}\int_0^t f(\tau,i(\tau))d\tau+\frac{G\alpha}{M(\alpha)}\int_0^t i(\tau)dB(\tau)$$
$$(21.109)$$

We define a mapping

$$|\Gamma(i(t))| = \left|\frac{1-\alpha}{M(\alpha)}f(t,i(t))+\frac{\alpha}{M(\alpha)}\int_0^t f(\tau,i(\tau))d\tau\right.$$
$$\left.+\frac{G\alpha}{M(\alpha)}\int_0^t i(\tau)dB(\tau)\right|$$
$$< \frac{1-\alpha}{M(\alpha)}\sup_{t\in[0,T]}|f(t,i(t))|+\frac{\alpha}{M(\alpha)}\int_0^t\sup_{\tau\in[0,t]}|f(\tau,i(\tau))|d\tau$$
$$+\frac{G\alpha}{M(\alpha)}\int_0^t\sup_{\tau\in[0,t]}|i(\tau)|dB(\tau)$$

$$
\begin{aligned}
&< \frac{1-\alpha}{M(\alpha)}\|f(.,i)\|_\infty + T\frac{\alpha}{M(\alpha)}\|f(.,i)\|_\infty + GL\|i\|_\infty \\
&< M_1 + M_2 + M_3 \\
&< M
\end{aligned}
$$

We have

$$
\begin{aligned}
|\Gamma(i) - \Gamma(i_1)| &= \left| \frac{1-\alpha}{M(\alpha)}(f(t,i(t)) - f(t,i_1(t))) \right. \\
&\quad + \frac{\alpha}{M(\alpha)} \int_0^t (f(\tau,i(\tau)) - f(\tau,i_1(\tau))) \, d\tau \\
&\quad \left. + \frac{G\alpha}{M(\alpha)} \int_0^t (i(\tau) - i_1(\tau)) \, dB(\tau) \right| \\
&< \frac{1-\alpha}{M(\alpha)} |f(t,i(t)) - f(t,i_1(t))| \\
&\quad + \frac{\alpha}{M(\alpha)} \int_0^t |f(\tau,i(\tau)) - f(\tau,i_1(\tau))| \, d\tau \\
&\quad + \frac{G\alpha}{M(\alpha)} \int_0^t |i(\tau) - i_1(\tau)| \, dB(\tau) \\
&< \frac{1-\alpha}{M(\alpha)} \sup_{t\in[0,T]} |f(t,i(t)) - f(t,i_1(t))| \\
&\quad + \frac{\alpha}{M(\alpha)} \int_0^t \sup_{\tau\in[0,t]} |f(\tau,i(\tau)) - f(\tau,i_1(\tau))| \, d\tau \\
&\quad + \frac{G\alpha}{M(\alpha)} \int_0^t \sup_{\tau\in[0,t]} |i(\tau) - i_1(\tau)| \, dB(\tau) \\
&< \frac{1-\alpha}{M(\alpha)} \frac{R}{A} \|i(t) - i_1(t)\|_\infty \\
&\quad + \frac{\alpha}{M(\alpha)} T\frac{R}{A} \|i(t)) - i_1(t)\|_\infty + GL\|i(t) - i_1(t)\|_\infty \\
&< \left(\frac{1-\alpha}{M(\alpha)} \frac{R}{A} + \frac{\alpha}{M(\alpha)} T\frac{R}{A} + GL \right) \|i(t) - i_1(t)\|_\infty \\
&< L_1 \|i(t) - i_1(t)\|_\infty
\end{aligned}
$$

We assume that $B(t)$ is Lipschitzian on $[0,T]$ with constant L, then

$$
\begin{aligned}
|\Gamma(i(t))| &= \left| \frac{1-\alpha}{M(\alpha)} f(t,i(t)) + \frac{\alpha}{M(\alpha)} \int_0^t f(\tau,i(\tau)) d\tau \right. \\
&\quad \left. + \frac{G\alpha}{M(\alpha)} \int_0^t i(\tau) dB(\tau) \right| \\
&< \frac{1-\alpha}{M(\alpha)} |f(t,i(t))|
\end{aligned}
$$

$$+\frac{\alpha}{M(\alpha)}\int_0^t |f(\tau,i(\tau))|\,d\tau$$

$$+\frac{G\alpha}{M(\alpha)}\int_0^t |i(\tau)|\,dB(\tau)$$

$$<\quad \frac{1-\alpha}{M(\alpha)}\sup_{l\in[0,t]}|f(l,i(l))|$$

$$+\frac{\alpha}{M(\alpha)}\int_0^t \sup_{l\in[0,\tau]}|f(l,i(l))|\,d\tau$$

$$+\frac{G\alpha}{M(\alpha)}\int_0^t \sup_{l\in[0,\tau]}|i(l)|\,dB(\tau)$$

This shows that our mapping has a unique solution. We now present the numerical solution of the equation

$$i(t_{n+1}) = i(0)+\frac{1-\alpha}{M(\alpha)}f(t_n,i(t_n))+\frac{\alpha}{M(\alpha)}\int_0^{t_{n+1}} f(\tau,i(\tau))\,d\tau$$

$$+\frac{G\alpha}{M(\alpha)}\int_0^{t_{n+1}} i(\tau)\,dB(\tau)$$

and

$$i(t_n) = i(0)+\frac{1-\alpha}{M(\alpha)}f(t_{n-1},i(t_{n-1}))+\frac{\alpha}{M(\alpha)}\int_0^{t_n} f(\tau,i(\tau))\,d\tau$$

$$+\frac{G\alpha}{M(\alpha)}\int_0^{t_n} i(\tau)\,dB(\tau)$$

Thus, we reach

$$i(t_{n+1}) = i(n)+\frac{1-\alpha}{M(\alpha)}\left(f(t_n,i(t_n))-f(t_{n-1},i(t_{n-1}))\right)$$

$$+\frac{\alpha}{M(\alpha)}\int_{t_n}^{t_{n+1}} f(\tau,i(\tau))\,d\tau+\frac{G\alpha}{M(\alpha)}\int_{t_n}^{t_{n+1}} i(\tau)\,dB(\tau)$$

$$= i(n)+\frac{1-\alpha}{M(\alpha)}\left(f(t_n,i(t_n))-f(t_{n-1},i(t_{n-1}))\right)$$

$$+\frac{\alpha}{M(\alpha)}\sum_{j=2}^{n}\int_{t_j}^{t_{j+1}} f(\tau,i(\tau))\,d\tau+\frac{G\alpha}{M(\alpha)}\sum_{j=2}^{n}\int_{t_j}^{t_{j+1}} i(\tau)\,dB(\tau)$$

$$= i(n)+\frac{1-\alpha}{M(\alpha)}\left(f(t_n,i(t_n))-f(t_{n-1},i(t_{n-1}))\right)$$

$$+\frac{\alpha}{M(\alpha)}\sum_{j=2}^{n}\left[-(\Delta t)\frac{4}{3}f(t_{j-1},i^{j-1})+(\Delta t)\frac{5}{12}f(t_{j-2},i^{j-2})\right.$$

$$+(\Delta t)\frac{23}{12}f(t_j, i^j)\Bigg]$$

$$+\frac{G\alpha}{M(\alpha)}\sum_{j=2}^{n}i(t_j)\left(B(t_{j+1}-B(t_j))\right)$$

Then, we obtain

$$
\begin{aligned}
i(t_{n+1}) &= i(n)+\frac{1-\alpha}{M(\alpha)}\left(\left(-\frac{R}{A}i(t_n)-\frac{1}{AC}\int_0^{t_n}i(y)dy+\frac{1}{A}V(t_n)\right)\right.\\
&\quad \left.-\left(-\frac{R}{A}i(t_{n-1})-\frac{1}{AC}\int_0^{t_{n-1}}i(y)dy+\frac{1}{A}V(t_{n-1})\right)\right)\\
&\quad +\frac{\alpha}{M(\alpha)}\sum_{j=2}^{n}\left[-(\Delta t)\frac{4}{3}\left(-\frac{R}{A}i(t_{j-1})-\frac{1}{AC}\int_0^{t_{j-1}}i(y)dy+\frac{1}{A}V(t_{j-1})\right)\right.\\
&\quad +(\Delta t)\frac{5}{12}\left(-\frac{R}{A}i(t_{j-2})-\frac{1}{AC}\int_0^{t_{j-2}}i(y)dy+\frac{1}{A}V(t_{j-2})\right)\\
&\quad \left.+(\Delta t)\frac{23}{12}\left(-\frac{R}{A}i(t_j)-\frac{1}{AC}\int_0^{t_j}i(y)dy+\frac{1}{A}V(t_j)\right)\right]\\
&\quad +\frac{G\alpha}{M(\alpha)}\sum_{j=2}^{n}i(t_j)\left(B(t_{j+1}-B(t_j))\right)
\end{aligned}
$$

We consider the following differential equation

$$\frac{di}{dt}=-\frac{R}{L}i(t)+\frac{1}{L}V(t)=f(t,i(t)) \qquad (21.110)$$

We convert the above differential equation to the fractional stochastic differential equation as: In Mittag-Leffler case we have

$$
\begin{aligned}
i(t)-i(0) &= \frac{1-\alpha}{AB(\alpha)}f(t,i(t))+\frac{\alpha}{AB(\alpha)\Gamma(\alpha)}\int_0^t f(\tau,i(\tau))(t-\tau)^{\alpha-1}d\tau\\
&\quad +\frac{G\alpha}{\Gamma(\alpha)AB(\alpha)}\int_0^t i(\tau)(t-\tau)^{\alpha-1}dB(\tau)
\end{aligned}
$$

We define a mapping

$$
\begin{aligned}
|\Gamma(i(t))| &= \left|\frac{1-\alpha}{AB(\alpha)}f(t,i(t))+\frac{\alpha}{AB(\alpha)\Gamma(\alpha)}\int_0^t f(\tau,i(\tau))(t-\tau)^{\alpha-1}d\tau\right.\\
&\quad \left.+\frac{G\alpha}{\Gamma(\alpha)AB(\alpha)}\int_0^t i(\tau)(t-\tau)^{\alpha-1}dB(\tau)\right|\\
&< \frac{1-\alpha}{AB(\alpha)}|f(t,i(t))|\\
&\quad +\frac{\alpha}{AB(\alpha)\Gamma(\alpha)}\int_0^t |f(\tau,i(\tau))|(t-\tau)^{\alpha-1}d\tau
\end{aligned}
$$

$$+ \frac{G\alpha}{\Gamma(\alpha)AB(\alpha)} \int_0^t |i(\tau)| (t-\tau)^{\alpha-1} dB(\tau)$$

$$< \frac{1-\alpha}{AB(\alpha)} \sup_{t\in[0,T]} |f(t,i(t))|$$

$$+ \frac{\alpha}{AB(\alpha)\Gamma(\alpha)} \int_0^t \sup_{\tau\in[0,t]} |f(\tau,i(\tau))| (t-\tau)^{\alpha-1} d\tau$$

$$+ \frac{G\alpha}{\Gamma(\alpha)AB(\alpha)} \int_0^t \sup_{\tau\in[0,t]} |i(\tau)| (t-\tau)^{\alpha-1} dB(\tau)$$

$$< \frac{1-\alpha}{AB(\alpha)} \|f(t,i(t))\|_\infty$$

$$+ \frac{T^\alpha \alpha}{AB(\alpha)\Gamma(\alpha+1)} \|f(t,i(t))\|_\infty$$

$$+ \frac{GL\alpha}{\Gamma(\alpha)AB(\alpha)} \|i(t)\|_\infty$$

$$< M_1 + M_2 + M_3 = M$$

We have

$$|\Gamma(i) - \Gamma(i_1)| = \left| \frac{1-\alpha}{AB(\alpha)} (f(t,i(t)) - f(t,i_1(t))) \right.$$

$$+ \frac{\alpha}{AB(\alpha)\Gamma(\alpha)} \int_0^t (f(\tau,i(\tau)) - f(\tau,i_1(\tau))) (t-\tau)^{\alpha-1} d\tau$$

We assume that $B(t)$ is Lipschitzian on $[0,T]$ with constant L, then

$$|\Gamma(i(t))| = \left| \frac{1-\alpha}{AB(\alpha)} f(t,i(t)) + \frac{\alpha}{AB(\alpha)\Gamma(\alpha)} \int_0^t f(\tau,i(\tau))(t-\tau)^{\alpha-1} d\tau \right.$$

$$\left. + \frac{G\alpha}{\Gamma(\alpha)AB(\alpha)} \int_0^t i(\tau)(t-\tau)^{\alpha-1} dB(\tau) \right|$$

$$< \frac{1-\alpha}{AB(\alpha)} |f(t,i(t))|$$

$$+ \frac{\alpha}{AB(\alpha)\Gamma(\alpha)} \int_0^t |f(\tau,i(\tau))| (t-\tau)^{\alpha-1} d\tau$$

$$+ \frac{G\alpha}{\Gamma(\alpha)AB(\alpha)} \int_0^t |i(\tau)| (t-\tau)^{\alpha-1} dB(\tau)$$

$$< \frac{1-\alpha}{AB(\alpha)} \sup_{l\in[0,t]} |f(l,i(l))|$$

$$+ \frac{\alpha}{AB(\alpha)\Gamma(\alpha)} \int_0^t \sup_{l\in[0,\tau]} |f(l,i(l))| (t-\tau)^{\alpha-1} d\tau$$

$$+ \frac{G\alpha}{\Gamma(\alpha)AB(\alpha)} \int_0^t \sup_{l\in[0,\tau]} |i(l)| (t-\tau)^{\alpha-1} dB(\tau)$$

This shows that our mapping has a unique solution. We now present the numerical solution of the equation

$$
\begin{aligned}
i(t_{n+1}) =&\, i(0) + \frac{1-\alpha}{AB(\alpha)} f(t_n, i(t_n)) + \frac{\alpha}{AB(\alpha)\Gamma(\alpha))} \int_0^{t_{n+1}} f(\tau, i(\tau))(t-\tau)^{\alpha-1} d\tau \\
&+ \frac{G\alpha}{\Gamma(\alpha)AB(\alpha)} \int_0^{t_{n+1}} i(\tau)(t-\tau)^{\alpha-1} dB(\tau) \\
=&\, i(0) + \frac{1-\alpha}{AB(\alpha)} f(t_n, i(t_n)) + \frac{\alpha}{AB(\alpha)\Gamma(\alpha))} \sum_{j=0}^{n} \int_{t_j}^{t_{j+1}} f(\tau, i(\tau))(t-\tau)^{\alpha-1} d\tau \\
&+ \frac{G\alpha}{\Gamma(\alpha)AB(\alpha)} \sum_{j=0}^{n} \int_{t_j}^{t_{j+1}} i(\tau)(t-\tau)^{\alpha-1} dB(\tau) \\
=&\, i(0) + \frac{(\Delta t)^\alpha}{AB(\alpha)\Gamma(\alpha+1)} \sum_{j=2}^{n} f(t_{j-2}, i^{j-2}) \left[(n-j+1)^\alpha - (n-j)^\alpha \right] \\
&+ \frac{\alpha(\Delta t)^\alpha}{AB(\alpha)\Gamma(\alpha+2)} \sum_{j=2}^{n} \left[f(t_{j-1}, i^{j-1}) - f(t_{j-2}, i^{j-2}) \right] \\
&\times \left[(n-j+1)^\alpha (n-j+3+2\alpha) - (n-j)^\alpha)(n-j+3+3\alpha) \right] \\
&+ \frac{(\Delta t)^\alpha \alpha}{2AB(\alpha)\Gamma(\alpha+3)} \left[f(t_j, i^j) - 2f(t_{j-1}, i^{j-1}) + f(t_{j-2}, i^{j-2}) \right] \\
&\times \left[(n-j+1)^\alpha \left(2(n-j)^2 + (3\alpha+10)(n-j) + 2\alpha^2 + 9\alpha + 12 \right) \right. \\
&\left. - (n-j)^\alpha \left(2(n-j)^2 + (5\alpha+10)(n-j) + 6\alpha^2 + 18\alpha + 12 \right) \right] \\
&+ \frac{G\alpha(\Delta t)^\alpha}{AB(\alpha)\Gamma(\alpha+1)} \sum_{j=0}^{n} i(t_j) \left[(n-j+1)^\alpha - (n-j)^\alpha \right] (B(t_{j+1}) - B(t_j))
\end{aligned}
$$

Then, we obtain

$$
\begin{aligned}
i(t_{n+1}) =&\, i(0) + \frac{\alpha(\Delta t)^\alpha}{AB(\alpha)\Gamma(\alpha+1)} \sum_{j=2}^{n} \left(-\frac{R}{L} i(t_{j-2}) + \frac{1}{L} V(t_{j-2}) \right) \\
&\times \left[(n-j+1)^\alpha - (n-j)^\alpha \right] \\
&+ \frac{\alpha(\Delta t)^\alpha}{AB(\alpha)\Gamma(\alpha+2)} \sum_{j=2}^{n} \left[\left(-\frac{R}{L} i(t_{j-1}) + \frac{1}{L} V(t_{j-1}) \right) \right. \\
&\left. - \left(-\frac{R}{L} i(t_{j-2}) + \frac{1}{L} V(t_{j-2}) \right) \right] \\
&\times \left[(n-j+1)^\alpha (n-j+3+2\alpha) - (n-j^\alpha)(n-j+3+3\alpha) \right] \\
&+ \frac{\alpha(\Delta t)^\alpha}{2AB(\alpha)\Gamma(\alpha+3)} \left[\left(-\frac{R}{L} i(t_j) + \frac{1}{L} V(t_j) \right) \right. \\
&\left. -2 \left(-\frac{R}{L} i(t_{j-1}) + \frac{1}{L} V(t_{j-1}) \right) + \left(-\frac{R}{L} i(t_{j-2}) + \frac{1}{L} V(t_{j-2}) \right) \right]
\end{aligned}
$$

$$\times \left[(n-j+1)^\alpha \left(2(n-j)^2+(3\alpha+10)(n-j)+2\alpha^2+9\alpha+12\right)\right.$$
$$\left.-(n-j)^\alpha \left(2(n-j)^2+(5\alpha+10)(n-j)+6\alpha^2+18\alpha+12\right)\right]$$
$$+\frac{G\alpha(\Delta t)^\alpha}{AB(\alpha)\Gamma(\alpha+1)}\sum_{j=0}^{n} i(t_j)\left[(n-j+1)^\alpha-(n-j)^\alpha\right]\left(B(t_{j+1})-B(t_j)\right)$$

We consider the following differential equation

$$\frac{di}{dt}=-\frac{R}{A}i(t)-\frac{1}{AC}\int_0^t i(y)dy+\frac{1}{A}V(t)=f(t,i(t)) \qquad (21.111)$$

We convert the above differential equation to the fractional stochastic differential equation as: In Mittag-Leffler case we have

$$i(t)-i(0) = \frac{1-\alpha}{AB(\alpha)}f(t,i(t))+\frac{\alpha}{AB(\alpha)\Gamma(\alpha))}\int_0^t f(\tau,i(\tau))(t-\tau)^{\alpha-1}d\tau$$
$$+\frac{G\alpha}{\Gamma(\alpha)AB(\alpha)}\int_0^t i(\tau)(t-\tau)^{\alpha-1}dB(\tau)$$

We define a mapping

$$|\Gamma(i(t))| = \left|\frac{1-\alpha}{AB(\alpha)}f(t,i(t))+\frac{\alpha}{AB(\alpha)\Gamma(\alpha))}\int_0^t f(\tau,i(\tau))(t-\tau)^{\alpha-1}d\tau\right.$$
$$\left.+\frac{G\alpha}{\Gamma(\alpha)AB(\alpha)}\int_0^t i(\tau)(t-\tau)^{\alpha-1}dB(\tau)\right|$$
$$< \frac{1-\alpha}{AB(\alpha)}|f(t,i(t))|$$
$$+\frac{\alpha}{AB(\alpha)\Gamma(\alpha))}\int_0^t |f(\tau,i(\tau))|(t-\tau)^{\alpha-1}d\tau$$
$$+\frac{G\alpha}{\Gamma(\alpha)AB(\alpha)}\int_0^t |i(\tau)|(t-\tau)^{\alpha-1}dB(\tau)$$
$$< \frac{1-\alpha}{AB(\alpha)}\sup_{t\in[0,T]}|f(t,i(t))|$$
$$+\frac{\alpha}{AB(\alpha)\Gamma(\alpha))}\int_0^t \sup_{\tau\in[0,t]}|f(\tau,i(\tau))|(t-\tau)^{\alpha-1}d\tau$$
$$+\frac{G\alpha}{\Gamma(\alpha)AB(\alpha)}\int_0^t \sup_{\tau\in[0,t]}|i(\tau)|(t-\tau)^{\alpha-1}dB(\tau)$$
$$< \frac{1-\alpha}{AB(\alpha)}\|f(t,i(t))\|_\infty$$
$$+\frac{T^\alpha\alpha}{AB(\alpha)\Gamma(\alpha+1))}\|f(t,i(t))\|_\infty$$

$$+\frac{GL\alpha}{\Gamma(\alpha)AB(\alpha)}\|i(t)\|_\infty$$
$$<\quad M_1+M_2+M_3=M$$

We have

$$|\Gamma(i)-\Gamma(i_1)|=\left|\frac{1-\alpha}{AB(\alpha)}\left(f(t,i(t))-f(t,i_1(t))\right)\right.$$

$$\left|\frac{\alpha}{AB(\alpha)\Gamma(\alpha)}\int_0^t\left(f(\tau,i(\tau))-f(\tau,i_1(\tau))\right)(t-\tau)^{\alpha-1}d\tau\right.$$

$$\left.+\frac{G\alpha}{\Gamma(\alpha)AB(\alpha)}\int_0^t\left(i(\tau)-i_1(\tau)\right)(t-\tau)^{\alpha-1}dB(\tau)\right|$$

$$<\frac{1-\alpha}{AB(\alpha)}|f(t,i(t))-f(t,i_1(t))|$$

$$+\frac{\alpha}{AB(\alpha)\Gamma(\alpha)}\int_0^t|f(\tau,i(\tau))-f(\tau,i_1(\tau))|(t-\tau)^{\alpha-1}d\tau$$

$$+\frac{G\alpha}{\Gamma(\alpha)AB(\alpha)}\int_0^t|i(\tau)-i_1(\tau)|(t-\tau)^{\alpha-1}dB(\tau)$$

$$<\frac{1-\alpha}{AB(\alpha)}\sup_{t\in[0,T]}|f(t,i(t))-f(t,i_1(t))|$$

$$+\frac{\alpha}{AB(\alpha)\Gamma(\alpha)}\int_0^t\sup_{\tau\in[0,t]}|f(\tau,i(\tau))-f(\tau,i_1(\tau))|(t-\tau)^{\alpha-1}d\tau$$

$$+\frac{G\alpha}{\Gamma(\alpha)AB(\alpha)}\int_0^t\sup_{\tau\in[0,t]}|i(\tau)-i_1(\tau)|(t-\tau)^{\alpha-1}dB(\tau)$$

$$<\frac{1-\alpha}{AB(\alpha)}\frac{R}{A}\|i(t)-i_1(t)\|_\infty$$

$$+\frac{\alpha}{AB(\alpha)\Gamma(\alpha)}\frac{R}{A}\frac{T^\alpha}{\alpha}\|i(t)-i_1(t)\|_\infty$$

$$+\frac{GL\alpha}{\Gamma(\alpha)AB(\alpha)}\|i(t)-i_1(t)\|_\infty$$

$$<\left(\frac{1-\alpha}{AB(\alpha)}\frac{R}{A}+\frac{\alpha}{AB(\alpha)\Gamma(\alpha)}\frac{R}{A}\frac{T^\alpha}{\alpha}+\frac{GL\alpha}{\Gamma(\alpha)AB(\alpha)}\right)\|i(t)-i_1(t)\|_\infty$$

$$<L_1\|i(t)-i_1(t)\|_\infty$$

We assume that $B(t)$ is Lipschitzian on $[0,T]$ with constant L, then

$$|\Gamma(i(t))|=\left|\frac{1-\alpha}{AB(\alpha)}f(t,i(t))+\frac{\alpha}{AB(\alpha)\Gamma(\alpha)}\int_0^t f(\tau,i(\tau))(t-\tau)^{\alpha-1}d\tau\right.$$

$$\left.+\frac{G\alpha}{\Gamma(\alpha)AB(\alpha)}\int_0^t i(\tau)(t-\tau)^{\alpha-1}dB(\tau)\right|$$

$$< \frac{1-\alpha}{AB(\alpha)}|f(t,i(t))|$$

$$+\frac{\alpha}{AB(\alpha)\Gamma(\alpha)}\int_0^t |f(\tau,i(\tau))|(t-\tau)^{\alpha-1}d\tau$$

$$+\frac{G\alpha}{\Gamma(\alpha)AB(\alpha)}\int_0^t |i(\tau)|(t-\tau)^{\alpha-1}dB(\tau)$$

$$< \frac{1-\alpha}{AB(\alpha)}\sup_{l\in[0,t]}|f(l,i(l))|$$

$$+\frac{\alpha}{AB(\alpha)\Gamma(\alpha)}\int_0^t \sup_{l\in[0,\tau]}|f(l,i(l))|(t-\tau)^{\alpha-1}d\tau$$

$$+\frac{G\alpha}{\Gamma(\alpha)AB(\alpha)}\int_0^t \sup_{l\in[0,\tau]}|i(l)|(t-\tau)^{\alpha-1}dB(\tau)$$

This shows that our mapping has a unique solution. We now present the numerical solution of the equation

$$i(t_{n+1})=i(0)+\frac{1-\alpha}{AB(\alpha)}f(t_n,i(t_n))+\frac{\alpha}{AB(\alpha)\Gamma(\alpha)}\int_0^{t_{n+1}} f(\tau,i(\tau))(t-\tau)^{\alpha-1}d\tau$$

$$+\frac{G\alpha}{\Gamma(\alpha)AB(\alpha)}\int_0^{t_{n+1}} i(\tau)(t-\tau)^{\alpha-1}dB(\tau)$$

$$=i(0)+\frac{1-\alpha}{AB(\alpha)}f(t_n,i(t_n))+\frac{\alpha}{AB(\alpha)\Gamma(\alpha)}\sum_{j=0}^n\int_{t_j}^{t_{j+1}} f(\tau,i(\tau))(t-\tau)^{\alpha-1}d\tau$$

$$+\frac{G\alpha}{\Gamma(\alpha)AB(\alpha)}\sum_{j=0}^n\int_{t_j}^{t_{j+1}} i(\tau)(t-\tau)^{\alpha-1}dB(\tau)$$

$$=i(0)+\frac{(\Delta t)^\alpha}{AB(\alpha)\Gamma(\alpha+1)}\sum_{j=2}^n f(t_{j-2},i^{j-2})[(n-j+1)^\alpha-(n-j)^\alpha]$$

$$+\frac{\alpha(\Delta t)^\alpha}{AB(\alpha)\Gamma(\alpha+2)}\sum_{j=2}^n [f(t_{j-1},i^{j-1})-f(t_{j-2},i^{j-2})]$$

$$\times [(n-j+1)^\alpha(n-j+3+2\alpha)-(n-j)^\alpha)(n-j+3+3\alpha)]$$

$$+\frac{(\Delta t)^\alpha\alpha}{2AB(\alpha)\Gamma(\alpha+3)}[f(t_j,i^j)-2f(t_{j-1},i^{j-1})+f(t_{j-2},i^{j-2})]$$

$$\times [(n-j+1)^\alpha(2(n-j)^2+(3\alpha+10)(n-j)+2\alpha^2+9\alpha+12)$$

$$-(n-j)^\alpha(2(n-j)^2+(5\alpha+10)(n-j)+6\alpha^2+18\alpha+12)]$$

$$+\frac{G\alpha(\Delta t)^\alpha}{AB(\alpha)\Gamma(\alpha+1)}\sum_{j=0}^n i(t_j)[(n-j+1)^\alpha-(n-j)^\alpha](B(t_{j+1})-B(t_j))$$

Then, we obtain

$$
i(t_{n+1}) = i(0) + \frac{\alpha(\Delta t)^\alpha}{AB(\alpha)\Gamma(\alpha+1)} \sum_{j=2}^{n} \left(-\frac{R}{A}i(t_{j-2}) - \frac{1}{AC}\int_0^{t_{j-2}} i(y)dy + \frac{1}{A}V(t_{j-2}) \right)
$$

$$
\times \left[(n-j+1)^\alpha - (n-j)^\alpha \right]
$$

$$
+ \frac{\alpha(\Delta t)^\alpha}{AB(\alpha)\Gamma(\alpha+2)} \sum_{j=2}^{n} \left[\left(-\frac{R}{A}i(t_{j-1}) - \frac{1}{AC}\int_0^{t_{j-1}} i(y)dy + \frac{1}{A}V(t_{j-1}) \right) \right.
$$

$$
\left. - \left(-\frac{R}{A}i(t_{j-2}) - \frac{1}{AC}\int_0^{t_{j-2}} i(y)dy + \frac{1}{A}V(t_{j-2}) \right) \right]
$$

$$
\times \left[(n-j+1)^\alpha(n-j+3+2\alpha) - (n-j^\alpha)(n-j+3+3\alpha) \right]
$$

$$
+ \frac{\alpha(\Delta t)^\alpha}{2AB(\alpha)\Gamma(\alpha+3)} \left[\left(-\frac{R}{A}i(t_j) - \frac{1}{AC}\int_0^{t_j} i(y)dy + \frac{1}{A}V(t_j) \right) \right.
$$

$$
- 2\left(-\frac{R}{A}i(t_{j-1}) - \frac{1}{AC}\int_0^{t_{j-1}} i(y)dy + \frac{1}{A}V(t_{j-1}) \right)
$$

$$
\left. + \left(-\frac{R}{A}i(t_{j-2}) - \frac{1}{AC}\int_0^{t_{j-2}} i(y)dy + \frac{1}{A}V(t_{j-2}) \right) \right]
$$

$$
\times \left[(n-j+1)^\alpha \left(2(n-j)^2 + (3\alpha+10)(n-j) + 2\alpha^2 + 9\alpha + 12 \right) \right.
$$

$$
\left. - (n-j)^\alpha \left(2(n-j)^2 + (5\alpha+10)(n-j) + 6\alpha^2 + 18\alpha + 12 \right) \right]
$$

$$
+ \frac{G\alpha(\Delta t)^\alpha}{AB(\alpha)\Gamma(\alpha+1)} \sum_{j=0}^{n} i(t_j)\left[(n-j+1)^\alpha - (n-j)^\alpha \right] \left(B(t_{j+1}) - B(t_j) \right)
$$

We consider the following differential equation

$$
\frac{di}{dt} = -\frac{R_1}{A}i(t) + \frac{R_1}{AR_2}V(t) + C\frac{d^2V(t)}{dt^2} + \left(\frac{CR_1}{A} + \frac{1}{R_2} \right)\frac{dV(t)}{dt} = f(t,i(t)) \quad (21.112)
$$

We convert the above differential equation to the fractional stochastic differential equation as: In Mittag-Leffler case we have

$$
i(t) - i(0) = \frac{1-\alpha}{AB(\alpha)}f(t,i(t)) + \frac{\alpha}{AB(\alpha)\Gamma(\alpha))}\int_0^t f(\tau,i(\tau))(t-\tau)^{\alpha-1}d\tau
$$

$$
+ \frac{G\alpha}{\Gamma(\alpha)AB(\alpha)}\int_0^t i(\tau)(t-\tau)^{\alpha-1}dB(\tau)
$$

We define a mapping

$$
\begin{aligned}
|\Gamma(i(t))| &= \left| \frac{1-\alpha}{AB(\alpha)} f(t,i(t)) + \frac{\alpha}{AB(\alpha)\Gamma(\alpha)} \int_0^t f(\tau,i(\tau))(t-\tau)^{\alpha-1} d\tau \right. \\
&\quad \left. + \frac{G\alpha}{\Gamma(\alpha)AB(\alpha)} \int_0^t i(\tau)(t-\tau)^{\alpha-1} dB(\tau) \right| \\
&< \frac{1-\alpha}{AB(\alpha)} |f(t,i(t))| \\
&\quad + \frac{\alpha}{AB(\alpha)\Gamma(\alpha)} \int_0^t |f(\tau,i(\tau))| (t-\tau)^{\alpha-1} d\tau \\
&\quad + \frac{G\alpha}{\Gamma(\alpha)AB(\alpha)} \int_0^t |i(\tau)| (t-\tau)^{\alpha-1} dB(\tau) \\
&< \frac{1-\alpha}{AB(\alpha)} \sup_{t\in[0,T]} |f(t,i(t))| \\
&\quad + \frac{\alpha}{AB(\alpha)\Gamma(\alpha)} \int_0^t \sup_{\tau\in[0,t]} |f(\tau,i(\tau))| (t-\tau)^{\alpha-1} d\tau \\
&\quad + \frac{G\alpha}{\Gamma(\alpha)AB(\alpha)} \int_0^t \sup_{\tau\in[0,t]} |i(\tau)| (t-\tau)^{\alpha-1} dB(\tau) \\
&< \frac{1-\alpha}{AB(\alpha)} \|f(t,i(t))\|_\infty \\
&\quad + \frac{T^\alpha \alpha}{AB(\alpha)\Gamma(\alpha+1)} \|f(t,i(t))\|_\infty \\
&\quad + \frac{GL\alpha}{\Gamma(\alpha)AB(\alpha)} \|i(t)\|_\infty \\
&< M_1 + M_2 + M_3 = M
\end{aligned}
$$

We have

$$
\begin{aligned}
|\Gamma(i) - \Gamma(i_1)| &= \left| \frac{1-\alpha}{AB(\alpha)} (f(t,i(t)) - f(t,i_1(t))) \right. \\
&\quad + \frac{\alpha}{AB(\alpha)\Gamma(\alpha)} \int_0^t (f(\tau,i(\tau)) - f(\tau,i_1(\tau)))(t-\tau)^{\alpha-1} d\tau \\
&\quad \left. + \frac{G\alpha}{\Gamma(\alpha)AB(\alpha)} \int_0^t (i(\tau) - i_1(\tau))(t-\tau)^{\alpha-1} dB(\tau) \right| \\
&< \frac{1-\alpha}{AB(\alpha)} |f(t,i(t)) - f(t,i_1(t))| \\
&\quad + \frac{\alpha}{AB(\alpha)\Gamma(\alpha)} \int_0^t |f(\tau,i(\tau)) - f(\tau,i_1(\tau))| (t-\tau)^{\alpha-1} d\tau \\
&\quad + \frac{G\alpha}{\Gamma(\alpha)AB(\alpha)} \int_0^t |i(\tau) - i_1(\tau)| (t-\tau)^{\alpha-1} dB(\tau)
\end{aligned}
$$

$$< \frac{1-\alpha}{AB(\alpha)} \sup_{t \in [0,T]} |f(t, i(t)) - f(t, i_1(t))|$$

$$+ \frac{\alpha}{AB(\alpha)\Gamma(\alpha)} \int_0^t \sup_{\tau \in [0,t]} |f(\tau, i(\tau)) - f(\tau, i_1(\tau))| (t - \tau)^{\alpha - 1} d\tau$$

$$+ \frac{G\alpha}{\Gamma(\alpha)AB(\alpha)} \int_0^t \sup_{\tau \in [0,t]} |i(\tau) - i_1(\tau)| (t - \tau)^{\alpha - 1} dB(\tau)$$

$$< \frac{1-\alpha}{AB(\alpha)} \frac{R_1}{A} \|i(t) - i_1(t)\|_\infty$$

$$+ \frac{\alpha}{AB(\alpha)\Gamma(\alpha)} \frac{R_1}{A} \frac{T^\alpha}{\alpha} \|i(t) - i_1(t)\|_\infty$$

$$+ \frac{GL\alpha}{\Gamma(\alpha)AB(\alpha)} \|i(t) - i_1(t)\|_\infty$$

$$< \left(\frac{1-\alpha}{AB(\alpha)} \frac{R_1}{A} + \frac{\alpha}{AB(\alpha)\Gamma(\alpha)} \frac{R_1}{A} \frac{T^\alpha}{\alpha} + \frac{GL\alpha}{\Gamma(\alpha)AB(\alpha)} \right) \|i(t) - i_1(t)\|_\infty$$

$$< L_1 \|i(t) - i_1(t)\|_\infty$$

We assume that $B(t)$ is Lipschitzian on $[0,T]$ with constant L, then

$$|\Gamma(i(t))| = \left| \frac{1-\alpha}{AB(\alpha)} f(t, i(t)) + \frac{\alpha}{AB(\alpha)\Gamma(\alpha)} \int_0^t f(\tau, i(\tau))(t - \tau)^{\alpha - 1} d\tau \right.$$

$$\left. + \frac{G\alpha}{\Gamma(\alpha)AB(\alpha)} \int_0^t i(\tau)(t - \tau)^{\alpha - 1} dB(\tau) \right|$$

$$< \frac{1-\alpha}{AB(\alpha)} |f(t, i(t))|$$

$$+ \frac{\alpha}{AB(\alpha)\Gamma(\alpha)} \int_0^t |f(\tau, i(\tau))| (t - \tau)^{\alpha - 1} d\tau$$

$$+ \frac{G\alpha}{\Gamma(\alpha)AB(\alpha)} \int_0^t |i(\tau)| (t - \tau)^{\alpha - 1} dB(\tau)$$

$$< \frac{1-\alpha}{AB(\alpha)} \sup_{l \in [0,t]} |f(l, i(l))|$$

$$+ \frac{\alpha}{AB(\alpha)\Gamma(\alpha)} \int_0^t \sup_{l \in [0,\tau]} |f(l, i(l))| (t - \tau)^{\alpha - 1} d\tau$$

$$+ \frac{G\alpha}{\Gamma(\alpha)AB(\alpha)} \int_0^t \sup_{l \in [0,\tau]} |i(l)| (t - \tau)^{\alpha - 1} dB(\tau)$$

This shows that our mapping has a unique solution. We now present the numerical solution of the equation

$$i(t_{n+1}) = i(0) + \frac{1-\alpha}{AB(\alpha)}f(t_n, i(t_n)) + \frac{\alpha}{AB(\alpha)\Gamma(\alpha)}\int_0^{t_{n+1}} f(\tau, i(\tau))(t-\tau)^{\alpha-1}d\tau$$

$$+ \frac{G\alpha}{\Gamma(\alpha)AB(\alpha)}\int_0^{t_{n+1}} i(\tau)(t-\tau)^{\alpha-1}dB(\tau)$$

$$= i(0) + \frac{1-\alpha}{AB(\alpha)}f(t_n, i(t_n)) + \frac{\alpha}{AB(\alpha)\Gamma(\alpha)}\sum_{j=0}^n\int_{t_j}^{t_{j+1}} f(\tau, i(\tau))(t-\tau)^{\alpha-1}d\tau$$

$$+ \frac{G\alpha}{\Gamma(\alpha)AB(\alpha)}\sum_{j=0}^n\int_{t_j}^{t_{j+1}} i(\tau)(t-\tau)^{\alpha-1}dB(\tau)$$

$$= i(0) + \frac{(\Delta t)^\alpha}{AB(\alpha)\Gamma(\alpha+1)}\sum_{j=2}^n f(t_{j-2}, i^{j-2})[(n-j+1)^\alpha - (n-j)^\alpha]$$

$$+ \frac{\alpha(\Delta t)^\alpha}{AB(\alpha)\Gamma(\alpha+2)}\sum_{j=2}^n [f(t_{j-1}, i^{j-1}) - f(t_{j-2}, i^{j-2})]$$

$$\times [(n-j+1)^\alpha(n-j+3+2\alpha) - (n-j)^\alpha(n-j+3+3\alpha)]$$

$$+ \frac{(\Delta t)^\alpha\alpha}{2AB(\alpha)\Gamma(\alpha+3)}[f(t_j, i^j) - 2f(t_{j-1}, i^{j-1}) + f(t_{j-2}, i^{j-2})]$$

$$\times [(n-j+1)^\alpha(2(n-j)^2 + (3\alpha+10)(n-j) + 2\alpha^2 + 9\alpha + 12)$$

$$-(n-j)^\alpha(2(n-j)^2 + (5\alpha+10)(n-j) + 6\alpha^2 + 18\alpha + 12)]$$

$$+ \frac{G\alpha(\Delta t)^\alpha}{AB(\alpha)\Gamma(\alpha+1)}\sum_{j=0}^n i(t_j)[(n-j+1)^\alpha - (n-j)^\alpha](B(t_{j+1}) - B(t_j))$$

Then, we obtain

$$i(t_{n+1}) = i(0) + \frac{\alpha(\Delta t)^\alpha}{AB(\alpha)\Gamma(\alpha+1)}\sum_{j=2}^n\left(-\frac{R_1}{A}i(t_{j-2}) + \frac{R_1}{AR_2}V(t_{j-2}) + C\frac{d^2V(t_{j-2})}{dt^2}\right.$$

$$\left.+ \left(\frac{CR_1}{A} + \frac{1}{R_2}\right)\frac{dV(t_{j-2})}{dt}\right)[(n-j+1)^\alpha - (n-j)^\alpha]$$

$$+ \frac{\alpha(\Delta t)^\alpha}{AB(\alpha)\Gamma(\alpha+2)}\sum_{j=2}^n\left[\left(-\frac{R_1}{A}i(t_{j-1}) + \frac{R_1}{AR_2}V(t_{j-1}) + C\frac{d^2V(t_{j-1})}{dt^2}\right.\right.$$

$$\left.+ \left(\frac{CR_1}{A} + \frac{1}{R_2}\right)\frac{dV(t_{j-1})}{dt}\right)$$

$$\left.- \left(-\frac{R_1}{A}i(t_{j-2}) + \frac{R_1}{AR_2}V(t_{j-2}) + C\frac{d^2V(t_{j-2})}{dt^2} + \left(\frac{CR_1}{A} + \frac{1}{R_2}\right)\frac{dV(t_{j-2})}{dt}\right)\right]$$

$$\times [(n-j+1)^\alpha(n-j+3+2\alpha) - (n-j^\alpha)(n-j+3+3\alpha)]$$

$$+ \frac{\alpha(\Delta t)^\alpha}{2AB(\alpha)\Gamma(\alpha+3)}\left[\left(-\frac{R_1}{A}i(t_j) + \frac{R_1}{AR_2}V(t_j) + C\frac{d^2V(t_j)}{dt^2}\right.\right.$$

$$+ \left(\frac{CR_1}{A} + \frac{1}{R_2} \right) \frac{dV(t_j)}{dt} \right)$$

$$-2 \left(-\frac{R_1}{A} i(t_{j-1}) + \frac{R_1}{AR_2} V(t_{j-1}) + C \frac{d^2V(t_{j-1})}{dt^2} + \left(\frac{CR_1}{A} + \frac{1}{R_2} \right) \frac{dV(t_{j-1})}{dt} \right)$$

$$+ \left(-\frac{R_1}{A} i(t_{j-2}) + \frac{R_1}{AR_2} V(t_{j-2}) + C \frac{d^2V(t_{j-2})}{dt^2} + \left(\frac{CR_1}{A} + \frac{1}{R_2} \right) \frac{dV(t_{j-2})}{dt} \right) \right]$$

$$\times \left[(n-j+1)^\alpha \left(2(n-j)^2 + (3\alpha+10)(n-j) + 2\alpha^2 + 9\alpha + 12 \right) \right.$$

$$\left. - (n-j)^\alpha \left(2(n-j)^2 + (5\alpha+10)(n-j) + 6\alpha^2 + 18\alpha + 12 \right) \right]$$

$$+ \frac{G\alpha(\Delta t)^\alpha}{AB(\alpha)\Gamma(\alpha+1)} \sum_{j=0}^{n} i(t_j) \left[(n-j+1)^\alpha - (n-j)^\alpha \right] \left(B(t_{j+1}) - B(t_j) \right)$$

We consider the following differential equation

$$\frac{di}{dt} = -RC \frac{d^2 i(t)}{dt^2} + \frac{1}{L} V(t) + \frac{RC}{L} \frac{dV(t)}{dt} = f(t, i(t)) \qquad (21.113)$$

We convert the above differential equation to the fractional stochastic differential equation as: In Mittag-Leffler case we have

$$i(t) - i(0) = \frac{1-\alpha}{AB(\alpha)} f(t, i(t)) + \frac{\alpha}{AB(\alpha)\Gamma(\alpha)} \int_0^t f(\tau, i(\tau))(t-\tau)^{\alpha-1} d\tau$$

$$+ \frac{G\alpha}{\Gamma(\alpha)AB(\alpha)} \int_0^t i(\tau)(t-\tau)^{\alpha-1} dB(\tau)$$

We define a mapping

$$|\Gamma(i(t))| = \left| \frac{1-\alpha}{AB(\alpha)} f(t, i(t)) + \frac{\alpha}{AB(\alpha)\Gamma(\alpha)} \int_0^t f(\tau, i(\tau))(t-\tau)^{\alpha-1} d\tau \right.$$

$$\left. + \frac{G\alpha}{\Gamma(\alpha)AB(\alpha)} \int_0^t i(\tau)(t-\tau)^{\alpha-1} dB(\tau) \right|$$

$$< \frac{1-\alpha}{AB(\alpha)} |f(t, i(t))|$$

$$+ \frac{\alpha}{AB(\alpha)\Gamma(\alpha)} \int_0^t |f(\tau, i(\tau))| (t-\tau)^{\alpha-1} d\tau$$

$$+ \frac{G\alpha}{\Gamma(\alpha)AB(\alpha)} \int_0^t |i(\tau)| (t-\tau)^{\alpha-1} dB(\tau)$$

$$< \frac{1-\alpha}{AB(\alpha)} \sup_{t \in [0,T]} |f(t, i(t))|$$

$$+ \frac{\alpha}{AB(\alpha)\Gamma(\alpha)} \int_0^t \sup_{\tau \in [0,t]} |f(\tau, i(\tau))| (t-\tau)^{\alpha-1} d\tau$$

$$+ \frac{G\alpha}{\Gamma(\alpha)AB(\alpha)} \int_0^t \sup_{\tau \in [0,t]} |i(\tau)| (t-\tau)^{\alpha-1} dB(\tau)$$

$$< \frac{1-\alpha}{AB(\alpha)} \|f(t,i(t))\|_\infty$$

$$+ \frac{T^\alpha \alpha}{AB(\alpha)\Gamma(\alpha+1))} \|f(t,i(t))\|_\infty$$

$$+ \frac{GL\alpha}{\Gamma(\alpha)AB(\alpha)} \|i(t)\|_\infty$$

$$< M_1 + M_2 + M_3 = M$$

We have

$$|\Gamma(i) - \Gamma(i_1)| = \left| \frac{1-\alpha}{AB(\alpha)} (f(t,i(t)) - f(t,i_1(t))) \right.$$

$$+ \frac{\alpha}{AB(\alpha)\Gamma(\alpha))} \int_0^t (f(\tau,i(\tau)) - f(\tau,i_1(\tau)))(t-\tau)^{\alpha-1} d\tau$$

$$+ \left. \frac{G\alpha}{\Gamma(\alpha)AB(\alpha)} \int_0^t (i(\tau) - i_1(\tau))(t-\tau)^{\alpha-1} dB(\tau) \right|$$

$$< \frac{1-\alpha}{AB(\alpha)} |f(t,i(t)) - f(t,i_1(t))|$$

$$+ \frac{\alpha}{AB(\alpha)\Gamma(\alpha))} \int_0^t |f(\tau,i(\tau)) - f(\tau,i_1(\tau))|(t-\tau)^{\alpha-1} d\tau$$

$$+ \frac{G\alpha}{\Gamma(\alpha)AB(\alpha)} \int_0^t |i(\tau) - i_1(\tau)|(t-\tau)^{\alpha-1} dB(\tau)$$

$$< \frac{1-\alpha}{AB(\alpha)} \sup_{t\in[0,T]} |f(t,i(t)) - f(t,i_1(t))|$$

$$+ \frac{\alpha}{AB(\alpha)\Gamma(\alpha))} \int_0^t \sup_{\tau\in[0,t]} |f(\tau,i(\tau)) - f(\tau,i_1(\tau))|(t-\tau)^{\alpha-1} d\tau$$

$$+ \frac{G\alpha}{\Gamma(\alpha)AB(\alpha)} \int_0^t \sup_{\tau\in[0,t]} |i(\tau) - i_1(\tau)|(t-\tau)^{\alpha-1} dB(\tau)$$

$$< \frac{1-\alpha}{AB(\alpha)} RC \|i(t) - i_1(t)\|_\infty$$

$$+ \frac{\alpha}{AB(\alpha)\Gamma(\alpha))} RC \frac{T^\alpha}{\alpha} \|i(t) - i_1(t)\|_\infty$$

$$+ \frac{GL\alpha}{\Gamma(\alpha)AB(\alpha)} \|i(t) - i_1(t)\|_\infty$$

$$< \left(\frac{1-\alpha}{AB(\alpha)} RC + \frac{\alpha}{AB(\alpha)\Gamma(\alpha))} RC \frac{T^\alpha}{\alpha} + \frac{GL\alpha}{\Gamma(\alpha)AB(\alpha)} \right) \|i(t) - i_1(t)\|_\infty$$

$$< L_1 \|i(t) - i_1(t)\|_\infty$$

We assume that $B(t)$ is Lipschitzian on $[0,T]$ with constant L, then

$$
\begin{aligned}
|\Gamma(i(t))| &= \left| \frac{1-\alpha}{AB(\alpha)} f(t,i(t)) + \frac{\alpha}{AB(\alpha)\Gamma(\alpha)} \int_0^t f(\tau,i(\tau))(t-\tau)^{\alpha-1}d\tau \right. \\
&\left. + \frac{G\alpha}{\Gamma(\alpha)AB(\alpha)} \int_0^t i(\tau)(t-\tau)^{\alpha-1}dB(\tau) \right| \\
&< \frac{1-\alpha}{AB(\alpha)} |f(t,i(t))| \\
&+ \frac{\alpha}{AB(\alpha)\Gamma(\alpha)} \int_0^t |f(\tau,i(\tau))| (t-\tau)^{\alpha-1}d\tau \\
&+ \frac{G\alpha}{\Gamma(\alpha)AB(\alpha)} \int_0^t |i(\tau)| (t-\tau)^{\alpha-1}dB(\tau) \\
&< \frac{1-\alpha}{AB(\alpha)} \sup_{l\in[0,t]} |f(l,i(l))| \\
&+ \frac{\alpha}{AB(\alpha)\Gamma(\alpha)} \int_0^t \sup_{l\in[0,\tau]} |f(l,i(l))| (t-\tau)^{\alpha-1}d\tau \\
&+ \frac{G\alpha}{\Gamma(\alpha)AB(\alpha)} \int_0^t \sup_{l\in[0,\tau]} |i(l)| (t-\tau)^{\alpha-1}dB(\tau)
\end{aligned}
$$

This shows that our mapping has a unique solution. We now present the numerical solution of the equation

$$
\begin{aligned}
i(t_{n+1}) =&\, i(0) + \frac{1-\alpha}{AB(\alpha)} f(t_n,i(t_n)) + \frac{\alpha}{AB(\alpha)\Gamma(\alpha)} \int_0^{t_{n+1}} f(\tau,i(\tau))(t-\tau)^{\alpha-1}d\tau \\
&+ \frac{G\alpha}{\Gamma(\alpha)AB(\alpha)} \int_0^{t_{n+1}} i(\tau)(t-\tau)^{\alpha-1}dB(\tau) \\
=&\, i(0) + \frac{1-\alpha}{AB(\alpha)} f(t_n,i(t_n)) + \frac{\alpha}{AB(\alpha)\Gamma(\alpha)} \sum_{j=0}^n \int_{t_j}^{t_{j+1}} f(\tau,i(\tau))(t-\tau)^{\alpha-1}d\tau \\
&+ \frac{G\alpha}{\Gamma(\alpha)AB(\alpha)} \sum_{j=0}^n \int_{t_j}^{t_{j+1}} i(\tau)(t-\tau)^{\alpha-1}dB(\tau) \\
=&\, i(0) + \frac{(\Delta t)^\alpha}{AB(\alpha)\Gamma(\alpha+1)} \sum_{j=2}^n f(t_{j-2},i^{j-2})[(n-j+1)^\alpha - (n-j)^\alpha] \\
&+ \frac{\alpha(\Delta t)^\alpha}{AB(\alpha)\Gamma(\alpha+2)} \sum_{j=2}^n [f(t_{j-1},i^{j-1}) - f(t_{j-2},i^{j-2})] \\
&\times [(n-j+1)^\alpha(n-j+3+2\alpha) - (n-j)^\alpha(n-j+3+3\alpha)] \\
&+ \frac{(\Delta t)^\alpha \alpha}{2AB(\alpha)\Gamma(\alpha+3)} [f(t_j,i^j) - 2f(t_{j-1},i^{j-1}) + f(t_{j-2},i^{j-2})] \\
&\times [(n-j+1)^\alpha (2(n-j)^2 + (3\alpha+10)(n-j) + 2\alpha^2 + 9\alpha + 12)
\end{aligned}
$$

$$-(n-j)^\alpha \left(2(n-j)^2 + (5\alpha+10)(n-j) + 6\alpha^2 + 18\alpha + 12\right)\big]$$

$$+\frac{G\alpha(\Delta t)^\alpha}{AB(\alpha)\Gamma(\alpha+1)} \sum_{j=0}^n i(t_j)\left[(n-j+1)^\alpha - (n-j)^\alpha\right]\left(B(t_{j+1}) - B(t_j)\right)$$

Then, we obtain

$$i(t_{n+1}) = i(0) + \frac{\alpha(\Delta t)^\alpha}{AB(\alpha)\Gamma(\alpha+1)} \sum_{j=2}^n \left(-RC\frac{d^2i(t_{j-2})}{dt^2} + \frac{1}{L}V(t_{j-2}) + \frac{RC}{L}\frac{dV(t_{j-2})}{dt}\right)$$

$$\times \left[(n-j+1)^\alpha - (n-j)^\alpha\right]$$

$$+\frac{\alpha(\Delta t)^\alpha}{AB(\alpha)\Gamma(\alpha+2)} \sum_{j=2}^n \left[\left(-RC\frac{d^2i(t_{j-1})}{dt^2} + \frac{1}{L}V(t_{j-1}) + \frac{RC}{L}\frac{dV(t_{j-1})}{dt}\right)\right.$$

$$\left.-\left(-RC\frac{d^2i(t_{j-2})}{dt^2} + \frac{1}{L}V(t_{j-2}) + \frac{RC}{L}\frac{dV(t_{j-2})}{dt}\right)\right]$$

$$\times \left[(n-j+1)^\alpha(n-j+3+2\alpha) - (n-j^\alpha)(n-j+3+3\alpha)\right]$$

$$+\frac{\alpha(\Delta t)^\alpha}{2AB(\alpha)\Gamma(\alpha+3)} \left[\left(-RC\frac{d^2i(t_j)}{dt^2} + \frac{1}{L}V(t_j) + \frac{RC}{L}\frac{dV(t_j)}{dt}\right)\right.$$

$$-2\left(-RC\frac{d^2i(t_{j-1})}{dt^2} + \frac{1}{L}V(t_{j-1}) + \frac{RC}{L}\frac{dV(t_{j-1})}{dt}\right)$$

$$\left.+\left(-RC\frac{d^2i(t_{j-2})}{dt^2} + \frac{1}{L}V(t_{j-2}) + \frac{RC}{L}\frac{dV(t_{j-2})}{dt}\right)\right]$$

$$\times \left[(n-j+1)^\alpha \left(2(n-j)^2 + (3\alpha+10)(n-j) + 2\alpha^2 + 9\alpha + 12\right)\right.$$

$$\left.-(n-j)^\alpha \left(2(n-j)^2 + (5\alpha+10)(n-j) + 6\alpha^2 + 18\alpha + 12\right)\right]$$

$$+\frac{G\alpha(\Delta t)^\alpha}{AB(\alpha)\Gamma(\alpha+1)} \sum_{j=0}^n i(t_j)\left[(n-j+1)^\alpha - (n-j)^\alpha\right]\left(B(t_{j+1}) - B(t_j)\right)$$

We consider the following differential equation

$$\frac{dV}{dt} = -\frac{1}{C}i(t) - \frac{1}{R_1C}V(t) + A\frac{d^2i(t)}{dt^2} + \left(R_2 + \frac{L}{R_1C}\right)\frac{di(t)}{dt} + \frac{R_1+R_2}{R_1C}i(t) = f(t,V(t))$$

(21.114)

We convert the above differential equation to the fractional stochastic differential equation as: In Mittag-Leffler case we have

$$V(t) - V(0) = \frac{1-\alpha}{AB(\alpha)}f(t,V(t)) + \frac{\alpha}{AB(\alpha)\Gamma(\alpha)}\int_0^t f(\tau,V(\tau))(t-\tau)^{\alpha-1}d\tau$$

$$+\frac{G\alpha}{\Gamma(\alpha)AB(\alpha)}\int_0^t V(\tau)(t-\tau)^{\alpha-1}dB(\tau)$$

We define a mapping

$$
\begin{aligned}
|\Gamma(V(t))| \quad = \quad & \left| \frac{1-\alpha}{AB(\alpha)} f(t,V(t)) + \frac{\alpha}{AB(\alpha)\Gamma(\alpha)} \int_0^t f(\tau,V(\tau))(t-\tau)^{\alpha-1} d\tau \right. \\
& \left. + \frac{G\alpha}{\Gamma(\alpha)AB(\alpha)} \int_0^t V(\tau)(t-\tau)^{\alpha-1} dB(\tau) \right| \\
< \quad & \frac{1-\alpha}{AB(\alpha)} |f(t,V(t))| \\
& + \frac{\alpha}{AB(\alpha)\Gamma(\alpha)} \int_0^t |f(\tau,V(\tau))| (t-\tau)^{\alpha-1} d\tau \\
& + \frac{G\alpha}{\Gamma(\alpha)AB(\alpha)} \int_0^t |V(\tau)| (t-\tau)^{\alpha-1} dB(\tau) \\
< \quad & \frac{1-\alpha}{AB(\alpha)} \sup_{t\in[0,T]} |f(t,V(t))| \\
& + \frac{\alpha}{AB(\alpha)\Gamma(\alpha)} \int_0^t \sup_{\tau\in[0,t]} |f(\tau,V(\tau))| (t-\tau)^{\alpha-1} d\tau \\
& + \frac{G\alpha}{\Gamma(\alpha)AB(\alpha)} \int_0^t \sup_{\tau\in[0,t]} |V(\tau)| (t-\tau)^{\alpha-1} dB(\tau) \\
< \quad & \frac{1-\alpha}{AB(\alpha)} \|f(t,V(t))\|_\infty \\
& + \frac{T^\alpha \alpha}{AB(\alpha)\Gamma(\alpha+1))} \|f(t,V(t))\|_\infty \\
& + \frac{GL\alpha}{\Gamma(\alpha)AB(\alpha)} \|V(t)\|_\infty \\
< \quad & M_1 + M_2 + M_3 = M
\end{aligned}
$$

We have

$$
\begin{aligned}
& |\Gamma(V) - \Gamma(V_1)| \\
= \quad & \left| \frac{1-\alpha}{AB(\alpha)} (f(t,V(t)) - f(t,V_1(t))) \right. \\
& + \frac{\alpha}{AB(\alpha)\Gamma(\alpha)} \int_0^t (f(\tau,V(\tau)) - f(\tau,V_1(\tau))) (t-\tau)^{\alpha-1} d\tau \\
& \left. + \frac{G\alpha}{\Gamma(\alpha)AB(\alpha)} \int_0^t (V(\tau) - V_1(\tau)) (t-\tau)^{\alpha-1} dB(\tau) \right| \\
< \quad & \frac{1-\alpha}{AB(\alpha)} |f(t,V(t)) - f(t,V_1(t))| \\
& + \frac{\alpha}{AB(\alpha)\Gamma(\alpha)} \int_0^t |f(\tau,V(\tau)) - f(\tau,V_1(\tau))| (t-\tau)^{\alpha-1} d\tau \\
& + \frac{G\alpha}{\Gamma(\alpha)AB(\alpha)} \int_0^t |V(\tau) - V_1(\tau)| (t-\tau)^{\alpha-1} dB(\tau)
\end{aligned}
$$

$$< \frac{1-\alpha}{AB(\alpha)} \sup_{t\in[0,T]} |f(t,V(t)) - f(t,V_1(t))|$$

$$+ \frac{\alpha}{AB(\alpha)\Gamma(\alpha)} \int_0^t \sup_{\tau\in[0,t]} |f(\tau,V(\tau)) - f(\tau,V_1(\tau))| (t-\tau)^{\alpha-1} d\tau$$

$$+ \frac{G\alpha}{\Gamma(\alpha)AB(\alpha)} \int_0^t \sup_{\tau\in[0,t]} |V(\tau) - V_1(\tau)| (t-\tau)^{\alpha-1} dB(\tau)$$

$$< \frac{1-\alpha}{AB(\alpha)} \frac{1}{R_1 C} \|V(t) - V_1(t)\|_\infty$$

$$+ \frac{\alpha}{AB(\alpha)\Gamma(\alpha)} \frac{1}{R_1 C} \frac{T^\alpha}{\alpha} \|V(t) - V_1(t)\|_\infty$$

$$+ \frac{GL\alpha}{\Gamma(\alpha)AB(\alpha)} \|V(t) - V_1(t)\|_\infty$$

$$< \left(\frac{1-\alpha}{AB(\alpha)} \frac{1}{R_1 C} + \frac{\alpha}{AB(\alpha)\Gamma(\alpha)} \frac{1}{R_1 C} \frac{T^\alpha}{\alpha} + \frac{GL\alpha}{\Gamma(\alpha)AB(\alpha)} \right) \|V(t) - V_1(t)\|_\infty$$

$$< L_1 \|V(t) - V_1(t)\|_\infty$$

We assume that $B(t)$ is Lipschitzian on $[0,T]$ with constant L, then

$$
\begin{aligned}
|\Gamma(V(t))| \quad = \quad & \left| \frac{1-\alpha}{AB(\alpha)} f(t,V(t)) + \frac{\alpha}{AB(\alpha)\Gamma(\alpha)} \int_0^t f(\tau,V(\tau))(t-\tau)^{\alpha-1} d\tau \right. \\
& \left. + \frac{G\alpha}{\Gamma(\alpha)AB(\alpha)} \int_0^t V(\tau)(t-\tau)^{\alpha-1} dB(\tau) \right| \\
< \quad & \frac{1-\alpha}{AB(\alpha)} |f(t,V(t))| \\
& + \frac{\alpha}{AB(\alpha)\Gamma(\alpha)} \int_0^t |f(\tau,V(\tau))| (t-\tau)^{\alpha-1} d\tau \\
& + \frac{G\alpha}{\Gamma(\alpha)AB(\alpha)} \int_0^t |V(\tau)| (t-\tau)^{\alpha-1} dB(\tau) \\
< \quad & \frac{1-\alpha}{AB(\alpha)} \sup_{l\in[0,t]} |f(l,V(l))| \\
& + \frac{\alpha}{AB(\alpha)\Gamma(\alpha)} \int_0^t \sup_{l\in[0,\tau]} |f(l,V(l))| (t-\tau)^{\alpha-1} d\tau \\
& + \frac{G\alpha}{\Gamma(\alpha)AB(\alpha)} \int_0^t \sup_{l\in[0,\tau]} |V(l)| (t-\tau)^{\alpha-1} dB(\tau)
\end{aligned}
$$

This shows that our mapping has a unique solution. We now present the numerical solution of the equation

$$
\begin{aligned}
& V(t_{n+1}) \\
&= V(0) + \frac{1-\alpha}{AB(\alpha)} f(t_n, V(t_n)) + \frac{\alpha}{AB(\alpha)\Gamma(\alpha)} \int_0^{t_{n+1}} f(\tau, V(\tau))(t-\tau)^{\alpha-1} d\tau \\
&\quad + \frac{G\alpha}{\Gamma(\alpha)AB(\alpha)} \int_0^{t_{n+1}} V(\tau)(t-\tau)^{\alpha-1} dB(\tau) \\
&= V(0) + \frac{1-\alpha}{AB(\alpha)} f(t_n, V(t_n)) + \frac{\alpha}{AB(\alpha)\Gamma(\alpha)} \sum_{j=0}^{n} \int_{t_j}^{t_{j+1}} f(\tau, V(\tau))(t-\tau)^{\alpha-1} d\tau \\
&\quad + \frac{G\alpha}{\Gamma(\alpha)AB(\alpha)} \sum_{j=0}^{n} \int_{t_j}^{t_{j+1}} V(\tau)(t-\tau)^{\alpha-1} dB(\tau) \\
&= V(0) + \frac{(\Delta t)^{\alpha}}{AB(\alpha)\Gamma(\alpha+1)} \sum_{j=2}^{n} f(t_{j-2}, V^{j-2})[(n-j+1)^{\alpha} - (n-j)^{\alpha}] \\
&\quad + \frac{\alpha(\Delta t)^{\alpha}}{AB(\alpha)\Gamma(\alpha+2)} \sum_{j=2}^{n} [f(t_{j-1}, V^{j-1}) - f(t_{j-2}, V^{j-2})] \\
&\quad \times [(n-j+1)^{\alpha}(n-j+3+2\alpha) - (n-j)^{\alpha}(n-j+3+3\alpha)] \\
&\quad + \frac{(\Delta t)^{\alpha}\alpha}{2AB(\alpha)\Gamma(\alpha+3)} [f(t_j, V^j) - 2f(t_{j-1}, V^{j-1}) + f(t_{j-2}, V^{j-2})] \\
&\quad \times [(n-j+1)^{\alpha}\left(2(n-j)^2 + (3\alpha+10)(n-j) + 2\alpha^2 + 9\alpha + 12\right) \\
&\quad - (n-j)^{\alpha}\left(2(n-j)^2 + (5\alpha+10)(n-j) + 6\alpha^2 + 18\alpha + 12\right)] \\
&\quad + \frac{G\alpha(\Delta t)^{\alpha}}{AB(\alpha)\Gamma(\alpha+1)} \sum_{j=0}^{n} V(t_j)[(n-j+1)^{\alpha} - (n-j)^{\alpha}] \left(B(t_{j+1}) - B(t_j)\right)
\end{aligned}
$$

Then, we obtain

$$
\begin{aligned}
V(t_{n+1}) &= V(0) + \frac{\alpha(\Delta t)^{\alpha}}{AB(\alpha)\Gamma(\alpha+1)} \sum_{j=2}^{n} \left(-\frac{1}{C} i(t_{j-2}) - \frac{1}{R_1 C} V(t_{j-2}) + A \frac{d^2 i(t_{j-2})}{dt^2} \right. \\
&\quad \left. + \left(R_2 + \frac{L}{R_1 C}\right) \frac{di(t_{j-2})}{dt} + \frac{R_1 + R_2}{R_1 C} i(t_{j-2}) \right) [(n-j+1)^{\alpha} - (n-j)^{\alpha}] \\
&\quad + \frac{\alpha(\Delta t)^{\alpha}}{AB(\alpha)\Gamma(\alpha+2)} \sum_{j=2}^{n} \left[\left(-\frac{1}{C} i(t_{j-1}) - \frac{1}{R_1 C} V(t_{j-1}) + A \frac{d^2 i(t_{j-1})}{dt^2} \right. \right. \\
&\quad \left. + \left(R_2 + \frac{L}{R_1 C}\right) \frac{di(t_{j-1})}{dt} + \frac{R_1 + R_2}{R_1 C} i(t_{j-1}) \right) \\
&\quad - \left(-\frac{1}{C} i(t_{j-2}) - \frac{1}{R_1 C} V(t_{j-2}) + A \frac{d^2 i(t_{j-2})}{dt^2} \right)
\end{aligned}
$$

$$+\left(R_2+\frac{L}{R_1C}\right)\frac{di(t_{j-2})}{dt}+\frac{R_1+R_2}{R_1C}i(t_{j-2})\bigg)\bigg]$$

$$\times\left[(n-j+1)^\alpha(n-j+3+2\alpha)-(n-j^\alpha)(n-j+3+3\alpha)\right]$$

$$+\frac{\alpha(\Delta t)^\alpha}{2AB(\alpha)\Gamma(\alpha+3)}\bigg[\left(-\frac{1}{C}i(t_j)-\frac{1}{R_1C}V(t_j)+A\frac{d^2i(t_j)}{dt^2}\right.$$

$$+\left(R_2+\frac{L}{R_1C}\right)\frac{di(t_j)}{dt}+\frac{R_1+R_2}{R_1C}i(t_j)\bigg)$$

$$-\left(-\frac{1}{C}i(t_{j-1})-\frac{1}{R_1C}V(t_{j-1})+A\frac{d^2i(t_{j-1})}{dt^2}\right.$$

$$+\left(R_2+\frac{L}{R_1C}\right)\frac{di(t_{j-1})}{dt}+\frac{R_1+R_2}{R_1C}i(t_{j-1})\bigg)$$

$$+\left(-\frac{1}{C}i(t_{j-2})-\frac{1}{R_1C}V(t_{j-2})+A\frac{d^2i(t_{j-2})}{dt^2}\right.$$

$$+\left(R_2+\frac{L}{R_1C}\right)\frac{di(t_{j-2})}{dt}+\frac{R_1+R_2}{R_1C}i(t_{j-2})\bigg)\bigg]$$

$$\times\left[(n-j+1)^\alpha\left(2(n-j)^2+(3\alpha+10)(n-j)+2\alpha^2+9\alpha+12\right)\right.$$

$$-(n-j)^\alpha\left(2(n-j)^2+(5\alpha+10)(n-j)+6\alpha^2+18\alpha+12\right)\bigg]$$

$$+\frac{G\alpha(\Delta t)^\alpha}{AB(\alpha)\Gamma(\alpha+1)}\sum_{j=0}^{n}i(t_j)\left[(n-j+1)^\alpha-(n-j)^\alpha\right]\left(B(t_{j+1})-B(t_j)\right)$$

We consider the following differential equation

$$\frac{dV_2}{dt}=-\frac{R_1+R_2}{R_1R_2(C_1+C_2)}V_2(t)+\frac{C_1}{C_1+C_2}\frac{dV_1(t)}{dt}+\frac{1}{R_1(C_1+C_2)}V_1(t)=f(t,V_2(t))$$

(21.115)

We convert the above differential equation to the fractional stochastic differential equation as: In Mittag-Leffler case we have

$$V_2(t)-V_2(0)=\frac{1-\alpha}{AB(\alpha)}f(t,V_2(t))+\frac{\alpha}{AB(\alpha)\Gamma(\alpha)}\int_0^t f(\tau,V_2(\tau))(t-\tau)^{\alpha-1}d\tau$$

$$+\frac{G\alpha}{\Gamma(\alpha)AB(\alpha)}\int_0^t V_2(\tau)(t-\tau)^{\alpha-1}dB(\tau)$$

We define a mapping

$$
\begin{aligned}
|\Gamma(V_2(t))| \quad = \quad & \left| \frac{1-\alpha}{AB(\alpha)} f(t,V_2(t)) + \frac{\alpha}{AB(\alpha)\Gamma(\alpha)} \int_0^t f(\tau,V_2(\tau))(t-\tau)^{\alpha-1} d\tau \right. \\
& \left. + \frac{G\alpha}{\Gamma(\alpha)AB(\alpha)} \int_0^t V_2(\tau)(t-\tau)^{\alpha-1} dB(\tau) \right| \\
< \quad & \frac{1-\alpha}{AB(\alpha)} |f(t,V_2(t))| \\
& + \frac{\alpha}{AB(\alpha)\Gamma(\alpha)} \int_0^t |f(\tau,V_2(\tau))| (t-\tau)^{\alpha-1} d\tau \\
& + \frac{G\alpha}{\Gamma(\alpha)AB(\alpha)} \int_0^t |V_2(\tau)| (t-\tau)^{\alpha-1} dB(\tau) \\
< \quad & \frac{1-\alpha}{AB(\alpha)} \sup_{t\in[0,T]} |f(t,V_2(t))| \\
& + \frac{\alpha}{AB(\alpha)\Gamma(\alpha)} \int_0^t \sup_{\tau\in[0,t]} |f(\tau,V_2(\tau))| (t-\tau)^{\alpha-1} d\tau \\
& + \frac{G\alpha}{\Gamma(\alpha)AB(\alpha)} \int_0^t \sup_{\tau\in[0,t]} |V_2(\tau)| (t-\tau)^{\alpha-1} dB(\tau) \\
< \quad & \frac{1-\alpha}{AB(\alpha)} \|f(t,V_2(t))\|_\infty \\
& + \frac{T^\alpha \alpha}{AB(\alpha)\Gamma(\alpha+1)} \|f(t,V_2(t))\|_\infty \\
& + \frac{GL\alpha}{\Gamma(\alpha)AB(\alpha)} \|V_2(t)\|_\infty \\
< \quad & M_1 + M_2 + M_3 = M
\end{aligned}
$$

We have

$$
\begin{aligned}
|\Gamma(V_2) - \Gamma(V_{2_1})| = & \left| \frac{1-\alpha}{AB(\alpha)} \left(f(t,V_2(t)) - f(t,V_{2_1}(t)) \right) \right. \\
& + \frac{\alpha}{AB(\alpha)\Gamma(\alpha)} \int_0^t \left(f(\tau,V_2(\tau)) - f(\tau,V_{2_1}(\tau)) \right)(t-\tau)^{\alpha-1} d\tau \\
& \left. + \frac{G\alpha}{\Gamma(\alpha)AB(\alpha)} \int_0^t \left(V_2(\tau) - V_{2_1}(\tau) \right)(t-\tau)^{\alpha-1} dB(\tau) \right| \\
< & \frac{1-\alpha}{AB(\alpha)} |f(t,V_2(t)) - f(t,V_{2_1}(t))| \\
& + \frac{\alpha}{AB(\alpha)\Gamma(\alpha)} \int_0^t |f(\tau,V_2(\tau)) - f(\tau,V_{2_1}(\tau))| (t-\tau)^{\alpha-1} d\tau \\
& + \frac{G\alpha}{\Gamma(\alpha)AB(\alpha)} \int_0^t |V_2(\tau) - V_{2_1}(\tau)| (t-\tau)^{\alpha-1} dB(\tau)
\end{aligned}
$$

$$< \frac{1-\alpha}{AB(\alpha)} \sup_{t\in[0,T]} |f(t,V_2(t)) - f(t,V_{2_1}(t))|$$

$$+ \frac{\alpha}{AB(\alpha)\Gamma(\alpha)} \int_0^t \sup_{\tau\in[0,t]} |f(\tau,V_2(\tau)) - f(\tau,V_{2_1}(\tau))| (t-\tau)^{\alpha-1} d\tau$$

$$+ \frac{G\alpha}{\Gamma(\alpha)AB(\alpha)} \int_0^t \sup_{\tau\in[0,t]} |V_2(\tau) - V_{2_1}(\tau)| (t-\tau)^{\alpha-1} dB(\tau)$$

$$< \frac{1-\alpha}{AB(\alpha)} \frac{R_1+R_2}{R_1R_2(C_1+C_2)} \|V_2(t) - V_{2_1}(t)\|_\infty$$

$$+ \frac{\alpha}{AB(\alpha)\Gamma(\alpha)} \frac{R_1+R_2}{R_1R_2(C_1+C_2)} \frac{T^\alpha}{\alpha} \|V_2(t) - V_{2_1}(t)\|_\infty$$

$$+ \frac{GL\alpha}{\Gamma(\alpha)AB(\alpha)} \|V_2(t) - V_{2_1}(t)\|_\infty$$

$$< \left(\frac{1-\alpha}{AB(\alpha)} \frac{R_1+R_2}{R_1R_2(C_1+C_2)} + \frac{\alpha}{AB(\alpha)\Gamma(\alpha)} \frac{R_1+R_2}{R_1R_2(C_1+C_2)} \frac{T^\alpha}{\alpha} \right.$$

$$\left. + \frac{GL\alpha}{\Gamma(\alpha)AB(\alpha)} \right) \|V_2(t) - V_{2_1}(t)\|_\infty$$

$$< L_1 \|V(t) - V_1(t)\|_\infty$$

We assume that $B(t)$ is Lipschitzian on $[0,T]$ with constant L, then

$$|\Gamma(V_2(t))| = \left| \frac{1-\alpha}{AB(\alpha)} f(t,V_2(t)) + \frac{\alpha}{AB(\alpha)\Gamma(\alpha)} \int_0^t f(\tau,V_2(\tau))(t-\tau)^{\alpha-1} d\tau \right.$$

$$\left. + \frac{G\alpha}{\Gamma(\alpha)AB(\alpha)} \int_0^t V_2(\tau)(t-\tau)^{\alpha-1} dB(\tau) \right|$$

$$< \frac{1-\alpha}{AB(\alpha)} |f(t,V_2(t))|$$

$$+ \frac{\alpha}{AB(\alpha)\Gamma(\alpha)} \int_0^t |f(\tau,V_2(\tau))| (t-\tau)^{\alpha-1} d\tau$$

$$+ \frac{G\alpha}{\Gamma(\alpha)AB(\alpha)} \int_0^t |V_2(\tau)| (t-\tau)^{\alpha-1} dB(\tau)$$

$$< \frac{1-\alpha}{AB(\alpha)} \sup_{l\in[0,t]} |f(l,V_2(l))|$$

$$+ \frac{\alpha}{AB(\alpha)\Gamma(\alpha)} \int_0^t \sup_{l\in[0,\tau]} |f(l,V_2(l))| (t-\tau)^{\alpha-1} d\tau$$

$$+ \frac{G\alpha}{\Gamma(\alpha)AB(\alpha)} \int_0^t \sup_{l\in[0,\tau]} |V_2(l)| (t-\tau)^{\alpha-1} dB(\tau)$$

This shows that our mapping has a unique solution. We now present the numerical solution of the equation

$$V_2(t_{n+1}) = V_2(0) + \frac{1-\alpha}{AB(\alpha)} f(t_n, V_2(t_n))$$

$$+ \frac{\alpha}{AB(\alpha)\Gamma(\alpha)} \int_0^{t_{n+1}} f(\tau, V_2(\tau))(t-\tau)^{\alpha-1} d\tau$$

$$+ \frac{G\alpha}{\Gamma(\alpha)AB(\alpha)} \int_0^{t_{n+1}} V_2(\tau)(t-\tau)^{\alpha-1} dB(\tau)$$

$$= V_2(0) + \frac{1-\alpha}{AB(\alpha)} f(t_n, V_2(t_n))$$

$$+ \frac{\alpha}{AB(\alpha)\Gamma(\alpha)} \sum_{j=0}^{n} \int_{t_j}^{t_{j+1}} f(\tau, V_2(\tau))(t-\tau)^{\alpha-1} d\tau$$

$$+ \frac{G\alpha}{\Gamma(\alpha)AB(\alpha)} \sum_{j=0}^{n} \int_{t_j}^{t_{j+1}} V_2(\tau)(t-\tau)^{\alpha-1} dB(\tau)$$

$$= V_2(0) + \frac{(\Delta t)^\alpha}{AB(\alpha)\Gamma(\alpha+1)} \sum_{j=2}^{n} f(t_{j-2}, V_2^{j-2}) [(n-j+1)^\alpha - (n-j)^\alpha]$$

$$+ \frac{\alpha(\Delta t)^\alpha}{AB(\alpha)\Gamma(\alpha+2)} \sum_{j=2}^{n} \left[f(t_{j-1}, V_2^{j-1}) - f(t_{j-2}, V_2^{j-2}) \right]$$

$$\times [(n-j+1)^\alpha(n-j+3+2\alpha) - (n-j)^\alpha(n-j+3+3\alpha)]$$

$$+ \frac{(\Delta t)^\alpha \alpha}{2AB(\alpha)\Gamma(\alpha+3)} \left[f(t_j, V_2^j) - 2f(t_{j-1}, V_2^{j-1}) + f(t_{j-2}, V_2^{j-2}) \right]$$

$$\times [(n-j+1)^\alpha (2(n-j)^2 + (3\alpha+10)(n-j) + 2\alpha^2 + 9\alpha + 12)$$

$$- (n-j)^\alpha (2(n-j)^2 + (5\alpha+10)(n-j) + 6\alpha^2 + 18\alpha + 12)]$$

$$+ \frac{G\alpha(\Delta t)^\alpha}{AB(\alpha)\Gamma(\alpha+1)} \sum_{j=0}^{n} V(t_j) [(n-j+1)^\alpha - (n-j)^\alpha] (B(t_{j+1}) - B(t_j))$$

Then, we obtain

$$V_2(t_{n+1}) = V_2(0) + \frac{\alpha(\Delta t)^\alpha}{AB(\alpha)\Gamma(\alpha+1)} \sum_{j=2}^{n} \left(-\frac{R_1+R_2}{R_1R_2(C_1+C_2)} V_2(t_{j-2}) \right.$$

$$+ \frac{C_1}{C_1+C_2} \frac{dV_1(t_{j-2})}{dt} + \frac{1}{R_1(C_1+C_2)} V_1(t_{j-2}) \bigg)$$

$$\times [(n-j+1)^\alpha - (n-j)^\alpha]$$

$$+ \frac{\alpha(\Delta t)^\alpha}{AB(\alpha)\Gamma(\alpha+2)} \sum_{j=2}^{n} \left[\left(-\frac{R_1+R_2}{R_1R_2(C_1+C_2)} V_2(t_{j-1}) \right. \right.$$

$$+ \frac{C_1}{C_1+C_2} \frac{dV_1(t_{j-1})}{dt} + \frac{1}{R_1(C_1+C_2)} V_1(t_{j-1}) \bigg)$$

$$- \left(-\frac{R_1+R_2}{R_1R_2(C_1+C_2)} V_2(t_{j-2}) + \frac{C_1}{C_1+C_2} \frac{dV_1(t_{j-2})}{dt} \right.$$

$$+ \frac{1}{R_1(C_1+C_2)} V_1(t_{j-2}) \bigg) \bigg]$$

$$\times \left[(n-j+1)^\alpha(n-j+3+2\alpha) - (n-j^\alpha)(n-j+3+3\alpha) \right]$$

$$+ \frac{\alpha(\Delta t)^\alpha}{2AB(\alpha)\Gamma(\alpha+3)} \left[\left(-\frac{R_1+R_2}{R_1R_2(C_1+C_2)} V_2(t_j) \right. \right.$$

$$+ \frac{C_1}{C_1+C_2} \frac{dV_1(t_j)}{dt} + \frac{1}{R_1(C_1+C_2)} V_1(t_j) \bigg)$$

$$-2 \left(-\frac{R_1+R_2}{R_1R_2(C_1+C_2)} V_2(t_{j-1}) + \frac{C_1}{C_1+C_2} \frac{dV_1(t_{j-1})}{dt} \right.$$

$$+ \frac{1}{R_1(C_1+C_2)} V_1(t_{j-1}) \bigg)$$

$$+ \left(-\frac{R_1+R_2}{R_1R_2(C_1+C_2)} V_2(t_{j-2}) + \frac{C_1}{C_1+C_2} \frac{dV_1(t_{j-2})}{dt} \right.$$

$$+ \frac{1}{R_1(C_1+C_2)} V_1(t_{j-2}) \bigg) \bigg]$$

$$\times \left[(n-j+1)^\alpha \left(2(n-j)^2 + (3\alpha+10)(n-j) + 2\alpha^2 + 9\alpha + 12 \right) \right.$$

$$\left. - (n-j)^\alpha \left(2(n-j)^2 + (5\alpha+10)(n-j) + 6\alpha^2 + 18\alpha + 12 \right) \right]$$

$$+ \frac{G\alpha(\Delta t)^\alpha}{AB(\alpha)\Gamma(\alpha+1)} \sum_{j=0}^{n} i(t_j) \left[(n-j+1)^\alpha - (n-j)^\alpha \right] \left(B(t_{j+1}) - B(t_j) \right)$$

We consider the following differential equation

$$\frac{dV_2}{dt} = -4 \times 10^6 V_2(t) - 4 \times 10^7 V_1(t) = f(t, V_2(t)) \tag{21.116}$$

We convert the above differential equation to the fractional stochastic differential equation as: In Mittag-Leffler case we have

$$V_2(t) - V_2(0) = \frac{1-\alpha}{AB(\alpha)} f(t, V_2(t)) + \frac{\alpha}{AB(\alpha)\Gamma(\alpha)} \int_0^t f(\tau, V_2(\tau))(t-\tau)^{\alpha-1} d\tau$$

$$+ \frac{G\alpha}{\Gamma(\alpha)AB(\alpha)} \int_0^t V_2(\tau)(t-\tau)^{\alpha-1} dB(\tau)$$

We define a mapping

$$|\Gamma(V_2(t))| = \left| \frac{1-\alpha}{AB(\alpha)} f(t, V_2(t)) + \frac{\alpha}{AB(\alpha)\Gamma(\alpha)} \int_0^t f(\tau, V_2(\tau))(t-\tau)^{\alpha-1} d\tau \right.$$

$$\left. + \frac{G\alpha}{\Gamma(\alpha)AB(\alpha)} \int_0^t V_2(\tau)(t-\tau)^{\alpha-1} dB(\tau) \right|$$

$$< \frac{1-\alpha}{AB(\alpha)} |f(t, V_2(t))|$$

$$+ \frac{\alpha}{AB(\alpha)\Gamma(\alpha))} \int_0^t |f(\tau, V_2(\tau))| \, (t-\tau)^{\alpha-1} d\tau$$

$$+ \frac{G\alpha}{\Gamma(\alpha)AB(\alpha)} \int_0^t |V_2(\tau)| \, (t-\tau)^{\alpha-1} dB(\tau)$$

$$< \frac{1-\alpha}{AB(\alpha)} \sup_{t\in[0,T]} |f(t, V_2(t))|$$

$$+ \frac{\alpha}{AB(\alpha)\Gamma(\alpha))} \int_0^t \sup_{\tau\in[0,t]} |f(\tau, V_2(\tau))| \, (t-\tau)^{\alpha-1} d\tau$$

$$+ \frac{G\alpha}{\Gamma(\alpha)AB(\alpha)} \int_0^t \sup_{\tau\in[0,t]} |V_2(\tau)| \, (t-\tau)^{\alpha-1} dB(\tau)$$

$$< \frac{1-\alpha}{AB(\alpha)} \|f(t, V_2(t))\|_\infty$$

$$+ \frac{T^\alpha \alpha}{AB(\alpha)\Gamma(\alpha+1))} \|f(t, V_2(t))\|_\infty$$

$$+ \frac{GL\alpha}{\Gamma(\alpha)AB(\alpha)} \|V_2(t)\|_\infty$$

$$< M_1 + M_2 + M_3 = M$$

We have

$$|\Gamma(V_2) - \Gamma(V_{2_1})| = \left| \frac{1-\alpha}{AB(\alpha)} (f(t, V_2(t)) - f(t, V_{2_1}(t))) \right.$$

$$+ + \frac{\alpha}{AB(\alpha)\Gamma(\alpha))} \int_0^t (f(\tau, V_2(\tau)) - f(\tau, V_{2_1}(\tau))) \, (t-\tau)^{\alpha-1} d\tau$$

$$+ \frac{G\alpha}{\Gamma(\alpha)AB(\alpha)} \int_0^t (V_2(\tau) - V_{2_1}(\tau)) \, (t-\tau)^{\alpha-1} dB(\tau) \Big|$$

$$< \frac{1-\alpha}{AB(\alpha)} |f(t, V_2(t)) - f(t, V_{2_1}(t))|$$

$$+ \frac{\alpha}{AB(\alpha)\Gamma(\alpha))} \int_0^t |f(\tau, V_2(\tau)) - f(\tau, V_{2_1}(\tau))| \, (t-\tau)^{\alpha-1} d\tau$$

$$+ \frac{G\alpha}{\Gamma(\alpha)AB(\alpha)} \int_0^t |V_2(\tau) - V_{2_1}(\tau)| \, (t-\tau)^{\alpha-1} dB(\tau)$$

$$< \frac{1-\alpha}{AB(\alpha)} \sup_{t\in[0,T]} |f(t, V_2(t)) - f(t, V_{2_1}(t))|$$

$$+ \frac{\alpha}{AB(\alpha)\Gamma(\alpha))} \int_0^t \sup_{\tau\in[0,t]} |f(\tau, V_2(\tau)) - f(\tau, V_{2_1}(\tau))| \, (t-\tau)^{\alpha-1} d\tau$$

$$+ \frac{G\alpha}{\Gamma(\alpha)AB(\alpha)} \int_0^t \sup_{\tau\in[0,t]} |V_2(\tau) - V_{2_1}(\tau)| \, (t-\tau)^{\alpha-1} dB(\tau)$$

$$< \frac{1-\alpha}{AB(\alpha)} 4 \times 10^6 \|V_2(t) - V_{2_1}(t)\|_\infty$$

$$+\frac{\alpha}{AB(\alpha)\Gamma(\alpha))}4\times10^6\frac{T^\alpha}{\alpha}\|V_2(t)-V_{2_1}(t)\|_\infty$$

$$+\frac{GL\alpha}{\Gamma(\alpha)AB(\alpha)}\|V_2(t)-V_{2_1}(t)\|_\infty$$

$$<\left(\frac{1-\alpha}{AB(\alpha)}4\times10^6+\frac{\alpha}{AB(\alpha)\Gamma(\alpha))}4\times10^6\frac{T^\alpha}{\alpha}\right.$$

$$\left.+\frac{GL\alpha}{\Gamma(\alpha)AB(\alpha)}\right)\|V_2(t)-V_{2_1}(t)\|_\infty$$

$$<L_1\|V(t)-V_1(t)\|_\infty$$

We assume that $B(t)$ is Lipschitzian on $[0,T]$ with constant L, then

$$
\begin{aligned}
|\Gamma(V_2(t))| &= \left|\frac{1-\alpha}{AB(\alpha)}f(t,V_2(t))+\frac{\alpha}{AB(\alpha)\Gamma(\alpha))}\int_0^t f(\tau,V_2(\tau))(t-\tau)^{\alpha-1}d\tau\right. \\
&\qquad\left.+\frac{G\alpha}{\Gamma(\alpha)AB(\alpha)}\int_0^t V_2(\tau)(t-\tau)^{\alpha-1}dB(\tau)\right| \\
&< \frac{1-\alpha}{AB(\alpha)}|f(t,V_2(t))| \\
&\qquad+\frac{\alpha}{AB(\alpha)\Gamma(\alpha))}\int_0^t|f(\tau,V_2(\tau))|(t-\tau)^{\alpha-1}d\tau \\
&\qquad+\frac{G\alpha}{\Gamma(\alpha)AB(\alpha)}\int_0^t|V_2(\tau)|(t-\tau)^{\alpha-1}dB(\tau) \\
&< \frac{1-\alpha}{AB(\alpha)}\sup_{l\in[0,t]}|f(l,V_2(l))| \\
&\qquad+\frac{\alpha}{AB(\alpha)\Gamma(\alpha))}\int_0^t\sup_{l\in[0,\tau]}|f(l,V_2(l))|(t-\tau)^{\alpha-1}d\tau \\
&\qquad+\frac{G\alpha}{\Gamma(\alpha)AB(\alpha)}\int_0^t\sup_{l\in[0,\tau]}|V_2(l)|(t-\tau)^{\alpha-1}dB(\tau)
\end{aligned}
$$

This shows that our mapping has a unique solution. We now present the numerical solution of the equation

$$
\begin{aligned}
V_2(t_{n+1}) &=V_2(0)+\frac{1-\alpha}{AB(\alpha)}f(t_n,V_2(t_n)) \\
&\qquad+\frac{\alpha}{AB(\alpha)\Gamma(\alpha))}\int_0^{t_{n+1}}f(\tau,V_2(\tau))(t-\tau)^{\alpha-1}d\tau \\
&\qquad+\frac{G\alpha}{\Gamma(\alpha)AB(\alpha)}\int_0^{t_{n+1}}V_2(\tau)(t-\tau)^{\alpha-1}dB(\tau) \\
&=V_2(0)+\frac{1-\alpha}{AB(\alpha)}f(t_n,V_2(t_n))
\end{aligned}
$$

$$+ \frac{\alpha}{AB(\alpha)\Gamma(\alpha))} \sum_{j=0}^{n} \int_{t_j}^{t_{j+1}} f(\tau, V_2(\tau))(t-\tau)^{\alpha-1} d\tau$$

$$+ \frac{G\alpha}{\Gamma(\alpha)AB(\alpha)} \sum_{j=0}^{n} \int_{t_j}^{t_{j+1}} V_2(\tau)(t-\tau)^{\alpha-1} dB(\tau)$$

$$= V_2(0) + \frac{(\Delta t)^\alpha}{AB(\alpha)\Gamma(\alpha+1)} \sum_{j=2}^{n} f(t_{j-2}, V_2^{j-2}) \left[(n-j+1)^\alpha - (n-j)^\alpha \right]$$

$$+ \frac{\alpha(\Delta t)^\alpha}{AB(\alpha)\Gamma(\alpha+2)} \sum_{j=2}^{n} \left[f(t_{j-1}, V_2^{j-1}) - f(t_{j-2}, V_2^{j-2}) \right]$$

$$\times \left[(n-j+1)^\alpha (n-j+3+2\alpha) - (n-j)^\alpha)(n-j+3+3\alpha) \right]$$

$$+ \frac{(\Delta t)^\alpha \alpha}{2AB(\alpha)\Gamma(\alpha+3)} \left[f(t_j, V_2^j) - 2f(t_{j-1}, V_2^{j-1}) + f(t_{j-2}, V_2^{j-2}) \right]$$

$$\times \left[(n-j+1)^\alpha \left(2(n-j)^2 + (3\alpha+10)(n-j) + 2\alpha^2 + 9\alpha + 12 \right) \right.$$

$$\left. -(n-j)^\alpha \left(2(n-j)^2 + (5\alpha+10)(n-j) + 6\alpha^2 + 18\alpha + 12 \right) \right]$$

$$+ \frac{G\alpha(\Delta t)^\alpha}{AB(\alpha)\Gamma(\alpha+1)} \sum_{j=0}^{n} V(t_j) \left[(n-j+1)^\alpha - (n-j)^\alpha \right] \left(B(t_{j+1}) - B(t_j) \right)$$

Then, we obtain

$$V_2(t_{n+1}) = V_2(0) + \frac{\alpha(\Delta t)^\alpha}{AB(\alpha)\Gamma(\alpha+1)} \sum_{j=2}^{n} \left(-4 \times 10^6 V_2(t_{j-2}) - 4 \times 10^7 V_1(t_{j-2}) \right)$$

$$\times \left[(n-j+1)^\alpha - (n-j)^\alpha \right]$$

$$+ \frac{\alpha(\Delta t)^\alpha}{AB(\alpha)\Gamma(\alpha+2)} \sum_{j=2}^{n} \left[\left(-4 \times 10^6 V_2(t_{j-1}) - 4 \times 10^7 V_1(t_{j-1}) \right) \right.$$

$$\left. - \left(-4 \times 10^6 V_2(t_{j-2}) - 4 \times 10^7 V_1(t_{j-2}) \right) \right]$$

$$\times \left[(n-j+1)^\alpha (n-j+3+2\alpha) - (n-j^\alpha)(n-j+3+3\alpha) \right]$$

$$+ \frac{\alpha(\Delta t)^\alpha}{2AB(\alpha)\Gamma(\alpha+3)} \left[\left(-4 \times 10^6 V_2(t_j) - 4 \times 10^7 V_1(t_j) \right) \right.$$

$$-2 \left(-4 \times 10^6 V_2(t_{j-1}) - 4 \times 10^7 V_1(t_{j-1}) \right)$$

$$\left. + \left(-4 \times 10^6 V_2(t_{j-2}) - 4 \times 10^7 V_1(t_{j-2}) \right) \right]$$

$$\times \left[(n-j+1)^\alpha \left(2(n-j)^2 + (3\alpha+10)(n-j) + 2\alpha^2 + 9\alpha + 12 \right) \right.$$

$$\left. -(n-j)^\alpha \left(2(n-j)^2 + (5\alpha+10)(n-j) + 6\alpha^2 + 18\alpha + 12 \right) \right]$$

$$+ \frac{G\alpha(\Delta t)^\alpha}{AB(\alpha)\Gamma(\alpha+1)} \sum_{j=0}^{n} i(t_j) \left[(n-j+1)^\alpha - (n-j)^\alpha \right] \left(B(t_{j+1}) - B(t_j) \right)$$

We consider the following differential equation

$$\frac{dV_2}{dt} = -\frac{R_1+R_2}{R_1R_2C}V_2(t) + \frac{1}{R_1C_1}V_1(t) + \frac{dV_1(t)}{dt} = f(t,V_2(t)) \qquad (21.117)$$

We convert the above differential equation to the fractional stochastic differential equation as: In Mittag-Leffler case we have

$$\begin{aligned} V_2(t) - V_2(0) &= \frac{1-\alpha}{AB(\alpha)}f(t,V_2(t)) + \frac{\alpha}{AB(\alpha)\Gamma(\alpha)}\int_0^t f(\tau,V_2(\tau))(t-\tau)^{\alpha-1}d\tau \\ &+ \frac{G\alpha}{\Gamma(\alpha)AB(\alpha)}\int_0^t V_2(\tau)(t-\tau)^{\alpha-1}dB(\tau) \end{aligned}$$

We define a mapping

$$\begin{aligned} |\Gamma(V_2(t))| &= \left| \frac{1-\alpha}{AB(\alpha)}f(t,V_2(t)) + \frac{\alpha}{AB(\alpha)\Gamma(\alpha)}\int_0^t f(\tau,V_2(\tau))(t-\tau)^{\alpha-1}d\tau \right. \\ &\quad \left. + \frac{G\alpha}{\Gamma(\alpha)AB(\alpha)}\int_0^t V_2(\tau)(t-\tau)^{\alpha-1}dB(\tau) \right| \\ &< \frac{1-\alpha}{AB(\alpha)}|f(t,V_2(t))| \\ &\quad + \frac{\alpha}{AB(\alpha)\Gamma(\alpha)}\int_0^t |f(\tau,V_2(\tau))|(t-\tau)^{\alpha-1}d\tau \\ &\quad + \frac{G\alpha}{\Gamma(\alpha)AB(\alpha)}\int_0^t |V_2(\tau)|(t-\tau)^{\alpha-1}dB(\tau) \\ &< \frac{1-\alpha}{AB(\alpha)}\sup_{t\in[0,T]}|f(t,V_2(t))| \\ &\quad + \frac{\alpha}{AB(\alpha)\Gamma(\alpha)}\int_0^t \sup_{\tau\in[0,t]}|f(\tau,V_2(\tau))|(t-\tau)^{\alpha-1}d\tau \\ &\quad + \frac{G\alpha}{\Gamma(\alpha)AB(\alpha)}\int_0^t \sup_{\tau\in[0,t]}|V_2(\tau)|(t-\tau)^{\alpha-1}dB(\tau) \\ &< \frac{1-\alpha}{AB(\alpha)}\|f(t,V_2(t))\|_\infty \\ &\quad + \frac{T^\alpha\alpha}{AB(\alpha)\Gamma(\alpha+1)}\|f(t,V_2(t))\|_\infty \\ &\quad + \frac{GL\alpha}{\Gamma(\alpha)AB(\alpha)}\|V_2(t)\|_\infty \\ &< M_1 + M_2 + M_3 = M \end{aligned}$$

We have

$$|\Gamma(V_2) - \Gamma(V_{2_1})| = \left| \frac{1-\alpha}{AB(\alpha)} \left(f(t, V_2(t)) - f(t, V_{2_1}(t)) \right) \right.$$

$$+ \frac{\alpha}{AB(\alpha)\Gamma(\alpha)} \int_0^t \left(f(\tau, V_2(\tau)) - f(\tau, V_{2_1}(\tau)) \right) (t-\tau)^{\alpha-1} d\tau$$

$$+ \left. \frac{G\alpha}{\Gamma(\alpha)AB(\alpha)} \int_0^t \left(V_2(\tau) - V_{2_1}(\tau) \right) (t-\tau)^{\alpha-1} dB(\tau) \right|$$

$$< \frac{1-\alpha}{AB(\alpha)} |f(t, V_2(t)) - f(t, V_{2_1}(t))|$$

$$+ \frac{\alpha}{AB(\alpha)\Gamma(\alpha)} \int_0^t |f(\tau, V_2(\tau)) - f(\tau, V_{2_1}(\tau))| (t-\tau)^{\alpha-1} d\tau$$

$$+ \frac{G\alpha}{\Gamma(\alpha)AB(\alpha)} \int_0^t |V_2(\tau) - V_{2_1}(\tau)| (t-\tau)^{\alpha-1} dB(\tau)$$

$$< \frac{1-\alpha}{AB(\alpha)} \sup_{t \in [0,T]} |f(t, V_2(t)) - f(t, V_{2_1}(t))|$$

$$+ \frac{\alpha}{AB(\alpha)\Gamma(\alpha)} \int_0^t \sup_{\tau \in [0,t]} |f(\tau, V_2(\tau)) - f(\tau, V_{2_1}(\tau))| (t-\tau)^{\alpha-1} d\tau$$

$$+ \frac{G\alpha}{\Gamma(\alpha)AB(\alpha)} \int_0^t \sup_{\tau \in [0,t]} |V_2(\tau) - V_{2_1}(\tau)| (t-\tau)^{\alpha-1} dB(\tau)$$

$$< \frac{1-\alpha}{AB(\alpha)} \frac{R_1+R_2}{R_1 R_2 C} \|V_2(t) - V_{2_1}(t)\|_\infty$$

$$+ \frac{\alpha}{AB(\alpha)\Gamma(\alpha)} \frac{R_1+R_2}{R_1 R_2 C} \frac{T^\alpha}{\alpha} \|V_2(t) - V_{2_1}(t)\|_\infty$$

$$+ \frac{GL\alpha}{\Gamma(\alpha)AB(\alpha)} \|V_2(t) - V_{2_1}(t)\|_\infty$$

$$< \left(\frac{1-\alpha}{AB(\alpha)} \frac{R_1+R_2}{R_1 R_2 C} + \frac{\alpha}{AB(\alpha)\Gamma(\alpha)} \frac{R_1+R_2}{R_1 R_2 C} \frac{T^\alpha}{\alpha} \right.$$

$$+ \left. \frac{GL\alpha}{\Gamma(\alpha)AB(\alpha)} \right) \|V_2(t) - V_{2_1}(t)\|_\infty$$

$$< L_1 \|V(t) - V_1(t)\|_\infty$$

We assume that $B(t)$ is Lipschitzian on $[0, T]$ with constant L, then

$$|\Gamma(V_2(t))| = \left| \frac{1-\alpha}{AB(\alpha)} f(t, V_2(t)) + \frac{\alpha}{AB(\alpha)\Gamma(\alpha)} \int_0^t f(\tau, V_2(\tau))(t-\tau)^{\alpha-1} d\tau \right.$$

$$+ \left. \frac{G\alpha}{\Gamma(\alpha)AB(\alpha)} \int_0^t V_2(\tau)(t-\tau)^{\alpha-1} dB(\tau) \right|$$

$$< \frac{1-\alpha}{AB(\alpha)} |f(t, V_2(t))|$$

$$+\frac{\alpha}{AB(\alpha)\Gamma(\alpha))}\int_0^t |f(\tau,V_2(\tau))|\,(t-\tau)^{\alpha-1}d\tau$$

$$+\frac{G\alpha}{\Gamma(\alpha)AB(\alpha)}\int_0^t |V_2(\tau)|\,(t-\tau)^{\alpha-1}dB(\tau)$$

$$<\frac{1-\alpha}{AB(\alpha)}\sup_{l\in[0,t]}|f(l,V_2(l))|$$

$$+\frac{\alpha}{AB(\alpha)\Gamma(\alpha))}\int_0^t \sup_{l\in[0,\tau]}|f(l,V_2(l))|\,(t-\tau)^{\alpha-1}d\tau$$

$$+\frac{G\alpha}{\Gamma(\alpha)AB(\alpha)}\int_0^t \sup_{l\in[0,\tau]}|V_2(l)|\,(t-\tau)^{\alpha-1}dB(\tau)$$

This shows that our mapping has a unique solution. We now present the numerical solution of the equation

$$V_2(t_{n+1})=V_2(0)+\frac{1-\alpha}{AB(\alpha)}f(t_n,V_2(t_n))$$

$$+\frac{\alpha}{AB(\alpha)\Gamma(\alpha))}\int_0^{t_{n+1}} f(\tau,V_2(\tau))(t-\tau)^{\alpha-1}d\tau$$

$$+\frac{G\alpha}{\Gamma(\alpha)AB(\alpha)}\int_0^{t_{n+1}} V_2(\tau)(t-\tau)^{\alpha-1}dB(\tau)$$

$$=V_2(0)+\frac{1-\alpha}{AB(\alpha)}f(t_n,V_2(t_n))$$

$$+\frac{\alpha}{AB(\alpha)\Gamma(\alpha))}\sum_{j=0}^n\int_{t_j}^{t_{j+1}} f(\tau,V_2(\tau))(t-\tau)^{\alpha-1}d\tau$$

$$+\frac{G\alpha}{\Gamma(\alpha)AB(\alpha)}\sum_{j=0}^n\int_{t_j}^{t_{j+1}} V_2(\tau)(t-\tau)^{\alpha-1}dB(\tau)$$

$$=V_2(0)+\frac{(\Delta t)^\alpha}{AB(\alpha)\Gamma(\alpha+1)}\sum_{j=2}^n f(t_{j-2},V_2^{j-2})[(n-j+1)^\alpha-(n-j)^\alpha]$$

$$+\frac{\alpha(\Delta t)^\alpha}{AB(\alpha)\Gamma(\alpha+2)}\sum_{j=2}^n\left[f(t_{j-1},V_2^{j-1})-f(t_{j-2},V_2^{j-2})\right]$$

$$\times[(n-j+1)^\alpha(n-j+3+2\alpha)-(n-j)^\alpha)(n-j+3+3\alpha)]$$

$$+\frac{(\Delta t)^\alpha\alpha}{2AB(\alpha)\Gamma(\alpha+3)}\left[f(t_j,V_2^j)-2f(t_{j-1},V_2^{j-1})+f(t_{j-2},V_2^{j-2})\right]$$

$$\times[(n-j+1)^\alpha(2(n-j)^2+(3\alpha+10)(n-j)+2\alpha^2+9\alpha+12)$$

$$-(n-j)^\alpha(2(n-j)^2+(5\alpha+10)(n-j)+6\alpha^2+18\alpha+12)]$$

$$+\frac{G\alpha(\Delta t)^\alpha}{AB(\alpha)\Gamma(\alpha+1)}\sum_{j=0}^n V(t_j)[(n-j+1)^\alpha-(n-j)^\alpha]\,(B(t_{j+1})-B(t_j))$$

Then, we obtain

$$
\begin{aligned}
V_2(t_{n+1}) =\, & V_2(0) + \frac{\alpha(\Delta t)^\alpha}{AB(\alpha)\Gamma(\alpha+1)} \sum_{j=2}^{n} \left(-\frac{R_1+R_2}{R_1 R_2 C} V_2(t_{j-2}) \right.\\
& + \frac{1}{R_1 C_1} V_1(t_{j-2}) + \left.\frac{dV_1(t_{j-2})}{dt} \right)\\
& \times \left[(n-j+1)^\alpha - (n\quad j)^\alpha \right]\\
& + \frac{\alpha(\Delta t)^\alpha}{AB(\alpha)\Gamma(\alpha+2)} \sum_{j=2}^{n} \left[\left(-\frac{R_1+R_2}{R_1 R_2 C} V_2(t_{j-1}) \right.\right.\\
& + \frac{1}{R_1 C_1} V_1(t_{j-1}) + \left.\frac{dV_1(t_{j-1})}{dt} \right)\\
& - \left(-\frac{R_1+R_2}{R_1 R_2 C} V_2(t_{j-2}) + \frac{1}{R_1 C_1} V_1(t_{j-2}) + \left.\frac{dV_1(t_{j-2})}{dt} \right) \right]\\
& \times \left[(n-j+1)^\alpha (n-j+3+2\alpha) - (n-j^\alpha)(n-j+3+3\alpha) \right]\\
& + \frac{\alpha(\Delta t)^\alpha}{2AB(\alpha)\Gamma(\alpha+3)} \left[\left(-\frac{R_1+R_2}{R_1 R_2 C} V_2(t_j) \right.\right.\\
& + \frac{1}{R_1 C_1} V_1(t_j) + \left.\frac{dV_1(t_j)}{dt} \right)\\
& - 2 \left(-\frac{R_1+R_2}{R_1 R_2 C} V_2(t_{j-1}) + \frac{1}{R_1 C_1} V_1(t_{j-1}) + \frac{dV_1(t_{j-1})}{dt} \right)\\
& + \left(-\frac{R_1+R_2}{R_1 R_2 C} V_2(t_{j-2}) \right.\\
& + \frac{1}{R_1 C_1} V_1(t_{j-2}) + \left.\left.\frac{dV_1(t_{j-2})}{dt} \right) \right]\\
& \times \left[(n-j+1)^\alpha \left(2(n-j)^2 + (3\alpha+10)(n-j) + 2\alpha^2 + 9\alpha + 12 \right)\right.\\
& \left. - (n-j)^\alpha \left(2(n-j)^2 + (5\alpha+10)(n-j) + 6\alpha^2 + 18\alpha + 12 \right) \right]\\
& + \frac{G\alpha(\Delta t)^\alpha}{AB(\alpha)\Gamma(\alpha+1)} \sum_{j=0}^{n} i(t_j) \left[(n-j+1)^\alpha - (n-j)^\alpha \right] \left(B(t_{j+1}) - B(t_j) \right)
\end{aligned}
$$

We consider the following differential equation

$$
\frac{dV_2}{dt} = -\frac{1}{(R_1+R_2)C} V_2(t) + \frac{1}{(R_1+R_2)C} V_1(t) + \frac{R_2}{R_1+R_2} \frac{dV_1(t)}{dt} = f(t, V_2(t))
$$

$$(21.118)$$

We convert the above differential equation to the fractional stochastic differential equation as: In Mittag-Leffler case we have

$$V_2(t) - V_2(0) = \frac{1-\alpha}{AB(\alpha)} f(t, V_2(t)) + \frac{\alpha}{AB(\alpha)\Gamma(\alpha)} \int_0^t f(\tau, V_2(\tau))(t-\tau)^{\alpha-1} d\tau$$

$$+ \frac{G\alpha}{\Gamma(\alpha)AB(\alpha)} \int_0^t V_2(\tau)(t-\tau)^{\alpha-1} dB(\tau)$$

We define a mapping

$$|\Gamma(V_2(t))| = \left| \frac{1-\alpha}{AB(\alpha)} f(t, V_2(t)) + \frac{\alpha}{AB(\alpha)\Gamma(\alpha)} \int_0^t f(\tau, V_2(\tau))(t-\tau)^{\alpha-1} d\tau \right.$$

$$\left. + \frac{G\alpha}{\Gamma(\alpha)AB(\alpha)} \int_0^t V_2(\tau)(t-\tau)^{\alpha-1} dB(\tau) \right|$$

$$< \frac{1-\alpha}{AB(\alpha)} |f(t, V_2(t))|$$

$$+ \frac{\alpha}{AB(\alpha)\Gamma(\alpha)} \int_0^t |f(\tau, V_2(\tau))| (t-\tau)^{\alpha-1} d\tau$$

$$+ \frac{G\alpha}{\Gamma(\alpha)AB(\alpha)} \int_0^t |V_2(\tau)| (t-\tau)^{\alpha-1} dB(\tau)$$

$$< \frac{1-\alpha}{AB(\alpha)} \sup_{t \in [0,T]} |f(t, V_2(t))|$$

$$+ \frac{\alpha}{AB(\alpha)\Gamma(\alpha)} \int_0^t \sup_{\tau \in [0,t]} |f(\tau, V_2(\tau))| (t-\tau)^{\alpha-1} d\tau$$

$$+ \frac{G\alpha}{\Gamma(\alpha)AB(\alpha)} \int_0^t \sup_{\tau \in [0,t]} |V_2(\tau)| (t-\tau)^{\alpha-1} dB(\tau)$$

$$< \frac{1-\alpha}{AB(\alpha)} \|f(t, V_2(t))\|_\infty$$

$$+ \frac{T^\alpha \alpha}{AB(\alpha)\Gamma(\alpha+1)} \|f(t, V_2(t))\|_\infty$$

$$\vdash \frac{GL\alpha}{\Gamma(\alpha)AB(\alpha)} \|V_2(t)\|_\infty$$

$$< M_1 + M_2 + M_3 = M$$

We have

$$|\Gamma(V_2) - \Gamma(V_{2_1})| = \left| \frac{1-\alpha}{AB(\alpha)} (f(t, V_2(t)) - f(t, V_{2_1}(t))) \right.$$

$$+ \frac{\alpha}{AB(\alpha)\Gamma(\alpha)} \int_0^t (f(\tau, V_2(\tau)) - f(\tau, V_{2_1}(\tau))) (t-\tau)^{\alpha-1} d\tau$$

$$\left. + \frac{G\alpha}{\Gamma(\alpha)AB(\alpha)} \int_0^t (V_2(\tau) - V_{2_1}(\tau)) (t-\tau)^{\alpha-1} dB(\tau) \right|$$

$$< \frac{1-\alpha}{AB(\alpha)}|f(t,V_2(t))-f(t,V_{2_1}(t))|$$

$$+\frac{\alpha}{AB(\alpha)\Gamma(\alpha)}\int_0^t |f(\tau,V_2(\tau))-f(\tau,V_{2_1}(\tau))|(t-\tau)^{\alpha-1}d\tau$$

$$+\frac{G\alpha}{\Gamma(\alpha)AB(\alpha)}\int_0^t |V_2(\tau)-V_{2_1}(\tau)|(t-\tau)^{\alpha-1}dB(\tau)$$

$$< \frac{1-\alpha}{AB(\alpha)}\sup_{t\in[0,T]}|f(t,V_2(t))-f(t,V_{2_1}(t))|$$

$$+\frac{\alpha}{AB(\alpha)\Gamma(\alpha)}\int_0^t \sup_{\tau\in[0,t]}|f(\tau,V_2(\tau))-f(\tau,V_{2_1}(\tau))|(t-\tau)^{\alpha-1}d\tau$$

$$+\frac{G\alpha}{\Gamma(\alpha)AB(\alpha)}\int_0^t \sup_{\tau\in[0,t]}|V_2(\tau)-V_{2_1}(\tau)|(t-\tau)^{\alpha-1}dB(\tau)$$

$$< \frac{1-\alpha}{AB(\alpha)}\frac{1}{(R_1+R_2)C}\|V_2(t)-V_{2_1}(t)\|_\infty$$

$$+\frac{\alpha}{AB(\alpha)\Gamma(\alpha)}\frac{1}{(R_1+R_2)C}\frac{T^\alpha}{\alpha}\|V_2(t)-V_{2_1}(t)\|_\infty$$

$$+\frac{GL\alpha}{\Gamma(\alpha)AB(\alpha)}\|V_2(t)-V_{2_1}(t)\|_\infty$$

$$< \left(\frac{1-\alpha}{AB(\alpha)}\frac{1}{(R_1+R_2)C}+\frac{\alpha}{AB(\alpha)\Gamma(\alpha)}\frac{1}{(R_1+R_2)C}\frac{T^\alpha}{\alpha}\right.$$

$$\left.+\frac{GL\alpha}{\Gamma(\alpha)AB(\alpha)}\right)\|V_2(t)-V_{2_1}(t)\|_\infty$$

$$<L_1\|V(t)-V_1(t)\|_\infty$$

We assume that $B(t)$ is Lipschitzian on $[0,T]$ with constant L, then

$$|\Gamma(V_2(t))| = \left|\frac{1-\alpha}{AB(\alpha)}f(t,V_2(t))+\frac{\alpha}{AB(\alpha)\Gamma(\alpha)}\int_0^t f(\tau,V_2(\tau))(t-\tau)^{\alpha-1}d\tau\right.$$

$$\left.+\frac{G\alpha}{\Gamma(\alpha)AB(\alpha)}\int_0^t V_2(\tau)(t-\tau)^{\alpha-1}dB(\tau)\right|$$

$$< \frac{1-\alpha}{AB(\alpha)}|f(t,V_2(t))|$$

$$+\frac{\alpha}{AB(\alpha)\Gamma(\alpha)}\int_0^t |f(\tau,V_2(\tau))|(t-\tau)^{\alpha-1}d\tau$$

$$+\frac{G\alpha}{\Gamma(\alpha)AB(\alpha)}\int_0^t |V_2(\tau)|(t-\tau)^{\alpha-1}dB(\tau)$$

$$< \frac{1-\alpha}{AB(\alpha)}\sup_{l\in[0,t]}|f(l,V_2(l))|$$

$$+\frac{\alpha}{AB(\alpha)\Gamma(\alpha))}\int_0^t \sup_{l\in[0,\tau]} |f(l,V_2(l))|\,(t-\tau)^{\alpha-1}d\tau$$

$$+\frac{G\alpha}{\Gamma(\alpha)AB(\alpha)}\int_0^t \sup_{l\in[0,\tau]} |V_2(l)|\,(t-\tau)^{\alpha-1}dB(\tau)$$

This shows that our mapping has a unique solution. We now present the numerical solution of the equation

$$V_2(t_{n+1}) = V_2(0) + \frac{1-\alpha}{AB(\alpha)}f(t_n,V_2(t_n))$$

$$+\frac{\alpha}{AB(\alpha)\Gamma(\alpha))}\int_0^{t_{n+1}} f(\tau,V_2(\tau))(t-\tau)^{\alpha-1}d\tau$$

$$+\frac{G\alpha}{\Gamma(\alpha)AB(\alpha)}\int_0^{t_{n+1}} V_2(\tau)(t-\tau)^{\alpha-1}dB(\tau)$$

$$=V_2(0) + \frac{1-\alpha}{AB(\alpha)}f(t_n,V_2(t_n))$$

$$+\frac{\alpha}{AB(\alpha)\Gamma(\alpha))}\sum_{j=0}^n\int_{t_j}^{t_{j+1}} f(\tau,V_2(\tau))(t-\tau)^{\alpha-1}d\tau$$

$$+\frac{G\alpha}{\Gamma(\alpha)AB(\alpha)}\sum_{j=0}^n\int_{t_j}^{t_{j+1}} V_2(\tau)(t-\tau)^{\alpha-1}dB(\tau)$$

$$=V_2(0) + \frac{(\Delta t)^\alpha}{AB(\alpha)\Gamma(\alpha+1)}\sum_{j=2}^n f(t_{j-2},V_2^{j-2})[(n-j+1)^\alpha-(n-j)^\alpha]$$

$$+\frac{\alpha(\Delta t)^\alpha}{AB(\alpha)\Gamma(\alpha+2)}\sum_{j=2}^n\left[f(t_{j-1},V_2^{j-1})-f(t_{j-2},V_2^{j-2})\right]$$

$$\times[(n-j+1)^\alpha(n-j+3+2\alpha)-(n-j)^\alpha)(n-j+3+3\alpha)]$$

$$+\frac{(\Delta t)^\alpha\alpha}{2AB(\alpha)\Gamma(\alpha+3)}\left[f(t_j,V_2^j)-2f(t_{j-1},V_2^{j-1})+f(t_{j-2},V_2^{j-2})\right]$$

$$\times[(n-j+1)^\alpha\left(2(n-j)^2+(3\alpha+10)(n-j)+2\alpha^2+9\alpha+12\right)$$

$$-(n-j)^\alpha\left(2(n-j)^2+(5\alpha+10)(n-j)+6\alpha^2+18\alpha+12\right)]$$

$$+\frac{G\alpha(\Delta t)^\alpha}{AB(\alpha)\Gamma(\alpha+1)}\sum_{j=0}^n V(t_j)[(n-j+1)^\alpha-(n-j)^\alpha]\left(B(t_{j+1})-B(t_j)\right)$$

Then, we obtain

$$
\begin{aligned}
V_2(t_{n+1}) \;=\;& V_2(0) + \frac{\alpha(\Delta t)^\alpha}{AB(\alpha)\Gamma(\alpha+1)} \sum_{j=2}^{n}\left(-\frac{1}{(R_1+R_2)C}V_2(t_{j-2})\right.\\
&+\frac{1}{(R_1+R_2)C}V_1(t_{j-2}) + \frac{R_2}{R_1+R_2}\frac{dV_1(t_{j-2})}{dt}\Bigg)\\
&\times\left[(n-j+1)^\alpha - (n-j)^\alpha\right]\\
&+\frac{\alpha(\Delta t)^\alpha}{AB(\alpha)\Gamma(\alpha+2)} \sum_{j=2}^{n}\left[\left(-\frac{1}{(R_1+R_2)C}V_2(t_{j-1})\right.\right.\\
&+\frac{1}{(R_1+R_2)C}V_1(t_{j-1}) + \frac{R_2}{R_1+R_2}\frac{dV_1(t_{j-1})}{dt}\Bigg)\\
&-\left(-\frac{1}{(R_1+R_2)C}V_2(t_{j-2}) + \frac{1}{(R_1+R_2)C}V_1(t_{j-2})\right.\\
&\left.\left.+\frac{R_2}{R_1+R_2}\frac{dV_1(t_{j-2})}{dt}\right)\right]\\
&\times\left[(n-j+1)^\alpha(n-j+3+2\alpha) - (n-j^\alpha)(n-j+3+3\alpha)\right]\\
&+\frac{\alpha(\Delta t)^\alpha}{2AB(\alpha)\Gamma(\alpha+3)}\left[\left(-\frac{1}{(R_1+R_2)C}V_2(t_j)\right.\right.\\
&+\frac{1}{(R_1+R_2)C}V_1(t_j) + \frac{R_2}{R_1+R_2}\frac{dV_1(t_j)}{dt}\Bigg)\\
&-2\left(-\frac{1}{(R_1+R_2)C}V_2(t_{j-1}) + \frac{1}{(R_1+R_2)C}V_1(t_{j-1})\right.\\
&\left.+\frac{R_2}{R_1+R_2}\frac{dV_1(t_{j-1})}{dt}\right)\\
&+\left(-\frac{1}{(R_1+R_2)C}V_2(t_{j-2}) + \frac{1}{(R_1+R_2)C}V_1(t_{j-2})\right.\\
&\left.\left.+\frac{R_2}{R_1+R_2}\frac{dV_1(t_{j-2})}{dt}\right)\right]\\
&\times\left[(n-j+1)^\alpha\left(2(n-j)^2+(3\alpha+10)(n-j)+2\alpha^2+9\alpha+12\right)\right.\\
&\left.-(n-j)^\alpha\left(2(n-j)^2+(5\alpha+10)(n-j)+6\alpha^2+18\alpha+12\right)\right]\\
&+\frac{G\alpha(\Delta t)^\alpha}{AB(\alpha)\Gamma(\alpha+1)}\sum_{j=0}^{n} i(t_j)\left[(n-j+1)^\alpha - (n-j)^\alpha\right]\left(B(t_{j+1})-B(t_j)\right)
\end{aligned}
$$

We consider the following differential equation

$$
\frac{dV_2}{dt} = -\frac{1}{R_1 C_1}V_2(t) - \frac{1}{R_1 C_2}V_1(t) + \frac{R_2}{R_1}\frac{dV_1(t)}{dt} = f(t, V_2(t)) \qquad (21.119)
$$

We convert the above differential equation to the fractional stochastic differential equation as: In Mittag-Leffler case we have

$$
\begin{aligned}
V_2(t) - V_2(0) &= \frac{1-\alpha}{AB(\alpha)} f(t, V_2(t)) + \frac{\alpha}{AB(\alpha)\Gamma(\alpha)} \int_0^t f(\tau, V_2(\tau))(t-\tau)^{\alpha-1} d\tau \\
&\quad + \frac{G\alpha}{\Gamma(\alpha)AB(\alpha)} \int_0^t V_2(\tau)(t-\tau)^{\alpha-1} dB(\tau)
\end{aligned}
$$

We define a mapping

$$
\begin{aligned}
|\Gamma(V_2(t))| &= \left| \frac{1-\alpha}{AB(\alpha)} f(t, V_2(t)) + \frac{\alpha}{AB(\alpha)\Gamma(\alpha)} \int_0^t f(\tau, V_2(\tau))(t-\tau)^{\alpha-1} d\tau \right. \\
&\quad \left. + \frac{G\alpha}{\Gamma(\alpha)AB(\alpha)} \int_0^t V_2(\tau)(t-\tau)^{\alpha-1} dB(\tau) \right| \\
&< \frac{1-\alpha}{AB(\alpha)} |f(t, V_2(t))| \\
&\quad + \frac{\alpha}{AB(\alpha)\Gamma(\alpha)} \int_0^t |f(\tau, V_2(\tau))| (t-\tau)^{\alpha-1} d\tau \\
&\quad + \frac{G\alpha}{\Gamma(\alpha)AB(\alpha)} \int_0^t |V_2(\tau)| (t-\tau)^{\alpha-1} dB(\tau) \\
&< \frac{1-\alpha}{AB(\alpha)} \sup_{t\in[0,T]} |f(t, V_2(t))| \\
&\quad + \frac{\alpha}{AB(\alpha)\Gamma(\alpha)} \int_0^t \sup_{\tau\in[0,t]} |f(\tau, V_2(\tau))| (t-\tau)^{\alpha-1} d\tau \\
&\quad + \frac{G\alpha}{\Gamma(\alpha)AB(\alpha)} \int_0^t \sup_{\tau\in[0,t]} |V_2(\tau)| (t-\tau)^{\alpha-1} dB(\tau) \\
&< \frac{1-\alpha}{AB(\alpha)} \|f(t, V_2(t))\|_\infty \\
&\quad + \frac{T^\alpha \alpha}{AB(\alpha)\Gamma(\alpha+1)} \|f(t, V_2(t))\|_\infty \\
&\quad + \frac{GL\alpha}{\Gamma(\alpha)AB(\alpha)} \|V_2(t)\|_\infty \\
&< M_1 + M_2 + M_3 = M
\end{aligned}
$$

We have

$$
\begin{aligned}
|\Gamma(V_2) - \Gamma(V_{2_1})| &= \left| \frac{1-\alpha}{AB(\alpha)} (f(t, V_2(t)) - f(t, V_{2_1}(t))) \right. \\
&\quad + \frac{\alpha}{AB(\alpha)\Gamma(\alpha)} \int_0^t (f(\tau, V_2(\tau)) - f(\tau, V_{2_1}(\tau)))(t-\tau)^{\alpha-1} d\tau \\
&\quad \left. + \frac{G\alpha}{\Gamma(\alpha)AB(\alpha)} \int_0^t (V_2(\tau) - V_{2_1}(\tau))(t-\tau)^{\alpha-1} dB(\tau) \right|
\end{aligned}
$$

$$< \frac{1-\alpha}{AB(\alpha)} |f(t,V_2(t)) - f(t,V_{2_1}(t))|$$

$$+ \frac{\alpha}{AB(\alpha)\Gamma(\alpha)} \int_0^t |f(\tau,V_2(\tau)) - f(\tau,V_{2_1}(\tau))| (t-\tau)^{\alpha-1} d\tau$$

$$+ \frac{G\alpha}{\Gamma(\alpha)AB(\alpha)} \int_0^t |V_2(\tau) - V_{2_1}(\tau)| (t-\tau)^{\alpha-1} dB(\tau)$$

$$< \frac{1-\alpha}{AB(\alpha)} \sup_{t\in[0,T]} |f(t,V_2(t)) - f(t,V_{2_1}(t))|$$

$$+ \frac{\alpha}{AB(\alpha)\Gamma(\alpha))} \int_0^t \sup_{\tau\in[0,t]} |f(\tau,V_2(\tau)) - f(\tau,V_{2_1}(\tau))| (t-\tau)^{\alpha-1} d\tau$$

$$+ \frac{G\alpha}{\Gamma(\alpha)AB(\alpha)} \int_0^t \sup_{\tau\in[0,t]} |V_2(\tau) - V_{2_1}(\tau)| (t-\tau)^{\alpha-1} dB(\tau)$$

$$< \frac{1-\alpha}{AB(\alpha)} \frac{1}{R_1C_1} \|V_2(t) - V_{2_1}(t)\|_\infty$$

$$+ \frac{\alpha}{AB(\alpha)\Gamma(\alpha))} \frac{1}{R_1C_1} \frac{T^\alpha}{\alpha} \|V_2(t) - V_{2_1}(t)\|_\infty$$

$$+ \frac{GL\alpha}{\Gamma(\alpha)AB(\alpha)} \|V_2(t) - V_{2_1}(t)\|_\infty$$

$$< \left(\frac{1-\alpha}{AB(\alpha)} \frac{1}{R_1C_1} + \frac{\alpha}{AB(\alpha)\Gamma(\alpha))} \frac{1}{R_1C_1} \frac{T^\alpha}{\alpha} \right.$$

$$+ \left. \frac{GL\alpha}{\Gamma(\alpha)AB(\alpha)} \right) \|V_2(t) - V_{2_1}(t)\|_\infty$$

$$< L_1 \|V(t) - V_1(t)\|_\infty$$

We assume that $B(t)$ is Lipschitzian on $[0,T]$ with constant L, then

$$|\Gamma(V_2(t))| = \left| \frac{1-\alpha}{AB(\alpha)} f(t,V_2(t)) + \frac{\alpha}{AB(\alpha)\Gamma(\alpha))} \int_0^t f(\tau,V_2(\tau))(t-\tau)^{\alpha-1} d\tau \right.$$

$$\left. + \frac{G\alpha}{\Gamma(\alpha)AB(\alpha)} \int_0^t V_2(\tau)(t-\tau)^{\alpha-1} dB(\tau) \right|$$

$$< \frac{1-\alpha}{AB(\alpha)} |f(t,V_2(t))|$$

$$+ \frac{\alpha}{AB(\alpha)\Gamma(\alpha))} \int_0^t |f(\tau,V_2(\tau))| (t-\tau)^{\alpha-1} d\tau$$

$$+ \frac{G\alpha}{\Gamma(\alpha)AB(\alpha)} \int_0^t |V_2(\tau)| (t-\tau)^{\alpha-1} dB(\tau)$$

$$< \frac{1-\alpha}{AB(\alpha)} \sup_{l\in[0,t]} |f(l,V_2(l))|$$

$$+ \frac{\alpha}{AB(\alpha)\Gamma(\alpha))} \int_0^t \sup_{l\in[0,\tau]} |f(l,V_2(l))| (t-\tau)^{\alpha-1} d\tau$$

$$+\frac{G\alpha}{\Gamma(\alpha)AB(\alpha)}\int_0^t \sup_{l\in[0,\tau]}|V_2(l)|(t-\tau)^{\alpha-1}dB(\tau)$$

This shows that our mapping has a unique solution. We now present the numerical solution of the equation

$$V_2(t_{n+1})=V_2(0)+\frac{1-\alpha}{AB(\alpha)}f(t_n,V_2(t_n))$$

$$+\frac{\alpha}{AB(\alpha)\Gamma(\alpha)}\int_0^{t_{n+1}}f(\tau,V_2(\tau))(t-\tau)^{\alpha-1}d\tau$$

$$+\frac{G\alpha}{\Gamma(\alpha)AB(\alpha)}\int_0^{t_{n+1}}V_2(\tau)(t-\tau)^{\alpha-1}dB(\tau)$$

$$=V_2(0)+\frac{1-\alpha}{AB(\alpha)}f(t_n,V_2(t_n))$$

$$+\frac{\alpha}{AB(\alpha)\Gamma(\alpha)}\sum_{j=0}^n\int_{t_j}^{t_{j+1}}f(\tau,V_2(\tau))(t-\tau)^{\alpha-1}d\tau$$

$$+\frac{G\alpha}{\Gamma(\alpha)AB(\alpha)}\sum_{j=0}^n\int_{t_j}^{t_{j+1}}V_2(\tau)(t-\tau)^{\alpha-1}dB(\tau)$$

$$=V_2(0)+\frac{(\Delta t)^\alpha}{AB(\alpha)\Gamma(\alpha+1)}\sum_{j=2}^n f(t_{j-2},V_2^{j-2})[(n-j+1)^\alpha-(n-j)^\alpha]$$

$$+\frac{\alpha(\Delta t)^\alpha}{AB(\alpha)\Gamma(\alpha+2)}\sum_{j=2}^n\left[f(t_{j-1},V_2^{j-1})-f(t_{j-2},V_2^{j-2})\right]$$

$$\times[(n-j+1)^\alpha(n-j+3+2\alpha)-(n-j)^\alpha)(n-j+3+3\alpha)]$$

$$+\frac{(\Delta t)^\alpha\alpha}{2AB(\alpha)\Gamma(\alpha+3)}\left[f(t_j,V_2^j)-2f(t_{j-1},V_2^{j-1})+f(t_{j-2},V_2^{j-2})\right]$$

$$\times[(n-j+1)^\alpha\left(2(n-j)^2+(3\alpha+10)(n-j)+2\alpha^2+9\alpha+12\right)$$

$$-(n-j)^\alpha\left(2(n-j)^2+(5\alpha+10)(n-j)+6\alpha^2+18\alpha+12\right)]$$

$$+\frac{G\alpha(\Delta t)^\alpha}{AB(\alpha)\Gamma(\alpha+1)}\sum_{j=0}^n V(t_j)[(n-j+1)^\alpha-(n-j)^\alpha]\left(B(t_{j+1})-B(t_j)\right)$$

Then, we obtain

$$V_2(t_{n+1})=V_2(0)+\frac{\alpha(\Delta t)^\alpha}{AB(\alpha)\Gamma(\alpha+1)}\sum_{j=2}^n\left(-\frac{1}{R_1C_1}V_2(t_{j-2})-\frac{1}{R_1C_2}V_1(t_{j-2})\right.$$

$$\left.+\frac{R_2}{R_1}\frac{dV_1(t_{j-2})}{dt}\right)[(n-j+1)^\alpha-(n-j)^\alpha]$$

$$+\frac{\alpha(\Delta t)^\alpha}{AB(\alpha)\Gamma(\alpha+2)}\sum_{j=2}^n\left[\left(-\frac{1}{R_1C_1}V_2(t_{j-1})\right.\right.$$

$$-\frac{1}{R_1C_2}V_1(t_{j-1})+\frac{R_2}{R_1}\frac{dV_1(t_{j-1})}{dt}\Bigg)$$

$$-\left(-\frac{1}{R_1C_1}V_2(t_{j-2})-\frac{1}{R_1C_2}V_1(t_{j-2})+\frac{R_2}{R_1}\frac{dV_1(t_{j-2})}{dt}\right)\Bigg]$$

$$\times\left[(n-j+1)^\alpha(n-j+3+2\alpha)-(n-j^\alpha)(n-j+3+3\alpha)\right]$$

$$+\frac{\alpha(\Delta t)^\alpha}{2AB(\alpha)\Gamma(\alpha+3)}\left[\left(-\frac{1}{R_1C_1}V_2(t_j)-\frac{1}{R_1C_2}V_1(t_j)+\frac{R_2}{R_1}\frac{dV_1(t_j)}{dt}\right)\right.$$

$$-2\left(-\frac{1}{R_1C_1}V_2(t_{j-1})-\frac{1}{R_1C_2}V_1(t_{j-1})+\frac{R_2}{R_1}\frac{dV_1(t_{j-1})}{dt}\right)$$

$$+\left(-\frac{1}{R_1C_1}V_2(t_{j-2})-\frac{1}{R_1C_2}V_1(t_{j-2})+\frac{R_2}{R_1}\frac{dV_1(t_{j-2})}{dt}\right)\Bigg]$$

$$\times\left[(n-j+1)^\alpha\left(2(n-j)^2+(3\alpha+10)(n-j)+2\alpha^2+9\alpha+12\right)\right.$$

$$-(n-j)^\alpha\left(2(n-j)^2+(5\alpha+10)(n-j)+6\alpha^2+18\alpha+12\right)\Big]$$

$$+\frac{G\alpha(\Delta t)^\alpha}{AB(\alpha)\Gamma(\alpha+1)}\sum_{j=0}^{n}i(t_j)\left[(n-j+1)^\alpha-(n-j)^\alpha\right]\left(B(t_{j+1})-B(t_j)\right)$$

We consider the following differential equation

$$\frac{dV_2}{dt}=-\frac{1}{RC}V_2(t)+\frac{1}{2RC}V_1(t)-\frac{dV_1(t)}{dt}=f(t,V_2(t)) \qquad (21.120)$$

We convert the above differential equation to the fractional stochastic differential equation as: In Mittag-Leffler case we have

$$V_2(t)-V_2(0)=\frac{1-\alpha}{AB(\alpha)}f(t,V_2(t))+\frac{\alpha}{AB(\alpha)\Gamma(\alpha)}\int_0^t f(\tau,V_2(\tau))(t-\tau)^{\alpha-1}d\tau$$

$$+\frac{G\alpha}{\Gamma(\alpha)AB(\alpha)}\int_0^t V_2(\tau)(t-\tau)^{\alpha-1}dB(\tau)$$

We define a mapping

$$|\Gamma(V_2(t))|=\left|\frac{1-\alpha}{AB(\alpha)}f(t,V_2(t))+\frac{\alpha}{AB(\alpha)\Gamma(\alpha)}\int_0^t f(\tau,V_2(\tau))(t-\tau)^{\alpha-1}d\tau\right.$$

$$+\frac{G\alpha}{\Gamma(\alpha)AB(\alpha)}\int_0^t V_2(\tau)(t-\tau)^{\alpha-1}dB(\tau)\Bigg|$$

$$<\frac{1-\alpha}{AB(\alpha)}|f(t,V_2(t))|$$

$$+\frac{\alpha}{AB(\alpha)\Gamma(\alpha)}\int_0^t|f(\tau,V_2(\tau))|(t-\tau)^{\alpha-1}d\tau$$

$$+\frac{G\alpha}{\Gamma(\alpha)AB(\alpha)}\int_0^t|V_2(\tau)|(t-\tau)^{\alpha-1}dB(\tau)$$

$$
\begin{aligned}
&< \frac{1-\alpha}{AB(\alpha)} \sup_{t \in [0,T]} |f(t,V_2(t))| \\
&\quad + \frac{\alpha}{AB(\alpha)\Gamma(\alpha)} \int_0^t \sup_{\tau \in [0,t]} |f(\tau,V_2(\tau))| \, (t-\tau)^{\alpha-1} d\tau \\
&\quad + \frac{G\alpha}{\Gamma(\alpha)AB(\alpha)} \int_0^t \sup_{\tau \in [0,t]} |V_2(\tau)| \, (t-\tau)^{\alpha-1} dB(\tau) \\
&< \frac{1-\alpha}{AB(\alpha)} \|f(t,V_2(t))\|_\infty \\
&\quad + \frac{T^\alpha \alpha}{AB(\alpha)\Gamma(\alpha+1))} \|f(t,V_2(t))\|_\infty \\
&\quad + \frac{GL\alpha}{\Gamma(\alpha)AB(\alpha)} \|V_2(t)\|_\infty \\
&< M_1 + M_2 + M_3 = M
\end{aligned}
$$

We have

$$
\begin{aligned}
|\Gamma(V_2) - \Gamma(V_{2_1})| = &\left| \frac{1-\alpha}{AB(\alpha)} \left(f(t,V_2(t)) - f(t,V_{2_1}(t)) \right) \right. \\
&+ \frac{\alpha}{AB(\alpha)\Gamma(\alpha)} \int_0^t \left(f(\tau,V_2(\tau)) - f(\tau,V_{2_1}(\tau)) \right) (t-\tau)^{\alpha-1} d\tau \\
&\left. + \frac{G\alpha}{\Gamma(\alpha)AB(\alpha)} \int_0^t \left(V_2(\tau) - V_{2_1}(\tau) \right) (t-\tau)^{\alpha-1} dB(\tau) \right| \\
<&\frac{1-\alpha}{AB(\alpha)} |f(t,V_2(t)) - f(t,V_{2_1}(t))| \\
&+ \frac{\alpha}{AB(\alpha)\Gamma(\alpha)} \int_0^t |f(\tau,V_2(\tau)) - f(\tau,V_{2_1}(\tau))| \, (t-\tau)^{\alpha-1} d\tau \\
&+ \frac{G\alpha}{\Gamma(\alpha)AB(\alpha)} \int_0^t |V_2(\tau) - V_{2_1}(\tau)| \, (t-\tau)^{\alpha-1} dB(\tau) \\
<&\frac{1-\alpha}{AB(\alpha)} \sup_{t \in [0,T]} |f(t,V_2(t)) - f(t,V_{2_1}(t))| \\
&+ \frac{\alpha}{AB(\alpha)\Gamma(\alpha)} \int_0^t \sup_{\tau \in [0,t]} |f(\tau,V_2(\tau)) \quad f(\tau,V_{2_1}(\tau))| \, (t-\tau)^{\alpha-1} d\tau \\
&+ \frac{G\alpha}{\Gamma(\alpha)AB(\alpha)} \int_0^t \sup_{\tau \in [0,t]} |V_2(\tau) - V_{2_1}(\tau)| \, (t-\tau)^{\alpha-1} dB(\tau) \\
<&\frac{1-\alpha}{AB(\alpha)} \frac{1}{RC} \|V_2(t) - V_{2_1}(t)\|_\infty \\
&+ \frac{\alpha}{AB(\alpha)\Gamma(\alpha)} \frac{1}{RC} \frac{T^\alpha}{\alpha} \|V_2(t) - V_{2_1}(t)\|_\infty \\
&+ \frac{GL\alpha}{\Gamma(\alpha)AB(\alpha)} \|V_2(t) - V_{2_1}(t)\|_\infty
\end{aligned}
$$

$$< \left(\frac{1-\alpha}{AB(\alpha)} \frac{1}{RC} + \frac{\alpha}{AB(\alpha)\Gamma(\alpha)} \frac{1}{RC} \frac{T^{\alpha}}{\alpha} \right.$$

$$\left. + \frac{GL\alpha}{\Gamma(\alpha)AB(\alpha)} \right) \|V_2(t) - V_{2_1}(t)\|_{\infty}$$

$$< L_1 \|V(t) - V_1(t)\|_{\infty}$$

We assume that $B(t)$ is Lipschitzian on $[0, T]$ with constant L, then

$$|\Gamma(V_2(t))| = \left| \frac{1-\alpha}{AB(\alpha)} f(t, V_2(t)) + \frac{\alpha}{AB(\alpha)\Gamma(\alpha)} \int_0^t f(\tau, V_2(\tau))(t-\tau)^{\alpha-1} d\tau \right.$$

$$\left. + \frac{G\alpha}{\Gamma(\alpha)AB(\alpha)} \int_0^t V_2(\tau)(t-\tau)^{\alpha-1} dB(\tau) \right|$$

$$< \frac{1-\alpha}{AB(\alpha)} |f(t, V_2(t))|$$

$$+ \frac{\alpha}{AB(\alpha)\Gamma(\alpha)} \int_0^t |f(\tau, V_2(\tau))| (t-\tau)^{\alpha-1} d\tau$$

$$+ \frac{G\alpha}{\Gamma(\alpha)AB(\alpha)} \int_0^t |V_2(\tau)| (t-\tau)^{\alpha-1} dB(\tau)$$

$$< \frac{1-\alpha}{AB(\alpha)} \sup_{l \in [0,t]} |f(l, V_2(l))|$$

$$+ \frac{\alpha}{AB(\alpha)\Gamma(\alpha)} \int_0^t \sup_{l \in [0,\tau]} |f(l, V_2(l))| (t-\tau)^{\alpha-1} d\tau$$

$$+ \frac{G\alpha}{\Gamma(\alpha)AB(\alpha)} \int_0^t \sup_{l \in [0,\tau]} |V_2(l)| (t-\tau)^{\alpha-1} dB(\tau)$$

This shows that our mapping has a unique solution. We now present the numerical solution of the equation

$$V_2(t_{n+1}) = V_2(0) + \frac{1-\alpha}{AB(\alpha)} f(t_n, V_2(t_n))$$

$$+ \frac{\alpha}{AB(\alpha)\Gamma(\alpha)} \int_0^{t_{n+1}} f(\tau, V_2(\tau))(t-\tau)^{\alpha-1} d\tau$$

$$+ \frac{G\alpha}{\Gamma(\alpha)AB(\alpha)} \int_0^{t_{n+1}} V_2(\tau)(t-\tau)^{\alpha-1} dB(\tau)$$

$$= V_2(0) + \frac{1-\alpha}{AB(\alpha)} f(t_n, V_2(t_n))$$

$$+ \frac{\alpha}{AB(\alpha)\Gamma(\alpha)} \sum_{j=0}^{n} \int_{t_j}^{t_{j+1}} f(\tau, V_2(\tau))(t-\tau)^{\alpha-1} d\tau$$

$$+ \frac{G\alpha}{\Gamma(\alpha)AB(\alpha)} \sum_{j=0}^{n} \int_{t_j}^{t_{j+1}} V_2(\tau)(t-\tau)^{\alpha-1} dB(\tau)$$

$$= V_2(0) + \frac{(\Delta t)^{\alpha}}{AB(\alpha)\Gamma(\alpha+1)} \sum_{j=2}^{n} f(t_{j-2}, V_2^{j-2})\left[(n-j+1)^{\alpha} - (n-j)^{\alpha}\right]$$

$$+ \frac{\alpha(\Delta t)^{\alpha}}{AB(\alpha)\Gamma(\alpha+2)} \sum_{j=2}^{n} \left[f(t_{j-1}, V_2^{j-1}) - f(t_{j-2}, V_2^{j-2})\right]$$

$$\times \left[(n-j+1)^{\alpha}(n-j+3+2\alpha) - (n-j)^{\alpha}(n-j+3+3\alpha)\right]$$

$$+ \frac{(\Delta t)^{\alpha}\alpha}{2AB(\alpha)\Gamma(\alpha+3)} \left[f(t_j, V_2^{j}) - 2f(t_{j-1}, V_2^{j-1}) + f(t_{j-2}, V_2^{j-2})\right]$$

$$\times \left[(n-j+1)^{\alpha}\left(2(n-j)^2 + (3\alpha+10)(n-j) + 2\alpha^2 + 9\alpha + 12\right)\right.$$

$$\left.-(n-j)^{\alpha}\left(2(n-j)^2 + (5\alpha+10)(n-j) + 6\alpha^2 + 18\alpha + 12\right)\right]$$

$$+ \frac{G\alpha(\Delta t)^{\alpha}}{AB(\alpha)\Gamma(\alpha+1)} \sum_{j=0}^{n} V(t_j)\left[(n-j+1)^{\alpha} - (n-j)^{\alpha}\right]\left(B(t_{j+1}) - B(t_j)\right)$$

Then, we obtain

$$V_2(t_{n+1}) = V_2(0) + \frac{\alpha(\Delta t)^{\alpha}}{AB(\alpha)\Gamma(\alpha+1)} \sum_{j=2}^{n} \left(-\frac{1}{RC}V_2(t_{j-2}) + \frac{1}{2RC}V_1(t_{j-2}) - \frac{dV_1(t_{j-2})}{dt}\right)$$

$$\times \left[(n-j+1)^{\alpha} - (n-j)^{\alpha}\right]$$

$$+ \frac{\alpha(\Delta t)^{\alpha}}{AB(\alpha)\Gamma(\alpha+2)} \sum_{j=2}^{n} \left[\left(-\frac{1}{RC}V_2(t_{j-1}) + \frac{1}{2RC}V_1(t_{j-1}) - \frac{dV_1(t_{j-1})}{dt}\right)\right.$$

$$\left.-\left(-\frac{1}{RC}V_2(t_{j-2}) + \frac{1}{2RC}V_1(t_{j-2}) - \frac{dV_1(t_{j-2})}{dt}\right)\right]$$

$$\times \left[(n-j+1)^{\alpha}(n-j+3+2\alpha) - (n-j^{\alpha})(n-j+3+3\alpha)\right]$$

$$+ \frac{\alpha(\Delta t)^{\alpha}}{2AB(\alpha)\Gamma(\alpha+3)} \left[\left(-\frac{1}{RC}V_2(t_j) + \frac{1}{2RC}V_1(t_j) - \frac{dV_1(t_j)}{dt}\right)\right.$$

$$-2\left(-\frac{1}{RC}V_2(t_{j-1}) + \frac{1}{2RC}V_1(t_{j-1}) - \frac{dV_1(t_{j-1})}{dt}\right) -$$

$$\left.+\left(-\frac{1}{RC}V_2(t_{j-2}) + \frac{1}{2RC}V_1(t_{j-2}) - \frac{dV_1(t_{j-2})}{dt}\right)\right]$$

$$\times \left[(n-j+1)^{\alpha}\left(2(n-j)^2 + (3\alpha+10)(n-j) + 2\alpha^2 + 9\alpha + 12\right)\right.$$

$$\left.-(n-j)^{\alpha}\left(2(n-j)^2 + (5\alpha+10)(n-j) + 6\alpha^2 + 18\alpha + 12\right)\right]$$

$$+ \frac{G\alpha(\Delta t)^{\alpha}}{AB(\alpha)\Gamma(\alpha+1)} \sum_{j=0}^{n} i(t_j)\left[(n-j+1)^{\alpha} - (n-j)^{\alpha}\right]\left(B(t_{j+1}) - B(t_j)\right)$$

22 Numerical Simulations of Some Circuit Problems

Fractional differential operators based on the power-law functions have been used in this chapter to include in mathematical equations describing the dynamics of a circuit system and the effect of fading and crossover behaviors. These models are second-orders circuit systems on the one hand. On the other hand, models with classical integral were extended by replacing the classical integral with fractional integral including the Riemann-Liouville integral. Suitable numerical methods were used to provide numerical solutions to these modified models [88–91].

22.1 FIRST PROBLEM

We consider the following system:

$$\begin{cases} \frac{dV}{dt} = \frac{1}{C}i \\ iR + L\frac{di}{dt} + V = \frac{d^2V}{dt^2} + \frac{R}{L}\frac{dV}{dt} + \frac{1}{LC}V \end{cases}$$

We use

$$\frac{di}{dt} = C\frac{d^2V}{dt^2}$$

Then, we obtain

$$CR\frac{dV}{dt} + LC\frac{d^2V}{dt^2} + V = \frac{d^2V}{dt^2} + \frac{R}{L}\frac{dV}{dt} + \frac{1}{LC}V$$

$$(LC-1)\frac{d^2V}{dt^2} + \left(CR - \frac{R}{L}\right)\frac{dV}{dt} + \left(1 - \frac{1}{LC}\right)V = 0$$

$$\frac{d^2V}{dt^2} + c_1\frac{dV}{dt} + c_0V = 0$$

where

$$c_1 = \frac{CR - \frac{R}{L}}{LC-1}, \quad c_0 = \frac{1 - \frac{1}{LC}}{LC-1}$$

DOI: 10.1201/9781003359869-22

Then, we fractionalize the $\frac{dV}{dt}$ and obtain

$$\frac{d^2V}{dt^2} + c_1 \frac{d^\alpha V}{dt^\alpha} + c_0 V = 0$$

We demonstrate the numerical simulation of the solution of this problem for different values of α by Figure 22.1.

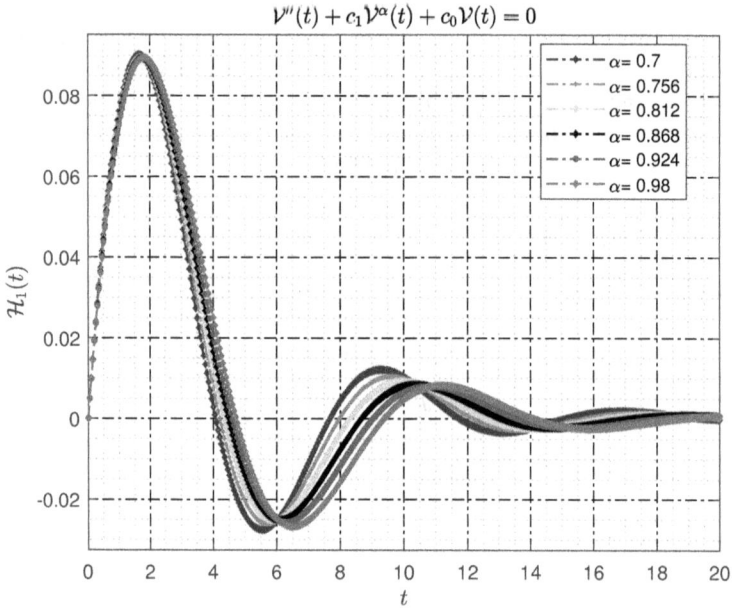

Figure 22.1 Numerical simulation of problem I for $c_1 = c_0 = 0.5$.

22.2   SECOND PROBLEM

We consider the following problem

$$\frac{V}{R} + \frac{1}{L} \int_0^t V(\tau) d\tau + C \frac{dV}{dt} = 0$$

We use

$$V = L \frac{di}{dt}$$

Then, we obtain

$$\frac{L}{R}\frac{di}{dt} + i(t) + CL\frac{d^2i}{dt^2} = 0$$

$$\frac{d^2i}{dt^2} + \frac{1}{RC}\frac{di}{dt} + \frac{1}{LC}i(t) = 0$$

Thus, we get

$$\frac{d^2i}{dt^2} + c_1\frac{di}{dt} + c_0 i(t) = 0$$

where

$$c_1 = \frac{1}{RC}, \quad c_0 = \frac{1}{LC}$$

Then, we fractionalize the $\frac{di}{dt}$ and obtain

$$\frac{d^2i}{dt^2} + c_1\frac{d^\alpha i}{dt^\alpha} + c_0 i(t) = 0$$

We demonstrate the numerical simulation of the solution of this problem for different values of α by Figure 22.2.

22.3 THIRD PROBLEM

We consider

$$\frac{V(t)}{R} + \frac{1}{L}\int_0^t V(\tau)d\tau + C\frac{dV}{dt} = \frac{d^2i}{dt^2} + \frac{1}{RC}\frac{di}{dt} + \frac{i}{CL}$$

We use

$$V = L\frac{di}{dt}$$

Then, we obtain

$$(LC-1)\frac{d^2}{dt^2} + \left(\frac{LC-1}{RC}\right)\frac{di}{dt} + \left(1 - \frac{1}{CL}\right)i$$

$$\frac{d^2}{dt^2} + c_1\frac{di}{dt} + c_0 i$$

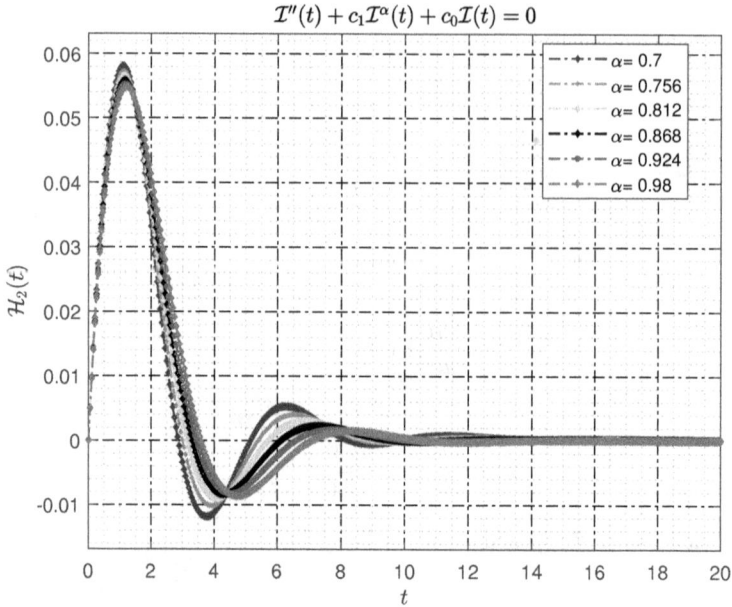

Figure 22.2   Numerical simulation of problem II for $c_1 = c_0 = 1$.

where

$$c_1 = \left(\frac{1}{RC}\right), \; c_0 = \left(\frac{1}{CL}\right)$$

Then, we fractionalize the $\frac{di}{dt}$ and obtain

$$\frac{d^2 i}{dt^2} + c_1 \frac{d^\alpha i}{dt^\alpha} + c_0 i(t) = 0$$

We demonstrate the numerical simulation of the solution of this problem for different values of α by Figure 22.3.

22.4   FOURTH PROBLEM

We consider

$$\frac{1}{L} \int_0^t V(\tau) d\tau + C\frac{dV}{dt} = \frac{i}{CL} + \frac{d^2 i}{dt^2}$$

We use

$$\mathcal{I}''(t) + c_1 \mathcal{I}^\alpha(t) + c_0 \mathcal{I}(t) = 0$$

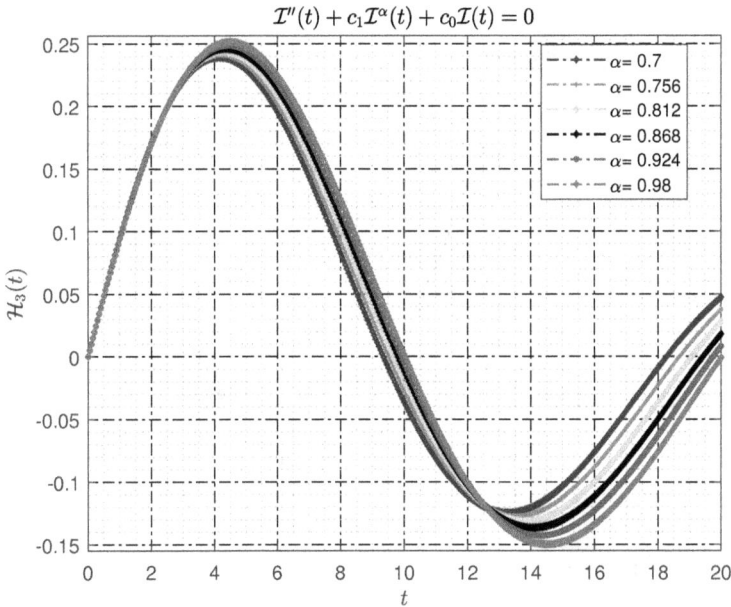

Figure 22.3 Numerical simulation of problem III for $c_1 = c_0 = 0.1$.

$$V(t) = L\frac{di}{dt}$$

Then, we obtain

$$\frac{1}{L}\int_0^t V(\tau)d\tau + C\frac{dV}{dt} = \frac{i}{CL} + \frac{d^2i}{dt^2}$$

$$(LC-1)\frac{d^2i}{dt^2} + \left(1 - \frac{1}{CL}\right)i = 0$$

$$\frac{d^2i}{dt^2} + \frac{1}{CL}i = 0$$

Then, we fractionalize the $\frac{d^2i}{dt^2}$ and obtain

$$\frac{d^\alpha i}{dt^\alpha} + \frac{1}{CL}i = 0$$

We demonstrate the numerical simulation of the solution of this problem for different values of α by Figure 22.4.

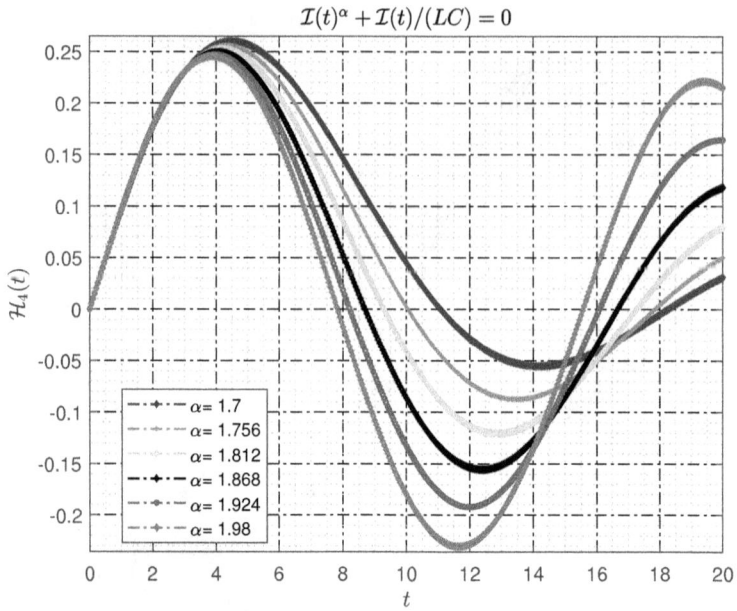

Figure 22.4   Numerical simulation of problem IV for $L = 2$ and $C = 3$.

23 Applications of General Integral Transform

In this chapter we apply the new general integral transform to obtain the transfer functions. The obtained transfer functions are new in the literature. Therefore, they will be very useful for mathematicians and engineers. We use power-law, exponential-decay, and Mittag-Leffler kernels. We present the applications of the circuit problems by the new general integral transform. In the circuit problems we check the effect of the three different kernels. We demonstrate the numerical simulations to prove the efficiency of the general integral transform. We use many integral transforms and obtain very interesting transfer functions.

An integral transform maps a function from its original function space into another function space by integration. Some of the features of the original function might be more easily characterized and manipulated than in the original function space. The transformed function can generally be mapped back to the original function space using the inverse transform. We have differential and integral operators with the different kernels [1–3, 5]. We have presented many useful integral transforms in this work. We have applied these transforms to many circuit problems in this paper.

Let $u(t)$ be an integrable function described for $t \geq 0$, $p(s) \neq 0$ and $q(s)$ are positive real functions. The general integral transform $T(s)$ of $u(t)$ has been presented by [8]:

$$T[u(t); s] = T(s) = p(s) \int_0^\infty u(t) \exp(-q(s)t)\, dt \qquad (23.1)$$

provided the integral exists for some $q(s)$.

This transform is very useful in many applications. We can obtain many integral transforms by choosing $p(s)$ and $q(s)$. We will obtain Mohand transform, Sawi transform, Elzaki transform, Aboodh transform, Pourreza transform, α integral Laplace transform, Kamal transform, G-transform, and Natural transform.

Linear physical system with one or multiple set of input and output can be presented by mathematical functions that depend on any of the outputs to any of the inputs. These functions are unique and are described relied on the systems governing equations. Transfer functions are described for any essentail set of input and output functions that may depend on the input and output together [7].

23.1 GENERAL INTEGRAL TRANSFORM

We define the following integral transforms that will be used in the following sections [8].

DOI: 10.1201/9781003359869-23

We have the definition of the Elzaki transform as [18]

$$E(f(t)) = s \int_0^\infty e^{-\frac{t}{s}} f(t)dt \tag{23.2}$$

We present the definition of the Aboodh transform as [19]

$$A(f(t)) = \frac{1}{s} \int_0^\infty \exp(-st)f(t)dt \tag{23.3}$$

We present the definition of the Pourreza transform as [20, 21]

$$P(f(t)) = s \int_0^\infty \exp(-s^2t)f(t)dt \tag{23.4}$$

We present the definition of the Mohand transform as [22]

$$R(f(t)) = s^2 \int_0^\infty \exp(-st)f(t)dt \tag{23.5}$$

We present the definition of the Sawi transform as [23]

$$Sa(f(t)) = \frac{1}{s^2} \int_0^\infty \exp(-\frac{t}{s})f(t)dt \tag{23.6}$$

We present the definition of Kamal transform as [24]

$$K(f(t)) = \int_0^\infty \exp\left(-\frac{t}{s}\right) f(t)dt \tag{23.7}$$

We present the definition of the G-transform as [25, 26]

$$G(f(t)) = s^\alpha \int_0^\infty \exp\left(-\frac{t}{s}\right) f(t)dt \tag{23.8}$$

We present the definition of the Natural transform as [27]

$$N(f(t)) = R(s,u) = \int_0^\infty \exp(-st)f(ut)dt \tag{23.9}$$

We consider t^{-a} and $\exp\left(-\frac{\alpha}{1-\alpha}\right)$.

23.1.1 MOHAND TRANSFORM

We find the Mohand transform of these functions as

$$M\left[t^{-\alpha}\right] = s^2 \int_0^\infty t^{-\alpha} \exp(-st)dt$$

If we use $st = u$ transform, then we will get

$$
\begin{aligned}
M\left[t^{-\alpha}\right] &= s^2 \int_0^\infty \left(\frac{u}{s}\right)^{-\alpha} \exp(-u)\frac{du}{s} \\
&= s^{\alpha+1} \int_0^\infty \exp(-u)u^{1-\alpha-1}du \\
&= s^{\alpha+1}\Gamma(1-\alpha)
\end{aligned}
$$

and

$$M\left[\exp\left(\frac{-\alpha}{1-\alpha}t\right)\right] = s^2\int_0^\infty \exp\left(\frac{-\alpha}{1-\alpha}t\right)\exp(-st)dt$$

$$= s^2\int_0^\infty \exp\left(-t\left(\frac{\alpha}{1-\alpha}\right)+s\right)dt$$

$$= s^2\left[\frac{1}{\frac{\alpha}{1-\alpha}+s}\right]$$

$$= \frac{s^2(1-\alpha)}{\alpha+s(1-\alpha)}$$

23.1.2 SAWI TRANSFORM

We find the Sawi transform of these functions as

$$Sa\left[t^{-\alpha}\right] = \frac{1}{s^2}\int_0^\infty \exp\left(\frac{-t}{s}\right)t^{-\alpha}dt$$

If we use $u = \frac{t}{s}$ transform, then we will get

$$Sa\left[t^{-\alpha}\right] = \frac{1}{s^2}\int_0^\infty \exp(-u)(su)^{-\alpha}sdu$$

$$= s^{-\alpha-1}\int_0^\infty \exp(-u)u^{1-\alpha-1}du$$

$$= s^{-\alpha-1}\Gamma(1-\alpha)$$

and

$$Sa\left[\exp\left(\frac{-\alpha}{1-\alpha}t\right)\right] = \frac{1}{s^2}\int_0^\infty \exp\left(\frac{-\alpha}{1-\alpha}t\right)\exp\left(\frac{-t}{s}\right)dt$$

$$= \frac{1}{s^2}\int_0^\infty \exp\left(-t\left(\frac{\alpha}{1-\alpha}+\frac{1}{s}\right)\right)dt$$

$$= \frac{1}{s^2}\frac{1}{\frac{\alpha}{1-\alpha}+\frac{1}{s}}$$

$$= \frac{s^{-1}(1-\alpha)}{\alpha s+1-\alpha}$$

23.1.3 ELZAKI TRANSFORM

We find the Elzaki transform of these functions as

$$E\left[t^{-\alpha}\right] = s\int_0^\infty t^{-\alpha}\exp\left(\frac{-t}{s}\right)dt$$

If we use $u = \frac{t}{s}$ transform, then we will get

$$
\begin{aligned}
E\left[t^{-\alpha}\right] &= s \int_0^\infty (us)^{-\alpha} \exp(-u)s\,du \\
&= s^{2-\alpha} \int_0^\infty \exp(-u)u^{-\alpha}\,du \\
&= s^{2-\alpha} \int_0^\infty \exp(-u)u^{1-\alpha-1} \\
&= s^{2-\alpha}\Gamma(1-\alpha)
\end{aligned}
$$

and

$$
\begin{aligned}
E\left[\exp\left(\frac{-\alpha}{1-\alpha}t\right)\right] &= s \int_0^\infty \exp\left(\frac{-\alpha}{1-\alpha}t\right)\exp\left(\frac{-t}{s}\right)dt \\
&= s \int_0^\infty \exp\left(-t\left(\frac{\alpha}{1-\alpha}+\frac{1}{s}\right)\right)dt \\
&= s\left(\frac{1}{\frac{\alpha}{1-\alpha}+\frac{1}{s}}\right) \\
&= \frac{s^2(1-\alpha)}{\alpha s + 1 - \alpha} \\
&= \frac{s^2(1-\alpha)}{\alpha s + 1 - \alpha}
\end{aligned}
$$

23.1.4 ABOODH TRANSFORM

We find the Aboodh transform of these functions as

$$
A\left[t^{-\alpha}\right] = \frac{1}{s}\int_0^\infty t^{-\alpha}\exp(-st)dt
$$

If we use $st = u$ transform, then we will get

$$
\begin{aligned}
A\left[t^{-\alpha}\right] &= \frac{1}{s}\int_0^\infty \left(\frac{u}{s}\right)^{-\alpha}\exp(-u)\frac{du}{s} \\
&= s^{\alpha-2}\int_0^\infty u^{-\alpha}\exp(-u)du \\
&= s^{\alpha-2}\int_0^\infty \exp(-u)u^{1-\alpha-1}du \\
&= s^{\alpha-2}\Gamma(1-\alpha)
\end{aligned}
$$

and

$$A\left[\exp\left(\frac{-\alpha}{1-\alpha}t\right)\right] = \frac{1}{s}\int_0^\infty \exp\left(\frac{-\alpha}{1-\alpha}t\right)\exp(-st)dt$$

$$= \frac{1}{s}\int_0^\infty \exp\left(-t\left(\frac{\alpha}{1-\alpha}+s\right)\right)dt$$

$$= \frac{1}{s}\left[\frac{1}{\frac{\alpha}{1-\alpha}+s}\right]$$

$$= \frac{1-\alpha}{s(\alpha+s(1-\alpha))}$$

23.1.5 POURREZA TRANSFORM

We find the Pourreza transform of these functions as

$$P\left[t^{-\alpha}\right] = s\int_0^\infty t^{-\alpha}\exp(-s^2t)dt$$

If we use $s^2t = u$ transform, then we will get

$$P\left[t^{-\alpha}\right] = s\int_0^\infty \left(\frac{u}{s^2}\right)^{-\alpha}\exp(-u)\frac{du}{s^2}$$

$$= s^{2\alpha-1}\int_0^\infty \exp(-u)u^{1-\alpha-1}du$$

$$= s^{2\alpha-1}\Gamma(1-\alpha)$$

and

$$P\left[\exp\left(\frac{-\alpha}{1-\alpha}t\right)\right] = s\int_0^\infty \exp\left(\frac{-\alpha}{1-\alpha}t\right)\exp(-s^2t)dt$$

$$= s\int_0^\infty \exp\left(-t\left(\frac{\alpha}{1-\alpha}+s^2\right)\right)dt$$

$$= s\left[\frac{1}{\frac{\alpha}{1-\alpha}+s^2}\right]$$

$$= \frac{s(1-\alpha)}{\alpha+s^2(1-\alpha)}$$

23.1.6 α INTEGRAL LAPLACE TRANSFORM

Since we define the fractional order with α we use β instead of α in the transform. We acquire the α integral Laplace transform of these functions as

$$L_\beta\left[t^{-\alpha}\right] = \int_0^\infty \exp\left(-s^{\frac{1}{\beta}}t\right)t^{-\alpha}dt$$

If we use $s^{\frac{1}{\beta}}t = u$ transform, then we will get

$$
\begin{aligned}
L_\beta\left[t^{-\alpha}\right] &= \int_0^\infty \exp(-u)\left(\frac{u}{s^{\frac{1}{\beta}}}\right)^{-\alpha}\frac{du}{s^{\frac{1}{\beta}}} \\
&= \int_0^\infty \frac{\exp(-u)u^{-\alpha}}{s^{\frac{-\alpha+1}{\beta}}}du \\
&= s^{\frac{\alpha-1}{\beta}}\int_0^\infty \exp(-u)u^{-\alpha-1}du \\
&= s^{\frac{\alpha-1}{\beta}}\Gamma(1-\alpha)
\end{aligned}
$$

and

$$
\begin{aligned}
L_\beta\left[\exp\left(\frac{-\alpha}{1-\alpha}t\right)\right] &= \int_0^\infty \exp\left(\frac{-\alpha}{1-\alpha}t\right)\exp\left(-s^{\frac{1}{k}}t\right)dt \\
&= \int_0^\infty \exp\left(-t\left(\frac{\alpha}{1-\alpha}+s^{\frac{1}{k}}\right)\right)dt \\
&= \frac{1}{\frac{\alpha}{1-\alpha}+s^{\frac{1}{k}}} \\
&= \frac{1-\alpha}{\alpha+s^{\frac{1}{k}}(1-\alpha)}
\end{aligned}
$$

23.1.7 KAMAL TRANSFORM

We find the Kamal transform of these functions as

$$
K\left[t^{-\alpha}\right] = \int_0^\infty t^{-\alpha}\exp\left(\frac{-t}{s}\right)dt
$$

If we use $\frac{t}{s} = u$ transform, then we will get

$$
\begin{aligned}
K\left[t^{-\alpha}\right] &= \int_0^\infty (us)^{-\alpha}\exp(-u)sdu \\
&= s^{1-\alpha}\int_0^\infty \exp(-u)u^{1-\alpha-1}du \\
&= s^{1-\alpha}\Gamma(1-\alpha)
\end{aligned}
$$

and

$$
\begin{aligned}
K\left[\exp\left(\frac{-\alpha}{1-\alpha}t\right)\right] &= \int_0^\infty \exp\left(\frac{-\alpha}{1-\alpha}t\right)\exp\left(\frac{-t}{s}\right)dt \\
&= \int_0^\infty \exp\left(-t\left(\frac{\alpha}{1-\alpha}+\frac{1}{s}\right)\right)dt \\
&= \frac{1}{\frac{\alpha}{1-\alpha}}+\frac{1}{s} = \frac{s(1-\alpha)}{\alpha s+1-\alpha} \\
&= \frac{s(1-\alpha)}{\alpha s+1-\alpha}
\end{aligned}
$$

23.1.8   G TRANSFORM

We find the G transform of these functions as

$$G\left[t^{-\alpha}\right] = s^k \int_0^\infty t^{-\alpha} \exp\left(\frac{-t}{s}\right) dt$$

If we use $\frac{t}{s} = u$ transform, then we will get

$$
\begin{aligned}
G\left[t^{-\alpha}\right] &= s^k \int_0^\infty (us)^{-\alpha} \exp(-u) s\, du \\
&= s^{k-\alpha+1} \int_0^\infty \exp(-u) u^{1-\alpha-1} du \\
&= s^{k-\alpha+1} \Gamma(1-\alpha)
\end{aligned}
$$

and

$$
\begin{aligned}
G\left[\exp\left(\frac{-\alpha}{1-\alpha t}\right)\right] &= s^k \int_0^\infty s^{\frac{-\alpha}{1-\alpha}t} \exp\left(\frac{-t}{s}\right) dt \\
&= s^k \int_0^\infty \exp\left(-t\left(\frac{\alpha}{1-\alpha} + \frac{1}{s}\right)\right) dt \\
&= s^k \frac{1}{\frac{\alpha}{1-\alpha} + \frac{1}{s}} \\
&= \frac{s^k s(1-\alpha)}{\alpha s + 1 - \alpha} \\
&= \frac{s^{k+1}(1-\alpha)}{\alpha s + 1 - \alpha}
\end{aligned}
$$

23.1.9   NATURAL TRANSFORM

We find the Natural transform of these functions as

$$
\begin{aligned}
N\left[t^{-\alpha}\right] &= \int_0^\infty (ut)^{-\alpha} \exp(-st) dt \\
&= \int_0^\infty u^{-\alpha} t^{-\alpha} e^{-st} dt \\
&= \int_0^\infty u^{-\alpha} \left(\frac{k}{s}\right)^{-\alpha} \exp(-k) \frac{dk}{s} \\
&= \frac{u^{-\alpha}}{s^{1-\alpha}} \int_0^\infty \exp(-k) k^{1-\alpha-1} dk \\
&= \frac{u^{-\alpha}}{s^{1-\alpha}} \Gamma(1-\alpha)
\end{aligned}
$$

and

$$
\begin{aligned}
N[f(t)] &= \int_0^\infty f(ut)\exp(-st)dt \\
&= \int_0^\infty \exp\left(\frac{-\alpha}{1-\alpha}ut\right)\exp(-st)dt \\
&= \int_0^\infty \exp\left(-t\left(\frac{\alpha}{1-\alpha}u+s\right)\right)dt \\
&= \frac{1}{-\left(\frac{\alpha}{1-\alpha}u+s\right)} \\
&= \frac{1-\alpha}{\alpha u+s(1-\alpha)}
\end{aligned}
$$

23.2 INTEGRAL TRANSFORMS OF SOME FRACTIONAL DIFFERENTIAL EQUATIONS

We obtain [4]

$$
{}_0^C D_t^\alpha u(t) = \frac{du(t)}{dt} * \frac{t^{-\alpha}}{\Gamma(1-\alpha)}. \tag{23.10}
$$

$$
{}_0^{CF} D_t^\alpha u(t) = \frac{du(t)}{dt} * \frac{M(\alpha)}{1-\alpha}\exp\left(-\frac{\alpha}{1-\alpha}t\right). \tag{23.11}
$$

$$
{}_0^{ABC} D_t^\alpha u(t) = \frac{du(t)}{dt} * \frac{AB(\alpha)}{1-\alpha}E_\alpha\left[\frac{-\alpha}{1-\alpha}t^\alpha\right]. \tag{23.12}
$$

We have the Elzaki transform of these derivatives as

$$
E\left({}_0^{CF}D_t^\alpha f\right) = \left(\frac{E(f)}{s^2}-f(0)\right)\frac{s^2 M(\alpha)}{\alpha(s-1)+1} \tag{23.13}
$$

$$
E\left({}_0^C D_t^\alpha f\right) = \frac{E(f)}{s^\alpha}-\frac{f(0)}{s^{\alpha-2}} \tag{23.14}
$$

$$
E\left({}_0^{ABC}D_t^\alpha f\right) = (E(f)-f(0))\frac{AB(\alpha)}{s^{-2}(1-\alpha)+\alpha s^{\alpha-2}} \tag{23.15}
$$

We have the Aboodh transform of these derivatives as

$$
A\left({}_0^{CF}D_t^\alpha f\right) = \left(sA(f)-\frac{f(0)}{s}\right)\frac{M(\alpha)}{s+\alpha-s\alpha} \tag{23.16}
$$

$$
A\left({}_0^C D_t^\alpha f\right) = (s^2 A(f)-f(0))s^{\alpha-2} \tag{23.17}
$$

$$
A\left({}_0^{ABC}D_t^\alpha f\right) = (s^2 A(f)-f(0))\frac{AB(\alpha)s^{\alpha-2}}{s^\alpha(1-\alpha)+\alpha} \tag{23.18}
$$

We have the Pourreza transform of these derivatives as

$$P\left({}_0^{CF}D_t^\alpha f\right) = (sP(f) - \sqrt{s}f(0))\frac{M(\alpha)}{s + \alpha - s\alpha} \tag{23.19}$$

$$P\left({}_0^{C}D_t^\alpha f\right) = (sP(f) - \sqrt{s}f(0))s^{\alpha-1} \tag{23.20}$$

$$P\left({}_0^{ABC}D_t^\alpha f\right) = (sP(f) - \sqrt{s}f(0))\frac{AB(\alpha)s^{\alpha-1}}{s^\alpha(1-\alpha) + \alpha} \tag{23.21}$$

We have the Mohand transform of these derivatives as

$$R\left({}_0^{CF}D_t^\alpha f\right) = (sR(f) - s^2 f(0))\frac{R(f)}{s + \alpha - s\alpha} \tag{23.22}$$

$$R\left({}_0^{C}D_t^\alpha f\right) = (\frac{R(f)}{s} - f(0))s^{\alpha+1} \tag{23.23}$$

$$R\left({}_0^{ABC}D_t^\alpha f\right) = \left(\frac{R(f)}{s} - f(0)\right)\frac{AB(\alpha)s^{\alpha+1}}{s^\alpha(1-\alpha) + \alpha} \tag{23.24}$$

We have the Sawi transform of these derivatives as

$$S\left({}_0^{CF}D_t^\alpha f\right) = (sS(f) - f(0))\frac{M(\alpha)}{s + \alpha - s\alpha} \tag{23.25}$$

$$S\left({}_0^{C}D_t^\alpha f\right) = \left(\frac{S(f)}{s} - f(0)\right)s^{\alpha+1} \tag{23.26}$$

$$S\left({}_0^{ABC}D_t^\alpha f\right) = \left(\frac{S(f)}{s} - f(0)\right)\frac{AB(\alpha)s^{\alpha+1}}{s^\alpha(1-\alpha) + \alpha} \tag{23.27}$$

We have the Kamal transform of these derivatives as

$$K\left({}_0^{CF}D_t^\alpha f\right) = (sK(f) - f(0))\frac{M(\alpha)}{s + \alpha - s\alpha} \tag{23.28}$$

$$K\left({}_0^{C}D_t^\alpha f\right) = (sK(f) - f(0))s^{\alpha-1} \tag{23.29}$$

$$K\left({}_0^{ABC}D_t^\alpha f\right) = (sK(f) - f(0))\frac{AB(\alpha)s^{\alpha-1}}{s^\alpha(1-\alpha) + \alpha} \tag{23.30}$$

We have the α-integral Laplace transform of these derivatives as

$$L_\beta\left({}_0^{CF}D_t^\alpha f\right) = \left(s^{\frac{1}{\beta}}f(t) - f(0)\right)\frac{M(\alpha)}{\alpha + s^{\frac{1}{\beta}}(1-\alpha)} \tag{23.31}$$

$$L_\beta\left({}_0^{C}D_t^\alpha f\right) = \left(s^{\frac{1}{\beta}}f(t) - f(0)\right)s^{\frac{\alpha-1}{\beta}} \tag{23.32}$$

$$L_\beta\left({}_0^{ABC}D_t^\alpha f\right) = \left(s^{\frac{1}{\beta}}f(t) - f(0)\right)\frac{AB(\alpha)s^{\frac{\alpha-1}{\beta}}}{(1-\alpha)^{\frac{\alpha}{\beta}} + \alpha} \tag{23.33}$$

Since the fractional order is α, we use β instead of α in the transform.

We have the G-Transform of these derivatives as

$$G\left({}^{CF}_0 D_t^\alpha f\right) = \left(sG(\alpha) - \frac{f(0)}{s^\alpha}\right) \tag{23.34}$$

$$G\left({}^C_0 D_t^\alpha f\right) = (s^{\alpha+1}G(f) - f(0))s^{-1} \tag{23.35}$$

$$G\left({}^{ABC}_0 D_t^\alpha f\right) = \frac{AB(\alpha)s^{-1}}{s^\alpha(1-\alpha)+\alpha} \tag{23.36}$$

We have the Natural transform of these derivatives as

$$N\left({}^{CF}_0 D_t^\alpha f\right) = \left(\frac{N(f)}{u} - f(0)\right)\frac{M(\alpha)}{s\alpha - s - u\alpha} \tag{23.37}$$

$$N\left({}^C_0 D_t^\alpha f\right) = \left(N(f) - \frac{f(0)}{u}\right)s^\alpha u^\alpha \tag{23.38}$$

$$N\left({}^{ABC}_0 D_t^\alpha f\right) = \frac{1}{u}(N(f) - f(0))\frac{AB(\alpha)s^{\alpha-1}}{s^\alpha(1-\alpha)+\alpha} \tag{23.39}$$

23.3   GENERAL TRANSFORM OF THE MITTAG-LEFFLER FUNCTIONS

We obtain the integral transforms of the Mittag-Leffler functions in this section.

23.3.1   ABOODH TRANSFORM

Theorem 23.1

We get the Aboodh transform of the Mittag-Leffler function by:

$$A\left[E_\alpha\left(-\frac{\alpha}{1-\alpha}t^\alpha\right)\right] = \frac{1}{s^2 + \frac{\alpha s^{2-\alpha}}{1-\alpha}} \tag{23.40}$$

where

$$\left|-\frac{\alpha}{1-\alpha}s^{-\alpha}\right| < 1$$

$$|s|^{-\alpha} < \frac{1-\alpha}{\alpha}$$

$$|s| < \left(\frac{1-\alpha}{\alpha}\right)^{-\frac{1}{\alpha}}$$

$$|s| < \left(\frac{\alpha}{1-\alpha}\right)^{\frac{1}{\alpha}}$$

$$a^2 + b^2 < \left(\frac{\alpha}{1-\alpha}\right)^{\frac{2}{\alpha}}$$

■

PROOF We have

$$A[t^\alpha] = \frac{1}{s}L[t^\alpha] \tag{23.41}$$

Then, we get

$$A\left[E_\alpha\left(-\frac{\alpha}{1-\alpha}t^\alpha\right)\right] = A\left[\sum_{k=0}^{\infty}\left(\frac{(-\frac{\alpha}{1-\alpha}t^\alpha)^k}{\Gamma(ak+1)}\right)\right]$$

$$= \sum_{k=0}^{\infty}\left(\frac{(-\frac{\alpha}{1-\alpha})^k}{\Gamma(ak+1)}\right)A[t^{ak}]$$

$$= \sum_{k=0}^{\infty}\left(\frac{(-\frac{\alpha}{1-\alpha})^k}{\Gamma(ak+1)}\right)\frac{1}{s}L[t^{ak}]$$

$$= \frac{1}{s}\sum_{k=0}^{\infty}\left(\frac{(-\frac{\alpha}{1-\alpha})^k}{\Gamma(ak+1)}\right)s^{-ak-1}\Gamma(ak+1)$$

$$= \frac{1}{s^2}\sum_{k=0}^{\infty}\left(-\frac{\alpha}{1-\alpha}s^{-\alpha}\right)^k$$

$$= \frac{1}{s^2}\left(\frac{1}{1+\frac{as^{-\alpha}}{1-\alpha}}\right)$$

$$= \frac{1}{s^2+\frac{as^{2-\alpha}}{1-\alpha}}$$

∎

23.3.2   MOHAND TRANSFORM

Theorem 23.2

We get the Mohand transform of the Mittag-Leffler function by:

$$M\left[E_\alpha\left(-\frac{\alpha}{1-\alpha}t^\alpha\right)\right] = \frac{1}{s^{-1}+\frac{as^{-1-\alpha}}{1-\alpha}} \tag{23.42}$$

where

$$\left| -\frac{\alpha}{1-\alpha} s^{-\alpha} \right| < 1$$

$$|s|^{-\alpha} < \frac{1-\alpha}{\alpha}$$

$$|s| < \left(\frac{1-\alpha}{\alpha} \right)^{-\frac{1}{\alpha}}$$

$$|s| < \left(\frac{\alpha}{1-\alpha} \right)^{\frac{1}{\alpha}}$$

$$a^2 + b^2 < \left(\frac{\alpha}{1-\alpha} \right)^{\frac{2}{\alpha}}$$

∎

PROOF We have

$$M[t^\alpha] = s^2 L[t^\alpha] \tag{23.43}$$

Then, we obtain

$$M\left[E_\alpha \left(-\frac{\alpha}{1-\alpha} t^\alpha \right) \right] = M\left[\sum_{k=0}^{\infty} \left(\frac{(-\frac{\alpha}{1-\alpha} t^\alpha)^k}{\Gamma(ak+1)} \right) \right]$$

$$= \sum_{k=0}^{\infty} \left(\frac{(-\frac{\alpha}{1-\alpha})^k}{\Gamma(ak+1)} \right) M[t^{ak}]$$

$$= \sum_{k=0}^{\infty} \left(\frac{(-\frac{\alpha}{1-\alpha})^k}{\Gamma(ak+1)} \right) s^2 L[t^{ak}]$$

$$= \sum_{k=0}^{\infty} \left(\frac{(-\frac{\alpha}{1-\alpha})^k}{\Gamma(ak+1)} \right) s^2 s^{-\alpha k-1} \Gamma(ak+1)$$

$$= s \sum_{k=0}^{\infty} \left(-\frac{\alpha}{1-\alpha} s^{-\alpha} \right)^k$$

$$= s \left(\frac{1}{1 + \frac{as^{-\alpha}}{1-\alpha}} \right)$$

$$= \frac{1}{s^{-1} + \frac{as^{-1-\alpha}}{1-\alpha}}$$

∎

23.3.3 SAWI TRANSFORM

Theorem 23.3

We get the Sawi transform of the Mittag-Leffler function by:

$$Sa\left[E_\alpha\left(-\frac{\alpha}{1-\alpha}t^\alpha\right)\right] = \frac{1}{s+\frac{\alpha s^{1+\alpha}}{1-\alpha}} \tag{23.44}$$

where

$$\left|-\frac{\alpha}{1-\alpha}s^\alpha\right| < 1$$

$$|s|^\alpha < \frac{1-\alpha}{\alpha}$$

$$|s| < \left(\frac{1-\alpha}{\alpha}\right)^{\frac{1}{\alpha}}$$

$$|s| < \left(\frac{1-\alpha}{\alpha}\right)^{\frac{2}{\alpha}}$$

$$a^2 + b^2 < \left(\frac{1-\alpha}{\alpha}\right)^{\frac{2}{\alpha}}$$

■

PROOF We have

$$Sa[t^\alpha] = s^{2\alpha}L[t^\alpha] \tag{23.45}$$

Then, we obtain

$$Sa\left[E_\alpha\left(-\frac{\alpha}{1-\alpha}t^\alpha\right)\right] = Sa\left[\sum_{k=0}^\infty \left(\frac{(-\frac{\alpha}{1-\alpha}t^\alpha)^k}{\Gamma(ak+1)}\right)\right]$$

$$= \sum_{k=0}^\infty \left(\frac{(-\frac{\alpha}{1-\alpha})^k}{\Gamma(ak+1)}\right)Sa[t^{ak}]$$

$$= \sum_{k=0}^\infty \left(\frac{(-\frac{\alpha}{1-\alpha})^k}{\Gamma(ak+1)}\right)s^{2\alpha k}L[t^{ak}]$$

$$= \sum_{k=0}^\infty \left(\frac{(-\frac{\alpha}{1-\alpha})^k}{\Gamma(ak+1)}\right)s^{2\alpha k}s^{-\alpha k-1}\Gamma(ak+1)$$

$$= \sum_{k=0}^\infty \left(-\frac{\alpha}{1-\alpha}\right)^k s^{\alpha k}s^{-1}$$

$$= \frac{1}{s}\sum_{k=0}^{\infty}\left(-\frac{\alpha}{1-\alpha}s^\alpha\right)^k$$

$$= \frac{1}{s}\left(\frac{1}{1+\frac{\alpha s^\alpha}{1-\alpha}}\right)$$

$$= \frac{1}{s+\frac{\alpha s^{1+\alpha}}{1-\alpha}}$$

$\blacksquare$

23.3.4   ELZAKI TRANSFORM

Theorem 23.4

We get the Elzaki transform of the Mittag-Leffler function by:

$$E\left[E_\alpha\left(-\frac{\alpha}{1-\alpha}t^\alpha\right)\right] = \frac{s^2}{1+\frac{\alpha s^\alpha}{1-\alpha}} \tag{23.46}$$

where

$$\left|-\frac{\alpha}{1-\alpha}s^\alpha\right| < 1$$

$$|s|^\alpha < \frac{1-\alpha}{\alpha}$$

$$|s| < \left(\frac{1-\alpha}{\alpha}\right)^{\frac{1}{\alpha}}$$

$$|s| < \left(\frac{1-\alpha}{\alpha}\right)^{\frac{2}{\alpha}}$$

$$a^2+b^2 < \left(\frac{1-\alpha}{\alpha}\right)^{\frac{2}{\alpha}}$$

$\blacksquare$

PROOF We have

$$E[t^\alpha] = s^{3+2\alpha}L[t^\alpha] \tag{23.47}$$

Then, we obtain

$$E\left[E_\alpha\left(-\frac{\alpha}{1-\alpha}t^\alpha\right)\right] = E\left[\sum_{k=0}^{\infty}\left(\frac{(-\frac{\alpha}{1-\alpha}t^\alpha)^k}{\Gamma(ak+1)}\right)\right]$$

$$= \sum_{k=0}^{\infty}\left(\frac{(-\frac{\alpha}{1-\alpha})^k}{\Gamma(ak+1)}\right)E[t^{ak}]$$

$$= \sum_{k=0}^{\infty}\left(\frac{(-\frac{\alpha}{1-\alpha})^k}{\Gamma(ak+1)}\right)s^{3+2\alpha k}L[t^{ak}]$$

$$= \sum_{k=0}^{\infty}\left(\frac{(-\frac{\alpha}{1-\alpha})^k}{\Gamma(ak+1)}\right)s^{3+2\alpha k}s^{-\alpha k-1}\Gamma(ak+1)$$

$$= \sum_{k=0}^{\infty}\left(-\frac{\alpha}{1-\alpha}\right)^k s^{\alpha k+2}$$

$$= s^2\sum_{k=0}^{\infty}\left(-\frac{\alpha}{1-\alpha}s^\alpha\right)^k$$

$$= s^2\left(\frac{1}{1+\frac{as^\alpha}{1-\alpha}}\right)$$

$$= \frac{s^2}{1+\frac{as^\alpha}{1-\alpha}}$$

∎

23.3.5 KAMAL TRANSFORM

Theorem 23.5

We get the Kamal transform of the Mittag-Leffler function by:

$$K\left[E_\alpha\left(-\frac{\alpha}{1-\alpha}t^\alpha\right)\right] = \frac{1}{s^{-1}+\frac{as^{\alpha-1}}{1-\alpha}} \tag{23.48}$$

where

$$\left| -\frac{\alpha}{1-\alpha} s^{\alpha} \right| < 1$$

$$|s|^{\alpha} < \frac{1-\alpha}{\alpha}$$

$$|s| < \left(\frac{1-\alpha}{\alpha} \right)^{\frac{1}{\alpha}}$$

$$|s| < \left(\frac{1-\alpha}{\alpha} \right)^{\frac{2}{\alpha}}$$

$$a^2 + b^2 < \left(\frac{1-\alpha}{\alpha} \right)^{\frac{2}{\alpha}}$$

∎

PROOF

We have

$$K[t^{\alpha}] = s^{2+2\alpha} L[t^{\alpha}] \tag{23.49}$$

Then, we obtain

$$K\left[E_{\alpha} \left(-\frac{\alpha}{1-\alpha} t^{\alpha} \right) \right] = K\left[\sum_{k=0}^{\infty} \left(\frac{(-\frac{\alpha}{1-\alpha} t^{\alpha})^k}{\Gamma(ak+1)} \right) \right]$$

$$= \sum_{k=0}^{\infty} \left(\frac{(-\frac{\alpha}{1-\alpha})^k}{\Gamma(ak+1)} \right) K[t^{ak}]$$

$$= \sum_{k=0}^{\infty} \left(\frac{(-\frac{\alpha}{1-\alpha})^k}{\Gamma(ak+1)} \right) s^{2+2\alpha k} L[t^{ak}]$$

$$= \sum_{k=0}^{\infty} \left(\frac{(-\frac{\alpha}{1-\alpha})^k}{\Gamma(ak+1)} \right) s^{2+2\alpha k} s^{-\alpha k-1} \Gamma(ak+1)$$

$$= s \sum_{k=0}^{\infty} \left(-\frac{\alpha}{1-\alpha} s^{\alpha} \right)^k$$

$$= s \left(\frac{1}{1+\frac{as^{\alpha}}{1-\alpha}} \right)$$

$$= \frac{s}{1+\frac{\alpha s^{\alpha}}{1-\alpha}}$$

$$= \frac{1}{s^{-1}+\frac{\alpha s^{\alpha-1}}{1-\alpha}}$$

23.3.6   POURREZA TRANSFORM

Theorem 23.6

We get the Pourreza transform of the Mittag-Leffler function by:

$$P\left[E_\alpha\left(-\frac{\alpha}{1-\alpha}t^\alpha\right)\right] = \frac{1}{s + \frac{\alpha s^{1-2\alpha}}{1-\alpha}} \tag{23.50}$$

where

$$\left|-\frac{\alpha}{1-\alpha}s^{-2\alpha}\right| < 1$$

$$|s|^{-2\alpha} < \frac{1-\alpha}{\alpha}$$

$$|s| < \left(\frac{1-\alpha}{\alpha}\right)^{\frac{1}{-2\alpha}}$$

$$|s| < \left(\frac{\alpha}{1-\alpha}\right)^{\frac{1}{2\alpha}}$$

$$a^2 + b^2 < \left(\frac{\alpha}{1-\alpha}\right)^{\frac{1}{\alpha}}$$

PROOF We have

$$P[t^\alpha] = s^{-\alpha}L[t^\alpha] \tag{23.51}$$

Then, we obtain

$$P\left[E_u\left(-\frac{\alpha}{1-\alpha}t^\alpha\right)\right] = P\left[\sum_{k=0}^{\infty}\left(\frac{(-\frac{\alpha}{1-\alpha}t^\alpha)^k}{\Gamma(ak+1)}\right)\right]$$

$$= \sum_{k=0}^{\infty}\left(\frac{(-\frac{\alpha}{1-\alpha})^k}{\Gamma(ak+1)}\right)P[t^{ak}]$$

$$= \sum_{k=0}^{\infty}\left(\frac{(-\frac{\alpha}{1-\alpha})^k}{\Gamma(ak+1)}\right)s^{-\alpha k}L[t^{ak}]$$

$$= \sum_{k=0}^{\infty} \left(\frac{(-\frac{\alpha}{1-\alpha})^k}{\Gamma(ak+1)} \right) s^{-2\alpha k} s^{-1} \Gamma(ak+1)$$

$$= \frac{1}{s} \sum_{k=0}^{\infty} \left(-\frac{\alpha}{1-\alpha} s^{-2\alpha} \right)^k$$

$$= \frac{1}{s} \left(\frac{1}{1 + \frac{as^{-2\alpha}}{1-\alpha}} \right)$$

$$= \frac{s^{-1}}{1 + \frac{as^{-2\alpha}}{1-\alpha}}$$

$$= \frac{1}{s + \frac{as^{1-2\alpha}}{1-\alpha}}$$

∎

23.3.7 α INTEGRAL LAPLACE TRANSFORM

Theorem 23.7

We use β instead of α in the transform. We obtain the α integral Laplace transform of the Mittag-Leffler function as

$$L_\beta \left[E_\alpha \left(-\frac{\alpha}{1-\alpha} t^\alpha \right) \right] = \frac{1}{s^{\frac{1}{\beta}} + \frac{as^{\frac{1}{\beta}(1-\alpha)}}{1-\alpha}} \tag{23.52}$$

where

$$\left| -\frac{\alpha}{1-\alpha} s^{-\frac{1}{\beta}\alpha} \right| < 1$$

$$|s|^{-\frac{1}{\beta}\alpha} < \frac{1-\alpha}{\alpha}$$

$$|s| < \left(\frac{1-\alpha}{\alpha} \right)^{-\frac{\beta}{\alpha}}$$

$$|s| < \left(\frac{\alpha}{1-\alpha} \right)^{\frac{\beta}{\alpha}}$$

$$a^2 + b^2 < \left(\frac{\alpha}{1-\alpha} \right)^{2\frac{\beta}{\alpha}}$$

∎

PROOF

We have

$$L_\beta[t^\alpha] = L[t^{\frac{1}{\beta}\alpha}] \tag{23.53}$$

Then, we have

$$L_\beta\left[E_\alpha\left(-\frac{\alpha}{1-\alpha}t^\alpha\right)\right] = L_\beta\left[\sum_{k=0}^\infty\left(\frac{(-\frac{\alpha}{1-\alpha}t^\alpha)^k}{\Gamma(ak+1)}\right)\right]$$

$$= \sum_{k=0}^\infty\left(\frac{(-\frac{\alpha}{1-\alpha})^k}{\Gamma(ak+1)}\right)L_\beta[t^{ak}]$$

$$= \sum_{k=0}^\infty\left(\frac{(-\frac{\alpha}{1-\alpha})^k}{\Gamma(ak+1)}\right)L[t^{\frac{1}{\beta}ak}]$$

$$= \sum_{k=0}^\infty\left(\frac{(-\frac{\alpha}{1-\alpha})^k}{\Gamma(ak+1)}\right)s^{\frac{1}{\beta}(-ak-1)}\Gamma(ak+1)$$

$$= s^{-\frac{1}{\beta}}\sum_{k=0}^\infty\left(-\frac{\alpha}{1-\alpha}s^{-\frac{1}{\beta}\alpha}\right)^k$$

$$= s^{-\frac{1}{\beta}}\left(\frac{1}{1+\frac{as^{-\frac{1}{\beta}\alpha}}{1-\alpha}}\right)$$

$$= \frac{1}{s^{\frac{1}{\beta}}+\frac{as^{\frac{1}{\beta}(1-\alpha)}}{1-\alpha}}$$

∎

23.3.8 G TRANSFORM

Theorem 23.8

We use β instead of α in the transform. We get the G transform of the Mittag-Leffler function by:

$$G\left[E_\alpha\left(-\frac{\alpha}{1-\alpha}t^\alpha\right)\right] = s^\beta\left(\frac{1}{s^{-1}+\frac{as^{\alpha-1}}{1-\alpha}}\right) \tag{23.54}$$

where

$$\left| -\frac{\alpha}{1-\alpha}s^{\alpha} \right| < 1$$

$$|s|^{\alpha} < \frac{1-\alpha}{\alpha}$$

$$|s| < \left(\frac{1-\alpha}{\alpha} \right)^{\frac{1}{\alpha}}$$

$$a^2 + b^2 < \left(\frac{1-\alpha}{\alpha} \right)^{\frac{2}{\alpha}}$$

■

PROOF We have

$$G[t^{\alpha}] = s^{\beta}K[t^{\alpha}], \quad K[t^{\alpha}] = s^{2+2\alpha}L[t^{\alpha}], \quad G[t^{\alpha}] = s^{\beta}s^{2+2\alpha}L[t^{\alpha}] \tag{23.55}$$

Then, we obtain

$$G\left[E_{\alpha}\left(-\frac{\alpha}{1-\alpha}t^{\alpha} \right) \right] = G\left[\sum_{k=0}^{\infty} \left(\frac{(-\frac{\alpha}{1-\alpha}t^{\alpha})^k}{\Gamma(ak+1)} \right) \right]$$

$$= \sum_{k=0}^{\infty} \left(\frac{(-\frac{\alpha}{1-\alpha})^k}{\Gamma(\alpha k+1)} \right) G[t^{ak}]$$

$$= \sum_{k=0}^{\infty} \left(\frac{(-\frac{\alpha}{1-\alpha})^k}{\Gamma(\alpha k+1)} \right) s^{\beta}s^{2+2\alpha}L[t^{ak}]$$

$$= \sum_{k=0}^{\infty} \left(\frac{(-\frac{\alpha}{1-\alpha})^k}{\Gamma(\alpha k+1)} \right) s^{\beta}s^{2+2\alpha k}s^{-\alpha k-1}\Gamma(\alpha k+1)$$

$$= s^{\beta}s^1 \sum_{k=0}^{\infty} \left(-\frac{\alpha}{1-\alpha}s^{\alpha} \right)^k$$

$$= ss^{\beta}\left(\frac{1}{1+\frac{as^{\alpha}}{1-\alpha}} \right)$$

$$= s^{\beta}\left(\frac{s}{1+\frac{as^{\alpha}}{1-\alpha}} \right)$$

$$= s^{\beta}\left(\frac{1}{s^{-1}+\frac{as^{\alpha-1}}{1-\alpha}} \right)$$

■

23.3.9   NATURAL TRANSFORM

Theorem 23.9

We get the Natural transform of the Mittag-Leffler function by:

$$N\left[E_\alpha\left(-\frac{\alpha}{1-\alpha}t^\alpha\right)\right] = \frac{1}{s + \frac{\alpha u^\alpha s^{1-\alpha}}{1-\alpha}} \tag{23.56}$$

where

$$\left|-\frac{\alpha}{1-\alpha}\left(\frac{s}{u}\right)^{-\alpha}\right| < 1$$

$$\left|\frac{s}{u}\right|^{-\alpha} < \frac{1-\alpha}{\alpha}$$

$$\left|\frac{s}{u}\right| < \left(\frac{1-\alpha}{\alpha}\right)^{\frac{-1}{\alpha}}$$

$$\left|\frac{s}{u}\right| < \left(\frac{\alpha}{1-\alpha}\right)^{\frac{1}{\alpha}}$$

$\blacksquare$

PROOF We have

$$R[s,u] = \frac{s}{u^2}A\left(\frac{s}{u}\right), \quad A(s) = \frac{1}{s}L(s), \quad R(s,u) = \frac{1}{u}L\left(\frac{s}{u}\right) \tag{23.57}$$

Then, we obtain

$$N\left[E_\alpha\left(-\frac{\alpha}{1-\alpha}\right)\right] = R\left[\sum_{k=0}^{\infty}\left(\frac{(-\frac{\alpha}{1-\alpha}t^\alpha)^k}{\Gamma(\alpha k+1)}\right)\right]$$

$$= \sum_{k=0}^{\infty}\left(\frac{(-\frac{\alpha}{1-\alpha})^k}{\Gamma(\alpha k+1)}\right)N[t^{\alpha k},u]$$

$$= \sum_{k=0}^{\infty}\left(\frac{(-\frac{\alpha}{1-\alpha})^k}{\Gamma(\alpha k+1)}\right)\frac{1}{u}L\left[\frac{t^{\alpha k}}{u^k}\right]$$

$$= \sum_{k=0}^{\infty}\left(\frac{(-\frac{\alpha}{1-\alpha})^k}{\Gamma(\alpha k+1)}\right)\frac{s^{-\alpha k-1}}{u^{-\alpha k-1}}\Gamma(\alpha k+1)$$

$$= \frac{1}{s}\left(\frac{1}{u}u\right)\sum_{k=0}^{\infty}\left(-\frac{\alpha}{1-\alpha}\left(\frac{s}{u}\right)^{-\alpha}\right)^k$$

$$= \frac{1}{s}\left(\frac{1}{1+\frac{\alpha\left(\frac{s}{u}\right)^{-\alpha}}{1-\alpha}}\right)$$

$$= \frac{1}{s+\frac{\alpha u^{\alpha} s^{1-\alpha}}{1-\alpha}}$$

■

23.4 GENERAL TRANSFORM OF THE EQUATIONS

We consider the following problems:

$$Df = \lambda f.$$

$$_0^{CF}D_t^{\alpha}f = \lambda f.$$

$$_0^C D_t^{\alpha}f = \lambda f.$$

$$_0^{ABC}D_t^{\alpha}f = \lambda f.$$

23.4.1 ELZAKI TRANSFORM

We give some examples of the Elzaki transform as

$$E(t^n) = n! s^{n+2}$$

$$E(\sin(at)) = \frac{as^3}{1+a^2s^2}$$

$$E(\cos(at)) = \frac{s^2}{1+a^2s^2}$$

We find the Elzaki transform of the above equations as

$$E(Df) = E(\lambda f)$$

$$\frac{E(f)}{s} - sf(0) = \lambda E(f)$$

$$E(f) = \frac{s^2 f(0)}{1-s\lambda}$$

$$E\left(_0^{CF}D_t^{\alpha}f\right) = E(\lambda f)$$

$$\frac{E(f)}{s^2} - f(0))\frac{s^2 M(\alpha)}{\alpha(s-1)+1} = \lambda E(f)$$

$$E(f) = \frac{f(0)s^2 M(\alpha)}{M(\alpha) - \lambda(\alpha s + 1 - \alpha)}$$

$$E({}_{0}^{C}D_t^{\alpha}f) = E(\lambda f)$$

$$\frac{E(f)}{s^{\alpha}} - \frac{f(0)}{s^{\alpha-2}} = \lambda E(f)$$

$$E(f) = \frac{f(0)}{1 - s^{\alpha}\lambda}$$

$$E({}_{0}^{ABC}D_t^{\alpha}f) = E(\lambda f)$$

$$(E(f) - f(0))\frac{AB(\alpha)}{s^{-2}(1-\alpha) + \alpha s^{\alpha-2}} = \lambda E(f)$$

$$E(f) = \frac{f(0)AB(\alpha)}{AB(\alpha) - \lambda(s^{-2}(1-\alpha) + \alpha s^{\alpha-2})}$$

23.4.2   ABOODH TRANSFORM

We give some examples of the Aboodh transform as

$$A(t^n) = \frac{n!}{s^n + 2}$$

$$A(\sin(at)) = \frac{a}{s(s^2 + a^2)}$$

$$A(\cos(at)) = \frac{1}{s^2 + a^2}$$

We get the Aboodh transform of the following equations by:

$$A(Df) = A(\lambda f)$$

$$sA(f) - \frac{f(0)}{s} = \lambda A(f)$$

$$A(f) = \frac{f(0)}{s(s - \lambda)}$$

$$A({}_{0}^{CF}D_t^{\alpha}f) = A(\lambda f)$$

$$\left(sA(f) - \frac{f(0)}{s}\right)\frac{M(\alpha)}{s + \alpha - s\alpha} = \lambda A(f)$$

$$A(f) = \frac{f(0)M(\alpha)}{sM(\alpha) - \lambda(s + \alpha - s\alpha)}$$

$$A\left({}_0^C D_t^\alpha f\right) = A(\lambda f)$$

$$\left(s^2 A(f) - f(0)\right) s^{\alpha-2} = \lambda A(f)$$

$$A(f) = \frac{f(0)s^{\alpha-2}}{s^\alpha - \lambda}$$

$$A\left({}_0^{ABC} D_t^\alpha f\right) = A(\lambda f)$$

$$\left(s^2 A(f) - f(0)\right) \frac{AB(\alpha)s^{\alpha-2}}{s^\alpha(1-\alpha)+\alpha} = \lambda A(f)$$

$$A(f) = \frac{f(0)AB(\alpha)s^{\alpha-2}}{AB(\alpha)s^\alpha - \lambda\left(s^\alpha(1-\alpha)+\alpha\right)}$$

23.4.3 POURREZA TRANSFORM

We give some examples of the Pourreza transform as

$$P(t^n) = \frac{n!s}{s^{2n+2}}$$

$$P(\sin(at)) = = \frac{as}{a^2 + s^4}$$

$$P(\cos(at)) = \frac{s^3}{s^4 + a^2}$$

We get the Pourreza transform of the following equations by:

$$P(Df) = P(\lambda f)$$

$$s^2 P(f) - sf(0) = \lambda P(f)$$

$$P(f) = \frac{sf(0)}{s^2 - \lambda}$$

$$P\left({}_0^{CF} D_t^\alpha f\right) = P(\lambda f)$$

$$\left(sP(f) - \sqrt{s}f(0)\right) \frac{M(\alpha)}{s+\alpha-s\alpha} = \lambda P(f)$$

$$P(f) = \frac{\sqrt{s}f(0)M(\alpha)}{sM(\alpha) - \lambda(s+\alpha-s\alpha)}$$

$$P\left({}_0^C D_t^\alpha f\right) = P(\lambda f)$$

$$\left(sP(f) - \sqrt{s}f(0)\right) s^{\alpha-1} = \lambda P(f)$$

$$P(f) = \frac{\sqrt{s}s^{\alpha-1}f(0)}{s^\alpha - \lambda}$$

$$P\left({}^{ABC}_0 D^\alpha_t f\right) = P(\lambda f)$$

$$\left(sP(\alpha) - \sqrt{s}f(0)\right)\frac{AB(\alpha)s^{\alpha-1}}{s^\alpha(1-\alpha)+\alpha} = \lambda P(f)$$

$$P(f) = \frac{\sqrt{s}f(0)AB(\alpha)s^{\alpha-1}}{AB(\alpha)s^\alpha - \lambda(s^\alpha(1-\alpha)+\alpha)}$$

23.4.4 MOHAND TRANSFORM

We give some examples of the Mohand transform as

$$R(t^n) = \frac{n!}{s^{n-1}}$$

$$R(\sin(at)) = \frac{as^2}{s^2+a^2}$$

$$R(\cos(at)) = \frac{s^3}{s^2+a^2}$$

We get the Mohand transform of the following equations by:

$$R(Df) = R(\lambda f)$$
$$sR(f) - s^2 f(0) = \lambda R(f)$$
$$R(f) = \frac{s^2 f(0)}{s-\lambda}$$

$$R\left({}^{CF}_0 D^\alpha_t f\right) = R(\lambda f)$$

$$\left(sR(f) - s^2 f(0)\right)\frac{R(f)}{s+\alpha - s\alpha} = \lambda R(f)$$

$$R(f) = \frac{s^2 f(0)M(\alpha)}{sM(\alpha) - \lambda(s+\alpha - s\alpha)}$$

$$R\left({}^C_0 D^\alpha_t f\right) = K(\lambda f)$$

$$\left(\frac{R(f)}{s} - f(0)\right)s^{\alpha+1} = \lambda R(f)$$

$$R(f) = \frac{f(0)s^{\alpha+1}}{s^\alpha - \lambda}$$

$$R\left({}_0^{ABC}D_t^\alpha f\right) \;=\; R(\lambda f)$$

$$\left(\frac{R(f)}{s} - f(0)\right)\frac{AB(\alpha)s^{\alpha+1}}{s^\alpha(1-\alpha)+\alpha} \;=\; \lambda R(f)$$

$$R(f) \;=\; \frac{f(0)AB(\alpha)s^{\alpha+1}}{AB(\alpha)s^\alpha - \lambda\left(s^\alpha(1-\alpha)+\alpha\right)}$$

23.4.5 SAWI TRANSFORM

We give some examples of the Sawi transform as

$$Sa(t^n) \;=\; s^{n-1}n!$$

$$Sa(\sin(at)) \;=\; \frac{a}{1+a^2s^2}$$

$$Sa(\cos(at)) \;=\; \frac{1}{s(1+a^2s^2)}$$

We get the Sawi transform of the following equations by:

$$Sa(Df) \;=\; Sa(\lambda f)$$

$$\frac{Sa(f)}{s} - \frac{f(0)}{s^2} \;=\; \lambda Sa(f)$$

$$Sa(f) \;=\; \frac{f(0)}{s^2(1-\lambda s)}$$

$$Sa\left({}_0^{CF}D_t^\alpha f\right) \;=\; Sa(\lambda f)$$

$$(sSa(f) - f(0))\frac{M(\alpha)}{s+\alpha-s\alpha} \;=\; \lambda Sa(f)$$

$$Sa(f) \;=\; \frac{f(0)M(\alpha)}{M(\alpha)s - \lambda(s+\alpha-s\alpha)}$$

$$Sa\left({}_0^C D_t^\alpha f\right) \;=\; = Sa(\lambda f)$$

$$\left(\frac{Sa(f)}{s} - f(0)\right)s^{\alpha+1} \;=\; \lambda Sa(f)$$

$$Sa(f) \;=\; \frac{f(0)s^{\alpha+1}}{s^\alpha - \lambda}$$

$$Sa\left({}_0^{ABC}D_t^\alpha f\right) = = Sa(\lambda f)$$

$$\left(\frac{Sa(f)}{s} - f(0)\right)\frac{AB(\alpha)s^{\alpha+1}}{s^\alpha(1-\alpha)+\alpha} = \lambda Sa(f)$$

$$Sa(f) = \frac{f(0)AB(\alpha)s^{\alpha-1}}{AB(\alpha)s^{\alpha-2}-\lambda(s^\alpha(1-\alpha)+\alpha)}$$

23.4.6 KAMAL TRANSFORM

We give some examples of the Kamal transform as

$$K(t^n) = n!s^{n+1}$$

$$K(\sin(at)) = \frac{as^2}{1+a^2s^2}$$

$$K(\cos(at)) = \frac{s}{1+a^2s^2}$$

We get the Kamal transform of the following equations by:

$$K(Df) = K(\lambda f)$$

$$\frac{K(f)}{s} - f(0) = \lambda K(f)$$

$$K(f) = \frac{f(0)s}{1-\lambda s}$$

$$K\left({}_0^{CF}D_t^\alpha f\right) = K(\lambda f)$$

$$(sK(f)-f(0))\frac{M(\alpha)}{s+\alpha-s\alpha} = \lambda K(f)$$

$$K(f) = \frac{f(0)M(\alpha)}{sM(\alpha)-\lambda(s+\alpha-s\alpha)}$$

$$K\left({}_0^C D_t^\alpha f\right) = K(\lambda f)$$

$$(sK(f)-f(0))s^{\alpha-1} - \lambda K(f)$$

$$K(f) = \frac{f(0)s^{\alpha-1}}{s-\lambda}$$

$$K\left({}^{ABC}_{0}D^{\alpha}_{t}f\right) = K(\lambda f)$$

$$(sK(f) - f(0))\frac{AB(\alpha)s^{\alpha-1}}{s^{\alpha}(1-\alpha)+\alpha} = \lambda K(f)$$

$$K(f) = \frac{f(0)AB(\alpha)s^{\alpha-1}}{AB(\alpha)s^{\alpha} - \lambda\left(s^{\alpha}(1-\alpha)+\alpha\right)}$$

23.4.7 G- TRANSFORM

We give some examples of the G-transform as

$$G(t^{n}) = n!s^{n+\alpha+1}$$

$$G(\sin(at)) = \frac{as^{\alpha+2}}{1+s^{2}a^{2}}$$

$$G(\cos(at)) = \frac{s^{\alpha+1}}{1+s^{2}a^{2}}$$

We find the G- transform of the following equations as

$$G(Df) = G(\lambda f)$$

$$\frac{G(f)}{s} - s^{\alpha}f(0) = \lambda G(f)$$

$$G(f) = \frac{s^{\alpha+1}f(0)}{1-s\lambda}$$

$$G\left({}^{CF}_{0}D^{\alpha}_{t}f\right) = G(\lambda f)$$

$$(sG(\alpha) - \frac{f(0)}{s^{\alpha}}) = \lambda G(f)$$

$$G(f) = \frac{f(0)M(\alpha)s^{-\alpha}}{sM(\alpha) - \lambda(s+\alpha-s\alpha)}$$

$$G\left({}^{C}_{0}D^{\alpha}_{t}f\right) = G(\lambda f)$$

$$(s^{\alpha+1}G(f) - f(0))s^{-1} = \lambda G(f)$$

$$G(f) = \frac{f(0)s^{-1}}{s^{\alpha}-\lambda}$$

$$G\left({}^{ABC}_0 D^\alpha_t f\right) = G(\lambda f)$$

$$(sG(f) - f(0))\frac{AB(\alpha)s^{-1}}{s^\alpha(1-\alpha)+\alpha} = \lambda G(f)$$

$$G(f) = \frac{f(0)AB(\alpha)s^{-1}}{AB(\alpha)-\lambda(s^\alpha(1-\alpha)+\alpha)}$$

23.4.8 NATURAL TRANSFORM

We give some examples of the Natural transform as

$$N(t^n) = \frac{u^n\Gamma(n+1)}{s^{n+1}}$$

$$N(\sin(at)) = \frac{au}{a^2u^2+s^2}$$

$$N(\cos(at)) = \frac{s}{a^2u^2+s^2}$$

We get the Natural transform of the following equations by:

$$N(Df) = N(\lambda f)$$

$$\frac{s}{u}N(f) - \frac{f(0)}{u} = \lambda N(f)$$

$$N(f) = \frac{f(0)}{s-\lambda u}$$

$$N\left({}^{CF}_0 D^\alpha_t f\right) = N(\lambda f)$$

$$\left(\frac{N(f)}{u} - f(0)\right)\frac{M(\alpha)}{s\alpha - s - u\alpha} = \lambda N(f)$$

$$N(f) = \frac{f(0)M(\alpha)u}{M(\alpha)-u\lambda(s\alpha-s-u\alpha)}$$

$$N\left({}^C_0 D^\alpha_t f\right) = N(\lambda f)$$

$$\left(N(f) - \frac{f(0)}{u}\right)s^\alpha u^\alpha = \lambda N(f)$$

$$N(f) = \frac{f(0)s^\alpha u^\alpha}{u(s^\alpha u^\alpha - \lambda)}$$

$$N\left({}_0^{ABC}D_t^\alpha f\right) = N(\lambda f)$$

$$\frac{1}{u}(N(f) - f(0))\frac{AB(\alpha)s^{\alpha-1}}{s^\alpha(1-\alpha)+\alpha} = \lambda N(f)$$

$$N(f) = \frac{f(0)AB(\alpha)s^{\alpha-1}}{AB(\alpha)s^{\alpha-1} - \lambda u(s^\alpha(1-\alpha)+\alpha)}$$

23.5 APPLICATIONS I

In this section we consider the circuit problems with different kernels.

$$_0^{CF}D_t^\alpha i(t) + \frac{1}{RC}\frac{di}{dt} = \frac{V(t)}{RCL} + \frac{1}{L}\frac{dV(t)}{dt} \qquad (23.58)$$

$$_0^{ABC}D_t^\alpha i(t) + \frac{1}{RC}\frac{di}{dt} = \frac{V(t)}{RCL} + \frac{1}{L}\frac{dV(t)}{dt} \qquad (23.59)$$

23.5.1 ELZAKI TRANSFORM

We get the Elzaki transform of these equations by:

$$E\left({}_0^{CF}D_t^\alpha i(t)\right) + E\left(\frac{1}{RC}\frac{di}{dt}\right) = E\left(\frac{V(t)}{RCL}\right) + E\left(\frac{1}{L}\frac{dV(t)}{dt}\right)$$

$$E\left(\frac{di(t)}{dt} * \frac{M(\alpha)}{1-\alpha}\exp\left(-\frac{\alpha}{1-\alpha}t\right)\right) + \frac{1}{RC}\left(\frac{E(i(t))}{s} - si(0)\right)$$

$$= \frac{1}{RCL}E(V(t)) + \frac{1}{L}\left(\frac{E(V(t))}{s} - sV(0)\right)$$

$$\frac{1}{s}E\left(\frac{di(t)}{dt}\right)\frac{M(\alpha)}{1-\alpha}E\left(\exp\left(-\frac{\alpha}{1-\alpha}t\right)\right) + \frac{1}{RC}\frac{E(i(t))}{s}$$

$$= \frac{1}{RCL}E(V(t)) + \frac{1}{L}\frac{E(V(t))}{s}$$

$$\frac{1}{s}\left(\frac{E(i(t))}{s} - si(0)\right)\frac{M(\alpha)}{1-\alpha}\left(\frac{s^2(1-\alpha)}{\alpha s-\alpha+1}\right) + \frac{1}{RC}\frac{E(i(t))}{s}$$

$$= \frac{1}{RCL}E(V(t)) + \frac{1}{L}\frac{E(V(t))}{s}$$

Then, we reach

$$\frac{E(V(t))}{E(i(t))} = \frac{(M(\alpha)RCs+\alpha s-\alpha+1)L}{(\alpha s-\alpha+1)(s+RC)}$$

and

$$E\left({}_0^{ABC}D_t^\alpha i(t)\right) + E\left(\frac{1}{RC}\frac{di}{dt}\right) = E\left(\frac{V(t)}{RCL}\right) + E\left(\frac{1}{L}\frac{dV(t)}{dt}\right)$$

$$E\left(\frac{di(t)}{dt} * \frac{AB(\alpha)}{1-\alpha}E_\alpha\left[\frac{-\alpha}{1-\alpha}t^\alpha\right]\right) + \frac{1}{RC}\left(\frac{E(i(t))}{s} - si(0)\right)$$

$$= \frac{1}{RCL}E(V(t)) + \frac{1}{L}\left(\frac{E(V(t))}{s} - sV(0)\right)$$

$$\frac{1}{s}E\left(\frac{di(t)}{dt}\right)\frac{AB(\alpha)}{1-\alpha}E\left(E_\alpha\left[\frac{-\alpha}{1-\alpha}t^\alpha\right]\right) + \frac{1}{RC}\frac{E(i(t))}{s}$$

$$= \frac{1}{RCL}E(V(t)) + \frac{1}{L}\frac{E(V(t))}{s}$$

$$\frac{1}{s}\left(\frac{E(i(t))}{s} - si(0)\right)\frac{AB(\alpha)}{1-\alpha}\left(\frac{s^2(1-\alpha)}{1-\alpha+\alpha s^\alpha}\right) + \frac{1}{RC}\frac{E(i(t))}{s}$$

$$= \frac{1}{RCL}E(V(t)) + \frac{1}{L}\frac{E(V(t))}{s}$$

Then, we reach

$$\frac{E(V(t))}{E(i(t))} = \frac{L(AB(\alpha)RCs + 1 - \alpha + \alpha s^\alpha)}{(1-\alpha+\alpha s^\alpha)(s+RC)}$$

23.5.2 ABOODH TRANSFORM

We get the Aboodh transform of these equations by:

$$A\left({}_0^{CF}D_t^\alpha i\right) + A\left(\frac{1}{RC}\frac{di}{dt}\right) = A\left(\frac{V(t)}{RCL}\right) + A\left(\frac{1}{L}\frac{dV(t)}{dt}\right)$$

$$A\left(\frac{di(t)}{dt} * \frac{M(\alpha)}{1-\alpha}\exp\left(-\frac{\alpha}{1-\alpha}t\right)\right) + \frac{1}{RC}\left(\frac{A(i(t))}{s} - si(0)\right)$$

$$= \frac{1}{RCL}A(V(t)) + \frac{1}{L}\left(\frac{A(V(t))}{s} - sV(0)\right)$$

$$sA\left(\frac{di(t)}{dt}\right)\frac{M(\alpha)}{1-\alpha}A\left(\exp\left(\frac{\alpha}{1-\alpha}t\right)\right) + \frac{1}{RC}\frac{A(i(t))}{s}$$

$$= \frac{1}{RCL}A(V(t)) + \frac{1}{L}\frac{A(V(t))}{s}$$

$$s\left(sA(i(t)) - \frac{i(0)}{s}\right)\frac{M(\alpha)}{1-\alpha}\left(\frac{1}{s}\left(\frac{1-\alpha}{s+\alpha-s\alpha}\right)\right) + \frac{1}{RC}\frac{A(i(t))}{s}$$

$$= \frac{1}{RCL}A(V(t)) + \frac{1}{L}\frac{A(V(t))}{s}$$

Then, we reach

$$\frac{A(V(t))}{A(i(t))} = \frac{L(RCs^2M(\alpha) + s + \alpha - s\alpha)}{(s+\alpha-s\alpha)(s+RC)}$$

and

$$A\left({}_0^{ABC}D_t^\alpha i(t)\right) + A\left(\frac{1}{RC}\frac{di}{dt}\right) = A\left(\frac{V(t)}{RCL}\right) + A\left(\frac{1}{L}\frac{dV(t)}{dt}\right)$$

$$A\left(\frac{di(t)}{dt} * \frac{AB(\alpha)}{1-\alpha}E_\alpha\left[\frac{-\alpha}{1-\alpha}t^\alpha\right]\right) + \frac{1}{RC}\left(\frac{A(i(t))}{s} - si(0)\right)$$

$$= \frac{1}{RCL}A(V(t)) + \frac{1}{L}\left(\frac{A(V(t))}{s} - sV(0)\right)$$

$$sA\left(\frac{di(t)}{dt}\right)\frac{AB(\alpha)}{1-\alpha}A\left(E_\alpha\left[\frac{-\alpha}{1-\alpha}t^\alpha\right]\right) + \frac{1}{RC}\frac{A(i(t))}{s}$$

$$= \frac{1}{RCL}A(V(t)) + \frac{1}{L}\frac{A(V(t))}{s}$$

$$s\left(sA(i(t)) - \frac{i(0)}{s}\right)\frac{AB(\alpha)}{1-\alpha}\frac{1-\alpha}{s^2(1-\alpha+\alpha s^{-\alpha})} + \frac{1}{RC}\frac{A(i(t))}{s}$$

$$= \frac{1}{RCL}A(V(t)) + \frac{1}{L}\frac{A(V(t))}{s}$$

Then, we reach

$$\frac{A(V(t))}{A(i(t))} = \frac{L(RCsAB(\alpha) + 1 - \alpha + \alpha s^{-\alpha})}{(1-\alpha+\alpha s^{-\alpha})(s+RC)}$$

23.5.3 POURREZA TRANSFORM

We get the Pourreza transform of these equations by:

$$HJ\left({}_0^{CF}D_t^\alpha i(t)\right) + HJ\left(\frac{1}{RC}\frac{di}{dt}\right) = HJ\left(\frac{V(t)}{RCL}\right) + HJ\left(\frac{1}{L}\frac{dV(t)}{dt}\right)$$

$$HJ\left(\frac{df(t)}{dt} * \frac{M(\alpha)}{1-\alpha}e^{(-\frac{\alpha}{1-\alpha}t)}\right) + \frac{1}{RC}\left(\frac{HJ(i(t))}{s} - si(0)\right)$$

$$= \frac{1}{RCL}HJ(V(t)) + \frac{1}{L}\left(\frac{HJ(V(t))}{s} - sV(0)\right)$$

$$\frac{1}{s}HJ\left(\frac{df(t)}{dt}\right)\frac{M(\alpha)}{1-\alpha}HJ\left(\exp\left((-\frac{\alpha}{1-\alpha}t)\right)\right) + \frac{1}{RC}\frac{A(i(t))}{s}$$

$$= \frac{1}{RCL}A(V(t)) + \frac{1}{L}\frac{A(V(t))}{s}$$

$$\frac{1}{s}\left(s^2HJ(i(t)) - si(0)\right)\frac{M(\alpha)}{1-\alpha}s\left(\frac{-1+\alpha}{s^2(-1+\alpha)-\alpha}\right) + \frac{1}{RC}\frac{A(i(t))}{s}$$

$$= \frac{1}{RCL}A(V(t)) + \frac{1}{L}\frac{A(V(t))}{s}$$

Thus, we obtain

$$\frac{HJ(V(t))}{HJ(i(t))} = \frac{s^2M(\alpha)RC + \alpha - s^2(1-\alpha)}{(\alpha - s^2(1-\alpha))(s+RC)}$$

and

$$HJ\left({}_{0}^{ABC}D_t^\alpha i(t)\right) + HJ\left(\frac{1}{RC}\frac{di}{dt}\right) = HJ\left(\frac{V(t)}{RCL}\right) + HJ\left(\frac{1}{L}\frac{dV(t)}{dt}\right)$$

$$HJ\left(\frac{di(t)}{dt} * \frac{AB(\alpha)}{1-\alpha}E_\alpha\left[\frac{-\alpha}{1-\alpha}t^\alpha\right]\right) + \frac{1}{RC}\left(\frac{HJ(i(t))}{s} - si(0)\right)$$

$$= \frac{1}{RCL}HJ(V(t)) + \frac{1}{L}\left(\frac{HJ(V(t))}{s} - sV(0)\right)$$

$$\frac{1}{s}HJ\left(\frac{di(t)}{dt}\right)\frac{AB(\alpha)}{1-\alpha}HJ\left(E_\alpha\left[\frac{-\alpha}{1-\alpha}t^\alpha\right]\right) + \frac{1}{RC}\frac{HJ(i(t))}{s}$$

$$= \frac{1}{RCL}HJ(V(t)) + \frac{1}{L}\frac{HJ(V(t))}{s}$$

$$\frac{1}{s}(s^2HJ(i(t)) - si(0))\frac{AB(\alpha)}{1-\alpha}\frac{1-\alpha}{s(1-\alpha+\alpha s^{-2\alpha})} + \frac{1}{RC}\frac{HJ(i(t))}{s}$$

$$= \frac{1}{RCL}HJ(V(t)) + \frac{1}{L}\frac{HJ(V(t))}{s}$$

Then, we get

$$\frac{HJ(V(t))}{HJ(i(t))} = \frac{(AB(\alpha)RCs + 1 - \alpha + \alpha s^{-2\alpha})L}{(1-\alpha+\alpha s^{-2\alpha})(s+RC)}$$

23.5.4 MOHAND TRANSFORM

We get the Mohand transform of these equations by:

$$M\left({}_0^{CF}D_t^\alpha i(t)\right) + \frac{1}{RC}\left(\frac{M(i(t))}{s} - si(0)\right)$$

$$= \frac{1}{RCL}M(V(t)) + \frac{1}{L}\left(\frac{M(V(t))}{s} - sV(0)\right)$$

$$M\left(\frac{di(t)}{dt} * \frac{M(\alpha)}{1-\alpha}e^{(-\frac{\alpha}{1-\alpha}t)}\right) + \frac{1}{RC}\left(\frac{M(i(t))}{s} - si(0)\right)$$

$$- \frac{1}{RCL}M(V(t)) + \frac{1}{L}\left(\frac{M(V(t))}{s} - sV(0)\right)$$

$$\frac{1}{s^2}M\left(\frac{di(t)}{dt}\right)\frac{M(\alpha)}{1-\alpha}M\left(\exp\left((-\frac{\alpha}{1-\alpha}t)\right)\right) + \frac{1}{RC}\frac{M(i(t))}{s}$$

$$= \frac{1}{RCL}M(V(t)) + \frac{1}{L}\frac{M(V(t))}{s}$$

$$\frac{1}{s^2}(sM(i(t)) - s^2i(0))\frac{M(\alpha)}{1-\alpha}s^2\frac{1-\alpha}{s+\alpha-s\alpha} + \frac{1}{RC}\frac{M(i(t))}{s}$$

$$= \frac{1}{RCL}M(V(t)) + \frac{1}{L}\frac{M(V(t))}{s}$$

Then, we obtain

$$\frac{M(V(t))}{M(i(t))} = \frac{(M(\alpha)s^2RC + s + \alpha - s\alpha)L}{(s + \alpha - s\alpha)(s + RC)}$$

and

$$M\left({}_0^{ABC}D_t^\alpha i(t)\right) + \frac{1}{RC}\left(\frac{M(i(t))}{s} - si(0)\right)$$

$$= \frac{1}{RCL}M(V(t)) + \frac{1}{L}\left(\frac{M(V(t))}{s} - sV(0)\right)$$

$$M\left(\frac{di(t)}{dt} * \frac{AB(\alpha)}{1-\alpha}E_\alpha\left[\frac{-\alpha}{1-\alpha}t^\alpha\right]\right) + \frac{1}{RC}\left(\frac{M(i(t))}{s} - si(0)\right)$$

$$= \frac{1}{RCL}M(V(t)) + \frac{1}{L}\left(\frac{M(V(t))}{s} - sV(0)\right)$$

$$\frac{1}{s^2}M\left(\frac{di(t)}{dt}\right)\frac{AB(\alpha)}{1-\alpha}M\left(E_\alpha\left[\frac{-\alpha}{1-\alpha}t^\alpha\right]\right) + \frac{1}{RC}\frac{M(i(t))}{s}$$

$$= \frac{1}{RCL}M(V(t)) + \frac{1}{L}\frac{M(V(t))}{s}$$

$$\frac{1}{s^2}(sM(i(t)) - s^2i(0))\frac{AB(\alpha)}{1-\alpha}s\frac{1-\alpha}{1-\alpha+\alpha s^{-\alpha}} + \frac{1}{RC}\frac{M(i(t))}{s}$$

$$= \frac{1}{RCL}M(V(t)) + \frac{1}{L}\frac{M(V(t))}{s}$$

Thus, we acquire

$$\frac{M(V(t))}{M(i(t))} = \frac{(AB(\alpha)RCs + 1 - \alpha + \alpha s^{-\alpha})L}{(1 - \alpha + \alpha s^{-\alpha})(s + RC)}$$

23.5.5 SAWI TRANSFORM

We get the Sawi transform of these equations by:

$$Sa\left({}_0^{CF}D_t^\alpha i(t)\right) + \frac{1}{RC}\left(\frac{Sa(i(t))}{s} - si(0)\right)$$

$$= \frac{1}{RCL}Sa(V(t)) + \frac{1}{L}\left(\frac{Sa(V(t))}{s} - sV(0)\right)$$

$$Sa\left(\frac{di(t)}{dt} * \frac{M(\alpha)}{1-\alpha}\exp\left((-\frac{\alpha}{1-\alpha}t)\right)\right) + \frac{1}{RC}\left(\frac{Sa(i(t))}{s} - si(0)\right)$$

$$= \frac{1}{RCL}Sa(V(t)) + \frac{1}{L}\left(\frac{Sa(V(t))}{s} - sV(0)\right)$$

$$s^2Sa\left(\frac{di(t)}{dt}\right)\frac{M(\alpha)}{1-\alpha}Sa\left(\exp\left(-\frac{\alpha}{1-\alpha}t\right)\right) + \frac{1}{RC}\frac{Sa(i(t))}{s}$$

$$= \frac{1}{RCL}Sa(V(t)) + \frac{1}{L}\frac{Sa(V(t))}{s}$$

$$s^2\left(\frac{Sa(i(t))}{s} - \frac{i(0)}{s^2}\right)\frac{M(\alpha)}{1-\alpha}\frac{1}{s^2}\frac{s(1-\alpha)}{1+\alpha(-1+s)} + \frac{1}{RC}\frac{Sa(i(t))}{s}$$

$$= \frac{1}{RCL}Sa(V(t)) + \frac{1}{L}\frac{Sa(V(t))}{s}$$

Then, we get

$$\frac{Sa(V(t))}{Sa(i(t))} = \frac{(M(\alpha)RCs+1+\alpha(-1+s))L}{(1+\alpha(-1+s))(s+RC)}$$

and

$$Sa\left({}_0^{ABC}D_t^\alpha i(t)\right) + \frac{1}{RC}\left(\frac{Sa(i(t))}{s} - si(0)\right)$$

$$= \frac{1}{RCL}Sa(V(t)) + \frac{1}{L}\left(\frac{Sa(V(t))}{s} - sV(0)\right)$$

$$Sa\left(\frac{di(t)}{dt} * \frac{AB(\alpha)}{1-\alpha}E_\alpha\left[\frac{-\alpha}{1-\alpha}t^\alpha\right]\right) + \frac{1}{RC}\left(\frac{Sa(i(t))}{s} - si(0)\right)$$

$$= \frac{1}{RCL}Sa(V(t)) + \frac{1}{L}\left(\frac{Sa(V(t))}{s} - sV(0)\right)$$

$$s^2Sa\left(\frac{di(t)}{dt}\right)\frac{AB(\alpha)}{1-\alpha}Sa\left(E_\alpha\left[\frac{-\alpha}{1-\alpha}t^\alpha\right]\right) + \frac{1}{RC}\frac{Sa(i(t))}{s}$$

$$= \frac{1}{RCL}Sa(V(t)) + \frac{1}{L}\frac{Sa(V(t))}{s}$$

$$s^2\left(\frac{Sa(i(t))}{s} - \frac{i(0)}{s^2}\right)\frac{AB(\alpha)}{1-\alpha}\frac{1}{s}\frac{1-\alpha}{1-\alpha+\alpha s^\alpha} + \frac{1}{RC}\frac{Sa(i(t))}{s}$$

$$= \frac{1}{RCL}Sa(V(t)) + \frac{1}{L}\frac{Sa(V(t))}{s}$$

Thus, we obtain

$$\frac{Sa(V(t))}{Sa(i(t))} = \frac{(AB(\alpha)RCs+1-\alpha+\alpha s^\alpha)L}{(1-\alpha+\alpha s^\alpha)(s+RC)}$$

23.5.6 KAMAL TRANSFORM

We get the Kamal transform of these equations by:

$$K\left({}_0^{CF}D_t^\alpha i(t)\right) + \frac{1}{RC}\left(\frac{K(i(t))}{s} - si(0)\right)$$

$$= \frac{1}{RCL}K(V(t)) + \frac{1}{L}\left(\frac{K(V(t))}{s} - sV(0)\right)$$

$$K\left(\frac{di(t)}{dt} * \frac{M(\alpha)}{1-\alpha}\exp\left((-\frac{\alpha}{1-\alpha}t)\right)\right) + \frac{1}{RC}\left(\frac{K(i(t))}{s} - si(0)\right)$$

$$= \frac{1}{RCL}K(V(t)) + \frac{1}{L}\left(\frac{K(V(t))}{s} - sV(0)\right)$$

$$K\left(\frac{di(t)}{dt}\right)\frac{M(\alpha)}{1-\alpha}K\left(\exp\left(\left(-\frac{\alpha}{1-\alpha}t\right)\right)\right) + \frac{1}{RC}\frac{K(i(t))}{s}$$

$$= \frac{1}{RCL}K(V(t)) + \frac{1}{L}\frac{K(V(t))}{s}$$

$$\left(\frac{K(i(t))}{s} - i(0)\right)\frac{M(\alpha)}{1-\alpha}s\frac{1-\alpha}{1+\alpha(-1+s)} + \frac{1}{RC}\frac{K(i(t))}{s}$$

$$= \frac{1}{RCL}K(V(t)) + \frac{1}{L}\frac{K(V(t)}{s}$$

Then, we obtain

$$\frac{K(V(t))}{K(i(t))} = \frac{(M(\alpha)RCs + 1 + \alpha(-1+s))L}{(1+\alpha(-1+s))(s+RC)}$$

and

$$K\left({}_0^{ABC}D_t^\alpha i(t)\right) + \frac{1}{RC}\left(\frac{K(i(t))}{s} - si(0)\right)$$

$$= \frac{1}{RCL}K(V(t)) + \frac{1}{L}\left(\frac{K(V(t))}{s} - sV(0)\right)$$

$$K\left(\frac{di(t)}{dt} * \frac{AB(\alpha)}{1-\alpha}E_\alpha\left[\frac{-\alpha}{1-\alpha}t^\alpha\right]\right) + \frac{1}{RC}\left(\frac{K(i(t))}{s} - si(0)\right)$$

$$= \frac{1}{RCL}K(V(t)) + \frac{1}{L}\left(\frac{K(V(t))}{s} - sV(0)\right)$$

$$K\left(\frac{di(t)}{dt}\right)\frac{AB(\alpha)}{1-\alpha}K\left(E_\alpha\left[\frac{-\alpha}{1-\alpha}t^\alpha\right]\right) + \frac{1}{RC}\frac{K(i(t))}{s}$$

$$= \frac{1}{RCL}K(V(t)) + \frac{1}{L}\frac{K(V(t))}{s}$$

$$\left(\frac{K(i(t))}{s} - i(0)\right)\frac{M(\alpha)}{1-\alpha}s\frac{1-\alpha}{1-\alpha+\alpha s^\alpha} + \frac{1}{RC}\frac{K(i(t))}{s}$$

$$= \frac{1}{RCL}K(V(t)) + \frac{1}{L}\frac{K(V(t)}{s}$$

Then, we get

$$\frac{K(V(t))}{K(i(t))} = \frac{(M(\alpha)RCs + 1 - \alpha + \alpha s^\alpha)L}{(1-\alpha+\alpha s^\alpha)(s+RC)}$$

23.5.7   G- TRANSFORM

We get the G- transform of these equations by:

$$G\left(_0^{CF}D_t^\alpha f\right) + \frac{1}{RC}\left(\frac{K(i(t))}{s} - si(0)\right)$$

$$= \frac{1}{RCL}K(V(t)) + \frac{1}{L}\left(\frac{K(V(t))}{s} - sV(0)\right)$$

$$G\left(\frac{df(t)}{dt} * \frac{M(\alpha)}{1-\alpha}\exp\left((-\frac{\alpha}{1-\alpha}t)\right)\right) + \frac{1}{RC}\left(\frac{K(i(t))}{s} - si(0)\right)$$

$$= \frac{1}{RCL}K(V(t)) + \frac{1}{L}\left(\frac{K(V(t))}{s} - sV(0)\right)$$

$$s^{-\beta}G\left(\frac{df(t)}{dt}\right)\frac{M(\alpha)}{1-\alpha}G\left(\exp\left((-\frac{\alpha}{1-\alpha}t)\right)\right) + \frac{1}{RC}\frac{K(i(t))}{s}$$

$$= \frac{1}{RCL}K(V(t)) + \frac{1}{L}\frac{K(V(t))}{s}$$

$$s^{-\beta}\left(\frac{G(f)}{s} - s^\beta f(0)\right)\frac{M(\alpha)}{1-\alpha}s^\beta\frac{s(1-\alpha)}{1+(-1+s)\alpha} + \frac{1}{RC}\frac{K(i(t))}{s}$$

$$= \frac{1}{RCL}K(V(t)) + \frac{1}{L}\frac{K(V(t)}{s}$$

Then, we obtain

$$\frac{G(V(t))}{G(i(t))} = \frac{(M(\alpha)RCs + 1 + (-1+s)\alpha)}{(1+(-1+s)\alpha)(s+RC)}$$

and

$$G\left(_0^{ABC}D_t^\alpha i(t)\right) + \frac{1}{RC}\left(\frac{G(i(t))}{s} - si(0)\right)$$

$$= \frac{1}{RCL}G(V(t)) + \frac{1}{L}\left(\frac{G(V(t))}{s} - sV(0)\right)$$

$$G\left(\frac{di(t)}{dt} * \frac{AB(\alpha)}{1-\alpha}E_\alpha\left[\frac{-\alpha}{1}t^\alpha\right]\right) + \frac{1}{RC}\left(\frac{G(i(t))}{s} - si(0)\right)$$

$$= \frac{1}{RCL}G(V(t)) + \frac{1}{L}(\frac{G(V(t))}{s} - sV(0))$$

$$s^{-\beta}G\left(\frac{di(t)}{dt}\right)\frac{AB(\alpha)}{1-\alpha}G\left(E_\alpha\left[\frac{-\alpha}{1-\alpha}t^\alpha\right]\right) + \frac{1}{RC}\frac{G(i(t))}{s}$$

$$= \frac{1}{RCL}G(V(t)) + \frac{1}{L}\frac{G(V(t))}{s}$$

$$s^{-\beta}\left(\frac{G(i(t))}{s} - s^\beta i(0)\right)\frac{AB(\alpha)}{1-\alpha}s^{\beta+1}\frac{1-\alpha}{1-\alpha+\alpha s^\alpha} + \frac{1}{RC}\frac{G(i(t))}{s}$$

$$\frac{1}{RCL}G(V(t)) + \frac{1}{L}\frac{G(V(t)}{s}$$

Thus, we reach

$$\frac{G(V(t))}{G(i(t))} = \frac{(AB(\alpha)RCs + 1 - \alpha + \alpha s^{\alpha})L}{(1 - \alpha + \alpha s^{\alpha})(s + RC)}$$

23.5.8 NATURAL TRANSFORM

We get the Natural transform of these equations by:

$$N\left({}^{CF}_{0}D^{\alpha}_{t}i(t)\right) + \frac{1}{RC}\left(\frac{N(i(t))}{s} - si(0)\right)$$

$$= \frac{1}{RCL}N(V(t)) + \frac{1}{L}\left(\frac{N(V(t))}{s} - sV(0)\right)$$

$$N\left(\frac{di(t)}{dt} * \frac{M(\alpha)}{1-\alpha}\exp\left(-\frac{\alpha}{1-\alpha}t\right)\right) + \frac{1}{RC}\left(\frac{N(i(t))}{s} - si(0)\right)$$

$$= \frac{1}{RCL}N(V(t)) + \frac{1}{L}(\frac{N(V(t))}{s} - sV(0))$$

$$uN\left(\frac{di(t)}{dt}\right)\frac{M(\alpha)}{1-\alpha}N\left(\exp\left((-\frac{\alpha}{1-\alpha}t)\right)\right) + \frac{1}{RC}(\frac{N(i(t))}{s} - si(0))$$

$$= \frac{1}{RCL}N(V(t)) + \frac{1}{L}(\frac{N(V(t))}{s} - sV(0))$$

$$u\left(\frac{s}{u}N(i(t)) - \frac{i(0)}{u}\right)\frac{M(\alpha)}{1-\alpha}\frac{-1+\alpha}{s(-1+\alpha)-u\alpha} + \frac{1}{RC}\frac{N(i(t))}{s}$$

$$= \frac{1}{RCL}N(V(t)) + \frac{1}{L}\frac{N(V(t))}{s}$$

Then, we obtain

$$\frac{N(V(t))}{N(i(t))} = \frac{(M(\alpha)RCs + u\alpha - s(-1+\alpha))L}{(u\alpha - s(-1+\alpha))(s + RC)}$$

and

$$N\left({}^{ABC}_{0}D^{\alpha}_{t}i(t)\right) + \frac{1}{RC}\left(\frac{N(i(t))}{s} - si(0)\right)$$

$$= \frac{1}{RCL}N(V(t)) + \frac{1}{L}(\frac{N(V(t))}{s} - sV(0))$$

$$N\left(\frac{di(t)}{dt} * \frac{AB(\alpha)}{1-\alpha}E_{\alpha}\left[\frac{-\alpha}{1-\alpha}t^{\alpha}\right]\right) + \frac{1}{RC}\left(\frac{N(i(t))}{s} - si(0)\right)$$

$$= \frac{1}{RCL}N(V(t)) + \frac{1}{L}\left(\frac{N(V(t))}{s} - sV(0)\right)$$

$$uN\left(\frac{di(t)}{dt}\right)\frac{AB(\alpha)}{1-\alpha}N\left(E_{\alpha}\left[\frac{-\alpha}{1-\alpha}t^{\alpha}\right]\right) + \frac{1}{RC}\left(\frac{N(i(t))}{s} - si(0)\right)$$

$$= \frac{1}{RCL}N(V(t)) + \frac{1}{L}\left(\frac{N(V(t))}{s} - sV(0)\right)$$

$$u\left(\frac{s}{u}N(i(t)) - \frac{i(0)}{u}\right)\frac{AB(\alpha)}{1-\alpha}\frac{1}{s}\frac{1-\alpha}{1-\alpha+\alpha(\frac{s}{u})^{-\alpha}} + \frac{1}{RC}\frac{N(i(t))}{s}$$

$$= \frac{1}{RCL}N(V(t)) + \frac{1}{L}\frac{N(V(t))}{s}$$

Then, we get

$$\frac{N(V(t))}{N(i(t))} = \frac{(AB(\alpha)RCs + 1 - \alpha + \alpha(\frac{s}{u})^{-\alpha})}{(1-\alpha+\alpha(\frac{s}{u})^{-\alpha})(s+RC)}$$

23.5.9   α INTEGRAL LAPLACE TRANSFORM

Since α is the fractional order of the equations, we use β instead of α in the transform. We get the α integral Laplace transform of these equations by:

$$L_\beta\left(^{CF}_0 D_t^\alpha i(t)\right) + \frac{1}{RC}\left(\frac{L_\beta(i(t))}{s} - si(0)\right)$$

$$= \frac{1}{RCL}L_\beta(V(t)) + \frac{1}{L}\left(\frac{L_\beta(V(t))}{s} - sV(0)\right)$$

$$L_\beta\left(\frac{di(t)}{dt} * \frac{M(\alpha)}{1-\alpha}\exp\left((-\frac{\alpha}{1-\alpha}t)\right)\right) + \frac{1}{RC}\left(\frac{L_\beta(i(t))}{s} - si(0)\right)$$

$$= \frac{1}{RCL}L_\beta(V(t)) + \frac{1}{L}\left(\frac{L_\beta(V(t))}{s} - sV(0)\right)$$

$$L_\beta\left(\frac{di(t)}{dt}\right)\frac{M(\alpha)}{1-\alpha}L_\beta\left(\exp\left(-\frac{\alpha}{1-\alpha}t\right)\right) + \frac{1}{RC}\left(\frac{L_\beta(i(t))}{s} - si(0)\right)$$

$$= \frac{1}{RCL}L_\beta(V(t)) + \frac{1}{L}\left(\frac{L_\beta(V(t))}{s} - sV(0)\right)$$

$$\left(s^{\frac{1}{\beta}}L_\beta(i(t)) - i(0)\right)\frac{M(\alpha)}{1-\alpha}\frac{1}{s^{\frac{1}{\beta}} + \frac{\alpha}{1-\alpha}} + \frac{1}{RC}\frac{L_\beta(i(t))}{s}$$

$$= \frac{1}{RCL}L_\beta(V(t)) + \frac{1}{L}\frac{L_\beta(V(t))}{s}$$

Then, we obtain

$$\frac{L_\beta(V(t))}{L_\beta(i(t))} - \frac{(M(\alpha)RCss^{\frac{1}{\beta}} + s^{\frac{1}{\beta}}(1-\alpha)+\alpha)L}{(s^{\frac{1}{\beta}}(1-\alpha)+\alpha)(s+RC)}$$

and

$$L_\beta \left(^{ABC}_0 D_t^\alpha i(t)\right) + \frac{1}{RC} \left(\frac{L_\beta(i(t))}{s} - si(0)\right)$$

$$= \frac{1}{RCL} L_\beta(V(t)) + \frac{1}{L}\left(\frac{L_\beta(V(t))}{s} - sV(0)\right)$$

$$L_\beta \left(\frac{di(t)}{dt} * \frac{AB(\alpha)}{1-\alpha} E_\alpha\left[\frac{-\alpha}{1-\alpha}t^\alpha\right]\right) + \frac{1}{RC}\left(\frac{L_\beta(i(t))}{s} - si(0)\right)$$

$$= \frac{1}{RCL} L_\beta(V(t)) + \frac{1}{L}\left(\frac{L_\beta(V(t))}{s} - sV(0)\right)$$

$$L_\beta\left(\frac{di(t)}{dt}\right)\frac{AB(\alpha)}{1-\alpha} L_\beta\left(E_\alpha\left[\frac{-\alpha}{1-\alpha}t^\alpha\right]\right) + \frac{1}{RC}\left(\frac{L_\beta(i(t))}{s} - si(0)\right)$$

$$= \frac{1}{RCL} L_\beta(V(t)) + \frac{1}{L}\left(\frac{L_\beta(V(t))}{s} - sV(0)\right)$$

$$\left(s^{\frac{1}{\beta}} L_\beta(i(t)) - i(0)\right)\frac{AB(\alpha)}{1-\alpha}\frac{1}{s^{\frac{1}{\beta}}}\frac{1-\alpha}{1-\alpha+\alpha s^{\frac{-\alpha}{\beta}}} + \frac{1}{RC}\frac{L_\beta(i(t))}{s}$$

$$= \frac{1}{RCL} L_\beta(V(t)) + \frac{1}{L}\frac{L_\beta(V(t))}{s}$$

Then, we reach

$$\frac{L_\beta(V(t))}{L_\beta(i(t))} = \frac{(AB(\alpha)RCs + 1 - \alpha + \alpha s^{-\frac{\alpha}{\beta}})L}{(1-\alpha+\alpha s^{-\frac{\alpha}{\beta}})(s+RC)}$$

23.6 APPLICATIONS II

We consider the following problem with Mittag-Leffler kernel in this section.

$$\frac{dV}{dt} = R_0^{ABC}D_t^\alpha i + \frac{1}{C}i \tag{23.60}$$

23.6.1 ELZAKI TRANSFORM

We obtain

$$E\left(\frac{dV}{dt}\right) = RE\left(^{ABC}_0 D_t^\alpha i\right) + E\left(\frac{1}{C}i\right)$$

$$\frac{E(v)}{s} - sv(0) = RE\left(\frac{di(t)}{dt} * \frac{AB(\alpha)}{1-\alpha} E_\alpha\left[\frac{-\alpha}{1-\alpha}t^\alpha\right]\right) + \frac{1}{C}E(i)$$

$$\frac{E(v)}{s} = R\frac{1}{s}E\left(\frac{di(t))}{dt}\right)\frac{AB(\alpha)}{1-\alpha}E\left(E_\alpha\left[\frac{-\alpha}{1-\alpha}t^\alpha\right]\right) + \frac{1}{C}E(i)$$

$$\frac{E(v)}{s} = R\frac{1}{s}\left(\frac{E(i)}{s} - si(0)\right)\frac{AB(\alpha)}{1-\alpha}\left(\frac{s^2(1-\alpha)}{1-\alpha+\alpha s^\alpha}\right) + \frac{1}{C}E(i)$$

$$\frac{E(v)}{s} = RE(i)\frac{AB(\alpha)}{1-\alpha+\alpha s^\alpha} + \frac{1}{C}E(i)$$

$$\frac{E(v)}{s} = E(i)\left(\frac{RAB(\alpha)}{1-\alpha+\alpha s^\alpha} + \frac{1}{C}\right)$$

Then, we get

$$\frac{E(i)}{E(v)} = \frac{(1-\alpha+\alpha s^{\alpha})C}{CsRAB(\alpha)+s(1-\alpha+\alpha s^{\alpha})}$$

23.6.2 ABOODH TRANSFORM

We obtain

$$A\left(\frac{dV}{dt}\right) = RA\left(_0^{ABC}D_t^{\alpha}i\right) + A\left(\frac{1}{C}i\right)$$

$$sA(v) - \frac{v(0)}{s} = RA\left(\frac{di(t)}{dt} * \frac{AB(\alpha)}{1-\alpha}E_\alpha\left[\frac{-\alpha}{1-\alpha}t^\alpha\right]\right) + \frac{1}{C}A(i)$$

$$sA(v) = RsA\left(\frac{di(t)}{dt}\right)\frac{AB(\alpha)}{1-\alpha}A\left(E_\alpha\left[\frac{-\alpha}{1-\alpha}t^\alpha\right]\right) + \frac{1}{C}A(i)$$

$$sA(v) = Rs\left(sA(i) - \frac{i(0)}{s}\right)\frac{AB(\alpha)}{1-\alpha}\left(\frac{1-\alpha}{s^2(1-\alpha+\alpha s^{-\alpha})}\right) + \frac{1}{C}A(i)$$

$$sA(v) = RA(i)\frac{AB(\alpha)}{1-\alpha+\alpha s^{-\alpha}} + \frac{1}{C}A(i)$$

$$sA(v) = A(i)\left(\frac{RAB(\alpha)}{1-\alpha+\alpha s^{-\alpha}} + \frac{1}{C}\right)$$

Then, we obtain

$$\frac{A(i)}{A(v)} = \frac{(1-\alpha+\alpha s^{-\alpha})Cs}{CRAB(\alpha)+1-\alpha+\alpha s^{-\alpha}}$$

23.6.3 POURREZA TRANSFORM

We obtain

$$HJ\left(\frac{dV}{dt}\right) = RHJ\left(_0^{ABC}D_t^{\alpha}i\right) + HJ(\frac{1}{C}i)$$

$$s^2HJ(v) - sv(0) = RHJ\left(\frac{di(t)}{dt} * \frac{AB(\alpha)}{1-\alpha}E_\alpha\left[\frac{-\alpha}{1-\alpha}t^\alpha\right]\right) + \frac{1}{C}HJ(i)$$

$$s^2HJ(v) = R\frac{1}{s}HJ\left(\frac{di(t)}{dt}\right)\frac{AB(\alpha)}{1-\alpha}HJ\left(E_\alpha\left[\frac{-\alpha}{1-\alpha}t^\alpha\right]\right) + \frac{1}{C}HJ(i)$$

$$s^2HJ(v) = R\frac{1}{s}\left(s^2HJ(i) - si(0)\right)\frac{AB(\alpha)}{1-\alpha}\left(\frac{1-\alpha}{s(1-\alpha+\alpha s^{-2\alpha})}\right) + \frac{1}{C}HJ(i)$$

$$s^2HJ(v) = RHJ(i)\frac{AB(\alpha)}{1-\alpha+\alpha s^{-2\alpha}} + \frac{1}{C}HJ(i)$$

$$s^2HJ(v) = HJ(i)\left(\frac{RAB(\alpha)}{1-\alpha+\alpha s^{-2\alpha}} + \frac{1}{C}\right)$$

Then, we get

$$\frac{HJ(i)}{HJ(v)} = \frac{(1-\alpha+\alpha s^{-2\alpha})Cs^2}{CRAB(\alpha)+1-\alpha+\alpha s^{-2\alpha}}$$

23.6.4 MOHAND TRANSFORM

We obtain

$$M\left(\frac{dV}{dt}\right) = RM\left({}^{ABC}_0 D^{\alpha}_t i\right) + M(\frac{1}{C}i)$$

$$sM(v) - s^2 v(0) = RM\left(\frac{di(t)}{dt} * \frac{AB(\alpha)}{1-\alpha} E_{\alpha}\left[\frac{-\alpha}{1-\alpha}t^{\alpha}\right]\right) + \frac{1}{C}M(i)$$

$$sM(v) = R\frac{1}{s^2}M\left(\frac{di(t)}{dt}\right)\frac{AB(\alpha)}{1-\alpha}M\left(E_{\alpha}\left[\frac{-\alpha}{1-\alpha}t^{\alpha}\right]\right) + \frac{1}{C}M(i)$$

$$sM(v) = R\frac{1}{s^2}(sM(i) - s^2 i(0))\frac{AB(\alpha)}{1-\alpha}\left(\frac{s(1-\alpha)}{1-\alpha+\alpha s^{-\alpha}}\right) + \frac{1}{C}M(i)$$

$$sM(v) = RM(i)\frac{AB(\alpha)}{1-\alpha+\alpha s^{-\alpha}} + \frac{1}{C}M(i)$$

$$sM(v) = M(i)\left(\frac{RAB(\alpha)}{1-\alpha+\alpha s^{-\alpha}} + \frac{1}{C}\right)$$

Then, we reach

$$\frac{M(i)}{M(v)} = \frac{(1-\alpha+\alpha s^{-\alpha})Cs}{CRAB(\alpha)+1-\alpha+\alpha s^{-\alpha}}$$

23.6.5 SAWI TRANSFORM

We have

$$Sa\left(\frac{dV}{dt}\right) = RSa\left({}^{ABC}_0 D^{\alpha}_t i\right) + Sa\left(\frac{1}{C}i\right)$$

$$\frac{Sa(v)}{s} - \frac{v(0)}{s^2} = RSa\left(\frac{di(t)}{dt} * \frac{AB(\alpha)}{1-\alpha} E_{\alpha}\left[\frac{-\alpha}{1-\alpha}t^{\alpha}\right]\right) + \frac{1}{C}Sa(i)$$

$$\frac{Sa(v)}{s} = Rs^2 Sa\left(\frac{di(t)}{dt}\right)\frac{AB(\alpha)}{1-\alpha}Sa\left(E_{\alpha}\left[\frac{-\alpha}{1-\alpha}t^{\alpha}\right]\right) + \frac{1}{C}Sa(i)$$

$$\frac{Sa(v)}{s} = Rs^2\left(\frac{Sa(i)}{s} - \frac{i(0)}{s^2}\right)\frac{AB(\alpha)}{1-\alpha}\left(\frac{1-\alpha}{s(1-\alpha+\alpha s^{\alpha})}\right) + \frac{1}{C}Sa(i)$$

$$\frac{Sa(v)}{s} = RSa(i)\frac{AB(\alpha)}{1-\alpha+\alpha s^{\alpha}} + \frac{1}{C}Sa(i)$$

$$\frac{Sa(v)}{s} = Sa(i)\left(\frac{RAB(\alpha)}{1-\alpha+\alpha s^{\alpha}} + \frac{1}{C}\right)$$

Then, we obtain

$$\frac{Sa(i)}{Sa(v)} = \frac{(1-\alpha+\alpha s^\alpha)C}{CsRAB(\alpha)+s(1-\alpha+\alpha s^\alpha)}$$

23.6.6 KAMAL TRANSFORM

We have

$$K\left(\frac{dV}{dt}\right) = RK\left({}_0^{ABC}D_t^\alpha i\right) + K\left(\frac{1}{C}i\right)$$

$$\frac{K(v)}{s} - v(0) = RK\left(\frac{di(t)}{dt} * \frac{AB(\alpha)}{1-\alpha}E_\alpha\left[\frac{-\alpha}{1-\alpha}t^\alpha\right]\right) + \frac{1}{C}K(i)$$

$$\frac{K(v)}{s} = RK\left(\frac{di(t)}{dt}\right)\frac{AB(\alpha)}{1-\alpha}K\left(E_\alpha\left[\frac{-\alpha}{1-\alpha}t^\alpha\right]\right) + \frac{1}{C}K(i)$$

$$\frac{K(v)}{s} = R\left(\frac{K(i)}{s} - i(0)\right)\frac{AB(\alpha)}{1-\alpha}\left(\frac{s(1-\alpha)}{1-\alpha+\alpha s^\alpha}\right) + \frac{1}{C}K(i)$$

$$\frac{K(v)}{s} = RK(i)\frac{AB(\alpha)}{1-\alpha+\alpha s^\alpha} + \frac{1}{C}K(i)$$

$$\frac{K(v)}{s} = K(i)\left(\frac{RAB(\alpha)}{1-\alpha+\alpha s^\alpha} + \frac{1}{C}\right)$$

Then, we obtain

$$\frac{K(i)}{K(v)} = \frac{(1-\alpha+\alpha s^\alpha)C}{CsRAB(\alpha)+s(1-\alpha+\alpha s^\alpha)}$$

23.6.7 G- TRANSFORM

We have

$$G\left(\frac{dV}{dt}\right) = RG\left({}_0^{ABC}D_t^\alpha i\right) + G\left(\frac{1}{C}i\right)$$

$$\frac{G(v)}{s} - s^\beta v(0) = RG\left(\frac{di(t)}{dt} * \frac{AB(\alpha)}{1-\alpha}E_\alpha\left[\frac{-\alpha}{1-\alpha}t^\alpha\right]\right) + \frac{1}{C}G(i)$$

$$\frac{G(v)}{s} = Rs^{-\beta}G\left(\frac{di(t)}{dt}\right)\frac{AB(\alpha)}{1-\alpha}G\left(E_\alpha\left[\frac{-\alpha}{1-\alpha}t^\alpha\right]\right) + \frac{1}{C}G(i)$$

$$\frac{G(v)}{s} = Rs^{\beta}\left(\frac{G(i)}{s} - s^\beta i(0)\right)\frac{AB(\alpha)}{1-\alpha}\left(\frac{s^\beta(1-\alpha)}{1-\alpha+\alpha s^u}\right) + \frac{1}{C}G(i)$$

$$\frac{G(v)}{s} = RG(i)\frac{AB(\alpha)}{1-\alpha+\alpha s^\alpha} + \frac{1}{C}G(i)$$

$$\frac{G(v)}{s} = G(i)\left(\frac{RAB(\alpha)}{1-\alpha+\alpha s^\alpha} + \frac{1}{C}\right)$$

Then, we have

$$\frac{G(i)}{G(v)} = \frac{(1-\alpha+\alpha s^{\alpha})C}{CsRAB(\alpha)+s(1-\alpha+\alpha s^{\alpha})}$$

23.6.8　NATURAL TRANSFORM

We have

$$N\left(\frac{dV}{dt}\right) = RN\left({}_{0}^{ADC}D_{t}^{u}i\right) + N\left(\frac{1}{C}i\right)$$

$$\frac{s}{u}N(v) - \frac{v(0)}{u} = RN\left(\frac{di(t)}{dt} * \frac{AB(\alpha)}{1-\alpha}E_{\alpha}\left[\frac{-\alpha}{1-\alpha}t^{\alpha}\right]\right) + \frac{1}{C}N(i)$$

$$\frac{s}{u}N(v) = RuN\left(\frac{di(t)}{dt}\right)\frac{AB(\alpha)}{1-\alpha}N\left(E_{\alpha}\left[\frac{-\alpha}{1-\alpha}t^{\alpha}\right]\right) + \frac{1}{C}N(i)$$

$$\frac{s}{u}N(v) = Ru\left(\frac{s}{u}N(i) - \frac{i(0)}{u}\right)\frac{AB(\alpha)}{1-\alpha}\left(\frac{1-\alpha}{s(1-\alpha+\alpha(\frac{s}{u})^{-\alpha})}\right) + \frac{1}{C}N(i)$$

$$\frac{s}{u}N(v) = RN(i)\frac{AB(\alpha)}{1-\alpha+\alpha\left(\frac{s}{u}\right)^{-\alpha}} + \frac{1}{C}N(i)$$

$$\frac{s}{u}N(v) = N(i)\left(\frac{RAB(\alpha)}{1-\alpha+\alpha(\frac{s}{u})^{-\alpha}} + \frac{1}{C}\right)$$

Then, we get

$$\frac{N(i)}{N(v)} = \frac{(1-\alpha+\alpha(\frac{s}{u})^{-\alpha}))Cs}{CuRAB(\alpha)+u(1-\alpha+\alpha(\frac{s}{u})^{-\alpha}))}$$

23.6.9　α INTEGRAL LAPLACE TRANSFORM

We have

$$L_{\alpha}\left(\frac{dv}{dt}\right) = RL_{\alpha}\left({}_{0}^{ABC}D_{t}^{\beta}i\right) + L_{\alpha}\left(\frac{1}{C}i\right)$$

$$s^{\frac{1}{\alpha}}L_{\alpha}(v) - v(0) = RL_{\alpha}\left(\frac{di(t)}{dt} * \frac{AB(\beta)}{1-\beta}E_{\beta}\left[\frac{-\beta}{1-\beta}t^{\beta}\right]\right) + \frac{1}{C}L_{\alpha}(i)$$

$$s^{\frac{1}{\alpha}}L_{\alpha}(v) = RL_{\alpha}\left(\frac{di(t)}{dt}\right)\frac{AB(\beta)}{1-\beta}L_{\alpha}\left(E_{\beta}\left[\frac{-\beta}{1-\beta}t^{\beta}\right]\right) + \frac{1}{C}L_{\alpha}(i)$$

$$s^{\frac{1}{\alpha}}L_{\alpha}(v) = R(s^{\frac{1}{\alpha}}L_{\alpha}(i) - i(0))\frac{AB(\beta)}{1-\beta}(\frac{1-\beta}{1-\beta+\beta s^{-\frac{\beta}{\alpha}}})s^{-\frac{1}{\alpha}} + \frac{1}{C}L_{\alpha}(i)$$

$$s^{\frac{1}{\alpha}}L_{\alpha}(v) = RL_{\alpha}(i)\frac{AB(\beta)}{1-\beta+\beta s^{-\frac{\beta}{\alpha}}} + \frac{1}{C}L_{\alpha}(i)$$

$$s^{\frac{1}{\alpha}}L_{\alpha}(v) = L_{\alpha}(i)\left(\frac{RAB(\beta)}{1-\beta+\beta s^{-\frac{\beta}{\alpha}}} + \frac{1}{C}\right)$$

Then, we obtain

$$\frac{L_\alpha(i)}{L_\alpha(v)} = \frac{(1-\beta+\beta s^{-\frac{\beta}{\alpha}})Cs^{\frac{1}{\alpha}}}{CRAB(\beta)+1-\beta+\beta s^{-\frac{\beta}{\alpha}}}$$

23.6.10 APPLICATIONS III

We consider the following circuit problems in this section.

$$\frac{dv}{dt} = R\frac{di}{dt} + \frac{1}{C}i \tag{23.61}$$

$$\frac{dE}{dt} = R\frac{di}{dt} + L\frac{d^2i}{dt^2} + \frac{1}{C}i \tag{23.62}$$

23.6.11 ELZAKI TRANSFORM

We find the Elzaki transform of these problems as

$$E\left[\frac{dv}{dt}\right] = E\left[R\frac{di}{dt}\right] + E\left[\frac{1}{C}i\right]$$

$$\frac{E(v)}{s} = R\frac{E(i)}{s} + \frac{1}{C}E(i)$$

$$\frac{E(v)}{s} = E(i)\left[\frac{R}{s} + \frac{1}{C}\right] = E(i)\left[\frac{RC+s}{sC}\right]$$

$$\frac{E(i)}{E(v)} = \frac{C}{RC+s}$$

and

$$\frac{dE}{dt} = R\frac{di}{dt} + L\frac{d^2i}{dt^2} + \frac{1}{C}i$$

$$E\left[\frac{dE}{dt}\right] = E\left[R\frac{di}{dt}\right] + E\left[L\frac{d^2i}{dt^2}\right] + E\left[\frac{1}{C}i\right]$$

$$\frac{E(E)}{s} = R\frac{E(i)}{s} + L\frac{E(i)}{s^2} + \frac{1}{C}E(i)$$

$$\frac{E(E)}{s} = E(i)\left[\frac{R}{s} + \frac{L}{s^2} + \frac{1}{C}\right]$$

$$\frac{E(E)}{s} = E(i)\left[\frac{sRC+LC+s^2}{s^2C}\right]$$

$$\frac{E(i)}{E(E)} = \frac{sC}{sRC+LC+s^2}$$

23.6.12   MOHAND TRANSFORM

We find the Mohand transform of these problems as

$$M\left[\frac{dv}{dt}\right] = M\left[R\frac{di}{dt}\right] + M\left[\frac{1}{C}i\right]$$

$$sM(v) = RsM(i) + \frac{1}{C}M(i)$$

$$sM(v) = M(i)\left[Rs + \frac{1}{C}\right]$$

$$sM(v) = M(i)\left[\frac{RsC+1}{C}\right]$$

$$\frac{M(i)}{M(v)} = \frac{Cs}{RCs+1}$$

and

$$M\left[\frac{dE}{dt}\right] = M\left[R\frac{di}{dt}\right] + M\left[L\frac{d^2i}{dt^2}\right] + M\left[\frac{1}{C}i\right]$$

$$\frac{M(E)}{s} = R\frac{M(i)}{s} + Ls^2 M(i) + \frac{1}{C}M(i)$$

$$\frac{M(E)}{s} = M(i)\left[\frac{R}{s} + \frac{Ls^2}{1} + \frac{1}{C}\right]$$

$$\frac{M(E)}{s} = M(i)\left[\frac{RC+Ls^3C+s}{sC}\right]$$

$$\frac{M(i)}{M(E)} = \frac{C}{RC+Ls^3C+s}$$

23.6.13   KAMAL TRANSFORM

We find the Kamal transform of these problems as

$$K\left[\frac{dv}{dt}\right] = K\left[R\frac{di}{dt}\right] + K\left[\frac{1}{C}i\right]$$

$$\frac{K(v)}{s} = R\frac{K(i)}{s} + \frac{1}{C}K(i)$$

$$\frac{K(v)}{s} = K(i)\left[\frac{R}{s} + \frac{1}{C}\right] = K(i)\left(\frac{RC+s}{sC}\right)$$

$$\frac{K(i)}{K(v)} = \frac{C}{RC+s}$$

and

$$K\left[\frac{dE}{dt}\right] = K\left[R\frac{di}{dt}\right] + K\left[L\frac{d^2i}{dt^2}\right] + K\left[\frac{1}{C}i\right]$$

$$\frac{K(E)}{s} = R\frac{K(i)}{s} + L\frac{1}{s^2}K(i) + \frac{1}{C}K(i)$$

$$\frac{K(E)}{s} = K(i)\left[\frac{R}{s} + \frac{L}{s^2} + \frac{1}{C}\right]$$

$$\frac{K(E)}{s} = K(i)\left(\frac{sCR + CL + s^2}{s^2C}\right)$$

$$\frac{K(i)}{K(E)} = \frac{sC}{sCR + CL + s^2}$$

23.6.14   ABOODH TRANSFORM

We find the Aboodh transform of these problems as

$$A\left[\frac{dv}{dt}\right] = A\left[R\frac{di}{dt}\right] + A\left[\frac{1}{C}i\right]$$

$$sA(v) = RsA(i) + \frac{1}{C}A(i)$$

$$sA(v) = A(i)\left(Rs + \frac{1}{C}\right)$$

$$sA(v) = A(i)\left(\frac{RsC + 1}{C}\right)$$

$$\frac{A(i)}{A(v)} = \frac{sC}{RsC + 1}$$

and

$$A\left[\frac{dE}{dt}\right] = A\left[R\frac{di}{dt}\right] + A\left[L\frac{d^2i}{dt^2}\right] + A\left[\frac{1}{C}i\right]$$

$$sA(E) = sRA(i) + Ls^2A(i) + \frac{1}{C}A(i)$$

$$sA(E) = A(i)\left[sR + Ls^2 + \frac{1}{C}\right]$$

$$sA(E) = A(i)\left[\frac{CsR + LCs^2 + 1}{C}\right]$$

$$\frac{A(i)}{A(E)} = \frac{Cs}{CsR + LCs^2 + 1}$$

23.6.15 SAWI TRANSFORM

We find the Sawi transform of these problems as

$$Sa\left[\frac{dv}{dt}\right] = Sa\left[R\frac{di}{dt}\right] + Sa\left[\frac{1}{C}i\right]$$

$$\frac{Sa(v)}{s} = R\frac{Sa(i)}{s} + \frac{1}{C}Sa(i)$$

$$\frac{Sa(v)}{s} = Sa(i)\left[\frac{R}{s} + \frac{1}{C}\right]$$

$$\frac{Sa(v)}{s} = Sa(i)\frac{(RC+s)}{sC}$$

$$\frac{Sa(i)}{Sa(v)} = \frac{C}{RC+s}$$

and

$$Sa\left[\frac{dE}{dt}\right] = Sa\left[R\frac{di}{dt}\right] + Sa\left[L\frac{d^2i}{dt^2}\right] + Sa\left[\frac{1}{C}i\right]$$

$$\frac{Sa(E)}{s} = R\frac{Sa(i)}{s} + L\frac{Sa(i)}{s^2} + \frac{1}{C}Sa(i)$$

$$\frac{Sa(E)}{s} = Sa(i)\left[\frac{R}{s} + \frac{L}{s^2} + \frac{1}{C}\right]$$

$$\frac{Sa(E)}{s} = Sa(i)\left(\frac{RsC+CL+s^2}{s^2C}\right)$$

$$\frac{Sa(i)}{Sa(E)} = \frac{sC}{RsC+CL+s^2}$$

23.6.16 α-INTEGRAL LAPLACE TRANSFORM

We get the α -Integral Laplace transform of these problems by:

$$L_\alpha\left[\frac{dv}{dt}\right] = L_\alpha\left[R\frac{di}{dt}\right] + L_\alpha\left[\frac{1}{C}i\right]$$

$$s^{\frac{1}{\alpha}}L_\alpha(v) = Rs^{\frac{1}{\alpha}}L_\alpha(i) + \frac{1}{C}L_\alpha(i)$$

$$s^{\frac{1}{\alpha}}L_\alpha(v) = L_\alpha(i)\left[Rs^{\frac{1}{\alpha}} + \frac{1}{C}\right]$$

$$s^{\frac{1}{\alpha}}L_\alpha(v) = L_\alpha(i)\frac{\left(CRs^{\frac{1}{\alpha}} + 1\right)}{C}$$

$$\frac{L_\alpha(i)}{L_\alpha(v)} = \frac{Cs^{\frac{1}{\alpha}}}{CRs^{\frac{1}{\alpha}} + 1}$$

and

$$L_\alpha\left[\frac{dE}{dt}\right]=L_\alpha\left[R\frac{di}{dt}\right]+L_\alpha\left[L\frac{d^2i}{dt^2}\right]+L_\alpha\left[\frac{1}{C}i\right]$$

$$s^{\frac{1}{\alpha}}L_\alpha(E)=s^{\frac{1}{\alpha}}RL_\alpha+Ls^{\frac{2}{\alpha}}L_\alpha(i)+\frac{1}{C}L_\alpha(i)$$

$$s^{\frac{1}{\alpha}}L_\alpha(E)=L_\alpha(i)\left[Rs^{\frac{1}{\alpha}}+Ls^{\frac{2}{\alpha}}+\frac{1}{C}\right]$$

$$\frac{L_\alpha(i)}{L_\alpha(E)}=\frac{s^{\frac{1}{\alpha}}}{Rs^{\frac{1}{\alpha}}+Ls^{\frac{2}{\alpha}}+\frac{1}{C}}=\frac{Cs^{\frac{1}{\alpha}}}{RCs^{\frac{1}{\alpha}}+LCs^{\frac{2}{\alpha}}+1}$$

23.6.17   G− TRANSFORM

We find the $G-$transform of these problems as

$$G\left[\frac{dv}{dt}\right]=G\left[R\frac{di}{dt}\right]+G\left[\frac{1}{C}i\right]$$

$$\frac{G(v)}{s}=R\frac{G(i)}{s}+\frac{1}{C}G(i)$$

$$\frac{G(v)}{s}=G(i)\left[\frac{R}{s}+\frac{1}{C}\right]$$

$$\frac{G(v)}{s}=G(i)\left(\frac{CR+s}{sC}\right)$$

$$\frac{G(i)}{G(v)}=\frac{C}{RC+s}$$

and

$$G\left[\frac{dE}{dt}\right]=G\left[R\frac{di}{dt}\right]+G\left[L\frac{d^2i}{dt^2}\right]+G\left[\frac{1}{C}i\right]$$

$$\frac{G(E)}{s}=R\frac{G(i)}{s}+L\frac{1}{s^2}G(i)+\frac{1}{C}G(i)$$

$$\frac{G(E)}{s}=G(i)\left[\frac{R}{s}+\frac{L}{s^2}+\frac{1}{C}\right]$$

$$\frac{G(E)}{s}=G(i)\left(\frac{sCR+CL+s^2}{s^2C}\right)$$

$$\frac{G(i)}{G(E)}=\frac{sC}{sCR+CL+s^2}$$

23.6.18   POURREZA TRANSFORM

We find the Pourreza transform of these problems as

$$H\left[\frac{dv}{dt}\right] = H\left[R\frac{di}{dt}\right] + H\left[\frac{1}{C}i\right]$$

$$s^2 H(v) = Rs^2 H(i) + \frac{1}{C}II(i)$$

$$s^2 H(v) = H(i)\left[Rs^2 + \frac{1}{C}\right]$$

$$s^2 H(v) = H(i)\left(\frac{RCs^2 + 1}{C}\right)$$

$$\frac{H(i)}{H(v)} = \frac{Cs^2}{RCs^2 + 1}$$

and

$$H\left[\frac{dE}{dt}\right] = H\left[R\frac{di}{dt}\right] + H\left[L\frac{d^2 i}{dt^2}\right] + H\left[\frac{1}{C}i\right]$$

$$s^2 H(E) = Rs^2 H(i) + Ls^4 H(i) + \frac{1}{C}H(i)$$

$$s^2 H(E) = H(i)\left[s^2 R + Ls^4 + \frac{1}{C}\right]$$

$$s^2 H(E) = H(i)\left(\frac{s^2 CR + LCs^4 + 1}{C}\right)$$

$$\frac{H(i)}{H(E)} = \frac{s^2 C}{s^2 CR + LCs^4 + 1}$$

23.6.19   NATURAL TRANSFORM

We find the Natural transform of these problems as

$$N\left[\frac{dv}{dt}\right] = N\left[R\frac{di}{dt}\right] + N\left[\frac{1}{C}i\right]$$

$$\frac{s}{u}N(v) = \frac{RsN(i)}{u} + \frac{1}{C}N(i)$$

$$\frac{s}{u}N(v) = N(i)\left[\frac{Rs}{u} + \frac{1}{C}\right]$$

$$\frac{s}{u}N(v) = N(i)\left[\frac{RsC + u}{uC}\right]$$

$$\frac{N(i)}{N(v)} = \frac{sC}{RsC + u}$$

and

$$N\left[\frac{dE}{dt}\right] = N\left[R\frac{di}{dt}\right] + N\left[L\frac{d^2i}{dt^2}\right] + N\left[\frac{1}{C}i\right]$$

$$\frac{s}{u}N(E) = R\frac{s}{u}N(i) + L\frac{s^2}{u^2}N(i) + \frac{1}{C}N(i)$$

$$\frac{s}{u}N(E) = N(i)\left[\frac{Rs}{u} + \frac{Ls^2}{u^2} + \frac{1}{C}\right]$$

$$\frac{s}{u}N(E) = N(i)\left[\frac{RsuC + LCs^2 + u^2}{u^2C}\right]$$

$$\frac{N(i)}{N(E)} = \frac{suC}{RsuC + LCs^2 + u^2}$$

23.6.20   APPLICATIONS IV

We consider the following circuit problem in this section.

$$\frac{d^2i(t)}{dt^2} + \frac{1}{RC}\frac{di}{dt} = \frac{V(t)}{RCL} + \frac{1}{L}\frac{dV(t)}{dt}$$

23.6.21   ELZAKI TRANSFORM

We find the Elzaki transform of the circuit problem as

$$E\left(\frac{d^2i(t)}{dt^2}\right) + \frac{1}{RC}E\left(\frac{di}{dt}\right) = E\left(\frac{V(t)}{RCL}\right) + E\left(\frac{1}{L}\frac{dV(t)}{dt}\right)$$

Then, we have

$$\left(\frac{E(i)}{s^2} - i(0) - si'(0)\right) + \frac{1}{RC}\left(\frac{E(i)}{s} - si(0)\right)$$

$$= \frac{E(V)}{RCL} + \frac{1}{L}\left(\frac{E(V)}{s} - sV(0)\right)$$

$$\frac{E(i)}{s^2} + \frac{E(i)}{sRC} = \frac{E(V)}{RCL} + \frac{E(V)}{sL}$$

$$E(i)\left(\frac{1}{s^2} + \frac{1}{sRC}\right) = E(V)\left(\frac{1}{RCL} + \frac{1}{sL}\right)$$

$$\frac{E(V)}{E(i)} = \frac{\left(\frac{1}{s^2} + \frac{1}{sRC}\right)}{\left(\frac{1}{RCL} + \frac{1}{sL}\right)}$$

Thus, we reach

$$\frac{E(V)}{E(i)} = \frac{L}{s}$$

23.6.22   ABOODH TRANSFORM

We find the Aboodh transform of the circuit problem as

$$A\left(\frac{d^2 i(t)}{dt^2}\right) + \frac{1}{RC}A\left(\frac{di}{dt}\right) = A\left(\frac{V(t)}{RCL}\right) + A\left(\frac{1}{L}\frac{dV(t)}{dt}\right)$$

$$\left(s^2 A(i) - i\,(0) - \frac{i'(0)}{s}\right) + \frac{1}{RC}\left(sA(i) - \frac{i(0)}{s}\right)$$

$$= \frac{A(V)}{RCL} + \frac{1}{L}\left(sA(V) - \frac{V(0)}{s}\right)$$

$$s^2 A(i) + \frac{sA(i)}{RC} = \frac{A(V)}{RCL} + \frac{sA(V)}{L}$$

$$A(i)\left(s^2 + \frac{s}{RC}\right) = A(V)\left(\frac{1}{RCL} + \frac{s}{L}\right)$$

$$\frac{A(V)}{A(i)} = \frac{\left(s^2 + \frac{s}{RC}\right)}{\left(\frac{1}{RCL} + \frac{s}{L}\right)} = \frac{Ls(sRC + 1)}{(sRC + 1)}$$

Then, we have

$$\frac{A(V)}{A(i)} = Ls$$

23.6.23   POURREZA TRANSFORM

We find the Pourreza transform of the circuit problem as

$$HJ\left(\frac{d^2 i(t)}{dt^2}\right) + \frac{1}{RC}HJ\left(\frac{di}{dt}\right) = HJ\left(\frac{V(t)}{RCL}\right) + HJ\left(\frac{1}{L}\frac{dV(t)}{dt}\right)$$

$$\left(sHJ(i) - \sqrt{s}\,i\,(0) - i'(0)\right) + \frac{1}{RC}\left(s^2 HJ(i) - si(0)\right)$$

$$= \frac{HJ(V)}{RCL} + \frac{1}{L}\left(s^2 HJ(V) - sV(0)\right)$$

$$sHJ(i) + \frac{s^2 HJ(i)}{RC} = \frac{HJ(V)}{RCL} + \frac{s^2 HJ(V)}{L}$$

$$HJ(i)\left(s + \frac{s^2}{RC}\right) = HJ(V)\left(\frac{1}{RCL} + \frac{s^2}{L}\right)$$

$$\frac{HJ(V)}{HJ(i)} = \frac{\left(s + \frac{s^2}{RC}\right)}{\left(\frac{1}{RCL} + \frac{s^2}{L}\right)}$$

Then, we have

$$\frac{HJ(V)}{HJ(i)} = \frac{Ls(RC + s)}{1 + s^2 RC}$$

23.6.24   MOHAND TRANSFORM

We find the Mohand transform of the circuit problem as

$$M\left(\frac{d^2 i(t)}{dt^2}\right) + \frac{1}{RC}M\left(\frac{di}{dt}\right) = M\left(\frac{V(t)}{RCL}\right) + M\left(\frac{1}{L}\frac{dV(t)}{dt}\right)$$

$$\left(s^2 M(i) - s^3 i(0) - s^2 i'(0)\right) + \frac{1}{RC}(sM(i) - s^2 i(0))$$

$$= \frac{M(V)}{RCL} + \frac{1}{L}(sM(V) - -s^2 V(0))$$

$$s^2 M(i) + \frac{sM(i)}{RC} = \frac{M(V)}{RCL} + \frac{sM(V)}{L}$$

$$M(i)\left(s^2 + \frac{s}{RC}\right) = M(V)\left(\frac{1}{RCL} + \frac{s}{L}\right)$$

$$\frac{M(V)}{M(i)} = \frac{\left(s^2 + \frac{s}{RC}\right)}{\left(\frac{1}{RCL} + \frac{s}{L}\right)} = \frac{Ls(sRC + 1)}{(sRC + 1)}$$

Then, we have

$$\frac{M(V)}{M(i)} = Ls$$

23.6.25   SAWI TRANSFORM

We find the Sawi transform of the circuit problem as

$$Sa\left(\frac{d^2 i(t)}{dt^2}\right) + \frac{1}{RC}Sa\left(\frac{di}{dt}\right) = Sa\left(\frac{V(t)}{RCL}\right) + Sa\left(\frac{1}{L}\frac{dV(t)}{dt}\right)$$

$$\left(\frac{Sa(i)}{s^2} - \frac{i(0)}{s^3} - \frac{i'(0)}{s^2}\right) + \frac{1}{RC}\left(\frac{Sa(i)}{s} - \frac{i(0)}{s^2}\right)$$

$$= \frac{Sa(V)}{RCL} + \frac{1}{L}\left(\frac{Sa(V)}{s} - \frac{V(0)}{s^2}\right)$$

$$\frac{Sa(i)}{s^2} + \frac{Sa(i)}{sRC} = \frac{Sa(V)}{RCL} + \frac{Su(V)}{sL}$$

$$Sa(i)\left(\frac{1}{s^2} + \frac{1}{sRC}\right) = Sa(V)\left(\frac{1}{RCL} + \frac{1}{sL}\right)$$

$$\frac{Sa(V)}{Sa(i)} = \frac{\left(\frac{1}{s^2} + \frac{1}{sRC}\right)}{\left(\frac{1}{RCL} + \frac{1}{sL}\right)}$$

Then, we get

$$\frac{Sa(V)}{Sa(i)} = \frac{L}{s}$$

23.6.26 KAMAL TRANSFORM

We find the Kamal transform of the circuit problem as

$$K\left(\frac{d^2i(t)}{dt^2}\right) + \frac{1}{RC}K\left(\frac{di}{dt}\right) = K\left(\frac{V(t)}{RCL}\right) + K\left(\frac{1}{L}\frac{dV(t)}{dt}\right)$$

$$\left(\frac{K(i)}{s^2} - \frac{i(0)}{s} - i'(0)\right) + \frac{1}{RC}\left(\frac{K(i)}{s} - i(0)\right)$$

$$= \frac{K(V)}{RCL} + \frac{1}{L}\left(\frac{K(V)}{s} - V(0)\right)$$

$$\frac{K(i)}{s^2} + \frac{K(i)}{sRC} = \frac{K(V)}{RCL} + \frac{K(V)}{sL}$$

$$K(i)\left(\frac{1}{s^2} + \frac{1}{sRC}\right) = K(V)\left(\frac{1}{RCL} + \frac{1}{sL}\right)$$

$$\frac{K(V)}{K(i)} = \frac{\left(\frac{1}{s^2} + \frac{1}{sRC}\right)}{\left(\frac{1}{RCL} + \frac{1}{sL}\right)}$$

Then, we get

$$\frac{K(V)}{K(i)} = \frac{L}{s}$$

23.6.27 G TRANSFORM

We find the G transform of the circuit problem as

$$G\left(\frac{d^2i(t)}{dt^2}\right) + \frac{1}{RC}G\left(\frac{di}{dt}\right) = G\left(\frac{V(t)}{RCL}\right) + G\left(\frac{1}{L}\frac{dV(t)}{dt}\right)$$

$$\left(\frac{G(i)}{s^2} - \frac{i(0)s^\beta}{s} - i'(0)s^\beta\right) + \frac{1}{RC}\left(\frac{G(i)}{s} - s^\beta i(0)\right)$$

$$= \frac{G(V)}{RCL} + \frac{1}{L}\left(\frac{G(V)}{s} - s^\beta V(0)\right)$$

$$\frac{G(i)}{s^2} + \frac{G(i)}{sRC} = \frac{G(V)}{RCL} + \frac{G(V)}{sL}$$

$$G(i)\left(\frac{1}{s^2} + \frac{1}{sRC}\right) = G(V)\left(\frac{1}{RCL} + \frac{1}{sL}\right)$$

$$\frac{G(V)}{G(i)} = \frac{\left(\frac{1}{s^2} + \frac{1}{sRC}\right)}{\left(\frac{1}{RCL} + \frac{1}{sL}\right)}$$

Then, we get

$$\frac{G(V)}{G(i)} = \frac{L}{s}$$

23.6.28   NATURAL TRANSFORM

We find the Natural transform of the circuit problem as

$$N\left(\frac{d^2 i(t)}{dt^2}\right) + \frac{1}{RC}N\left(\frac{di}{dt}\right) = N\left(\frac{V(t)}{RCL}\right) + N\left(\frac{1}{L}\frac{dV(t)}{dt}\right)$$

$$\left(\frac{s^2}{u^2}N(i) - i\,(0) - \frac{i'(0)}{u}\right) + \frac{1}{RC}\left(\frac{s}{u}N(i) - \frac{i(0)}{u}\right)$$

$$= \frac{N(V)}{RCL} + \frac{1}{L}\left(\frac{s}{u}N(V) - \frac{V(o)}{u}\right)$$

$$\left(\frac{s^2}{u^2}\right)N(i) + \frac{sN(i)}{uRC} = \frac{N(V)}{RCL} + \frac{sN(V)}{Lu}$$

$$N(i)\left(\frac{s^2}{u^2} + \frac{s}{uRC}\right) = N(V)\left(\frac{1}{RCL} + \frac{s}{Lu}\right)$$

$$\frac{N(V)}{N(i)} = \frac{\left(\frac{s^2}{u^2} + \frac{s}{uRC}\right)}{\left(\frac{1}{RCL} + \frac{s}{Lu}\right)} = \frac{Ls(sRC + u)}{u(sRC + u)}$$

Then, we get

$$\frac{N(V)}{N(i)} = \frac{Ls}{u}$$

23.6.29   α INTEGRAL LAPLACE TRANSFORM

We find the α integral Laplace transform of the circuit problem as

$$L_\alpha\left(\frac{d^2 i(t)}{dt^2}\right) + \frac{1}{RC}L_\alpha\left(\frac{di}{dt}\right) = L_\alpha\left(\frac{V(t)}{RCL}\right) + L_\alpha\left(\frac{1}{L}\frac{dV(t)}{dt}\right)$$

$$\left(s^{\frac{2}{\alpha}}L_\alpha(i) - s^{\frac{1}{\alpha}}i\,(0) - i'(0)\right) + \frac{1}{RC}\left(s^{\frac{1}{\alpha}}L_\alpha(i) - i\,(0)\right)$$

$$= \frac{L_\alpha(V)}{RCL} + \frac{1}{L}\left(s^{\frac{1}{\alpha}}L_\alpha(V) - V(0)\right)$$

$$s^{\frac{2}{\alpha}}L_\alpha(i) + \frac{s^{\frac{1}{\alpha}}L_\alpha(i)}{RC} = \frac{L_\alpha(V)}{RCL} + \frac{s^{\frac{1}{\alpha}}L_\alpha(V)}{L}$$

$$L_\alpha(i)\left(s^{\frac{2}{\alpha}} + \frac{s^{\frac{1}{\alpha}}}{RC}\right) = L_\alpha(V)\left(\frac{1}{RCL} + \frac{s^{\frac{1}{\alpha}}}{L}\right)$$

$$\frac{L_\alpha(V)}{L_\alpha(i)} = \frac{\left(s^{\frac{2}{\alpha}} + \frac{s^{\frac{1}{\alpha}}}{RC}\right)}{\left(\frac{1}{RCL} + \frac{s^{\frac{1}{\alpha}}}{L}\right)}$$

$$\frac{L_\alpha(V)}{L_\alpha(i)} = \frac{Ls^{\frac{1}{\alpha}}\left(s^{\frac{1}{\alpha}}RC + 1\right)}{\left(s^{\frac{1}{\alpha}}RC + 1\right)}$$

Then, we get

$$\frac{L\alpha(V)}{L\alpha(i)} = Ls^{\frac{1}{\alpha}}$$

23.7   APPLICATION V

We consider the following circuit problem in this section.

$$_0^C D_t^\alpha i(t) + \frac{1}{RC}\frac{di}{dt} = \frac{V(t)}{RCL} + \frac{1}{L}\frac{dV(t)}{dt}$$

23.7.1   ELZAKI TRANSFORM

We find the Elzaki transform of the circuit problem as

$$E(_0^C D_t^\alpha i(t)) + \frac{1}{RC}E\left(\frac{di}{dt}\right) = E\left(\frac{V(t)}{RCL}\right) + E\left(\frac{1}{L}\frac{dV(t)}{dt}\right)$$

$$s^{2-a}\left(\frac{E(i)}{s} - i(0) - si'(0)\right) + \frac{1}{RC}\left(\frac{E(i)}{s} - si(0)\right)$$

$$= \frac{E(V)}{RCL} + \frac{1}{L}\left(\frac{E(V)}{s} - sV(0)\right)$$

$$\frac{E(i)}{s^a} + \frac{E(i)}{sRC} = \frac{E(V)}{RCL} + \frac{E(V)}{sL}$$

$$E(i)\left(\frac{1}{s^a} + \frac{1}{sRC}\right) = E(V)\left(\frac{1}{RCL} + \frac{1}{sL}\right)$$

$$\frac{E(V)}{E(i)} = \frac{\left(\frac{1}{s^a} + \frac{1}{sRC}\right)}{\left(\frac{1}{RCL} + \frac{1}{sL}\right)}$$

$$\frac{E(V)}{E(i)} = \frac{L(s^{-a}sRC + 1)}{s + RC}$$

23.7.2   ABOODH TRANSFORM

We find the Aboodh transform of the circuit problem as

$$A(_0^C D_t^\alpha i(t)) + \frac{1}{RC}A\left(\frac{di}{dt}\right) = A\left(\frac{V(t)}{RCL}\right) + A\left(\frac{1}{L}\frac{dV(t)}{dt}\right)$$

$$s^{a-2}\left(s^2 A(i) - i(0) - \frac{i'(0)}{s}\right) + \frac{1}{RC}\left(sA(i) - \frac{i(0)}{s}\right)$$

$$= \frac{A(V)}{RCL} + \frac{1}{L}\left(sA(V) - \frac{V(0)}{s}\right)$$

$$s^a A(i) + \frac{sA(i)}{RC} = \frac{A(V)}{RCL} + \frac{sA(V)}{L}$$

$$A(i)\left(s^a + \frac{s}{RC}\right) = A(V)\left(\frac{1}{RCL} + \frac{s}{L}\right)$$

$$\frac{A(V)}{A(i)} = \frac{\left(s^a + \frac{s}{RC}\right)}{\left(\frac{1}{RCL} + \frac{s}{L}\right)}$$

$$\frac{A(V)}{A(i)} = \frac{L\left(s^a RC + s\right)}{1 + sRC}$$

23.7.3 POURREZA TRANSFORM

We find the Pourreza transform of the circuit problem as

$$HJ\left({}_{0}^{C}D_{t}^{a}i(t)\right) + \frac{1}{RC}HJ\left(\frac{di}{dt}\right) = HJ\left(\frac{V(t)}{RCL}\right) + HJ\left(\frac{1}{L}\frac{dV(t)}{dt}\right)$$

$$s^{2a-1}(sHJ(i) - \sqrt{s}\, i(0) - i'(0)) + \frac{1}{RC}(s^2 HJ(i) - si(0))$$

$$= \frac{HJ(V)}{RCL} + \frac{1}{L}(s^2 HJ(V) - sV(0))$$

$$s^{2a}HJ(i) + \frac{s^2 HJ(i)}{RC} = \frac{HJ(V)}{RCL} + \frac{s^2 HJ(V)}{L}$$

$$HJ(i)\left(s^{2a} + \frac{s^2}{RC}\right) = HJ(V)\left(\frac{1}{RCL} + \frac{s^2}{L}\right)$$

$$\frac{HJ(V)}{HJ(i)} = \frac{\left(s^{2a} + \frac{s^2}{RC}\right)}{\left(\frac{1}{RCL} + \frac{s^2}{L}\right)}$$

$$\frac{HJ(V)}{HJ(i)} = \frac{L(s^{2a}RC + s^2)}{1 + s^2 RC}$$

23.7.4 MOHAND TRANSFORM

We find the Mohand transform of the circuit problem as

$$M\left({}_{0}^{C}D_{t}^{a}i(t)\right) + \frac{1}{RC}M\left(\frac{di}{dt}\right) = M\left(\frac{V(t)}{RCL}\right) + M\left(\frac{1}{L}\frac{dV(t)}{dt}\right)$$

$$s^{a-2}(s^2 M(i) - s^3 i(0) - s^2 i'(0)) + \frac{1}{RC}(sM(i) - s^2 i(0))$$

$$= \frac{M(V)}{RCL} + \frac{1}{L}(sM(V) - s^2 V(0))$$

$$s^a M(i) + \frac{sM(i)}{RC} = \frac{M(V)}{RCL} + \frac{sM(V)}{L}$$

$$M(i)\left(s^a + \frac{s}{RC}\right) = M(V)\left(\frac{1}{RCL} + \frac{s}{L}\right)$$

$$\frac{M(V)}{M(i)} = \frac{\left(s^a + \frac{s}{RC}\right)}{\left(\frac{1}{RCL} + \frac{s}{L}\right)}$$

$$\frac{M(V)}{M(i)} = \frac{L(s^a RC + s)}{1 + sRC}$$

23.7.5 SAWI TRANSFORM

We find the Sawi transform of the circuit problem as

$$Sa(^C_0 D^a_t i(t)) + \frac{1}{RC} Sa\left(\frac{di}{dt}\right) = Sa\left(\frac{V(t)}{RCL}\right) + Sa\left(\frac{1}{L}\frac{dV(t)}{dt}\right)$$

$$s^{2-a}\left(\frac{Sa(i)}{s^2} - \frac{i(0)}{s^3} - \frac{i'(0)}{s^2}\right) + \frac{1}{RC}\left(\frac{Sa(i)}{s} - \frac{i(0)}{s^2}\right)$$

$$= \frac{Sa(V)}{RCL} + \frac{1}{L}\left(\frac{Sa(V)}{s} - \frac{V(0)}{s^2}\right)$$

$$s^{-a} Sa(i) + \frac{Sa(i)}{sRC} = \frac{Sa(V)}{RCL} + \frac{Sa(V)}{sL}$$

$$Sa(i)\left(\frac{1}{s^a} + \frac{1}{sRC}\right) = Sa(V)\left(\frac{1}{RCL} + \frac{1}{sL}\right)$$

$$\frac{Sa(V)}{Sa(i)} = \frac{\left(\frac{1}{s^a} + \frac{1}{sRC}\right)}{\left(\frac{1}{RCL} + \frac{1}{sL}\right)}$$

$$\frac{Sa(V)}{Sa(i)} = \frac{L(s^{-a}sRC + 1)}{s + RC}$$

23.7.6 KAMAL TRANSFORM

We find the Kamal transform of the circuit problem as

$$K\left(^C_0 D^a_t i(t)\right) + \frac{1}{RC}K\left(\frac{di}{dt}\right) = K\left(\frac{V(t)}{RCL}\right) + K\left(\frac{1}{L}\frac{dV(t)}{dt}\right)$$

$$s^{2-a}\left(\frac{K(i)}{s^2} - \frac{i(0)}{s} - i'(0)\right) + \frac{1}{RC}\left(\frac{K(i)}{s} - i(0)\right)$$

$$= \frac{K(V)}{RCL} + \frac{1}{L}\left(\frac{K(V)}{s} - V(0)\right)$$

$$\frac{K(i)}{s^a} + \frac{K(i)}{sRC} = \frac{K(V)}{RCL} + \frac{K(V)}{sL}$$

$$K(i)\left(\frac{1}{s^a} + \frac{1}{sRC}\right) = K(V)\left(\frac{1}{RCL} + \frac{1}{sL}\right)$$

$$\frac{K(V)}{K(i)} = \frac{\left(\frac{1}{s^a} + \frac{1}{sRC}\right)}{\left(\frac{1}{RCL} + \frac{1}{sL}\right)}$$

$$\frac{K(V)}{K(i)} = \frac{L(s^{-a}sRC + 1)}{s + RC}$$

23.7.7 G TRANSFORM

We find the G transform of the circuit problem as

$$G\left({}_{0}^{C}D_{t}^{a}i(t)\right) + \frac{1}{RC}G\left(\frac{di}{dt}\right) = G\left(\frac{V(t)}{RCL}\right) + G\left(\frac{1}{L}\frac{dV(t)}{dt}\right)$$

$$s^{2-a}\left(\frac{G(i)}{s^2} - \frac{i(0)s^\beta}{s} - i'(0)s^\beta\right) + \frac{1}{RC}\left(\frac{G(i)}{s} - s^\beta i(0)\right)$$

$$= \frac{G(V)}{RCL} + \frac{1}{L}\left(\frac{G(V)}{s} - s^\beta V(0)\right)$$

$$\frac{G(i)}{s^a} + \frac{G(i)}{sRC} = \frac{G(V)}{RCL} + \frac{G(V)}{sL}$$

$$(i)\left(\frac{1}{s^a} + \frac{1}{sRC}\right) = G(V)\left(\frac{1}{RCL} + \frac{1}{sL}\right)$$

$$\frac{G(V)}{G(i)} = \frac{\left(\frac{1}{s^a} + \frac{1}{sRC}\right)}{\left(\frac{1}{RCL} + \frac{1}{sL}\right)}$$

$$\frac{G(V)}{G(i)} = \frac{L(s^{-a}sRC+1)}{s+RC}$$

23.7.8 NATURAL TRANSFORM

We find the Natural transform of the circuit problem as

$$N\left({}_{0}^{C}D_{t}^{a}i(t)\right) + \frac{1}{RC}N\left(\frac{di}{dt}\right) = N\left(\frac{V(t)}{RCL}\right) + N\left(\frac{1}{L}\frac{dV(t)}{dt}\right)$$

$$\frac{s^{-2+a}}{u^{-2+a}}\left(\frac{s^2}{u^2}N(i) - i(0) - \frac{i'(0)}{u}\right)$$

$$+ \frac{1}{RC}\left(\frac{s}{u}N(i) - \frac{i(0)}{u}\right)$$

$$- \frac{N(V)}{RCL} + \frac{1}{L}\left(\frac{s}{u}N(V) - \frac{V(o)}{u}\right)$$

$$\frac{s^a}{u^a}N(i) + \frac{sN(i)}{uRC} = \frac{N(V)}{RCL} + \frac{sN(V)}{Lu}$$

$$N(i)\left(\frac{s^a}{u^a} + \frac{s}{uRC}\right) = N(V)\left(\frac{1}{RCL} + \frac{s}{Lu}\right)$$

$$\frac{N(V)}{N(i)} = \frac{\left(\frac{s^a}{u^a} + \frac{s}{uRC}\right)}{\left(\frac{1}{RCL} + \frac{s}{Lu}\right)}$$

$$\frac{N(V)}{N(i)} = \frac{L(s^a uRC + su^a)}{u^a(u + sRC)}$$

23.7.9　α INTEGRAL LAPLACE TRANSFORM

We find the α integral Laplace transform of the circuit problem as

$$L_\alpha\left({}_0^C D_t^\beta i(t)\right) + \frac{1}{RC} L_\alpha\left(\frac{di}{dt}\right) = L_\alpha\left(\frac{V(t)}{RCL}\right) + L_\alpha\left(\frac{1}{L}\frac{dV(t)}{dt}\right)$$

$$s^{-\frac{1}{\beta}+1}\left(s^{\frac{2}{\alpha}} L_\alpha(i) - s^{\frac{1}{\alpha}} i(0) - i'(0)\right)$$

$$+ \frac{1}{RC}\left(s^{\frac{1}{\alpha}} L_\alpha(i) - i(0)\right)$$

$$= \frac{L_\alpha(V)}{RCL} + \frac{1}{L}\left(s^{\frac{1}{\alpha}} L_\alpha(V) - V(0)\right)$$

$$s^{\frac{1}{\alpha}-\frac{1}{\beta}+1} L_\alpha(i) + \frac{s^{\frac{1}{\alpha}} L_\alpha(i)}{RC} = \frac{L_\alpha(V)}{RCL} + \frac{s^{\frac{1}{\alpha}} L_\alpha(V)}{L}$$

$$L_\alpha(i)\left(s^{\frac{1}{\alpha}-\frac{1}{\beta}+1} + \frac{s^{\frac{1}{\alpha}}}{RC}\right) = L_\alpha(V)\left(\frac{1}{RCL} + \frac{s^{\frac{1}{\alpha}}}{L}\right)$$

$$\frac{L_\alpha(V)}{L_\alpha(i)} = \frac{\left(s^{\frac{1}{\alpha}-\frac{1}{\beta}+1} + \frac{s^{\frac{1}{\alpha}}}{RC}\right)}{\left(\frac{1}{RCL} + \frac{s^{\frac{1}{\alpha}}}{L}\right)}$$

$$\frac{L_\alpha(V)}{L_\alpha(i)} = \frac{L\left(s^{\frac{1}{\alpha}-\frac{1}{\beta}+1} RC + s^{\frac{1}{\alpha}}\right)}{1 + s^{\frac{1}{\alpha}} RC}$$

23.8　APPLICATION VI

We consider the following circuit model in this section.

$$_0^{ABC} D_t^a V = R\frac{di}{dt} + \frac{1}{C} i \tag{23.63}$$

23.8.1　ELZAKI TRANSFORM

We obtain the Elzaki transform of the circuit model as

$$E\left(_0^{ABC} D_t^a V\right) = E\left(R\frac{di}{dt}\right) + E\left(\frac{1}{C} i\right)$$

$$E\left(\frac{dV}{dt} * \frac{AB(a)}{1-a} E_a\left(-\frac{a}{1-a} t^a\right)\right) = R\left(\frac{E(i)}{s} - si(0)\right) + \frac{1}{C} E(i)$$

$$\frac{1}{s} E\left(\frac{dV}{dt}\right) \frac{AB(a)}{1-a} E\left(E_a\left(-\frac{a}{1-a} t^a\right)\right) = \frac{R}{s} E(i) + \frac{1}{C} E(i)$$

$$\frac{1}{s}\left(\frac{E(V)}{s} - sV(0)\right) \frac{AB(a)}{1-a} \frac{s^2(1-a)}{1-a+as^a} = E(i)\left(\frac{R}{s} + \frac{1}{C}\right)$$

$$\frac{E(v)AB(a)}{1-a+as^a} = E(i)\left(\frac{RC+s}{sC}\right)$$

$$\frac{E(v)AB(a)}{1-a+as^a} = \frac{(RC+s)E(i)}{sC}$$

$$\frac{E(i)}{E(V)} = \frac{sCAB(a)}{(RC+s)(1-a+as^a)}$$

23.8.2 ABOODH TRANSFORM

We obtain the Aboodh transform of the circuit model as

$$A(^{ABC}_0 D^a_t V) = A\left(R\frac{di}{dt}\right) + A\left(\frac{1}{C}i\right)$$

$$sA\left(\frac{dV}{dt} * \frac{AB(a)}{1-a}E_a\left(-\frac{a}{1-a}t^a\right)\right) = R\left(sA(i) - \frac{i(0)}{s}\right) + \frac{1}{C}A(i)$$

$$sA\left(\frac{dV}{dt}\right)\frac{AB(a)}{1-a}A\left(E_a\left(-\frac{a}{1-a}t^a\right)\right) = RsA(i) + \frac{1}{C}A(i)$$

$$s\left(sA(V) - \frac{V(0)}{s}\right)\frac{AB(a)}{1-a}\frac{(1-a)}{s^2(1-a+as^{-a})} = A(i)\left(Rs + \frac{1}{C}\right)$$

$$\frac{A(v)AB(a)}{1-a+as^{-a}} = A(i)\left(\frac{RsC+1}{C}\right)$$

$$\frac{A(v)AB(a)}{1-a+as^{-a}} = \frac{(RsC+1)A(i)}{C}$$

$$\frac{A(i)}{A(V)} = \frac{CAB(a)}{(RsC+1)(1-a+as^{-a})}$$

23.8.3 POURREZA TRANSFORM

We obtain the Pourreza transform of the circuit model as

$$HJ(^{ABC}_0 D^a_t V) - HJ\left(R\frac{di}{dt}\right) + HJ\left(\frac{1}{C}i\right)$$

$$HJ\left(\frac{dV}{dt} * \frac{AB(a)}{1-a}E_a\left(-\frac{a}{1-a}t^a\right)\right) = R(s^2 HJ(i) - si(0)) + \frac{1}{C}HJ(i)$$

$$\frac{1}{s}HJ\left(\frac{dV}{dt}\right)\frac{AB(a)}{1-a}HJ\left(E_a\left(-\frac{a}{1-a}t^a\right)\right) = HJ(i)\left(Rs^2 + \frac{1}{C}\right)$$

$$\frac{1}{s}(s^2 HJ(V) - sV(0))\frac{AB(a)}{1-a}\frac{(1-a)}{s(1-a+as^{-2a})} = HJ(i)\left(Rs^2 + \frac{1}{C}\right)$$

$$\frac{HJ(v)AB(a)}{(1-a+as^{-2a})} = HJ(i)\left(\frac{Rs^2 C+1}{C}\right)$$

$$\frac{HJ(v)AB(a)}{(1-a+as^{-2a})} = \frac{HJ(i)(Rs^2C+1)}{C}$$

$$\frac{HJ(i)}{HJ(V)} = \frac{CAB(a)}{(Rs^2C+1)(1-a+as^{-2a})}$$

23.8.4 MOHAND TRANSFORM

We obtain the Mohand transform of the circuit model as

$$M\left(^{ABC}_{0}D_t^a V\right) = M\left(R\frac{di}{dt}\right) + M\left(\frac{1}{C}i\right)$$

$$M\left(\frac{dV}{dt}*\frac{AB(a)}{1-a}E_a\left(-\frac{a}{1-a}t^a\right)\right) = R(sM(i)-s^2i(0)) + \frac{1}{C}M(i)$$

$$\frac{1}{s^2}M\left(\frac{dV}{dt}\right)\frac{AB(a)}{1-a}M\left(E_a\left(-\frac{a}{1-a}t^a\right)\right) = RsM(i) + \frac{1}{C}M(i)$$

$$\frac{1}{s^2}(sM(V)-s^2V(0))\frac{AB(a)}{1-a}\frac{s(1-a)}{1-a+as^{-a}} = RsM(i) + \frac{1}{C}M(i)$$

$$\frac{M(v)AB(a)}{1-a+as^{-a}} = M(i)\left(Rs+\frac{1}{C}\right)$$

$$\frac{M(v)AB(a)}{1-a+as^{-a}} = \frac{(RsC+1)M(i)}{C}$$

$$\frac{M(i)}{M(V)} = \frac{CAB(a)}{(RsC+1)(1-a+as^{-a})}$$

23.8.5 SAWI TRANSFORM

We obtain the Sawi transform of the circuit model as

$$Sa\left(^{ABC}_{0}D_t^a V\right) = Sa\left(R\frac{di}{dt}\right) + Sa\left(\frac{1}{C}i\right)$$

$$Sa\left(\frac{dV}{dt}*\frac{AB(a)}{1-a}E_a\left(-\frac{a}{1-a}t^a\right)\right) = R\left(\frac{Sa(i)}{s}-\frac{i(0)}{s^2}\right) + \frac{1}{C}Sa(i)$$

$$s^2Sa\left(\frac{dV}{dt}\right)\frac{AB(a)}{1-a}Sa\left(E_a\left(-\frac{a}{1-a}t^a\right)\right) = R\left(\frac{Sa(i)}{s}\right) + \frac{1}{C}Sa(i)$$

$$s^2\left(\frac{Sa(V)}{s}-\frac{V(0)}{s^2}\right)\frac{AB(a)}{1-a}\frac{(1-a)}{s(1-a+as^a)} = Sa(i)\left(\frac{R}{s}+\frac{1}{C}\right)$$

$$\frac{Sa(v)AB(a)}{1-a+as^a} = Sa(i)\left(\frac{RC+s}{sC}\right)$$

$$\frac{Sa(v)AB(a)}{1-a+as^a} = \frac{Sa(i)(RC+s)}{sC}$$

$$\frac{Sa(i)}{Sa(V)} = \frac{sCAB(a)}{(RC+s)(1-a+as^a)}$$

23.8.6 KAMAL TRANSFORM

We obtain the Kamal transform of the circuit model as

$$K(^{ABC}_0 D^a_t V) = K\left(R\frac{di}{dt}\right) + K\left(\frac{1}{C}i\right)$$

$$K\left(\frac{dV}{dt}*\frac{AB(a)}{1-a}E_a\left(-\frac{a}{1-a}t^a\right)\right) = R\left(\frac{K(i)}{s}-i(0)\right)+\frac{1}{C}K(i)$$

$$K\left(\frac{dV}{dt}\right)\frac{AB(a)}{1-a}K\left(E_a\left(-\frac{a}{1-a}t^a\right)\right) = \frac{R}{s}K(i)+\frac{1}{C}K(i)$$

$$\left(\frac{K(V)}{s}-V(0)\right)\frac{AB(a)}{1-a}\frac{s(1-a)}{1-a+as^a} = K(i)\left(\frac{R}{s}+\frac{1}{C}\right)$$

$$\frac{K(v)AB(a)}{1-a+as^a} = K(i)\left(\frac{RC+s}{sC}\right)$$

$$\frac{E(v)AB(a)}{1-a+as^a} = \frac{(RC+s)E(i)}{sC}$$

$$\frac{K(i)}{K(V)} = \frac{sCAB(a)}{(RC+s)(1-a+as^a)}$$

23.8.7 G TRANSFORM

We obtain the G transform of the circuit model as

$$G(^{ABC}_0 D^a_t V) = G\left(R\frac{di}{dt}\right) + G\left(\frac{1}{C}i\right)$$

$$G\left(\frac{dV}{dt}*\frac{AB(a)}{1-a}E_a\left(-\frac{a}{1-a}t^a\right)\right) = R\left(\frac{G(i)}{s}-s^\beta i(0)\right)+\frac{1}{C}E(i)$$

$$s^{-\beta}G\left(\frac{dV}{dt}\right)\frac{AB(a)}{1-a}G\left(E_a\left(-\frac{a}{1-a}t^a\right)\right) = \frac{R}{s}G(i)+\frac{1}{C}G(i)$$

$$s^{-\beta}\left(\frac{G(V)}{s}-s^\beta V(0)\right)\frac{AB(a)}{1-a}\frac{s^\beta(1-a)}{1-a+as^a} = G(i)\left(\frac{R}{s}+\frac{1}{C}\right)$$

$$\frac{G(V)AB(a)}{s(1-a+as^a)} = G(i)\left(\frac{RC+s}{sC}\right)$$

$$\frac{G(V)AB(a)}{s(1-a+as^a)} = \frac{(RC+s)G(i)}{sC}$$

$$\frac{G(i)}{G(V)} = \frac{CAB(a)}{(RC+s)(1-a+as^a)}$$

23.8.8   NATURAL TRANSFORM

We obtain the Natural transform of the circuit model as

$$N\left({}^{ABC}_0 D_t^a V\right) = N\left(R\frac{di}{dt}\right) + N\left(\frac{1}{C}i\right)$$

$$N\left(\frac{dV}{dt} * \frac{AB(a)}{1-a} E_a\left(-\frac{a}{1-a}t^a\right)\right) = R\left(\frac{s}{u}N(i) - \frac{i(o)}{u}\right) + \frac{1}{C}N(i)$$

$$uN\left(\frac{dV}{dt}\right)\frac{AB(a)}{1-a}N\left(E_a\left(-\frac{a}{1-a}t^a\right)\right) = \left(\frac{Rs}{u}\right)N(i) + \frac{1}{C}N(i)$$

$$u\left(\frac{s}{u}N(v) - \frac{v(o)}{u}\right)\frac{AB(a)}{1-a}\frac{(1-a)}{s(1-a+a\left(\frac{s}{u}\right)^{-a}} = N(i)\left(\frac{Rs}{u} + \frac{1}{C}\right)$$

$$\frac{N(v)AB(a)}{1-a+a\left(\frac{s}{u}\right)^{-a}} = N(i)\left(\frac{RsC+u}{uC}\right)$$

$$\frac{N(v)AB(a)}{1-a+a\left(\frac{s}{u}\right)^{-a}} = \frac{(RsC+u)N(i)}{uC}$$

$$\frac{N(i)}{N(V)} = \frac{uCAB(a)}{(RsC+u)\left(1-a+a\left(\frac{s}{u}\right)^{-a}\right)}$$

23.8.9 α INTEGRAL LAPLACE TRANSFORM

We obtain the α integral Laplace transform of the circuit model as

$$L_\alpha\left({}^{ABC}_0 D_t^\beta V\right) = L_\alpha\left(R\frac{di}{dt}\right) + L_\alpha\left(\frac{1}{C}i\right)$$

$$L_\alpha\left(\frac{dV}{dt} * \frac{AB(\beta)}{1-\beta}E_\beta\left(-\frac{\beta}{1-\beta}t^\beta\right)\right) = R\left(s^{\frac{1}{\alpha}}L_\alpha(i) - i(0)\right) + \frac{1}{C}L_\alpha(i)$$

$$L_\alpha\left(\frac{dV}{dt}\right)\frac{AB(\beta)}{1-\beta}L_\alpha\left(E_\beta\left(-\frac{\beta}{1-\beta}t^\beta\right)\right) = Rs^{\frac{1}{\alpha}}L_\alpha(i) + \frac{1}{C}L_\alpha(i)$$

$$\left(s^{\frac{1}{\alpha}}L_\alpha(V) - V(0)\right)\frac{AB(\beta)}{1-\beta}\frac{(1-\beta)}{1-\beta+\alpha s^{\frac{-\beta}{\alpha}}}s^{-\frac{1}{\alpha}} = \left(Rs^{\frac{1}{\alpha}} + \frac{1}{C}\right)L_\alpha(i)$$

$$\frac{L_\alpha(V)AB(\beta)}{\left(1-\beta+\alpha s^{\frac{-\beta}{\alpha}}\right)} = \left(Rs^{\frac{1}{\alpha}} + \frac{1}{C}\right)L_\alpha(i)$$

$$\frac{L_\alpha(V)AB(\beta)}{\left(1-\beta+\alpha s^{\frac{-\beta}{\alpha}}\right)} = \frac{L_\alpha(i)\left(Rs^{\frac{1}{\alpha}}C+1\right)}{C}$$

$$\frac{L_\alpha(i)}{L_\alpha(V)} = \frac{CAB(\beta)}{\left(Rs^{\frac{1}{\alpha}}C+1\right)\left(1-\beta+\alpha s^{\frac{-\beta}{\alpha}}\right)}$$

23.8.10 SIMULATIONS

We present the simulations of the obtained transfer functions in this section. We firstly consider the Eqs. (23.58)–(23.59). We apply many integral transforms (Elzaki transform, Aboodh transform, Pourreza transform, Mohand transform, Sawi transform, Kamal transform, G transform, α integral Laplace transform) to demonstrate the simulations of the transfer functions. In Figures 23.1–23.16, we choose our fractional order $\alpha = 0.1...0.9$ and $M(\alpha) = 1$, $R = 2$, $C = 3, \text{Ł} = 1$. In Figure 23.1, we demonstrate the transfer function of Eq. (23.58) by Elzaki transform. In Figure 23.2, we show the transfer function of Eq. (23.59) by Elzaki transform. In Figure 23.3, we present the transfer function of Eq. (23.58) by Aboodh transform. In Figure 23.4, we present the transfer function of Eq. (23.59) by Aboodh transform. In Figure 23.5, we give the transfer function of Eq. (23.58) by Pourreza transform. In Figure 23.6, we give the transfer function of Eq. (23.59) by Pourreza transform. We show the transfer function of Eq. (23.58) by Mohand transform for different values of fractional order in Figure 23.7. We demonstrate the transfer function of Eq. (23.59) by Mohand transform for different values of fractional order in Figure 23.8. We present the transfer function of Eq. (23.58) by Sawi transform for different values of fractional order in Figure 23.9. We show the transfer function of Eq. (23.59) by Sawi transform for different values of fractional order in Figure 23.10. We demonstrate the transfer function of Eq. (23.58) by Kamal transform for different values of fractional order in Figure 23.11. We show the transfer function of Eq. (23.59) by Kamal transform for different values of fractional order in Figure 23.12. We show the transfer function of Eq. (23.58) by G transform for different values of fractional order in Figure 23.13. We give the transfer function of Eq. (23.59) by G transform for different values of fractional order in Figure 23.14. We show the transfer function of Eq. (23.58) by α integral Laplace transform for different values of fractional order in Figure 23.15. We give the transfer function of Eq. (23.59) by α integral Laplace transform for different values of fractional order in Figure 23.16. We choose $\beta = 1$ in Figures 23.15–23.16.

Then, we consider the Eq. (23.60). We choose fractional order $\alpha = 0.2, 0.4, 0.6, 0.8$ and $R = 2$, $C = 3$ in Figures 23.17–23.24. We apply the integral transforms (Elzaki transform, Aboodh transform, Pourreza transform, Mohand transform, Sawi transform, Kamal transform, G transform, α integral Laplace transform) to show the simulations of the transfer functions. We can see the effect of the transforms and the effect of the fractional order in these figures.

Then, we consider Eqs. (23.61)–(23.62). We choose $R = 2$, $C = 3$, $L = 1$ in Figures 23.25–23.40. We apply the integral transforms (Elzaki transform, Aboodh transform, Pourreza transform, Mohand transform, Sawi transform, Kamal transform, G transform, α integral Laplace transform) to demonstrate the simulations of the transfer functions.

Then, we consider Eq. (23.63). We choose $R = 2$, $C = 3$, $L = 1$ in Figures 23.41–23.48. We apply the integral transforms (Elzaki transform, Aboodh transform, Pourreza transform, Mohand transform, Sawi transform, Kamal transform, G transform, α integral Laplace transform) to demonstrate the simulations of the transfer functions.

Then, we consider the Eq. (23.63). We choose $R = 2$, $C = 3$, $L = 1$ and $a = 0.3, 0.6, 0.9$ in Figures 23.49–23.58. We apply the integral transforms (Elzaki transform, Aboodh transform, Pourreza transform, Mohand transform, Sawi transform, Kamal transform, G transform, α integral Laplace transform) to present the simulations of the transfer functions.

Then, we consider the Eq. (23.63). We choose $R = 2$, $C = 3$ and $a = 0.15, 0.30, 0.45, 0.60, 0.75, 0.90$ in Figures 23.59–23.64. We apply the integral transforms (Elzaki transform, Aboodh transform, Pourreza transform, Mohand transform, Sawi transform, Kamal transform, G transform, α integral Laplace transform) to present the simulations of the transfer functions.

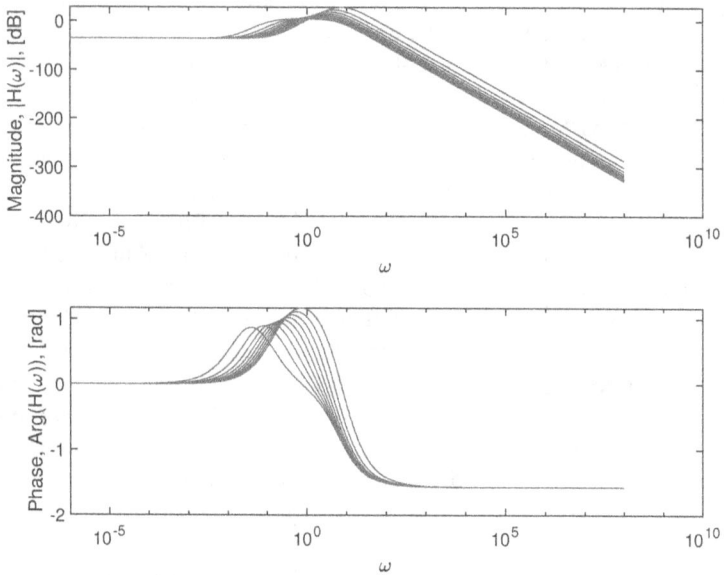

Figure 23.1   Elzaki transform of the Eq. (23.58).

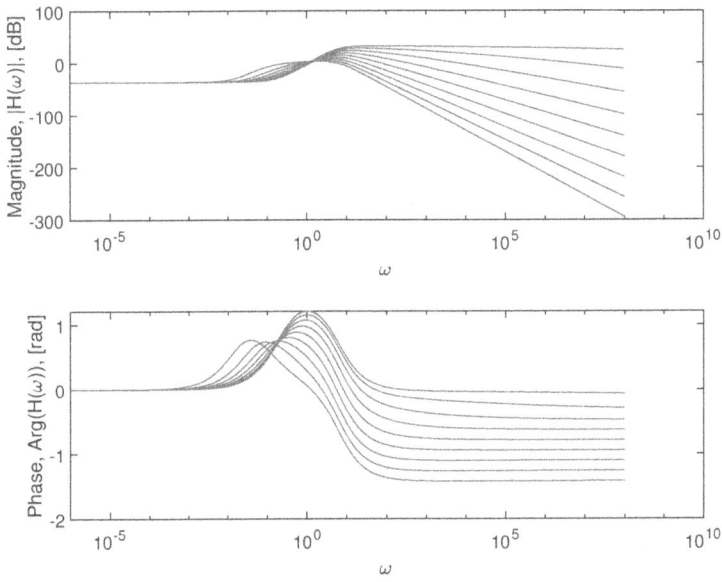

Figure 23.2 Elzaki transform of the Eq. (23.59).

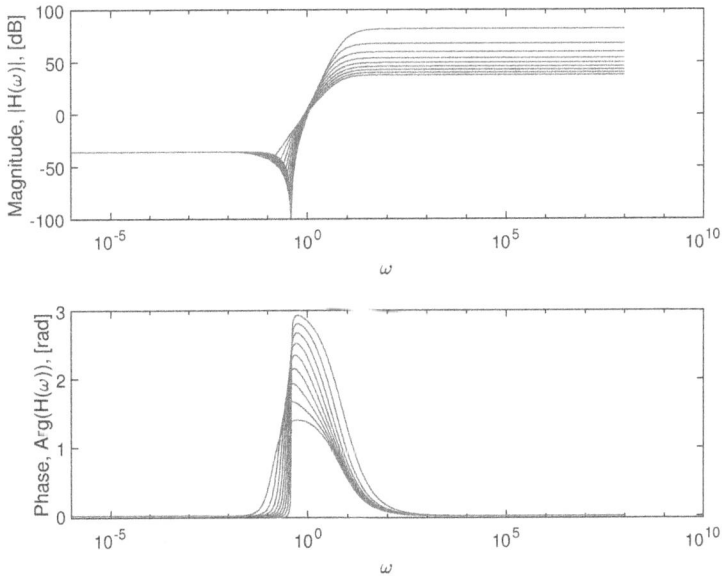

Figure 23.3 Aboodh transform of the Eq. (23.58).

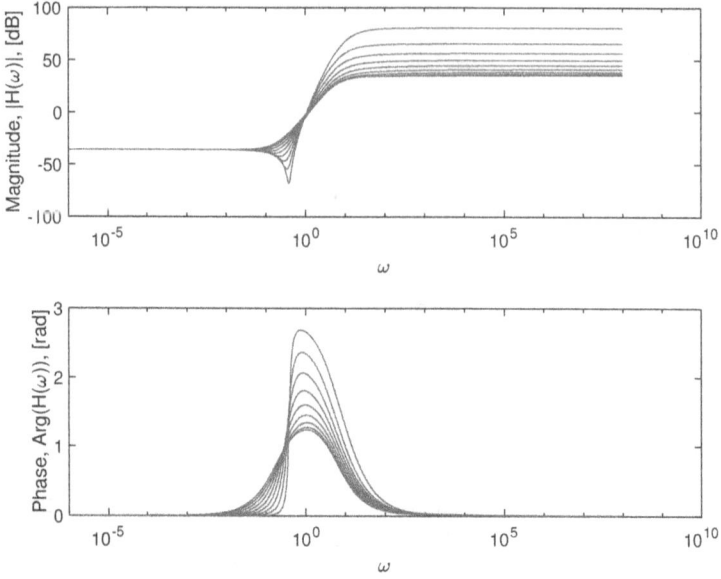

Figure 23.4   Aboodh transform of the Eq. (23.59).

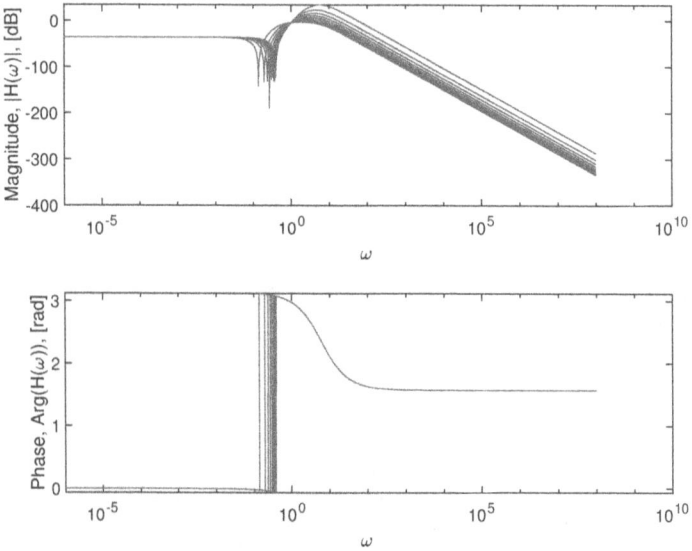

Figure 23.5   Pourreza transform of the Eq. (23.58).

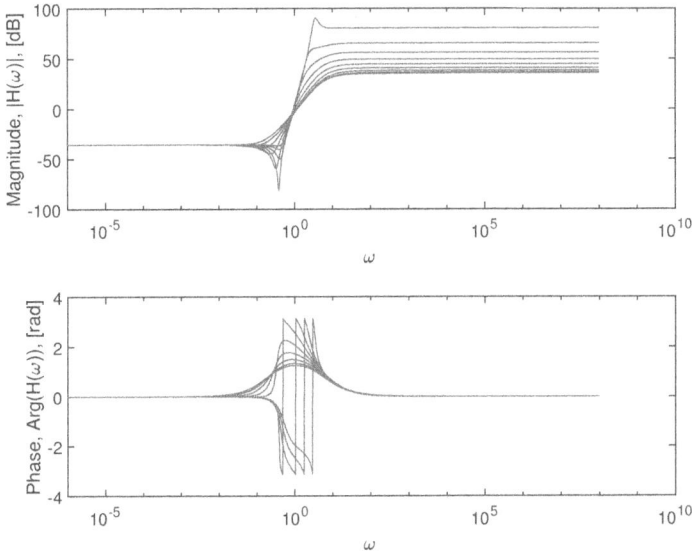

Figure 23.6 Pourreza transform of the Eq. (23.59).

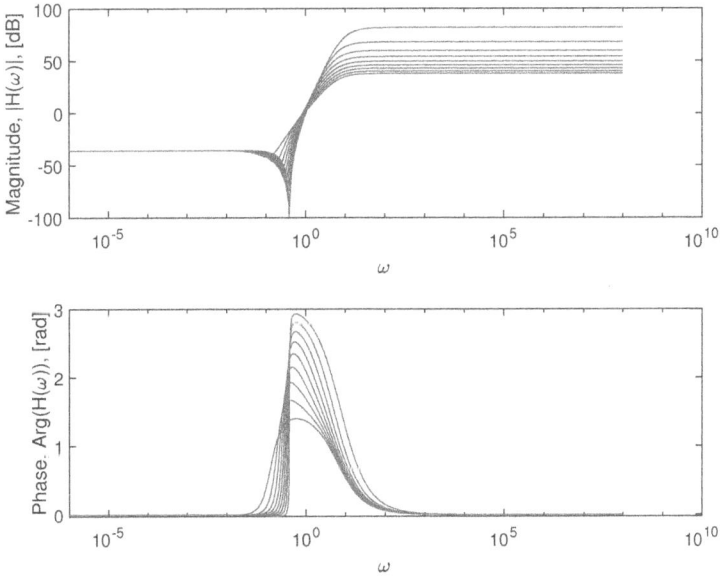

Figure 23.7 Mohand transform of the Eq. (23.58).

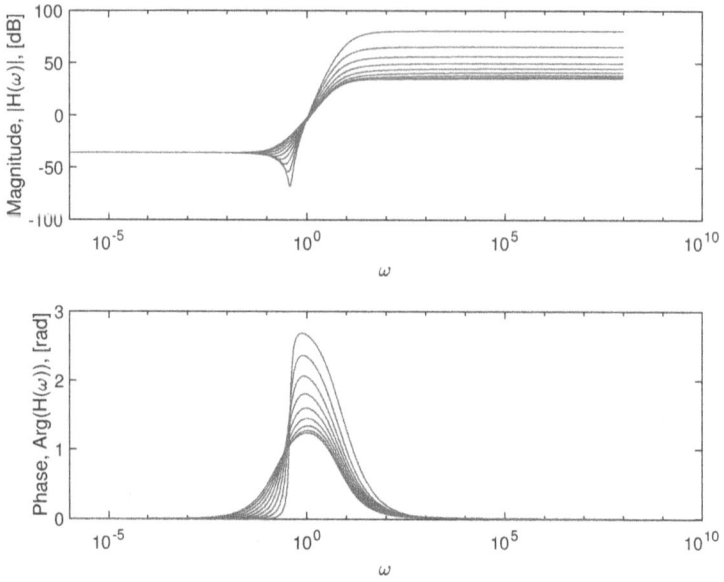

Figure 23.8   Mohand transform of the Eq. (23.59).

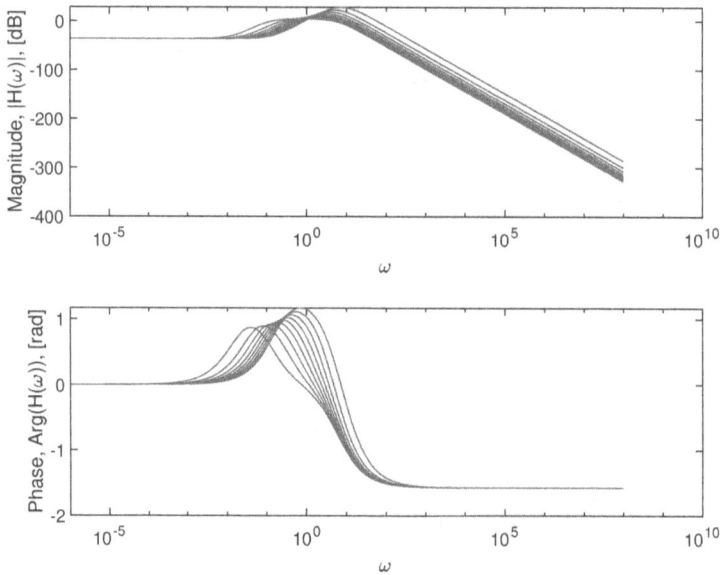

Figure 23.9   Sawi transform of the Eq. (23.58).

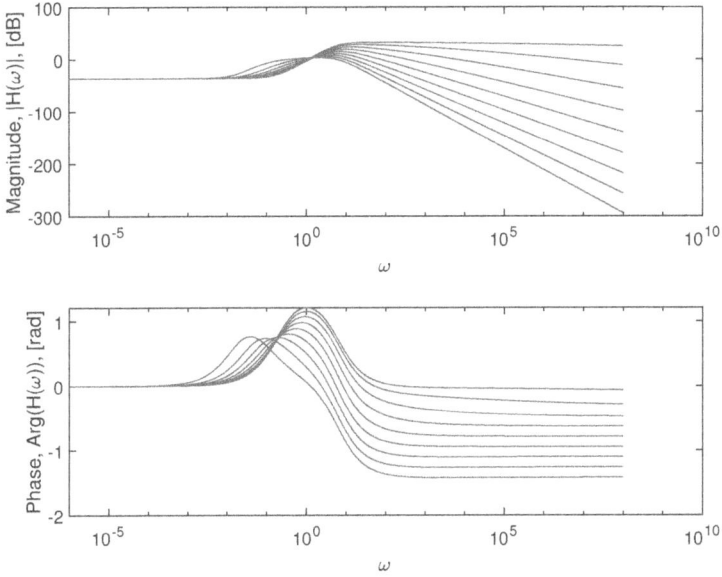

Figure 23.10   Sawi transform of the Eq. (23.59).

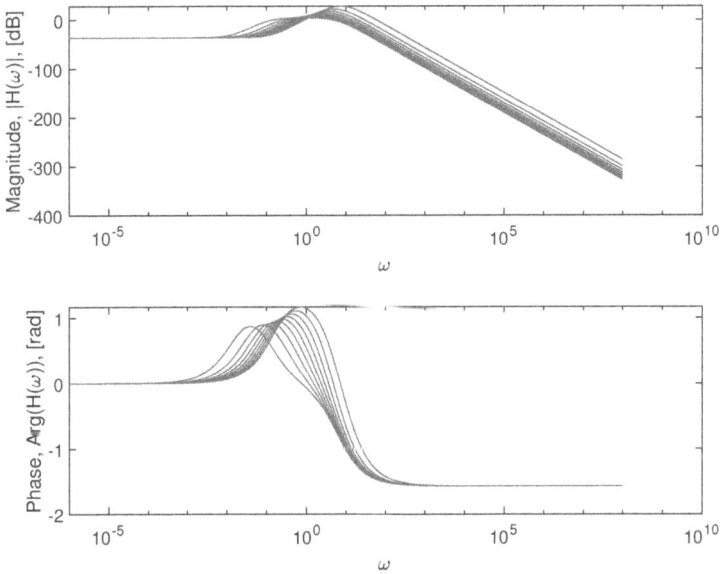

Figure 23.11   Kamal transform of the Eq. (23.58).

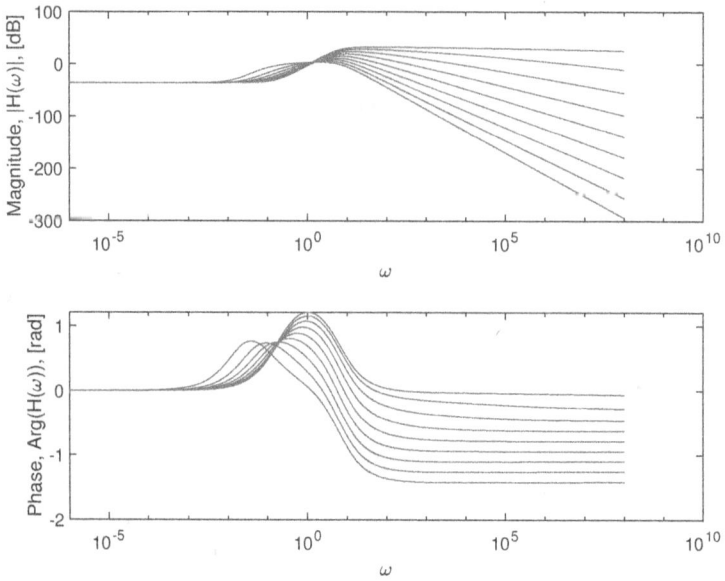

Figure 23.12   Kamal transform of the Eq. (23.59).

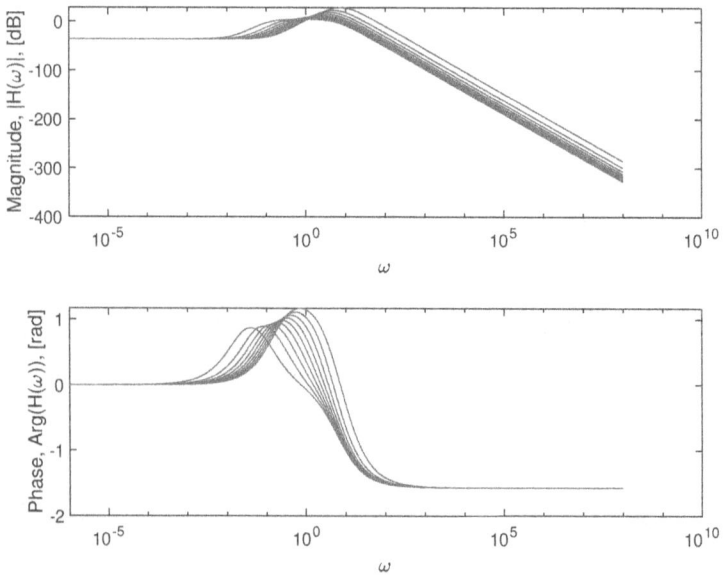

Figure 23.13   G transform of the Eq. (23.58).

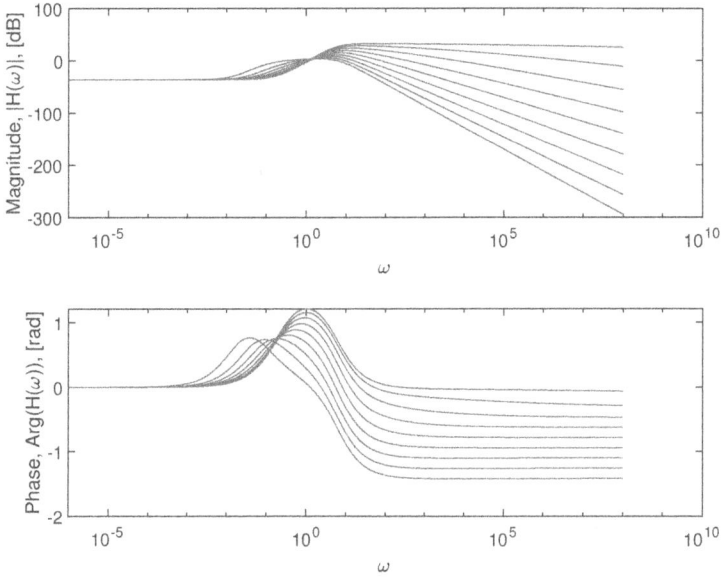

Figure 23.14   G transform of the Eq. (23.59).

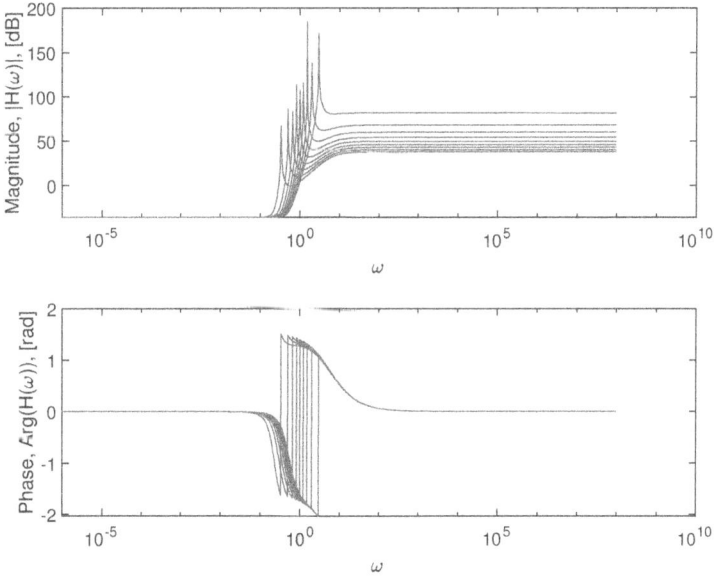

Figure 23.15 α integral Laplace transform of the Eq. (23.58).

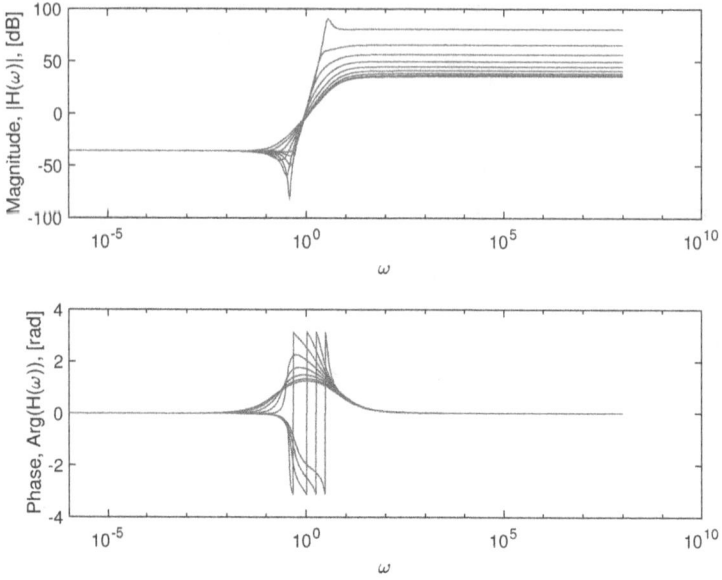

Figure 23.16 α integral Laplace transform of the Eq. (23.59).

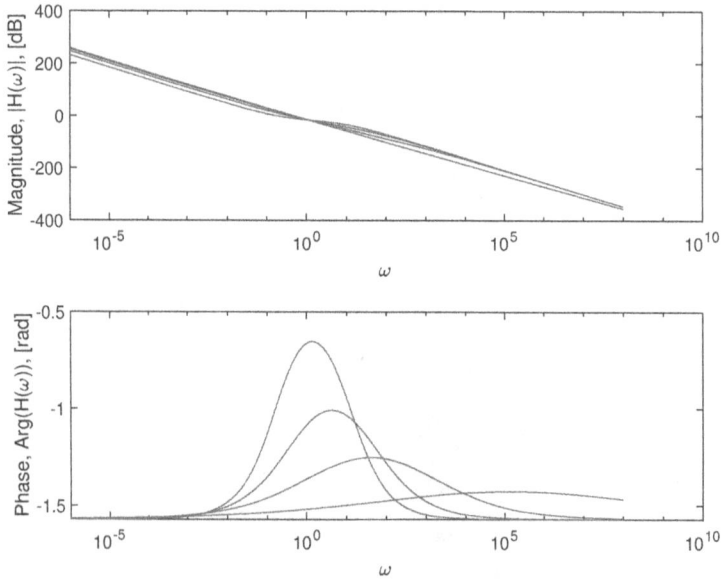

Figure 23.17   Elzaki transform of the Eq. (23.60).

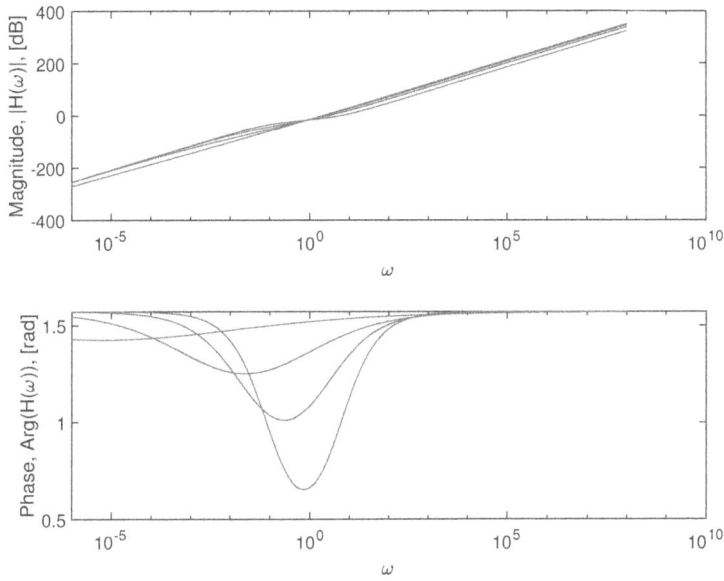

Figure 23.18   Aboodh transform of the Eq. (23.60).

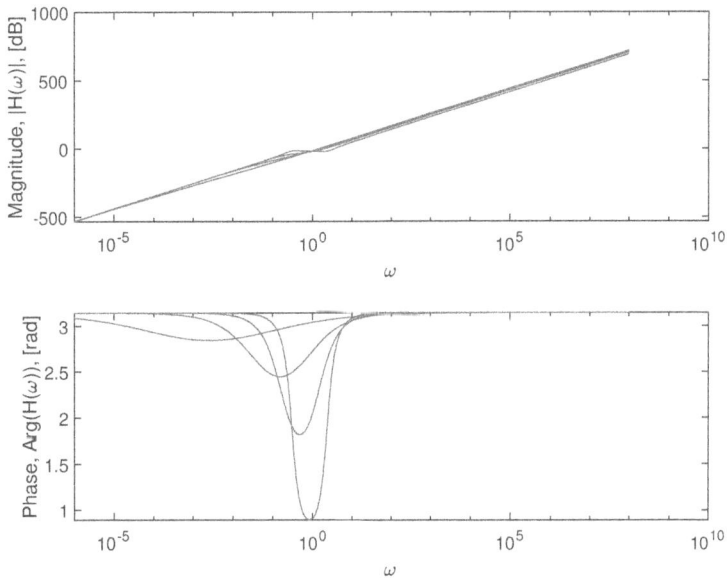

Figure 23.19   Pourreza transform of the Eq. (23.60).

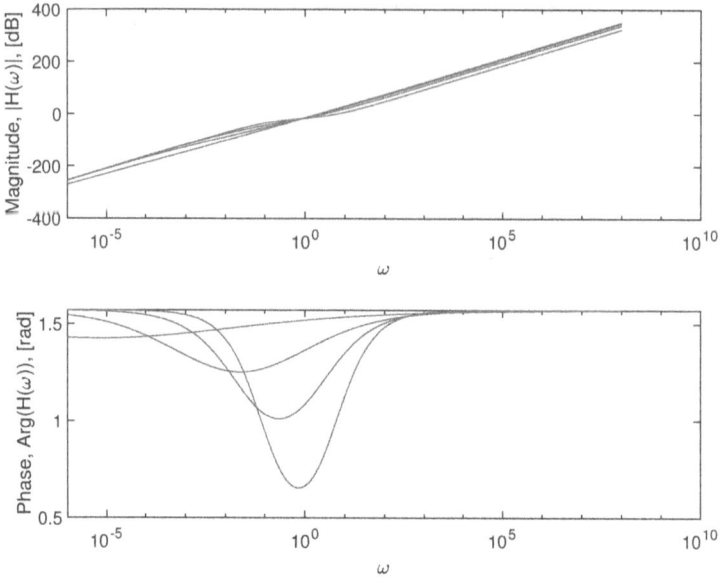

Figure 23.20 Mohand transform of the Eq. (23.60).

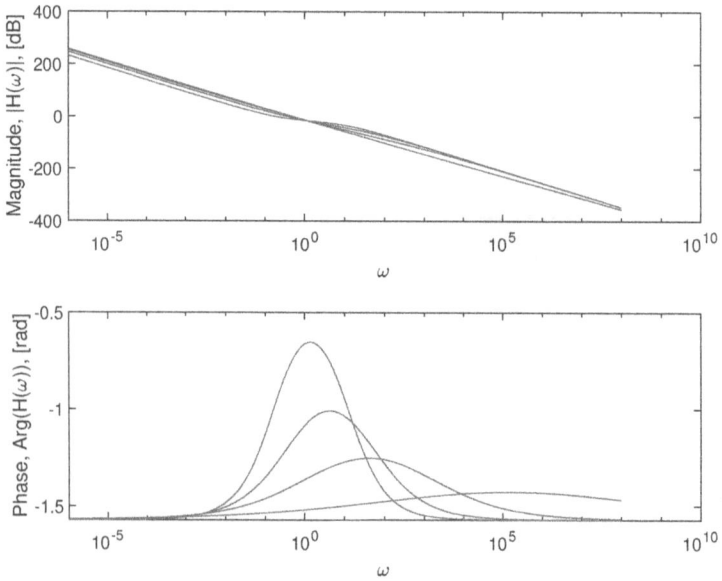

Figure 23.21 Sawi transform of the Eq. (23.60).

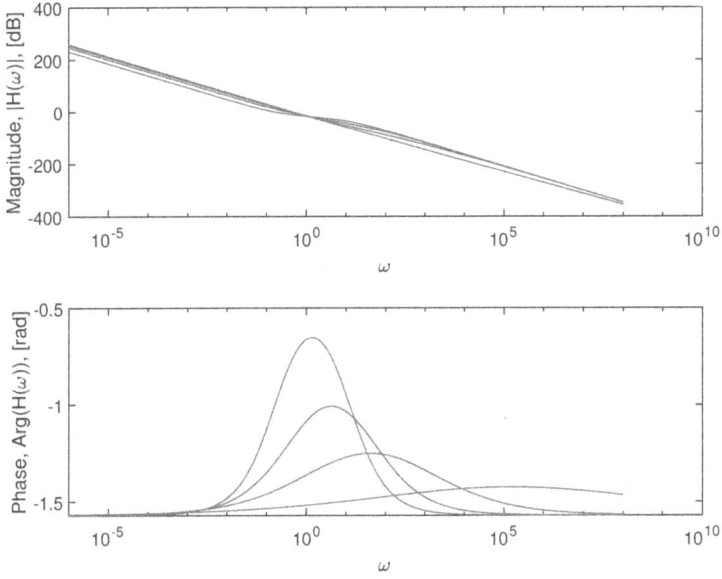

Figure 23.22 Kamal transform of the Eq. (23.60).

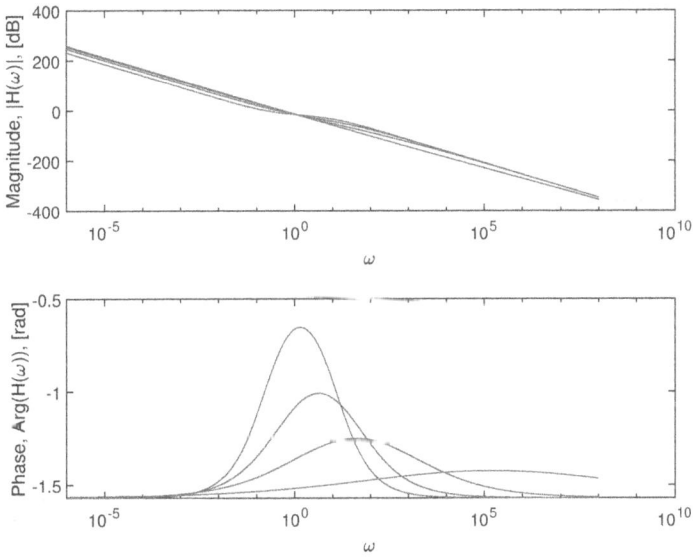

Figure 23.23 G transform of the Eq. (23.60).

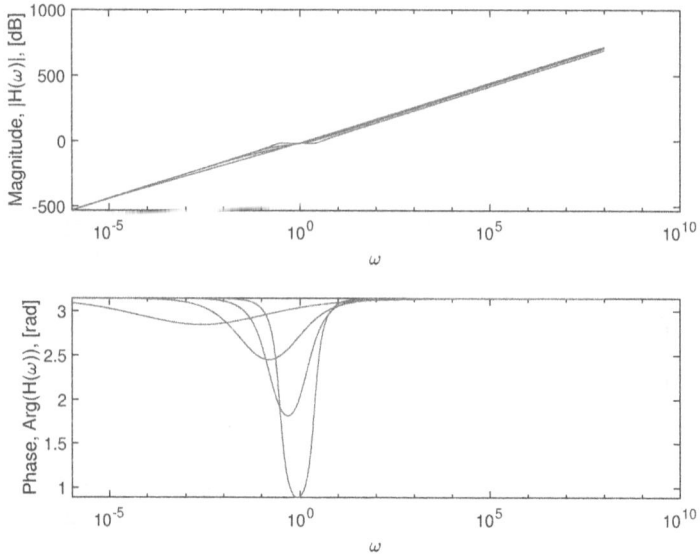

Figure 23.24 α integral Laplace transform of the Eq. (23.60) for $\alpha = 0.5$ and different values of β.

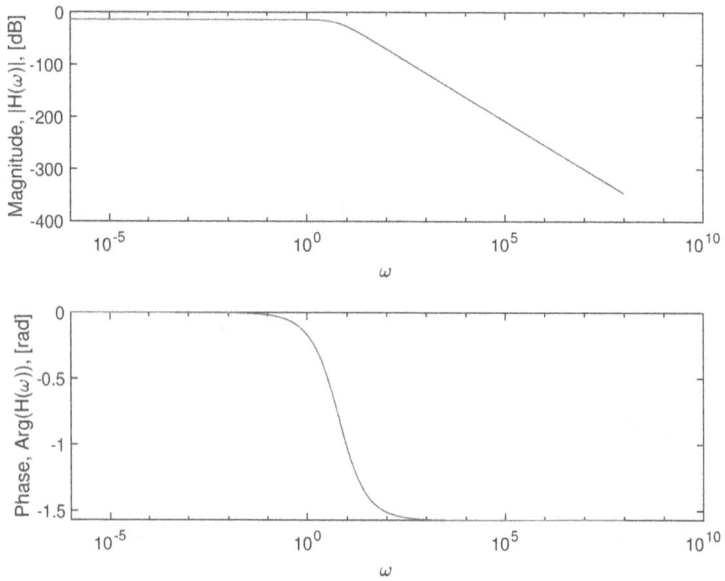

Figure 23.25 Elzaki transform of the Eq. (23.61).

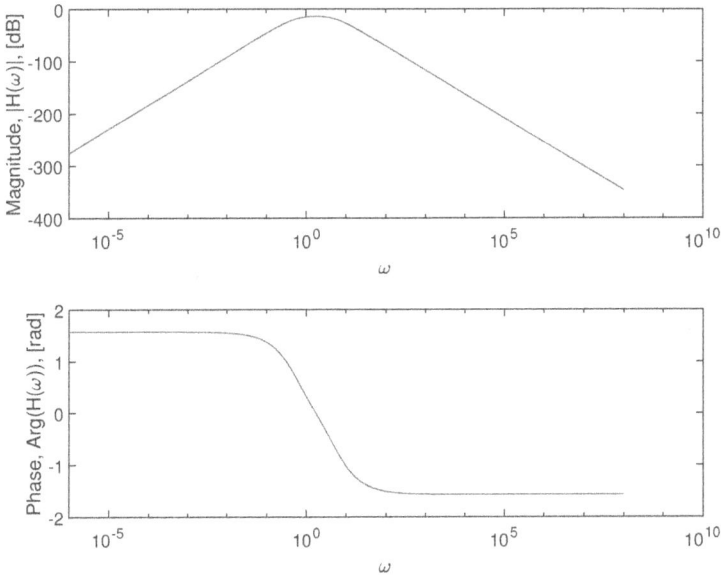

Figure 23.26 Elzaki transform of the Eq. (23.62).

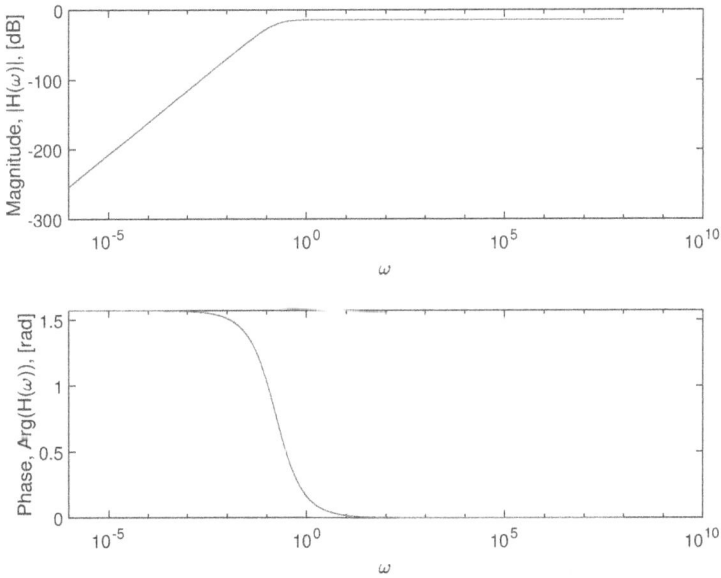

Figure 23.27 Mohand transform of the Eq. (23.61).

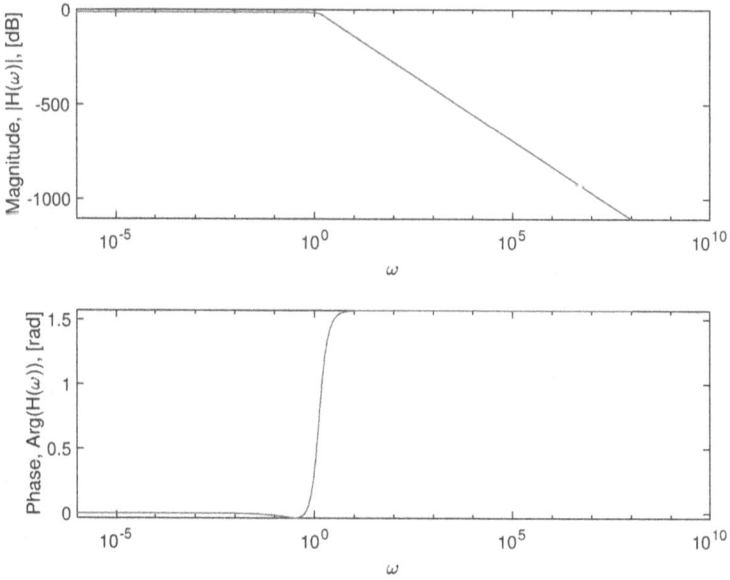

Figure 23.28 Mohand transform of the Eq. (23.62).

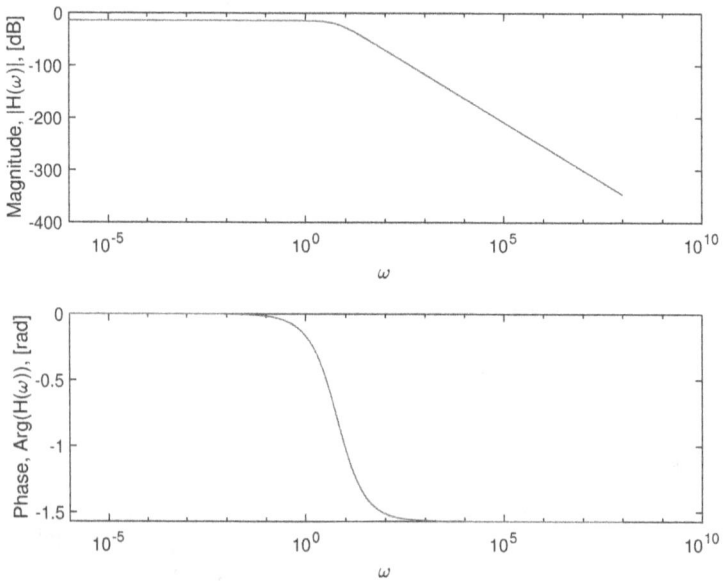

Figure 23.29 Kamal transform of the Eq. (23.61).

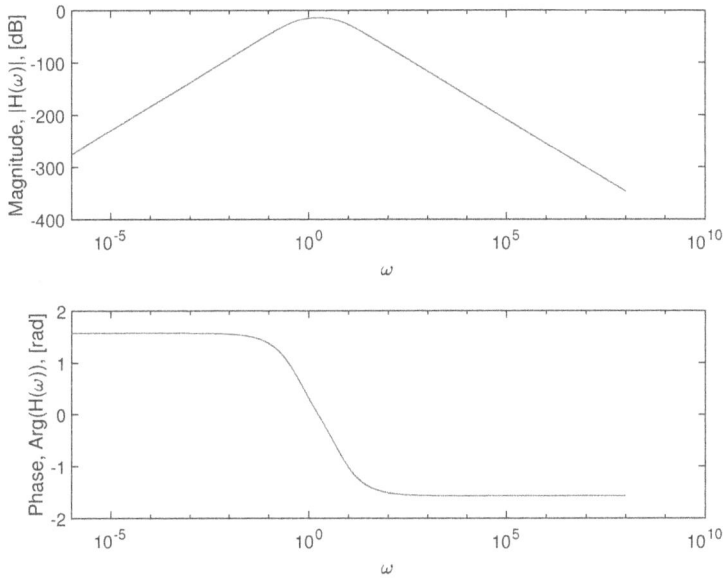

Figure 23.30 Kamal transform of the Eq. (23.62).

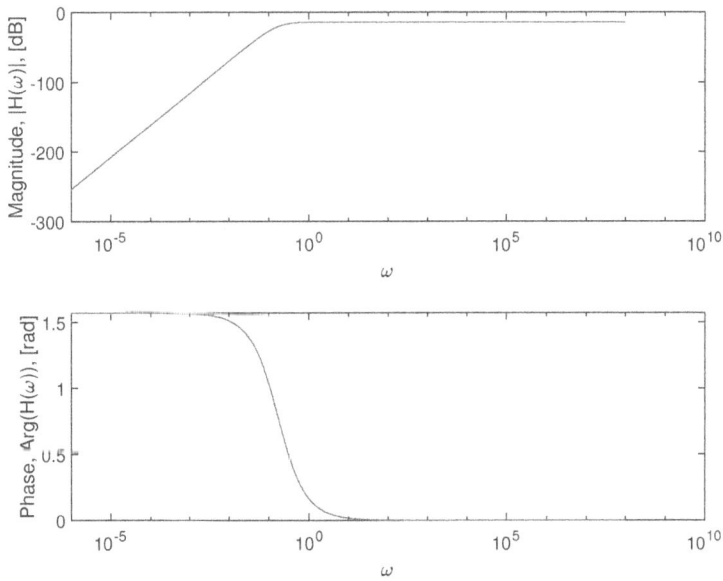

Figure 23.31 Aboodh transform of the Eq. (23.61).

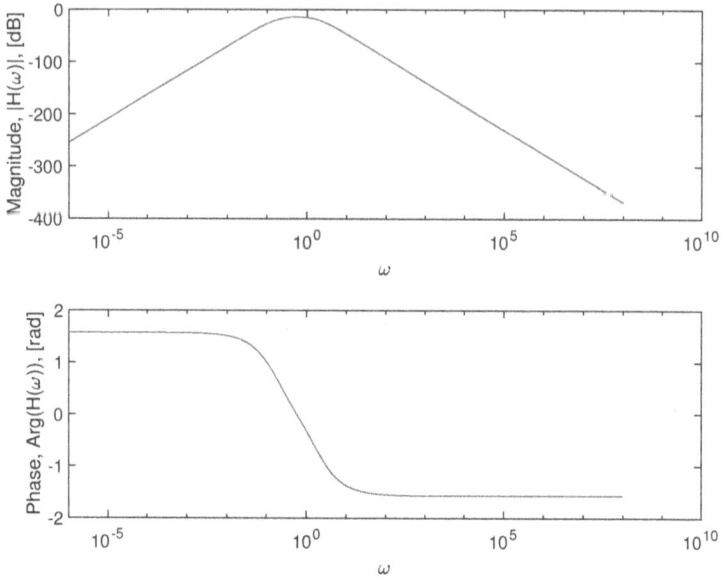

Figure 23.32 Aboodh transform of the Eq. (23.62).

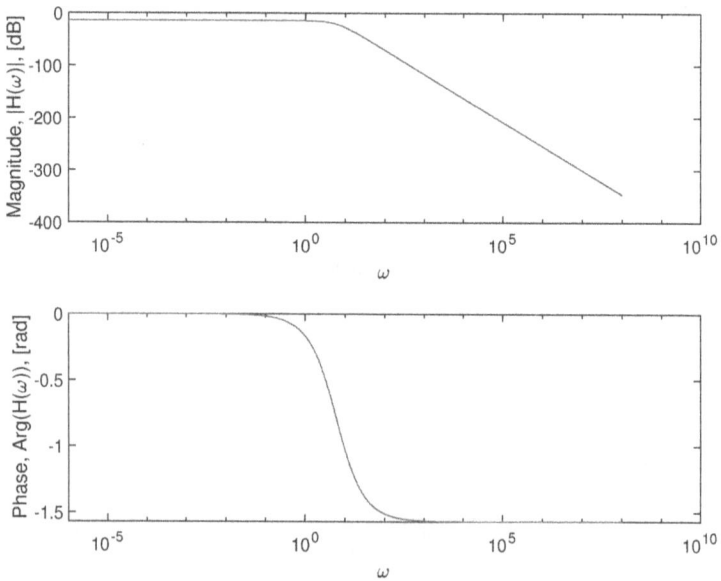

Figure 23.33 Sawi transform of the Eq. (23.61).

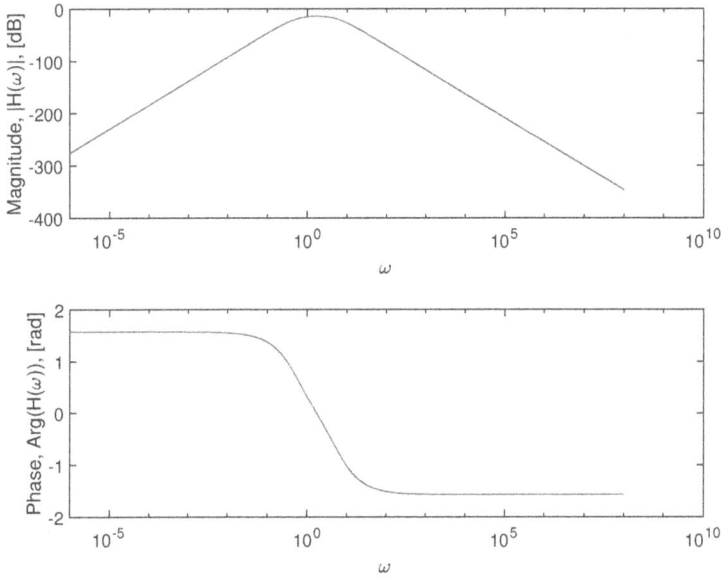

Figure 23.34 Sawi transform of the Eq. (23.62).

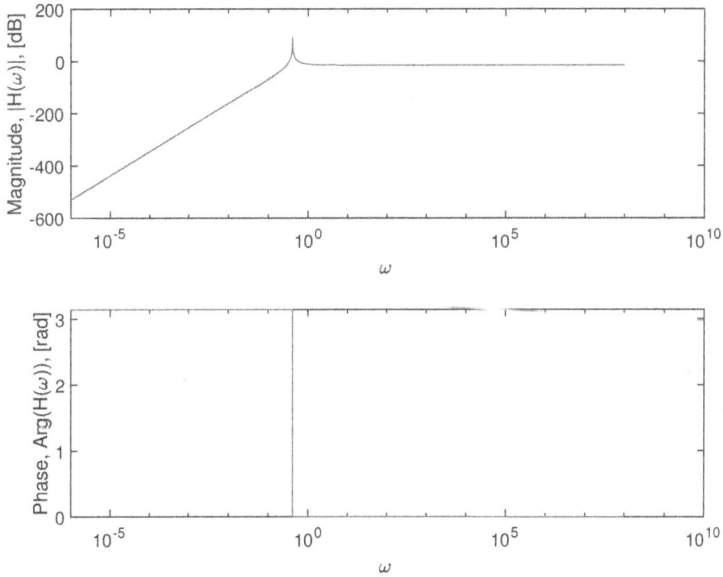

Figure 23.35 α integral Laplace transform of the Eq. (23.61) for $\alpha = 0.5$.

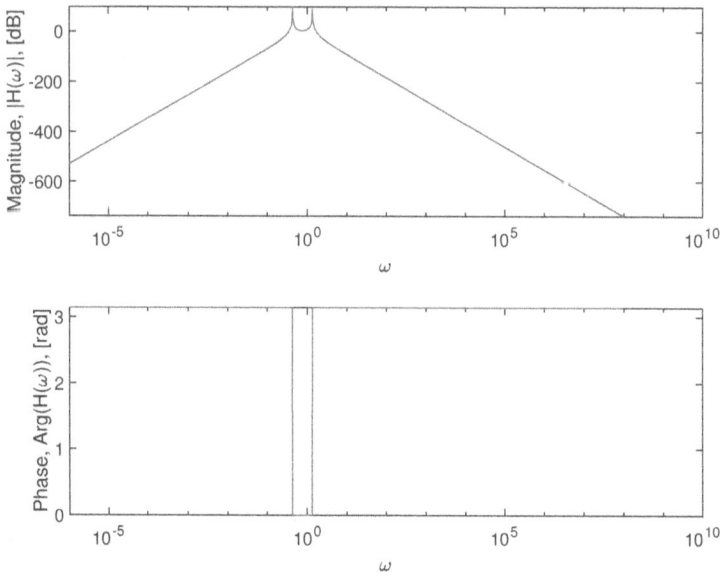

Figure 23.36 α integral Laplace transform of the Eq. (23.62) for $\alpha = 0.5$.

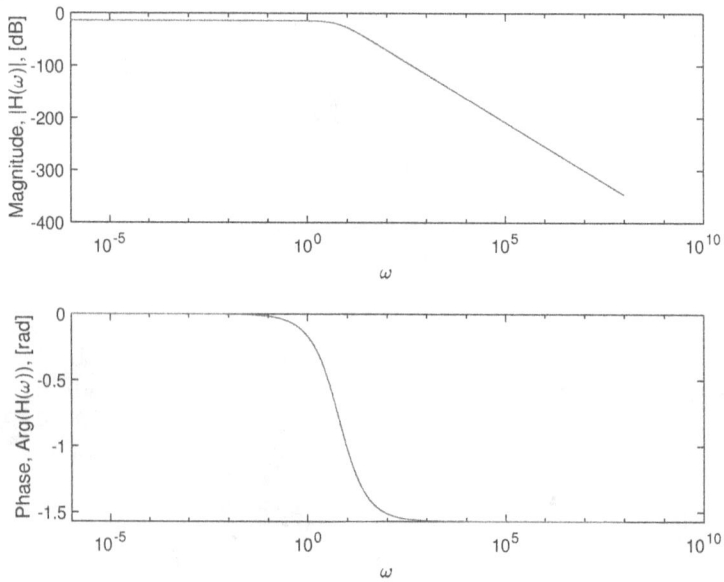

Figure 23.37 G transform of the Eq. (23.61).

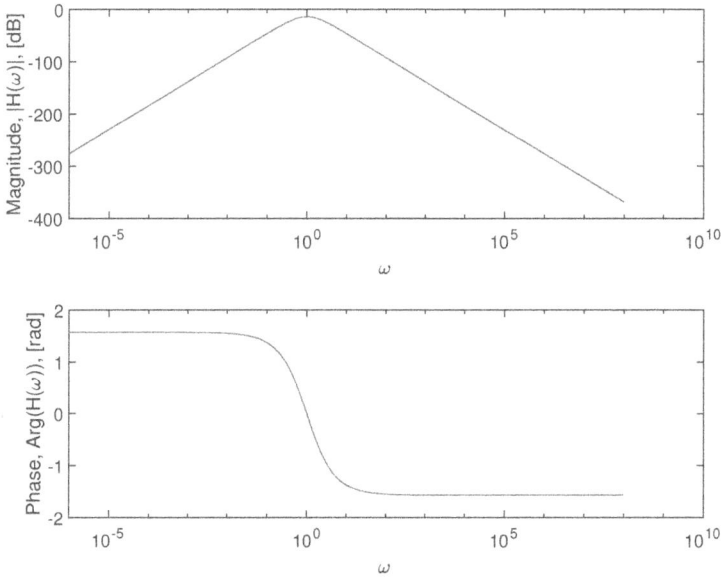

Figure 23.38 G transform of the Eq. (23.62).

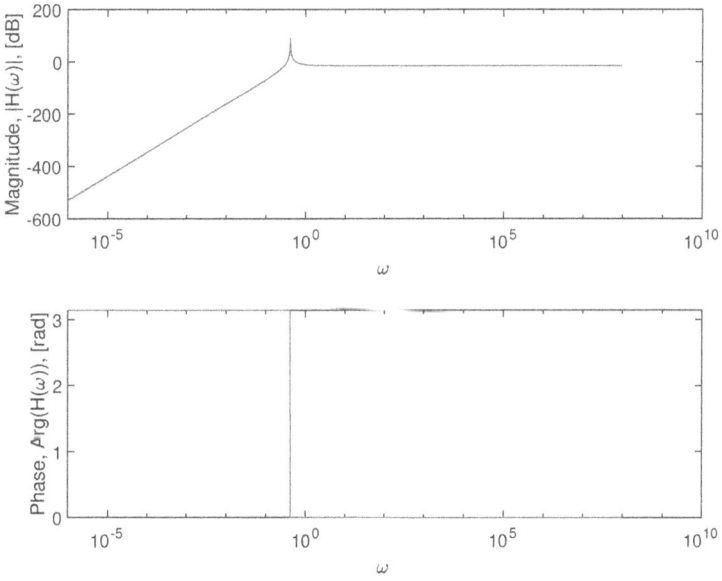

Figure 23.39 Pourreza transform of the Eq. (23.61).

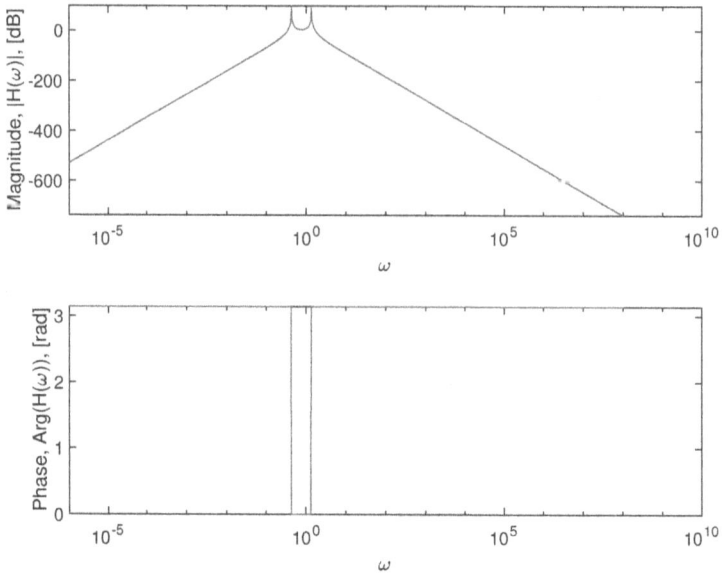

Figure 23.40 Pourreza transform of the Eq. (23.62).

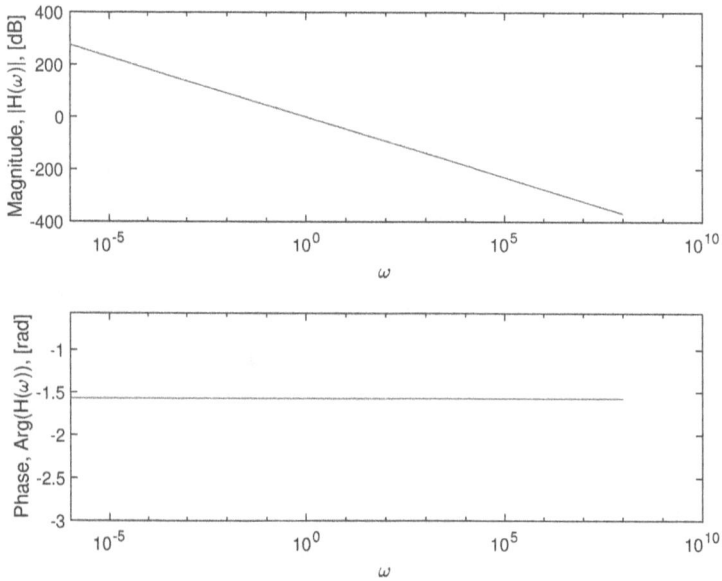

Figure 23.41 Elzaki transform of the Eq. (23.63).

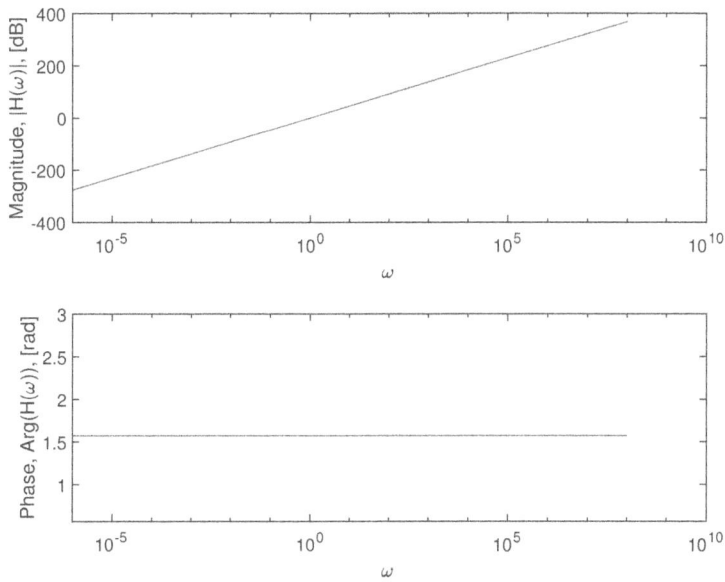

Figure 23.42   Aboodh transform of the Eq. (23.63).

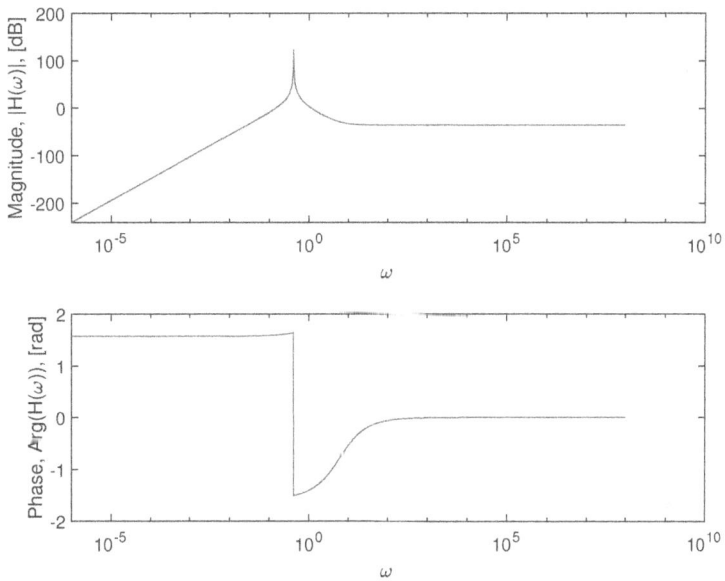

Figure 23.43   Pourreza transform of the Eq. (23.63).

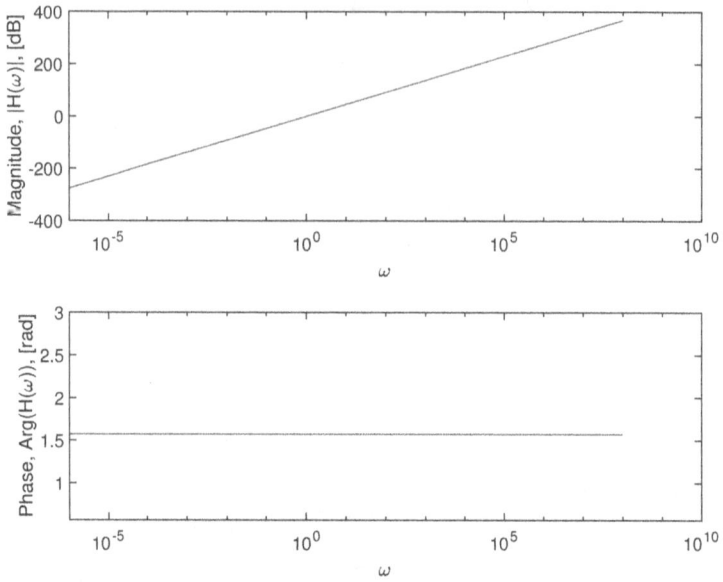

Figure 23.44 Mohand transform of the Eq. (23.63).

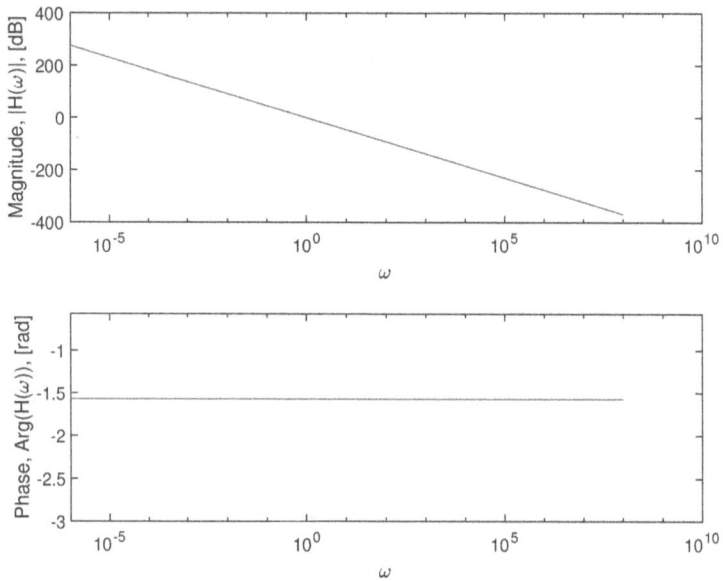

Figure 23.45 Sawi transform of the Eq. (23.63).

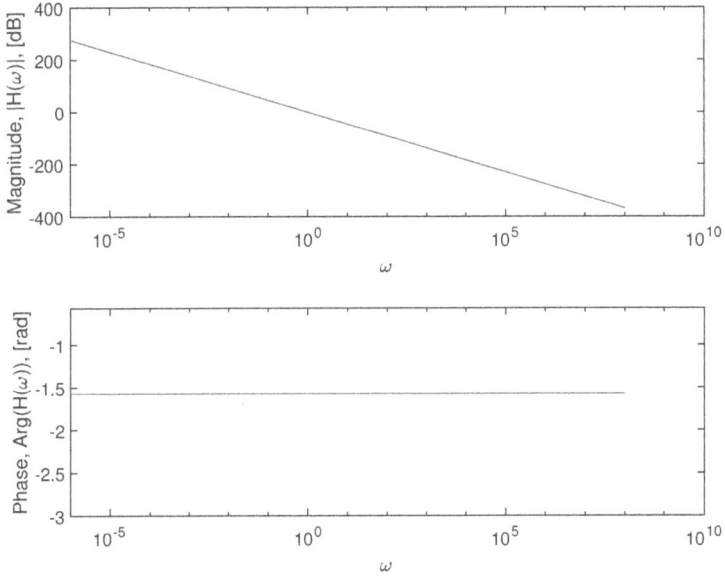

Figure 23.46   Kamal transform of the Eq. (23.63).

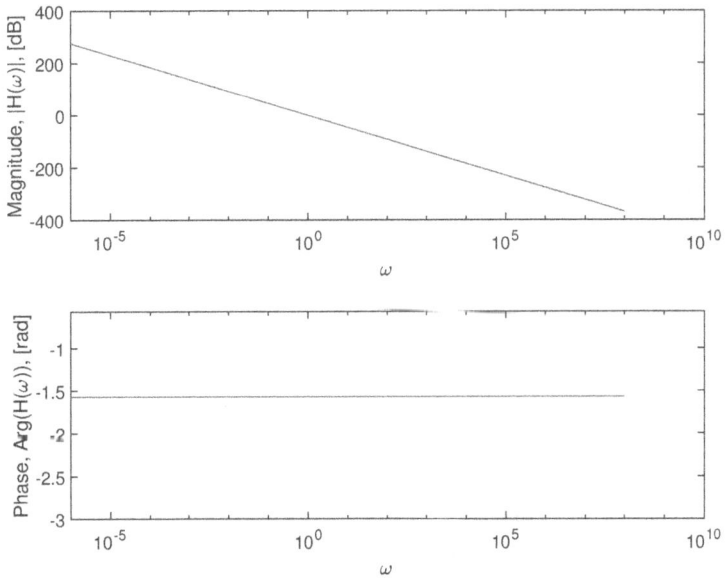

Figure 23.47   G transform of the Eq. (23.63).

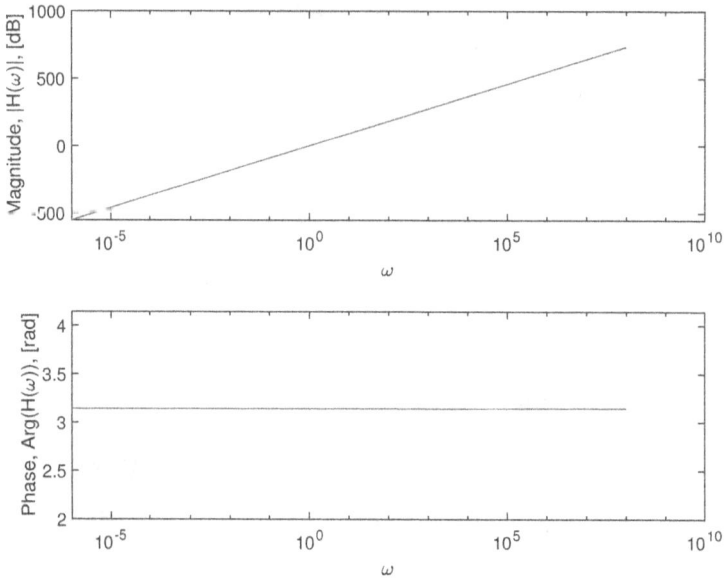

Figure 23.48 α integral Laplace transform of the Eq. (23.63) for $\alpha = 0.5$.

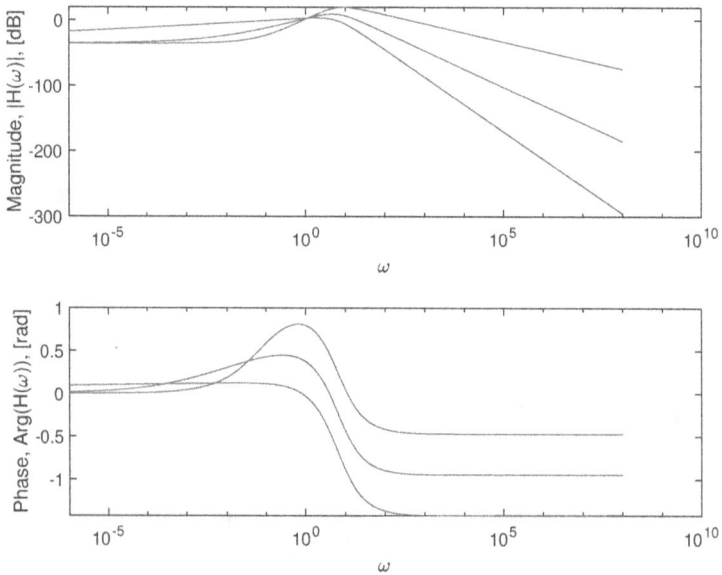

Figure 23.49 Elzaki transform of the Eq. (23.63).

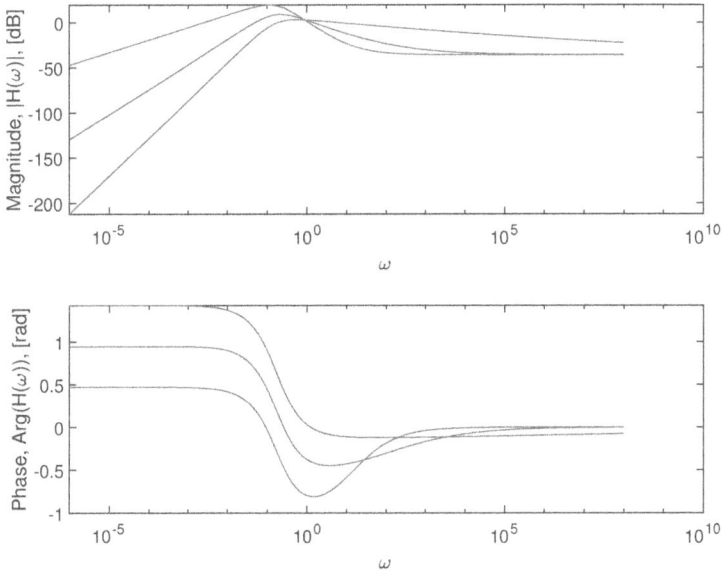

Figure 23.50   Aboodh transform of the Eq. (23.63).

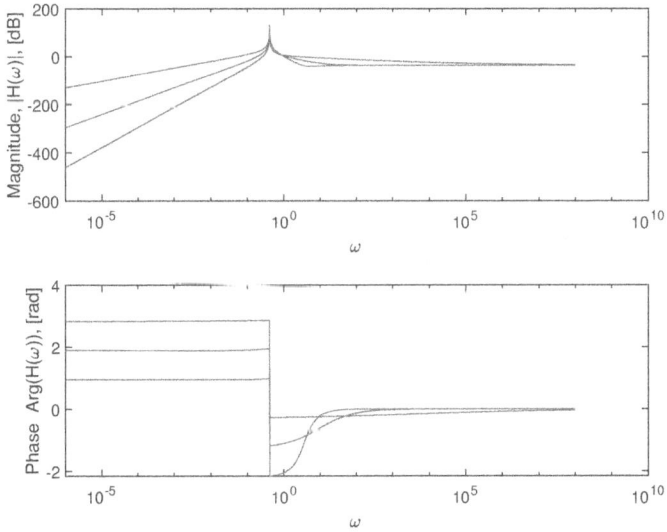

Figure 23.51   Pourreza transform of the Eq. (23.63).

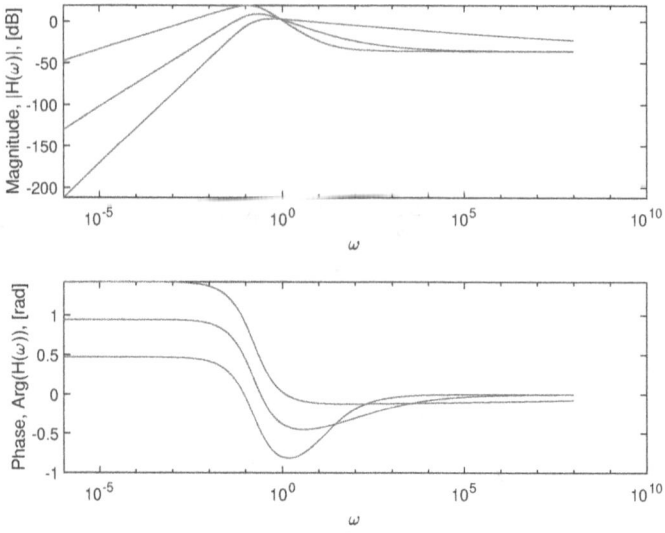

Figure 23.52 Mohand transform of the Eq. (23.63).

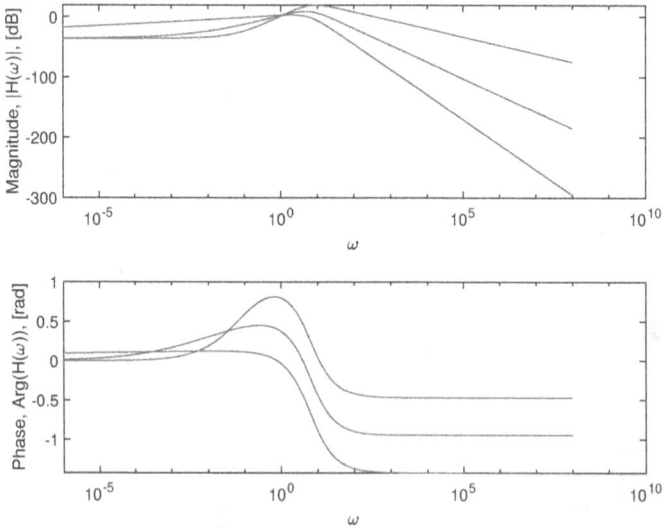

Figure 23.53 Sawi transform of the Eq. (23.63).

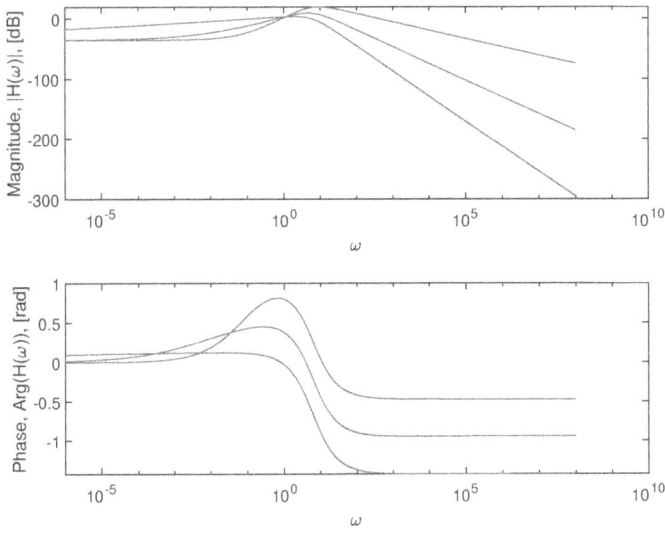

Figure 23.54   Kamal transform of the Eq. (23.63).

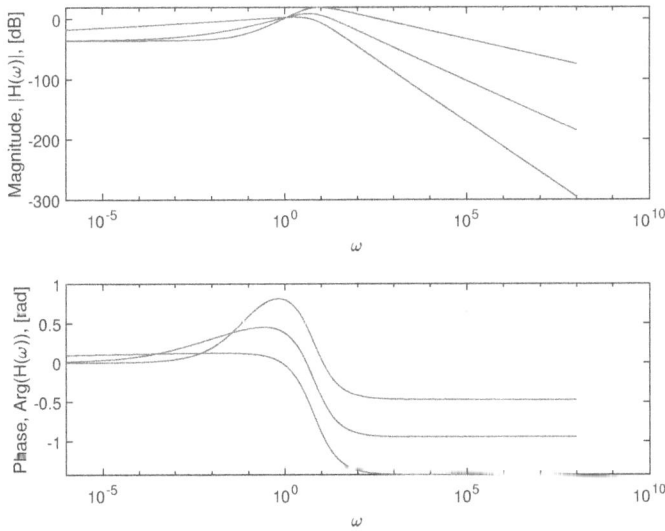

Figure 23.55   G transform of the Eq. (23.63).

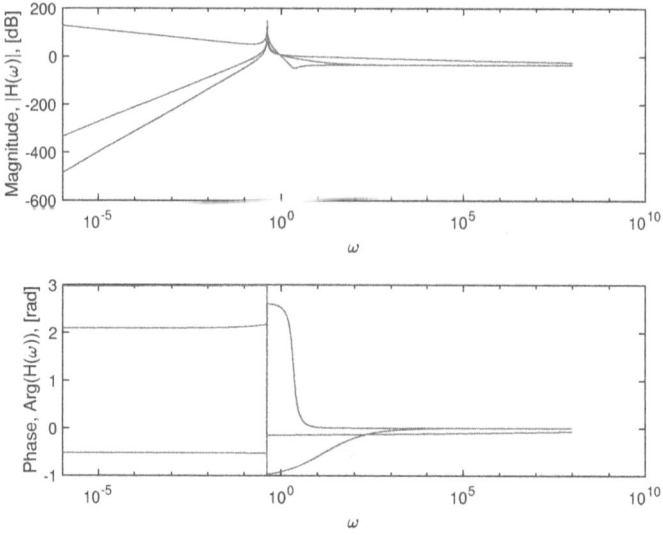

Figure 23.56 α integral Laplace transform of the Eq. (23.63) for $\alpha = 0.5$.

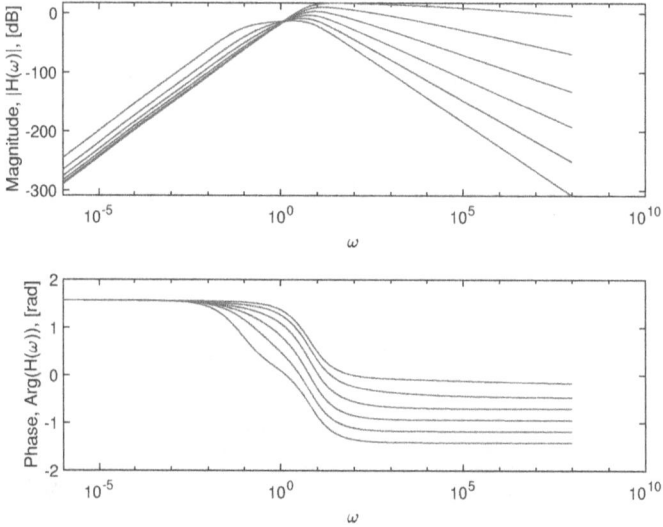

Figure 23.57 Elzaki transform of the Eq. (23.63).

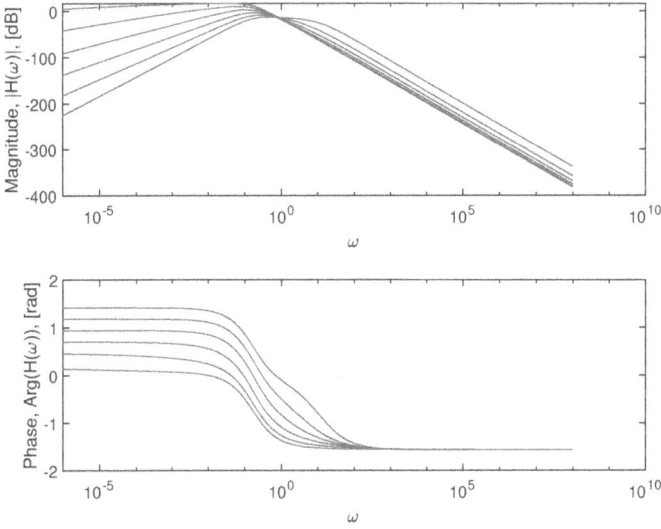

Figure 23.58 Aboodh transform of the Eq. (23.63).

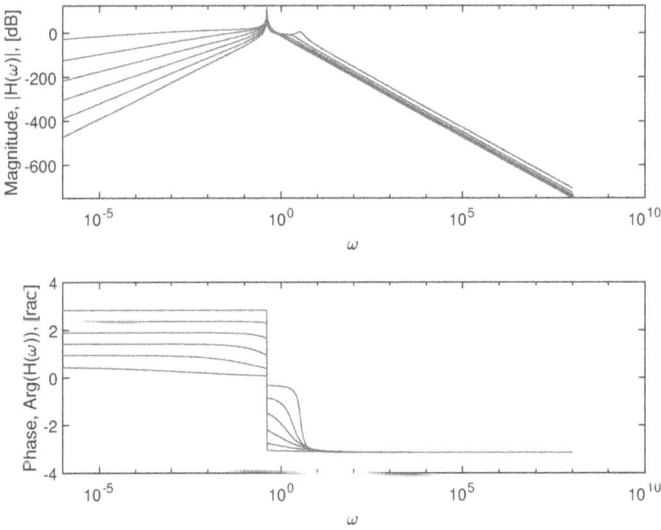

Figure 23.59 Pourreza transform of the Eq. (23.63).

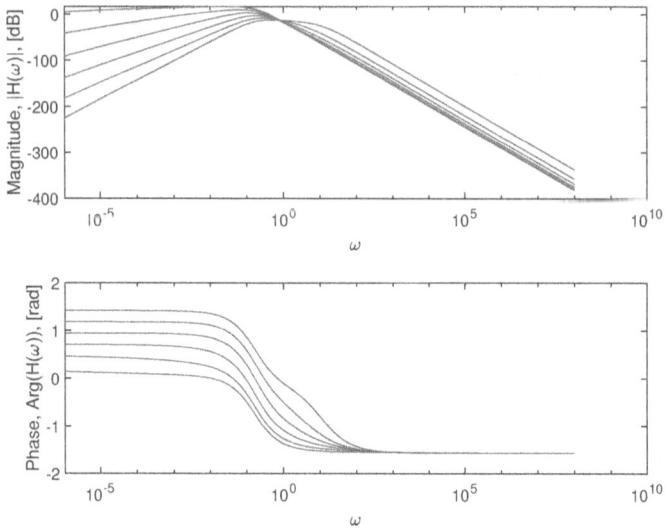

Figure 23.60 Mohand transform of the Eq. (23.63).

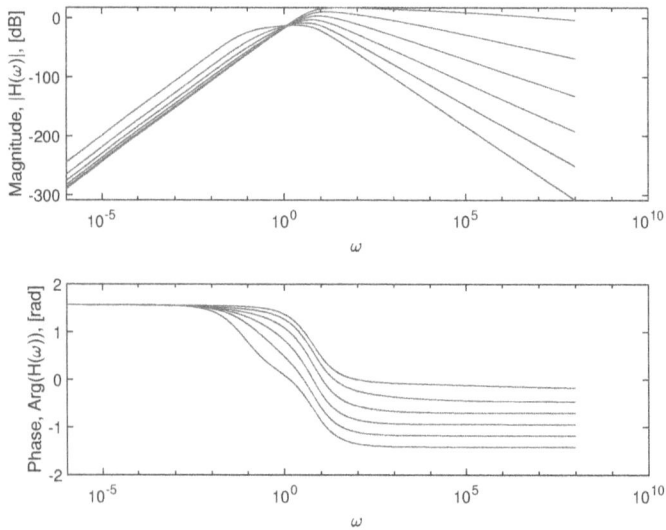

Figure 23.61 Sawi transform of the Eq. (23.63).

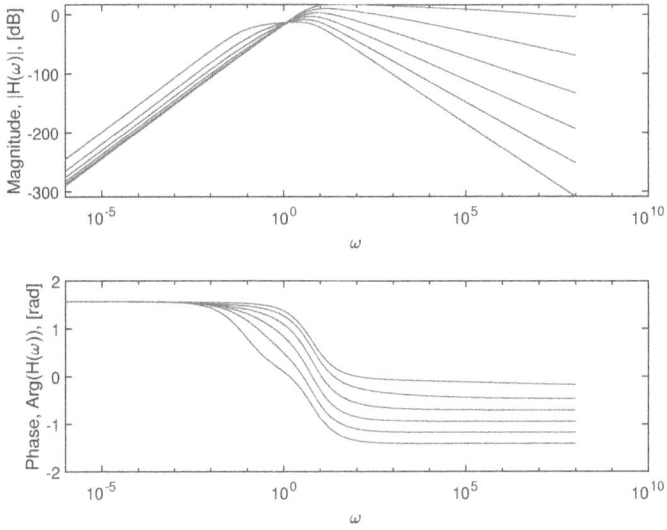

Figure 23.62   Kamal transform of the Eq. (23.63).

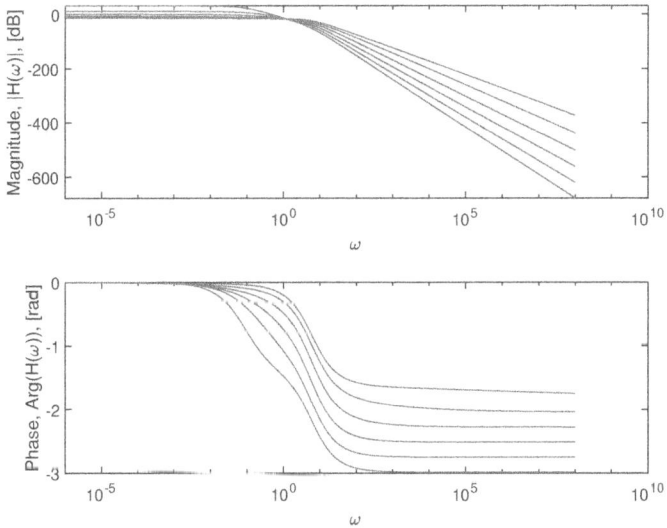

Figure 23.63   G transform of the Eq. (23.63).

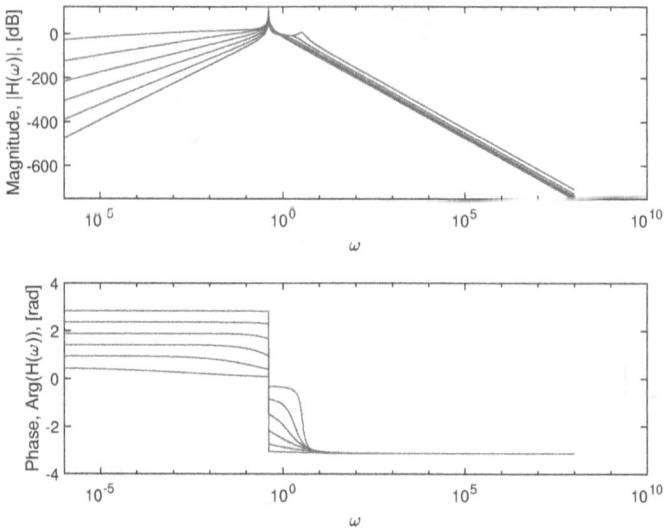

Figure 23.64 α integral Laplace transform of the Eq. (23.63).

Conclusion

In this book we investigated the general integral transform in detail. We applied many integral transforms to many circuit problems to obtain the transfer functions. We used different kernels (power-law, exponential-decay, Mittag-Leffler) in the models. We demonstrated the simulations to prove the efficiency of the proposed integral transforms. We discussed many useful applications. We obtained new and interesting transfer functions that will be useful for engineers.

References

1. J. Liouville. Mémoire sur le calcul des différentielles à indices quelconques. *Journal de l'École Polytechnique, Paris*, 13:71–162, 1832.
2. I. Podlubny, R. L. Magin, and I. Trymorush. Niels Henrik Abel and the birth of fractional calculus. *Fractional Calculus and Applied Analysis*, 20(5):1068–1075, 2017.
3. A. Atangana. Non validity of index law in fractional calculus: A fractional differential operator with Markovian and non-Markovian properties. *Physica A*, 505:688–706, 2018.
4. A. Atangana and A. Akgül. Can transfer function and Bode diagram be obtained from Sumudu transform. *Alexandria Engineering Journal*, 59(4):1971–1984, 2020.
5. A. Atangana and D. Baleanu. New fractional derivatives with nonlocal and non-singular kernel, theory and application to heat transfer model. *Thermal Science*, 20:763–769, 2016.
6. G. Doetsch. *Theorie und Anwendung der Laplacesche Transformation [Theory and Application of the Laplace Transform.* Berlin: Springer Translation, 1937.
7. A. Izadian. *Fundamentals of Modern Electric Circuit Analysis and Filter Synthesis: A Transfer Function Approach.* Cham, Switzerland: Springer Nature Switzerland AG, 2019.
8. H. Jafari. A new general integral transform for solving integral equations. *Journal of Advanced Research*, 32:133–138, 2021.
9. J. L. Schiff. *The Laplace Transform Theory and Applications, Series: Undergraduate Texts in Mathematics.* New York: Springer, 1999.
10. G. K. Watugala. Sumudu transform: A new integral transform to solve differential equations and control engineering problems. *International Journal of Mathematical Education in Science and Technology*, 24(1):35–43, 1993.
11. F. B. M. Belgacem, A. A. Karaballi, and S. L. Kalla. Analytical investigations of the Sumudu transform and applications to integral production equations. *Mathematical Problems in Engineering*, 3:103–118, 2003.
12. S. Weerakoon. Application of Sumudu transform to partial differential equations. *International Journal of Mathematical Education in Science and Technology*, 25(2):277–283, 1994.
13. M. Caputo. Linear model of dissipation whose Q is almost frequency independent-II. *Geophysical Journal of the Royal Astronomical Society*, 13:529–539, 1967.
14. D. Tavares, R. Almeida, and D. F. M. Torres. Caputo derivatives of fractional variable order: Numerical approximations. *Communications in Nonlinear Science and Numerical Simulation*, 35:69–87, 2016.
15. M. Caputo and M. Fabrizio. A new definition of fractional derivative without singular kernel. *Progress in Fractional Differentiation and Applications*, 2:73–85, 2015.
16. G. Hu, et al. Controlling spiral waves in a model of two-dimensional arrays of Chua's circuits. *Physical Review Letters*, 80(9):1884, 1998.
17. G. A. Leonov and N. V. Kuznetsov. Analytical-numerical methods for hidden attractors' localization: The 16th Hilbert problem, Aizerman and Kalman conjectures, and Chua circuits. In: *Numerical Methods for Differential Equations, Optimization, and Technological Problems.* Springer, Dordrecht, 41–64, 2013.

18. T. M. Elzaki, S. M. Elzaki, and E. M. A. Hilal. Elzaki and Sumudu transforms for solving, some differential equations. *Global Journal of Pure and Applied Mathematics*, 8(2):167–173, 2012.

19. K. S. Aboodh, The new integral transform "Aboodh Transform." *Global Journal of Pure and Applied Mathematics*, 9(1):35–43, 2013.

20. S. A. P. Ahmadi, H. Hosseinzadeh, and A. Y. Cherati. A new integral transform for solving higher order linear ordinary differential equations, *Nonlinear Dynamics and Systems Theory*, 19(2):243–252, 2019.

21. S. A. P. Ahmadı, H. Hosseinzadeh, and A. Y. Cherati. A new integral transform for solving higher order linear ordinary Laguerre and Hermite differential equations. *International Journal of Applied and Computational Mathematics*, 5:142, 2019.

22. M. M. Abdelrahim Mahgoub. The new integral transform "Mohand Transform." *Advances in Theoretical and Applied Mathematics*, 12(2):113–120, 2017.

23. M. M. Abdelrahim Mahgoub. The new integral transform "Sawi Transform." *Advances in Theoretical and Applied Mathematics*, 14(1):81–87, 2019.

24. H. Kamal and A. Sedeeg. The new integral transform "Kamal Transform." *Advances in Theoretical and Applied Mathematics*, 11(4):451–458, 2016.

25. H. Kim. On the form and properties of an integral transform with strength in integral transforms. *Far East Journal of Mathematical Sciences*, 102(11):2831–2844, 2017.

26. H. Kim, The intrinsic structure and properties of Laplace-typed integral transforms. *Mathematical Problems in Engineering*, 2017, Article ID 1762729, 8 pages.

27. Z. H. Khan and W. A. Khan. N-transform properties and applications. *NUST Journal of Engineering Science*,1(1):127–133, 2008.

28. K. F. Riley, M. P. Hobson, and S. J. Bence. *Mathematical Methods for Physics and Engineering (3rd ed.)*. Cambridge University Press, pp. 455, 2010. ISBN 978-0-521-86153-3.

29. J. J. Distefano, A. R. Stubberud, and I. J. Williams. *Feedback Systems and Control, Schaum's Outlines (2nd ed.)*. McGraw-Hill, p. 78, 1995, ISBN 978-0-07-017052-0.

30. H. Kober. On fractional integrals and derivatives. *The Quarterly Journal of Mathematics*, os-11(1):193–211, 1940.

31. J. Liouville. Mémoire sur quelques questions de géométrie et de mécanique, et sur un nouveau genre de calcul pour résoudre ces questions. *Journal de l'École Polytechnique, Paris*, 13:1–69, 1832.

32. B. Ross, A brief history and exposition of the fundamental theory of fractional calculus. Fractional Calculus and Its Applications. *Lecture Notes in Mathematics*, 457:1–36, 1975. doi:10.1007/BFb0067096.

33. L. Debnath. A brief historical introduction to fractional calculus. *International Journal of Mathematical Education in Science and Technology*, 35(4):87–501, 2004. doi:10.1080/00207390410001686571.

34. B. Girod, R. Rabenstein, and A. Stenger. *Signals and Systems, 2nd ed.*. Wiley, p. 50, 2001, ISBN 0-471-98800-6.

35. M. A. Laughton and D. F. Warne. *Electrical Engineer's Reference Book (16th ed.)*. Newnes, Burlington, USA, pp. 14/9–14/10, 2002.

36. M. E. Van Valkenburg. In memoriam: Hendrik W. Bode (1905–1982). *IEEE Transactions on Automatic Control*, AC-29(3):193–194, 1984

37. D. A. Mindell. *Between Human and Machine: Feedback, Control, and Computing before Cybernetics*. JHU Press, pp. 127–131, 2004. ISBN 0801880572.

38. S. Skogestad and I. Postlewaite. *Multivariable Feedback Control.* John Wiley & Sons Ltd., Chichester, West Sussex, England, 2005. ISBN 0-470-01167-X.

39. T. H. Lee. *The Design of CMOS Radio-Frequency Integrated Circuits (2nd ed.).* Cambridge University Press, Cambridge, UK, pp. 451–453, 2004.

40. H. Nyquist. Regeneration theory. *Bell System Technical Journal. USA: American Telephone and Telegraph Company (AT&T)*, 11(1):126–147, 1932.

41. U. A. Bakshi and A. V. Bakshi. *Circuit Analysis – II.* Technical Publications, 2009. ISBN 9788184315974.

42. P. Horowitz and W. Hill. *The Art of Electronics (3rd ed.).* Cambridge University Press, 2015. ISBN 0521809266.

43. K. V. Cartwright, E. Joseph, and E. J. Kaminsky. Finding the exact maximum impedance resonant frequency of a practical parallel resonant circuit without calculus. *The Technology Interface International Journal*, 11(1):26–34, 2010.

44. J. Blanchard. The history of electrical resonance. *Bell System Technical Journal*, 20(4):415, 1941.

45. Savary, Felix. Memoirs sur l'Aimentation. *Annales de Chimie et de Physique*, 34:5–37, 1827.

46. T. D. S. Hamilton. *Handbook of Linear Integrated Elecronics for Research.* V. K. McGraw-Hill, London, 1977.

47. M. A. Reddy. Operational amplifier circuits with variable shift and their application to high-Q active RC-filters and RC-oscillators, *IEEE Transactions on Circuits and Systems*, 23(6):384–389, 1976.

48. A. V. Oppenheim, A. S. Wilsky, and Ian T. Young. *Signals and Systems.* Prentice-Hall, Englewood Cliffs, NJ, 1983.

49. P. R. Geffe. Exact synthesis with real amplifiers. *IEEE Transactions on Circuits and Systems*, 21(3):369–376, 1974.

50. F. N. Trofimenkoff, D. H. Treleaven, and L. T. Bruton. Noise performanceor RC-active quadratic filter sections, *IEEE Transactions on Circuit Theory*, CT-20:524–532, 1973.

51. N. S. Nise. *Control Systems Engineering (4th ed.).* Wiley & Sons, 2004. ISBN 0-471-44577-0.

52. P. Horowitz and W. Hill. *The Art of Electronics (2nd ed.).* Cambridge University Press, 2001. ISBN 0-521-37095-7.

53. J. J. Cathey. *Electronic Devices and Circuits (Schaum's Outlines Series).* McGraw-Hill, 1988. ISBN 0-07-010274-0.

54 J. D. Irwin, and J. David. *Basic Engineering Circuit Analysis.* Wiley, 2006. ISBN 7-302-13021-3.

55. K. L. Kaiser. *Electromagnetic Compatibility Handbook.* CRC Press, 2004. ISBN 0-8493-2087-9.

56. J. W. Nilsson, S. A. Riedel, and A. Susan. *Electric Circuits.* Prentice Hall, 2008. ISBN 978-0-13-198925-2.

57. R. J Smith. *Circuits, Devices and Systems (International ed.).* Wiley, New York, p. 21, 1966.

58. D. P. Ellerman. Chapter 12: Parallel addition, series-parallel duality, and financial mathematics. *Intellectual Trespassing as a Way of Life: Essays in Philosophy, Economics, and Mathematics.* Rowman & Littlefield Publishers, Inc., 1995. ISBN 0-8476-7932-2.

59. T. Williams. *The Circuit Designer's Companion.* Butterworth-Heinemann, 2005. ISBN 0-7506-6370-7.

60. E. Kolářová. Modelling RL Electrical Circuits by Stochastic Diferential Equations, in

Proceedings of IEEE R8 International Conference on Computer as a Tool. Belgrade, pp. 1236–1238, 2005.

61. E. Kolářová. Statistical Estimates of Stochastic Solutions of RL Electrical Circuits, in *Proceedings of IEEE International Conference on Industrial Technology ICIT2006.* Mumbai, pp. 2546–2550, 2006.

62. E. Kolářová and L. Brančík. Vector Linear Stochastic Differential Equations and Their Applications to Electrical Networks, in *Proceedings of 35th International Conference on Telecommunications and Signal Processing TSP2012.* Prague, pp. 311–315, 2012.

63. L. Brančík. Time and Laplace-domain methods for MTL transient and sensitivity analysis. *COMPEL: The International Journal for Computation and Mathematics in Electrical and Electronic Engineering*, 30(4):1205–1223, 2011.

64. N. S. Patil, B. G. Gawalwad, and S. N. Sharma. A Random Input-Driven Resistor-Capacitor Series Circuit, in *Proceedings of 2011 International Conference on Recent Advancements in Electrical, Electronics and Control Engineering.* Sivakasi, pp. 100–103, 2011.

65. F. Rahman and N. Parisa. A stochastic perspective of RL electrical circuit using different noise terms. *COMPEL: The International Journal for Computation and Mathematics in Electrical and Electronic Engineering*, 30(2): 812–822, 2011.

66. U. M. Ascher and L. R. Petzold, Computer methods for ordinary differential equations and differential-algebraic equations. *SIAM: Society for Industrial and Applied Mathematics*, USA, 1998. ISBN 0-8987-1412-5.

67. L. Brančík. Modified technique of FFT-based numerical inversion of Laplace transforms with applications. *Przeglad Elektrotechniczny*, 83(11):53–56, 2007.

68. D. V. Ginste, D. De Zutter, D. Deschrijver, T. Dhaene, P. Manfredi, and F. Canavero, Stochastic modeling-based variability analysis of on-chip interconnects. *IEEE Transactions on Components, Packaging, and Manufacturing Technology*, 2(7):1182–1192, 2012.

69. C. R. Paul. *Analysis of Multiconductor Transmission Lines.* John Wiley & Sons, Inc., Hoboken, New Jersey, USA, 2008.

70. M. Cheng, K. T. Chau, C. C. Chan, E. Zhou, and X. Huang, Nonlinear varying-network magnetic circuit analysis for doubly salient permanentmagnet machines. *IEEE Transactions on Magnetics*, 36(1):339–348, 2000.

71. W. L. Soong, D. A. Staton, and T. J. E. Miller, Validation of lumpedcircuit and finite-element modeling of axially-laminated brushless machines, in *Proceedings of. 6th International. Conference on Electrical Machines and Drives*, pp. 85–90, 1993.

72. M. Toufik and A. Atangana. New numerical approximation of fractional derivative with non-local and non-singular kernel: Application to chaotic models. *European Physical Journal – Plus*, 132:444, 2017. https://doi.org/10.1140/epjp/i2017-11717-0.

73. T. Matsumoto. A chaotic attractor from Chua's circuit. *IEEE Transactions of Circuit System*, CAS-31(12):1055–1058, 1984.

74. T. T. Hartley, C. F. Lorenzo, and H. K. Qammer. Chaos in a fractional order Chua's system. *IEEE Transactions of Circuit and Systems, I: Fundamental Theory and Applications*, 42(8):485–490, 1995.

75. L. P. Chen, J. F. Qu, Y. Chai, R. Wu, and G. Qi. Synchronization of a class of fractional-order chaotic neural networks. *Entropy*, 15(8): 3265–3276, 2013.

76. A. G. Radwan, A. S. Elwakil, and A. M. Soliman. On the generalization of second-order filters to the fractional-order domain. *Journal of Circuits, Systems and Computers*, 18:361–386, 2009.

77. D. Xu, Z. Yang, and Y. Huang. Existence–uniqueness and continuation theorems for stochastic functional differential equations. *Journal of Differential Equations*, 245:1681–1703, 2008.

78. T. Caraballo, I. D. Chueshov, P. Marín-Rubio, and J. Real. Existence and asymptotic behaviour for stochastic heat equations with multiplicative noise in materials with memory. *Discrete and Continuous Dynamical Systems*, 18(2–3):253–270, 2007.

79. Z. Fan, M. Liu, and W. Cao. Existence and uniqueness of the solutions and convergence of semi-implicit Euler methods for stochastic pantograph equations. *Journal of Mathematical Analysis and Applications*, 325(2):1142–1159, 2007.

80. F. Jiang and Y. Shen. A note on the existence and uniqueness of mild solutions to neutral stochastic partial functional differential equations with non-Lipschitz coefficients. *Computers and Mathematics with Applications*, 61(6):1590–1594, 2011.

81. T. Taniguchi. The existence and uniqueness of energy solutions to local non-Lipschitz stochastic evolution equations. *Journal of Mathematical Analysis and Applications*, 360(1):245–253, 2009.

82. Heinrich, W. (1977), Arnold, L., *Stochastic Differential Equations, Theory and Applications*, New York. John Wiley & Sons. 1974. XVI, 228 S., £ 9.50 (engl. Übersetzung des deutschen Originals, R. Oldenbourg 1973). Z. angew. Math. Mech., 57: 271–271. https://doi.org/10.1002/zamm.19770570413

83. K. J. Åström. On a first order stochastic differential equation. *International Journal of Control*, 1(4): 301–326, 1965.

84. F. Battelli and M. Fečkan. Nonlinear RLC circuits and implicit ODEs. *Differential Integral Equations*, 27:671–690, 2014.

85. F. Battelli and M. Fečkan. On the existence of solutions connecting singularities in nonlinear RLC circuits. *Nonlinear Analysis*, 116:26–36, 2015.

86. L. O. Chua, Ch. A. Desoer, and E. S. Kuh. *Linear and Nonlinear Circuits*. McGraw-Hill, New York, 1987.

87. E. Gluskin, A nonlinear resistor and nonlinear inductor using a nonlinear capacitor. *Journal of the Franklin Institute*, 336:1035–1047, 1999.

88. R. Riaza. *Differential-Algebraic Systems: Analytical Aspects and Circuit Applications*. World Scientific Publishing Company Co. Pte. Ltd., Singapore, 2008.

89. J. Lambert. *Numerical Methods for Ordinary Differential Systems: The Initial Value Problem (1st ed)*. Wiley, West Sussex, 1991.

90. P. Rodrigues. *Computer-Aided Analysis of Nonlinear Microwave Circuits*. Artech House, Norwood MA, 1998.

91. J. Pedro and N. Carvalho. *Intermodulation Distortion in Microwave and Wireless Circuits (1st ed.)*. Artech House, Norwood MA, 2003.

92. H. Brachtendorf, G. Welsch, R. Laur, and A. Bunse-Gerstner. Numerical steady-state analysis of electronic circuits driven by multi-tone signals. *Electrical Engineering (Springer-Velag)*, 79(2):103–112, 1996. doi: 10.1007/BF01232919.

Index

For Product Safety Concerns and Information please contact our EU
representative GPSR@taylorandfrancis.com
Taylor & Francis Verlag GmbH, Kaufingerstraße 24, 80331 München, Germany

www.ingramcontent.com/pod-product-compliance
Lightning Source LLC
Chambersburg PA
CBHW060425220326
41598CB00021BA/2299